SOLAR AND STELLAR FLARES

SOLAR AND STELLAR FLARES

*Proceedings of the 104th Colloquium
of the International Astronomical Union
held in Stanford, California, August 15–19, 1988*

Edited by

BERNHARD M. HAISCH

Lockheed Palo Alto Research Laboratory, Palo Alto, California, U.S.A.

and

MARCELLO RODONÒ

Institute of Astronomy, University of Catania and Astrophysical Observatory,

Catania, Italy

Reprinted from Solar Physics, Volume 121, Nos 1–2, 1989

KLUWER ACADEMIC PUBLISHERS
DORDRECHT / BOSTON / LONDON

Library of Congress Cataloging-in-Publication Data

```
International Astronomical Union.  Colloquium (104th : 1988 :
Stanford, Calif.)
  Solar and stellar flares : proceedings of the 104th Colloquium of
the International Astronomical Union, held at Stanford, California,
August 15-19, 1988 / edited by Bernhard M. Haisch and Marcello
Rodonò.
     p.    cm.
  ISBN-13:978-94-010-6948-9       e-ISBN-13:978-94-009-1017-1
  DOI: 10.1007/978-94-009-1017-1
  1. Solar flares--Congresses.  2. Flare stars--Congresses.
I. Haisch, Bernhard M.  II. Rodonò, Marcello, 1941-   .  III. Title.
QB526.F6I57  1988
523.7'5--dc20
                                                      89-15599
```

Published by Kluwer Academic Publishers,
P.O. Box 17, 3300 AA Dordrecht, The Netherlands.

Kluwer Academic Publishers incorporates the publishing programmes of
D. Reidel, Martinus Nijhoff, Dr W. Junk and MTP Press.

Sold and distributed in the U.S.A. and Canada
by Kluwer Academic Publishers,
101 Philip Drive, Norwell, MA 02061, U.S.A.

In all other countries, sold and distributed
by Kluwer Academic Publishers Group,
P.O. Box 322, 3300 AH Dordrecht, The Netherlands.

Printed on acid free paper

Dedicated to:
The Solar Maximum Mission Project
and
The Flare Star Consortium

TABLE OF CONTENTS

(Solar and Stellar Flares)

EDITORIAL

Some years ago we considered extending the subject matter of *Solar Physics* to include the area of solar phenomena observed on other stars. In the years since we first discussed this, the solar-stellar connection has grown and considerable progress has been made in this field.

It seems to the Editors and to the Editorial Board that the scientific justification for including some solar-stellar papers in *Solar Physics* is more compelling now than ever. The two fields have grown closer together in recent years. Both solar and stellar physicists will benefit from the improved communication that will result from sharing the pages of a journal. It is clear that each group has much to learn from closer contact with the other.

Solar physics is a field which has connections both to stellar physics and to solar-terrestrial physics. Our journal already publishes papers in the solar-terrestrial field, and we view the proposed broadened scope as a symmetrical extension of this policy to the other end of the disciplinary spectrum.

A good opportunity presents itself now to begin such an expansion of our journal. The Editors have arranged for the publication in this volume of the proceedings of IAU Colloquium No. 104, 'Solar and Stellar Flares'. We intend to publish some invited reviews in the solar-stellar field in our journal in the near future. A letter to the international solar-stellar field has already invited contributed papers to the journal on this topic. We have recently expanded the subtitle of *Solar Physics* to include solar-stellar research, we have elected stellar physicists to three positions on the Editorial Board starting with the beginning of 1989, and we will continue such representation in future Board elections. It has been agreed between the three editors that papers dealing with solar-stellar research with be handled by Cornelis de Jager (Utrecht).

We now welcome contributed papers on the solar-stellar connection, and the Editors promise the solar-stellar community, as we promise the solar community, that we will adhere to the highest scientific and editorial standards. We intend to maintain a journal from which both communities can benefit, and we anticipate that a healthy synergism will develop within the pages of *Solar Physics*. Both fields will advance from a broader physical understanding of phenomena which are common to the two disciplines.

We welcome your comments and suggestions about these changes. We feel that this step will broaden and strengthen our journal and ultimately advance both of these closely related fields.

CORNELIS DE JAGER, ZDENĚK ŠVESTKA, and ROBERT F. HOWARD
Solar Physics Editors

Solar Physics **121:** ix, 1989.
© 1989 *Kluwer Academic Publishers.*

PREFACE

For almost three decades following the first flare detection on a dMe star in 1948, solar and stellar flare research more or less went their separate ways: various aspects of solar flares were observed but hardly ever white-light events, whereas on the stellar side optical flaring was virtually the only observable phenomenon. With the first X-ray detections of flares on UV Ceti, YZ CMi, and Proxima Centauri in the mid-1970's things changed dramatically and by the time of the 1982 Catania IAU Colloquium No. 71, *Activity in Red Dwarf Stars*, space observations of flare stars was an exciting topic. Nonetheless, participation at that meeting was still dominated by the stellar community. The Palo Alto IAU Colloquium No. 104, *Solar and Stellar Flares*, is the successor to the Catania meeting, and is the first major IAU conference to bring together solar and stellar topics and investigators on an even footing. More and more, solar and stellar researchers are speaking the same language and addressing the same problems; moreover there has been an increase in the number of investigators who actually do research on both sides. The Solar Maximum Mission, especially, seems to have been a catalyst for such 'crossover' activity.

This conference was four years in the planning, and thus it was with a certain measure of relief that we finally welcomed 200 scientists from 29 countries to Stanford University on 15–19 August, 1988. To bring this about required support of many sorts from many sources. The conference was co-hosted by three institutes: the Lockheed Palo Alto Research Laboratory, the University of Catania and Stanford University. We were fortunate in obtaining generous funding from the NASA Solar Maximum Mission project to organize the meeting, and for this we owe special gratitude to SMM XRP Principal Investigator, Dr Keith T. Strong, and SMM Project Scientist, Dr Joseph B. Gurman. As a result we were able to engage the outstanding logistical support of the SLW Associates, Susan Sweeney, Margaret London, and Nancy Winton, professional conference organizers. Inviting 200 guests to town for a week is no job for amateurs! Extensive additional funding for travel was also provided by NASA, and by the IAU, ESA, and COSPAR; this allowed us to support approximately half of the participants. The all important meeting bags were generously provided by Lockheed, and well supplied coffee breaks were paid for by the Stanford Solar Observatory, Lockheed and Kluwer Academic Publishers. Mailings and other miscellaneous logistics were underwritten by the Catania Astrophysical Observatory and Lockheed. The Scientific Organizing Committee was chaired by us with invaluable support from SOC member Peter Sturrock. Other members of the SOC were R. Bonnet, J. Butler, L. Cram, R. Gershberg, M. Giampapa, D. Gibson, C. de Jager, C. Jordan, M. Machado, M. Oda, E. Priest, and H. Zirin.

As promised, the skies were clear and the temperatures balmy every single day. Social events included a Sunday night reception, a Tuesday night wine and cheese in the Stanford Rodin Garden and a Thursday night banquet at the Stanford Faculty Club with a lecture by Dr Morris Berman. Most memorable, perhaps, was the Monday night

Solar Physics **121**: 1–2, 1989.
© 1989 *Kluwer Academic Publishers*.

concert at Dinkelspiel Auditorium 'An Evening of Songs and Arias' hosted by Dr Kip Cranna of San Francisco Opera, produced and directed by Elizabeth Tucker, and featuring soprano Ellie Holt Murray, mezzo-soprano Marsha Sims, tenor Richard Walker, and baritone David Taft Kekuewa, with piano accompaniment by Mark Haffner, staff coach for San Francisco Opera.

Two scientific themes clearly emerged from this conference: (1) the key to progress in flare research lies in a multispectral approach with as much temporal resolution as the photon fluxes allow; and (2) the key to understanding the physics lies in a dynamic interaction between solar and stellar investigations and investigators. During the eight sessions solar and stellar topics were balanced and intermixed in 33 invited and oral presentations. We are particularly pleased that these proceedings will be the springboard to publication of solar-stellar articles in the journal *Solar Physics*. In addition, 115 very exciting posters were also displayed and a companion volume containing many of these is available as a publication of the Catania Astrophysical Observatory.

We dedicate this book to the Solar Maximum Mission and to the Flare Star Consortium. To all our solar-stellar friends and colleagues: 'Thank you!'

BERNHARD M. HAISCH and MARCELLO RODONÒ
28 March, 1989

AN OVERVIEW OF SOLAR AND STELLAR FLARE RESEARCH

BERNHARD M. HAISCH

Div. 91–30, Bldg. 255, Lockheed Palo Alto Research Laboratory,
3251 Hanover St., Palo Alto, CA 94304, U.S.A.

Abstract. An overview of the many topics discussed at IAU Colloquium No. 104 is presented as an introduction to the Proceedings. Suggested areas for future research emerging from the conference are summarized.

1. Background

Solar and stellar flare research were carried out in remarkable isolation from each other until the mid-1970s. Solar flares were observed in Hα, and their radio, X-ray, and energetic particle outputs were studied, but until the recent initiation of a white light flare patrol program, only a few dozen white light flares had been observed in over a hundred years of solar observation. On the stellar side an almost exactly opposite situation prevailed: optical flaring was virtually the only observable phenomenon and was a frequent occurrence in the class of UV Ceti-type flare stars, a subset of the dMe and dKe population.

The stellar emphasis shifted considerably with the first X-ray detections of flares on UV Ceti, YZ CMi, and Proxima Centauri in 1974–1975 so that by the time of the 1982 Catania IAU Colloquium, *Activity in Red Dwarf Stars*, space observations of flare stars was an exciting topic. Nonetheless, participation at that meeting was mainly limited to the stellar community. Recently there has been an increase in the number of investigators who actually do research on both sides and in particular, the Solar Maximum Mission seems to have spurred quite a bit of research activity on the stellar side, such as in the application of solar flare magnetohydrodynamic loop models to stellar observations.

This colloquium was the first major meeting to bring together solar and stellar topics and investigators on an even footing. Approximately 200 scientists from 29 countries met for five days at Stanford University, and the following is an overview of what was said and learned in that exchange.

2. Where Do Flares Occur?

Although apparently brought about by photospheric motions, solar flares are primarily a coronal phenomenon involving magnetic structure interactions and reconfigurations. Unfortunately, very little hard data actually exists on the coronal magnetic field. Apart from field line modeling extrapolated from photospheric vector magnetograms, radio observations offer the best possibility at present of measuring field strengths, although this technique is intrinsically model-dependent. Zheleznyakov reported on the detection of radio emission at 1658 MHz interpreted as thermal cyclotron radiation and, using

Solar Physics **121**: 3–18, 1989.
© 1989 *Kluwer Academic Publishers.*

a model, a field strength of ≈ 200 G was derived for a coronal loop containing $T \sim 4 \times 10^6$ K plasma.

In the photosphere, a confusing variety of magnetic situations have been associated with flare production: flux emergence, decreasing flux, adjacent increasing *and* decreasing flux. Livi presented evidence that cancellation of flux is the common denominator herein, and that flares are centered around cancellation sites; that is to say, flares occur at or between opposite polarity features that are cancelling. It remains to be seen if cancellation always leads to flares and if the cancellation process can lead to storage of energy for subsequent release as a flare. Another flare-conducive condition is magnetic shear, and the relationship of a sheared magnetic configuration to bipolar cancellation is an open topic. Machado *et al.* (1988) have proposed that a flare basically consists of one bipole impacting one of more adjacent bipoles, with the X-ray emission concentrated in the impacting bipole at the onset; thereafter it may spread throughout the interacting bipole systems, but it does not concentrate at the impact point.

While individual magnetic bipoles on the Sun tend to last for about 2 weeks on average, regions of enhanced and persistent activity last for one to several solar rotations. Gaizauskas emphasized that it is not just the local conditions at the flare site that are important, but that 'multiple structures are involved in the energy release' which are not 'uncoupled from a global background'. This is clear from the many kinds of flare precursors: for example, homologous flares show that flare-producing stresses of similar sort can keep building up over some time and sympathetic flares demonstrate that flare-triggering excitations can travel between adjacent or even remote active regions, and on an even more global scale surviving for several solar rotations, there is the existence of clusters or nests of active region production and superactive regions giving rise to major flares. Unfortunately there is no precursor that *always* predicts a flare.

Progress on the stellar side has accelerated primarily as a result of taking a multi-spectral approach to flare observations, made possible recently by increased instrumental sensitivities and by the dogged persistence of teams of investigators willing to split up and travel to opposite ends of the Earth to simultaneously 'catch a flare' with different instruments. This has paid off slowly but surely though in making plausible physial parameters emerge from what are now many different spectral combinations of data.

On stars, of course, there is very little data concerning magnetic fields, but advances are taking place, in deconvolving field strengths from filling factors using detailed differential intercomparison of profiles (in unpolarized light) of magnetically sensitive (i.e., having a high Landé *g* factor) and insensitive lines which are otherwise similar in their formation, the so-called 'Robinson technique'; the classical method based on circular polarization line differences used on the Sun fails in the stellar case since, without spatial resolution, the overall north and south polarities cancel out leaving no measurable net magnetic flux to detect. The Robinson-type method has now been used to measure about 50 late-type stars. While this method in principle yields both an average field strength, B, and an average filling factor, f, there are important limitations and challengeable assumptions underlying this application: These were summarized by

Linsky (cf. also Hartmann, 1987). Perhaps the most important problem is that the stellar atmospheric structure is not the same inside and outside of magnetic flux tubes and starspots – sunspots are obviously much darker than the rest of the Sun – and this difference in 'visibility' may account for some of the anomalously high filling factors ($f \approx 0.9$) which have been found. Among the emerging trends noted by Linsky, an important one appears to be $\langle B \rangle \sim \langle P_g^{1/2} \rangle$, which is, however, far from being established at this point.

Concerning the types of stars that flare, as summarized by Pettersen, activity has been reported now along the entire Main Sequence and in many evolved stars, although the reality of a good number of the events may legitimately be questioned, since they are often only a single, weakly observed occurrence; one can also never be sure that the event did not occur on an undetected faint, red companion. Flaring T Tauri stars, and the several hundred flaring objects reported in Orion and in the Pleiades are definitely young; on the other hand, other stars observed to flare are well along in stellar evolution. In close binaries, it is possible that flare-like activity is related to mass transfer. To date, confirmed, solar-like flares have been seen only on stars having outer convection zones. It has been proposed that dynamo processes originate primarily in the radiative core/convection zone interface, suggesting that fully convective stars would manifest little activity. Stars fainter than $L = 3 \times 10^{31}$ ergs s^{-1} ($M = 0.3 \, M_\odot$) are fully convective, but Pettersen reports no evidence of a qualitative change in the flaring properties at this important breakpoint; rather he finds a continuous linear relationship between the maximum time-averaged flare luminosity, $\langle L_f \rangle$, for a given class of star (individual stars of a given class vary widely, of course) and the volume of the convection zone V_{conv}. (He claims a similar relation for maximum X-ray luminosity: $L_x \sim V_{conv}$).

There are now about 80 'classical' dKe/dMe red dwarf flare stars identified. Their ratio of time-average flare luminosity to bolometric luminosity can reach $\langle L_f \rangle / L_{bol} = 1-3 \times 10^{-3}$ and including other non-photospheric energy fluxes could result in a ratio on the order of 10^{-2} for the non-photospheric power fraction, whereas in the Sun this is certainly no more than $\sim 10^{-5}$. Mullan speculated that this ratio might climb high enough to alter the internal constitution and equilibrium of these stars from that predicted using standard models. The exact evolutionary status of these stars is uncertain, but in fact they all 'age' quite slowly, the Main-Sequence lifetimes being: $\approx 40 \times 10^9$ yr for a dK5, $\approx 70 \times 10^9$ yr for a dM0, and $\approx 270 \times 10^9$ yr for a dM5, i.e., longer than the present age of the Universe. Among evolved stars, the RS CVn binaries are a growing class of flaring objects, along with mass transfer systems of the Algol-, W UMa-, and FK Com-types. The cluster flare stars tend to be on or near the Main Sequence (cf. Mirzoyan and Ambaryan, 1988, for field and cluster flare star comparisons).

There are significant differences between RS CVn and dMe flares as summarized by Byrne. In general the RS CVn events can be up to two orders of magnitude more energetic than dMe flares; the RS CVn flares are also of much longer duration, with IUE ultraviolet line enhancements lasting up to 7 hours; and flare temperatures, derived from low resolution soft X-ray spectra, are found to be in the range $T = 1-5 \times 10^7$ K for the dMe stars, but can apparently range up to $T = 10^8$ K in RS CVn flares.

Radio flares have been observed in dMe and dKe stars, in RS CVn systems, pre-Main-Sequence stars and X-ray binaries as reviewed by Kuijpers. While radio emission is small in terms of the flare energy budget, it is of potentially tremendous use as a signature of physical processes; unfortunately, the interpretation of this is very model-dependent. Also, as pointed out by Kundu, while radio flares on RS CVn stars are probably due to a radiative mechanism identified and associated with solar flares – non-thermal gyrosynchrotron emission – flares on dMe stars appear to be due to an as yet unidentified coherent mechanism. Mullan addressed the question of whether the electron cyclotron maser might be the mechanism, and concluded that the requirements for producing this type of radiation are incompatible with stellar flare loop model parameters (which are of course derived from models). VLBI observations imply that the size of the RS CVn radio sources are comparable to the size of the star itself.

3. The Characteristics of Solar Flares

The different 'phases' of a solar flare are partially a problem of semantics, but also a genuine problem of identification and categorization of phenomena that 'always occur' in a flare – given an observation of sufficient sensitivity. Sturrock argued for four phases: (1) an *activation phase*; (2) an *impulsive phase*; (3) a *gradual phase* in which particle acceleration continues to take place; and (4) a *late phase* dominated by soft X-ray emission cooling the hot plasma, although even during this late phase there is likely to be some energy release (continued heating) taking place. Many other researchers however refer to the entire soft X-ray phase as the gradual phase, distinguishing only an impulsive and a gradual phase. On the stellar side, it has only been possible, because of wavelength and sensitivity limits, to observe the late phase in X-rays. However, the phenomenon of negative infrared preflaring could be an important manifestation of pre-impulsive phase activation resulting in the ejection of obscuring prominence-like material. In one study discussed by Gaizauskas, surging arches preceded just over half the observed flares. On the Sun a flare-preceding Coronal Mass Ejection may sometimes involve more kinetic energy than the 'flare itself'. The point is that mass motions caused by magnetic reconfiguration could be the real flare onset, or even the dominant mode of magnetic energy dissipation.

During the impulsive phase, energy release can attain a rate of 10^{30} ergs s^{-1}; moreover the length scales of the process involved are probably well below 1 arc sec (~ 700 km); and the time-scales range down to at least the $10-100$ ms regime as shown by hard X-ray bursts. Dennis summarized five possible mechanisms for transporting energy from the corona to loop footpoints: (1) thermal plasma with $T \geq 10^8$ K (i.e., so-called thermal model); (2) fast electrons with energies $\sim 10-100$ keV (i.e., so-called thick-target model, and at present the most 'popular'); (3) relativistic electrons; (4) protons with energies ≤ 1 MeV; and (5) protons with energies ≥ 1 MeV, the distinction between (2) and (3), and (4) and (5), relating to different resultant X-ray production mechanisms. The thick-target model in which electron beams produce heating and X-ray bremsstrahlung has been the most successful to date, although proton beams with energies of $\sim 100-1000$ keV are being seriously investigated as an alternative; both linear polarization of chromospheric lines and a red-shifted Lα line are predicted from this mechanism.

Thick-target interacting electron beams heat efficiently: the ratio of collisional losses to X-ray bremsstrahlung is $\sim 10^5$. On the other hand, due to the much lower energy per particle, heating by electron beams of necessity involves more particles, hence, more charge, than proton beams; and this means that such collective processes as, for example, a neutral return current need to be dealt with in that case. (Such a current can even arise in the case of conduction if a sufficiently intense heat flux is required; in which case classical conduction conditions no longer apply.) Yet another point is that the number of electrons accelerated exceeds the number initially in the coronal loop: where do they come from?

During the impulsive phase γ-rays are also observed and their spectra consist of: electron bremsstrahlung, nuclear lines, and pion decay emission, with the nuclear lines dominating in the 1–10 MeV range. The SMM γ-ray spectrometer (GRS) has observed over 150 flares as discussed by Rieger; the Japanese Hinotori satellite also observed solar flare γ-rays. An important observation is that the majority of γ-ray flares are seen at the limb; this supports the thick-target electron beam impulsive phase model since bremsstrahlung radiation is anisotropic and for an electron beam impinging vertically on the lower atmosphere, there would be a greater likelihood of seeing the resultant bremsstrahlung out the side ($\sim 90°$) of a limb flare than back out the top ($\sim 180°$) of a disk flare. (Compton scattering destroys this effect for hard X-ray bremsstrahlung.) Evidence for such a directivity also comes from analysis of 'stereoscopically observed flares' using ISEE-3 in conjunction with Pioneer Venus and from the determination of an apparent systematic increase of GRS flare luminosity toward the limb. Unfortunately, while the above observation fits nicely into a concept of beam heating, another γ-ray observation is puzzling: the fact that γ-rays attributed to pion decay persisted for 30 min following a 65 s impulsive phase flare indicating that ion acceleration somehow continued long after the impulsive phase in that event.

Four intriguing SMM results are: (1) that non-thermal broadening of soft X-ray lines indicative of random mass motions often takes place before the impulsive phase; (2) during the impulsive phase blue shifts appear (in 80% of BCS-observed M- and X-flares) provided that there is a high enough signal to temporally resolve this short-lived event; (3) blue-shifts of this sort are absent for flares past 60 deg in longitude; and (4) the densities derived from line ratios are in general much higher than those derived from emission measures, implying very small filling factors and, hence, very filamentary magnetic structure. These results could be mutually consistent if preflare activation is a process of many magnetic reconnections taking place in numerous adjacent filamentary loops, with this eventually leading to chromospheric evaporation and, hence, directed upflows into the now heated loops. Maximum upward velocities may reach 1000 km s^{-1}.

The soft X-ray phase typically manifests plasma temperatures, $T \approx 15-20 \times 10^6$ K. Early on it was thought that the time integral of the hard X-ray light curve equalled the current value of the soft X-ray light curve, which would imply that all energy input occurred in the impulsive phase with relatively slow eventual loss via soft X-ray cooling, but, in fact, the evidence for post-impulsive phase continued heating in many flares is now considerable.

Pallavicini, Serio, and Vaiana (1977) introduced the concept of two distinct types of flares: (1) small compact events ($E \sim 10^{30}$ ergs) on a short time-scale ($t \sim 10^3$ s) thought to occur in one or more loops without disrupting the basic magnetic configurations versus (2) more energetic ($E \sim 10^{32}$ ergs) and longer lived ($t \sim 10^4$ s) two-ribbon flares involving disruptive opening up of magnetic arcades. Švestka called these *confined flares* when the late phase involves only cooling and *dynamic flares* or *long-decay flares* when there is continued energy release. These categorizations are basically the same as the compact versus two-ribbon types, bearing in mind that *most* compact flares involve more than just a single loop, although there are, according to Švestka, indeed a few good single-loop flares. On the other hand there are flares which fit neither of the categories, or which show characteristics of both. Of course, many 'single-loop events' prove to be anything but simple when other diagnostic data in addition to images are available: Švestka cited examples in which filling factors may be as low as 0.01, indicative of highly filamentary structure, and other cases which clearly indicate the presence of both hot (10^7 K) and cool material (10^4 K) in the same flaring region – also a likely sign of filamented structure.

Dröge found that these two categories of flares can be distinguished by their electron energy spectra as measured *in situ* by spacecraft outside the Earth's geomagnetic field. The long duration events (two-ribbon flares presumably) give rise to electrons having a single power-law exponent in their ~ 0.1–100 MeV spectra suggesting a single acceleration mechanism; flares lasting < 1 hr generate more complex electron distributions.

The opening up and reconnection of field lines that makes a two-ribbon, or even 'n-ribbon' flare, is an important physical process that may go on for hours releasing energy and creating huge post-flare loops; soft X-ray loops grew for 11 hours in one Skylab event; HXIS observed similarly long-lived giant X-ray arches. Moreover, during this long period there must be continuous input of material from the footpoints to fill the loops, which means that there must be considerable heating at the footpoints to evaporate material into the coronal loops. Blue shifts indicating upflows on the order of 0.5–10 km s^{-1} have been observed which could supply the 10^{16}–10^{17} g required to fill dense ($n_e \sim 10^{12}$ cm^{-3}) post-flare loops; and in between the ribbons, redshifts indicative of cooling, falling matter have been observed. As the loops cool, they are also observed to shrink in size.

Mass motions associated with flares rarely exceed velocities of 20 km s^{-1} in chromospheric elements, but in the corona there is a wide range of velocities from, say, 20–2000 km s^{-1}. Martin placed the dynamic coronal components of flares into the following categories: erupting filaments, expanding clouds, flare-loops, flaring arches, and surges. Although involving material in the low corona, all of these are observable in Hα. Švestka pointed out that there are several examples of flares in which analysis shows that thermal energy in the late phase has gone more into mass motions rather than radiation or conduction. It has been proposed that Coronal Mass Ejection is a propagating compressive response to a flare; however the CME-flare relationship is not yet clear, and an up-to-date commentary on this issue may be found in Harrison and Sime (1989).

4. The Optical or White-Light Flare

In the dMe stars, optical flares are dominated by blue continuum radiation best seen in the U-band where the contrast with the photospheric continuum is greatest; U-band flares can be as bright as $\Delta U \approx 5$ mag. The general trend is for Balmer line enhancement to precede a continuum rise, but once started the continuum dominates over lines (H, He, Ca II) in the U- and B-bands; then after the continuum fades out, the enhanced emission lines remain prominent for some time. Overall, though, most of the energy is in the continuum. Bolometric energies involved can exceed 10^{37} ergs, although this is rare and from flare colors one infers, $E_{bol} \approx 6 \times E_{opt}$ and $E_{opt} \approx 3$–$5 \times E_U$. Other approximate relationships for stellar flares are: $E_{opt} = 10 \times E_{Balmer} \approx 30 \times E_{H\alpha}$. For the dMe stars, approximately equal amounts of energy are emitted in X-rays, the UV and the optical continuum. Tuominen reported on past and present efforts to detect polarization changes during flares, but to date no conclusive flare-related polarization effects have been measured.

There are tight correlations among flare contributions by the Balmer lines, chromospheric lines (Mg II), transition region lines and X-rays which, according to Shakhovskaya, do not depend on the energy of the individual flare nor upon the star. The following important statistical relation has been noted: a relation between optical flare energy and flare frequency which appears to be a power law for both the Sun and dMe stars: $v(E) \sim E^\beta$ with $\beta \approx 0.4$–1.4.

On the Sun, the white-light flare (WLF), reviewed by Neidig, represents the most extreme conditions found in flares and is the solar analog of the classical stellar flare. WLF emission appears as bright patches, waves or ribbons often containing brighter kernels ($< 3''$) and about 15 yr^{-1} are now seen using a small aperture patrol telescope. WLFs are probably not a special class of solar flare, rather just a matter of detection threshold: a GOES M2 flare or stronger in a large, active region would likely show white light flare kernels. WLFs are 'blue' like stellar flares, this due to Balmer continuum, at $T_{eff} \approx 10^4$ K, and the appearance of many short wavelength lines. The largest recently monitored WLF had peak power, $P = 2 \times 10^{29}$ ergs s^{-1} and $E > 3 \times 10^{31}$ ergs, which, however, is still 10^{-3} times fainter than the brightest of stellar flares. In the bright kernels of WLFs, optical continuum radiative losses exceed those in optical emission lines; in one case, $P_{cont} > 60 \times P_{H\alpha}$; allowing for other emission lines leads to the conclusion that perhaps 90% was continuum emission for this event. Filling factors are still unknown, of course and even a properly calibrated WLF spectrum has not yet been taken.

The WLF appears to originate above the $\tau = 1$ level, in the upper photosphere and/or chromosphere. Hydrogen recombination is responsible for much of the emission in flares showing Balmer jumps, and in flares of this type, $n_e > 10^{13}$ cm^{-3}. Some WLFs do not show Balmer jumps and in those cases the emission might be attributed to free-free, Paschen recombination or H$^-$ emission. The huge radiative losses of WLFs poses a major problem to mechanisms of energy transport: conductive heating from the corona would require such high temperature gradients that the emission measure in the WLF temperature interval would be too small; irradiation by soft X-rays is not feasible since $L_z < L_{opt}$.

Thez WLF power tracks the hard X-ray emission and this is the basis for theoretical investigations by Aboudarham and Henoux into how electron beams can bring about enough non-thermal hydrogen ionization – primarily in the low chromosphere – to account for the white light emission via recombination in the Paschen continuum. Further down in the atmosphere, at the temperature minimum and upper photosphere, the temperature is increased by about 240 K due to radiative heating by chromospheric continuum radiation, and as a result, n_e is increased leading to a significant increase in the H$^-$ population. Overall, these models show that white light emission can be explained by electron bombardment, but heating the upper photosphere directly by the electron beam is out of the question since $E > 900$ keV electrons would be necessary to penetrate to this depth.

Mullan addressed the issue of whether electron beam heating could also explain the properties of the stellar optical flare. In his analysis, the electron beam peaks in energy at ~ 20 keV, and the question then is whether there is enough energy flux of 20 keV electrons to penetrate effectively to the chromospheric level; there is a discrepancy in the assumed conditions for this analysis, since Aboudarham and Henoux showed that only $E > 70$ keV electrons actually reach the solar chromosphere. An 'electron beam flare' would also produce strong Hα Stark wings, whereas a beam stopped in the corona but creating a thermal conduction front propagating into the chromosphere, a 'thermal conduction flare', would have no Stark wings nor central reversal. Conditions on the Sun could go either way, but using the Proxima Centauri flare loops deduced by Haisch (1983) as an examplar, Mullan found that a 20 keV beam would not penetrate to the chromospheric level.

5. The Question of Microflaring

A major issue today is whether flaring and microflaring account for the heating of solar and stellar coronae. The balloon observations of Lin *et al.* (1984) are usually cited as an observational basis, since they were the first high resolution (~ 11 keV), hard X-ray (≥ 20 keV) detections of events having peak fluxes 10–100 times lower than normal flares at a rate (at least during the 141 min observing window) of once every 5 min somewhere on the Sun. However, related observations go back to *Skylab* in which the solar transition region shows continual small-scale activity as identified by Emslie and Noyes (1978) in EUV bursts seen simultaneously in several lines spanning the temperature range from the chromosphere to the corona. Bruner and Lites (1979) observed transient, red-shifted brightenings – such as intensity increases by as much as a factor of 5 in less than 30 s – in the transition region C IV line above active regions and sunspots by using the OSO-8 spectrometer, and these were subsequently also seen in HRTS spectra (Dere *et al.*, 1981). The SMM–UVSP has observed such burst activity in detail (Porter, Toomre, and Gebbie, 1984), and stochastic fluctuations were found on time-scales of a minute or less. Athay (1984) analyzed C IV enhancements in an active region, finding order-of-magnitude increases in the intensities along with high velocities during the rise phases. There also seem to be density enhancements during such bursts (Hayes and Shine, 1987). Moreover, bursts are not confined to active regions; Nishikawa (1986) found time-varying EUV sources in *Skylab* spectroheliograms of quiet-Sun network and cell interiors. Porter *et al.* (1987) showed that C IV brightenings in the quiet Sun occur directly over the neutral lines of small magnetic bipoles that, on the basis of 10830 Å dark point correspondences, are *possibly* X-ray bright points. These brightenings occur mostly in the network but sometimes in the cells. Individual impulsive brightenings have lifetimes of ∼ 10–40 s either as isolated events or as repeated fluctuations at a single site. Most recently, Haisch *et al.* (1988) actually observed the soft X-ray fluctuations in large active region coronal loops which, based on their size, ought to be the archetypes of steady, large-scale structures.

In Parker's view, flares, microflares and the X-ray corona all result from the same process: magnetic neutral point reconnection. Waves cannot heat the corona since there is no correlation between brightness and size of coronal structures as one would expect: the theoretical expectation is that the smaller the region, the smaller the fraction of the wave-power spectrum to do the heating... and this is not seen. Attention has focused on the continuous random displacement of loop footpoints by granular motions as the likely driving mechanism of coronal heating. The key point is that with the gas pressure being ∼ 10^{-2} of the magnetic pressure, any non-potential field must be force-free, i.e., $\nabla \times \mathbf{B} = \alpha\mathbf{B}$, but although α must be constant along a given field line, the physical helicity randomly goes in both directions and so 'internal tangential discontinuities develop across which the magnitude of the field is continuous but the direction changes discontinuously' and this results in dissipation, the characteristic scale of which Parker estimated from the characteristic scale of the footpoint shuffling process to be ∼ 10^{24} ergs, which he has labeled a *nanoflare*. In this view, a continuous succession of nanoflares produces the general glow of the individual X-ray loops. Moreover, it appears that a burst of reconnection may occur when a critical level of discontinuity is exceeded; the typical flare is then seen as a 'coordinated burst of nanoflares throughout a finite volume of field'. An important point is that while there may be a triggering site – probably where two loops interact – most of the energy released appears to come from throughout the loops involved. The closest we have come to seeing nanoflares is perhaps in the ∼ 10^4 decimetric spikes of ∼ 10^{26} ergs each observed (cf. Dennis' article) during some impulsive flares.

On the stellar side, it has been shown that there is a relationship between mean optical flare power and X-ray luminosity and this has been invoked by a number of investigators (present author included) as evidence that flaring is the source of coronal heating. Shakhovskaya pointed out that the flare power-frequency relation is relevant to this argument in that the microflaring (small E) end of the $v(E) = E^\beta$ relation will only contribute a small additional fraction (say 10%) to the observed $\langle L_f \rangle$ for the many stars having a relatively large β, which may not be enough to bring the ratio $\langle L_f \rangle / L_x$ up to unity. The problem is, of course, that neither $\langle L_f \rangle$ nor L_x are very precisely determined; moreover in the case of L_x the measurement only refers to a single epoch in virtually all cases and it is well established that the Sun varies by an order of magnitude in L_x over its cycle.

Interestingly, just as observations of the phenomenon of microflaring on the Sun are increasing in number, the stellar microflaring case is suffering some observational setbacks, one of which was presented at the conference by the Crimean Astrophysical Observatory group. They used the Soviet 6-m telescope in a mode which gave them the arrival time of each photon with a precision of 5×10^{-8} s; claiming a temporal resolution on the order of 3×10^{-7} s, they find no evidence of *optical* microflaring in timescales of 10^{-6}–10^{-1} s. However, Tovmasyan and Zalinyan (1988) have published observations of flares lasting less than 1 s. Then there is the conflicting evidence for a general level of variability in the X-ray flux of dMe stars: Ambruster, Sciortino, and Golub (1987) find evidence of such microflaring-like behaviour in Einstein observations; Collura, Pasquini, and Schmitt (1988) find no such evidence in Exosat observations! On the other hand the sought after solar analog of the stellar X-ray microflare – fluctuations of sufficient degree in soft X-ray flux to be observable if the Sun were an unresolved point source at stellar distances – has recently been found in the temporal variation of the large active region loops observed by Haisch *et al.* (1988) with SMM.

6. Theoretical Questions

Helioseismology now indicates that there is little or no vertical gradient in the solar angular velocity, Ω, which undermines the foundation of the $\alpha\omega$ dynamo concept; Ω varies primarily with latitude, and this together with the helicity of the convection generates vertically propagating dynamo waves which can still lead to bands $(2$–4×10^5 km wide) of azimuthal field $(3$–10×10^3 G) moving along the bottom of the convection zone $(2 \times 10^5$ km). A new effect identified recently by Parker is that such bands obstruct upward convective heat transport resulting in 'cool shadows' above the magnetic bands which in turn suppress the magnetic buoyancy. Underneath each band there is an accumulation of heat, however, which may via the Rayleigh–Taylor instability locally initiate thermal plumes buoyant enough to penetrate to the surface and thereby create active regions. This still does not explain why the field is so intensely concentrated into finely structured loops in which flares originate above the surface. The explanation of that may lie in the fact that such field concentration may represent an energy minimum condition, in that there is less obstruction of convective heat transport if the same amount of magnetic field is compressed into discrete fibrils than uniformly distributed. The physical mechanism for achieving this could be downdrafts in the fibrils. It appears likely at this point that the concentration of field into fibril structures is a near-surface phenomenon.

Having established (perhaps) how magnetic flux concentrations originate below the solar surface, their extension into the solar atmosphere and their energy-releasing interactions therein are the next considerations. In the solar atmosphere, magnetic reconnection at current sheets or in current-carrying arches is thought to be the mechanism responsible for the rapid release of energy. Two key questions are: (1) How are magnetic fields stressed on time-scales of hours to days and spatial scales ranging from network magnetic elements to complexes of activity; (2) which instability finally sets in to release some of this built-up energy? Sakurai discussed these issues and especially the difference between the termination of an equilibrium sequence in which the resulting perturbation is uni-direcitonal and an explosive instability in which the new configuration which a system moves toward is determined by initial infinitesimal perturbations. While the concept of 'loss of equilibrium' resulting in magnetic field eruptions is appealing, Sturrock argued that this does not, in fact, occur; he has found that 'the magnetic field develops in a well-behaved manner and shows no evidence of catastrophic behaviour'.

Coronal mass ejections could result from a purely MHD instability resulting from lengthwise shearing of the 'quonset hut' magnetic arcade spanning a neutral line. Beyond a critical shear, the field energy exceeds that of a configuration in which the field lines extend to infinity, and this results in a jump from the closed sheared to the open potential configuration. Coronal gas is ejected outward with the expanding field. Moore presented results showing that the decrease in the volume-integrated, filament-traced magnetic field appears to agree well with estimates of the total flare/CME energy suggesting that magnetic expansion is the primary source of energization.

The next question after that of energy release is, of course, 'How is this energy converted into particle beam acceleration?' and this question was addressed by both Cargill (who gave the invited presentation of Vlahos) and by Rieger. Vlahos pointed to three critical observations that suggest 'new thinking' on particle acceleration and transport in solar flares: (1) the hard X-ray microflaring discussed above; (2) short duration (≤ 100 ms) radio spikes in the 200–300 MHz band; and (3) the timing and duration of γ-ray observations. The excellent temporal correspondence between γ-rays and hard X-rays at the outset of the impulsive phase demonstrates that in a single step electrons can be accelerated impulsively to ~ 100 MeV. The explanatory requirements for any acceleration mechanism are thus several: it must account for the proper electron and proton acceleration time-scales, particle energy spectra, correct ratio of particles, and it should have a properly high efficiency (or else another energy sink or channel would need to be identified). Overall it appears that particles are accelerated to all energies almost simultaneously.

There are three generic types of particle acceleration mechanism: coherent, Fermi or stochastic, and shock wave. The first of these can result from the action of a DC electric field or a narrow band electromagnetic wave; the electric field can appear from magnetic reconnection or from double layers. Stochastic acceleration can result from Alfvén waves having wavelengths on the same scale as the particle gyroradii. The shock wave theory combines coherent and stochastic elements. None of these are yet 'on firm ground'. Vlahos and Cargill have explored the interesting consequences of many small, localized energy releases in a highly structured (fibrous) corona: nonlinear particle and shock interactions will heat and/or accelerate particles over a much larger volume. They emphasize the consequences of such global coupling and urge a de-emphasis on the single loop model.

7. Hydrodynamic Models

Hydrodynamic models have been developed which simulate the response of plasma confined within a rigid loop when subject to heating having a given temporal and spatial profile. Such models represent the thermal phase of compact flares in other words. These models are, of course, not restricted to the simple analytical approximations of purely radiative or purely conductive cooling. (Conduction really just redistributes the energy to some other place where it usually winds up being radiated away.) Peres described an application of one such model, which was first applied to SMM observations specifically to the light curves and time histories of the line profiles of soft X-ray lines. Such analysis can model the secondary effects of a flare – plasma motion, hard X-ray impulsive response, soft X-ray late-phase light curves, line profile responses – but not specifically address the primary causes of a flare. Such a model can be applied to a stellar flare X-ray and temperature light curve (but cf. below).

Poletto showed that a reconnection model of two-ribbon flares developed for interpretation of solar SMM data could be also successfully be applied to stellar X-ray light curves. In her model, the open magnetic field created by an eruption closes back to a lower, potential energy state; reconnection occurs at progressively higher levels in the corona corresponding to the rising post-flare loops. The time profile of the magnetic energy release is then compared to the X-ray light curve. An Exosat-observed flare on EQ Pegasi was interpreted in this way, although a major uncertainty is exactly what fraction of the released magnetic energy goes into the prolonged soft X-ray emission and how this fraction might change throughout the course of the flare.

Unfortunately neither of these analyses is unique: the X-ray event the author observed on Proxima Centauri with Einstein (Haisch et al., 1983) has since been successfully modeled both ways!: as a compact event by Reale et al. (1988) using the hydrodynamics simulation and as a two-ribbon event by Poletto, Pallavicini, and Kopp (1988) using their reconnection model! On the other hand, Schmitt presented a very reassuring result relating to stellar flare analysis: he analyzed a solar flare observed by Einstein via X-ray scattering from the Earth's upper atmosphere, analyzed it as one would an unresolved stellar flare, and then checked that analysis against SMM observations of the same event. In general, the basic physical parameters derived from the spatially unresolved IPC observation were verified.

Antonucci reviewed high temperature solar flare diagnostics and showed that significant spectral differences are predicted by different heating functions in the compact loop model, but because of count rate limitations of the SMM, these rapidly changing signatures cannot yet be satisfactorily resolved. Thus even on the Sun, with its high flux, time resolution is still a major limitation for identifying the initial processes of a flare.

As for the very concept of a flare loop model, the above notwithstanding, Emslie said forthrightly: '...the assumption of an isolated, rigid, unyielding, non-evolving field structure is a highly questionable one.' For example, gas pressures during a flare *can* reach ambient magnetic pressures; heating mechanisms *can* act primarily on electrons or, conversely, ions, etc.

8. For the Future

The following potpourri that emerged during the conference is offered as a collection of possibly useful ideas and hopefully stimulating suggestions.

Foing suggested a number of areas in which improved observations of stellar flare spectra can be expected to address specific outstanding problems of flare physics. For example, high time-resolution observations of Hα line profiles during the impulsive phase should be able to distinguish signatures of electron beam heating, proton beam heating and conductive heating. Major profile changes have already been measured: extremely broad Stark wings have been seen in some of the Balmer lines along with pronounced red asymmetries, and this is consistent with an electron beam-induced pressure wave propagating down into the chromosphere; Mg II line shifts have also been observed, but while the Ca II lines brighten they appear to not otherwise be significantly altered. Better high time-resolution observations should result in valuable new diagnostics.

Recently Canfield *et al.* (1987) evaluated the equality of upward momentum (derived from blue-shifted soft X-ray lines) and downward momentum (derived from redshifted Hα) in solar flares and found a rough equality. Mullan attempted to make similar arguments for stellar flares, but for lack of simultaneous data was forced to do this using heterogeneous observations. A significant discrepancy emerged: namely that the downward momentum appeared to exceed the upward momentum by as much as five orders of magnitude.

The measurement of electron densities at various temperatures using line ratios during flares has been attempted with IUE, but using HST this could become a powerful tool, since this will allow the derivation of stellar flare volumes.

IRAS data indicate that unusual infrared excesses may be present in the quiescent dMe spectrum. Mullan suggested that this could be a signature of synchrotron radiation suggesting the presence of relativistic electrons even in the quiet coronae of dMe stars.

It was proposed by Rao that the impulsive component of some 'superflares' has indeed already been observed – in the form of cosmic γ-ray bursts, and that the corresponding late-phase has been observed in the form of 'fast-transient X-ray' (FTX) events observed during the Ariel V and HEAO-1 sky surveys. Such superflares would be extremely energetic, 5–7 orders of magnitude beyond a solar flare, and he proposed that they take place in the inter-star region of an active binary.

Schmitt suggested using microwave bursts as proxies for the extremely faint expected stellar hard X-ray impulsive phase emission.

It appears that the transition region differential emission measure should decrease in response to conductive flare heating, since a steepening temperature gradient requirement more than compensates for the transition region 'moving to lower levels in the atmosphere'. But since UV line fluxes rise along with the hard X-ray burst and such bursts have also shown evidence of being co-spatial (UVSP and HXIS), this argues for electron beams as the preferred heating mechanism. Should we ever be in a position to observe stellar hard X-rays (perhaps by proxy as suggested above), such a UV/hard X-ray (anti-)correlation would be a diagnostic of the stellar flare heating mechanism.

Regarding the underlying origin of flares, Bumba and Hejna examined the large-scale distributions of magnetic field on the Sun and found a correlation of enhanced flare activity with the redistribution of global magnetic fields. They raised the possibility that the accumulation of magnetic energy in the coronal field configuration may be less important than direct input of magnetic energy into the atmosphere from below for flare production.

The question of periodicity in occurrence of flares is still wide open; there are conflicting claims. Finding a stellar analog of the 152-day solar periodicity would be an important discovery.

Zirin has stated that 'all flares are associated with filaments'; Sakurai discussed how a magnetic island formed in an arcade may undergo explosive kink instability, or, in observational terms, filament activation. The relationship of the evolution of the prominence magnetic field *and* the role of the cool material itself is an important area of future research. In the stellar case, negative, pre-flare dips in the photometry together with simultaneous high time-resolution line spectrometry may be the key to observing the mass motions that accompany/trigger a flare.

Establishing a relationship between change in magnetic energy of a flaring volume with the energy of a flare would be a tremendous step forward; to date only a qualitative relationship has been established, for example, the untwisting of chromospheric fibrils.

The possibility of spatially resolving flare sources using very large baseline interferometry is an exciting prospect; exploratory efforts have been carried out on RS CVn systems using microwave VLBI.

Although the flare is believed to be energized by magnetic reconnection, the alternative of double-layer formation should be considered.

Since the magnetic Zeeman broadening increases as λ^2, perfecting and using the Robinson technique on infrared lines holds much promise for observing B and f on stars.

9. Concluding Remarks

Having discussed these many new, exciting, and even exotic measurements and theoretical models, it is worthwhile to return for a moment to classical solar $H\alpha$ observations, as reviewed by Martin in her article on mass motions, since practically all aspects of a flare manifest themselves in some way and at some time in $H\alpha$. There are several parting thoughts worth keeping in mind, since they clearly exemplify the complexity of 'real flares'. (1) Chromospheric flares consist of a succession of elementary localized brightenings which shows that the chromospheric flare elements are individual sub-arc sec structures. (2) These fine-scale strutures evolve at different rates and with different phases, so that even in the 'overall' decay phase, newly forming chromospheric flare elements can appear; this points out a real problem in defining flare phases, because a whole flare does not form simultaneously. (3) Remote brightenings can take place up to a few $\times 10^5$ km from the primary flare site, and this demonstrates the long distance interconnectedness of magnetic structure. (4) Surges show material rising, stopping and falling back down, which is what one would expect for material confined by magnetic flux tubes, but the downflow is not always along precisely the same path as the upflow which probably demonstrates that the magnetic field can change significantly on this time-scale. (5) Flaring arches are similar to surges but appear to show a clump of material flowing upward, traversing an arch-like path, and descending to another point in the atmosphere, and this shows that material can indeed flow from one end of a loop to the other.

It is impossible to obtain observations of this sort for stars, but as summarized in the review by Burne, significant progress is being made in measuring mass motions associated with stellar flares through high time-resolution spectroscopy. Especially in the dMe stars, flare lines stand out prominently in the spectrum above the emission from the rest of the star, and so new high-technology techniques for obtaining time-honored spectral information hold considerable promise.

This was a highly productive conference, and I have no hesitation in saying that the four years (!) of preparation and planning by myself and my colleague, Prof. Marcello Rodono, that brought it about were well worth it in the end. Two themes clearly emerged: (1) the key to progress in flare research lies in a multispectral approach with as much temporal and spectral resolution as the photon fluxes allow; and (2) the key to understanding the physics lies in a dynamic interaction between solar and stellar investigations and investigators.

Acknowledgements

This work and much of the conference was funded by the NASA SMM project via contract NAS5–30431 to the Lockheed Palo Alto Research Laboratory; stellar flare research is supported in part by the Lockheed Independent Research Program. Too many people to even begin to name made IAU Colloquium No. 104 possible, but I would be remiss to not point out my gratitude to my co-chairman, Prof. Marcello Rodono, SOC member Prof. Peter Sturrock, SMM XRP Principal Investigator Dr Keith Strong, and SMM Project Scientist, Dr Joseph B. Gurman.

References

Ambruster, C. W., Sciortino, S., and Golub, L.: 1987, *Astrophys. J. Suppl.* **65**, 273.
Athay, R. G.: 1984, *Solar Phys.* **93**, 123.
Bruner, E. C. and Lites, B. W.: 1979, *Astrophys. J.* **228**, 322.
Canfield, R. C., Metcalf, T. R., Strong, K. T., and Zarro, D. M.: 1987, *Nature* **326**, 165.

Collura, A., Pasquini, L., and Schmitt, J. H. M. M.: 1988, *Astron. Astrophys.* **205**, 197.

Dere, K. P., Bartoe, J.-D., Brueckner, G. E., Dykton, M. D., and van Hoosier, M. E.: 1981, *Astrophys. J.* **249**, 333.

Emslie, A. G. and Noyes, R. W.: 1978, *Solar Phys.* **57**, 373.

Haisch, B. M.: *Activity in Red Dwarf Stars*, D. Reidel Publ. Co., Dordrecht, Holland, p. 255.

Haisch, B. M., Linsky, J. L., Bornmann, P. L., Stencel, R. E., Antiochos, S. K., Golub, L., and Vaiana, G. S.: 1983, *Astrophys. J.* **167**, 280.

Haisch, B. M., Strong, K. T., Harrison, R. A., and Gary, G. A.: 1988, *Astrophys. J. Suppl.* **68**, 371.

Harrison, R. A. and Sime, D. G.: 1989, *Astron. Astrophys.* **208**, 274.

Hartmann, L.: 1987, 'Cool Stars, Stellar Systems and the Sun', *Lecture Notes in Physics*, Vol. 291, Springer-Verlag, Berlin, p. 1.

Hayes, M. and Shine, R. A.: 1987, *Astrophys. J.* **312**, 943.

Lin, R. P., Schwartz, R. A., Kane, S. R., Pelling, R. M., and Hurley, K. C.: 1984, *Astrophys. J.* **283**, 421.

Machado, M., Moore, R. L., Hernandes, A. M., Rovira, M. G., Hagyard, M. J., and Smith, J. B.: 1988, *Astrophys. J.* **326**, 425.

Mirzoyan, L. V. and Ambaryan, V. V.: 1988, *Astrofizika* **28**, 375.

Nishikawa, T.: 1986, *Solar Phys.* **105**, 339.

Pallavicini, R., Serio, S., and Vaiana, G. S.: 1977, *Astrophys. J.* **247**, 692.

Poletto, G., Pallavicini, R., and Kopp, R. A.: 1988, *Astron. Astrophys.* **201**, 93.

Porter, J. G., Toomre, J., and Gebbie, K. B.: 1984, *Astrophys. J.* **283**, 879.

Porter, J. G. *et al.*: 1987, *Astrophys. J.* **323**, 380.

Reale, F. *et al.*: 1988, *Astrophys. J.* **328**, 256.

Tovmasyan, G. M. and Zalinyan, V. P.: 1988, *Astrofizika* **28**, 131.

ELECTRON BEAM AS ORIGIN OF WHITE-LIGHT
SOLAR FLARES

J. ABOUDARHAM and J. C. HENOUX

DASOP, Observatoire de Paris-Meudon, 92195 Meudon Principal Cedex, France

Abstract. We study the effect of chromospheric bombardment by an electron beam during solar flares. Using a semi-empirical flare model, we investigate energy balance at temperature minimum level and in the upper photosphere. We show that non-thermal hydrogen ionization (i.e., due to the electrons of the beam) leads to an increase of chromospheric hydrogen continuum emission, H⁻ population, and absorption of photospheric and chromospheric continuum radiation. So, the upper photosphere is radiatively heated by chromospheric continuum radiation produced by the beam. The effect of hydrogen ionization is an enhanced white-light emission both at chromospheric and photospheric level, due to Paschen and H⁻ continua emission, respectively. We then obtain white-light contrasts compatible with observations, obviously showing the link between white-light flares and atmospheric bombardment by electron beams.

1. Introduction

White-light flares (WLF) are characterized by an increase in emission in the visible range. It has been deduced from observations that the source of this emission is placed in the low atmosphere: low chromosphere and upper photosphere, showing a heating of these layers during the flare. It is important to determine whether this temperature increase can be explained by energy transport from the lower corona, or whether it requires a partial or total *in situ* energy release.

The two main contributors of white-light emission are negative hydrogen, originating in the upper photosphere (see Neidig, 1983, for instance), and Paschen recombination continuum from chromospheric origin.

Hudson (1972) suggested a link between WLF and atmospheric bombardment by beams, proton, or electron beams. But none of these seem to be able to explain the observed low atmosphere heating: protons, which could penetrate deeply in the atmosphere, must be over 100 MeV in order to reach the photosphere and then must emit γ-emission which is not systematically correlated with WLF; and electrons do not seem to be able to deposit enough energy in the upper photosphere. But, due to the lack of detailed calculations, it was impossible to remove doubt.

In Section 2, we shall show what makes the electron beam a good applicant for WLF, deduced from hard X-ray observations.

Section 3 is devoted to the detailed calculation concerning both an electron beam and the flaring atmosphere. We first derived the energy deposited by the beam along its way through the atmosphere. From this, we deduced hydrogen excitation and ionization rates due to the electrons of the beam, what we call 'non-thermal' collisional rates. The semi-empirical flare model F_2 (Machado *et al.*, 1980), giving temperature distribution versus column mass for a strong flare was used in order to estimate the thermal collisional excitation and ionization rates. The quiet-Sun model C (Vernazza, Avrett, and Loeser, 1981) was used as a reference, for the 'background' Sun, necessary in the calculation of the energy input present in the quiet-Sun conditions and of the contrast during the flare. We then studied the atmospheric response, solving radiative transfer, statistical equilibrium and electrical neutrality equations, leading to the determination of radiative losses and thermal (i.e., due to background electrons) excitation and ionization rates. In Section 4 we show that non-thermal rates should not be neglected. Their contribution leads to an increase in electron density which enhances Paschen continuum emission and H^- absorption. Consequently, the upper photosphere as represented by semi-empirical flare model F_2 is not in energetic equilibrium. When the initial total energy flux of the electron beam is 10^{12} ergs cm^{-2} s^{-1}, energy balance in the upper photosphere is reached by increasing the upper photosphere temperature by approximately 240 K.

This radiative heating of the low atmosphere during flares leads to a strong white-light contrast which can now account for WLF observations.

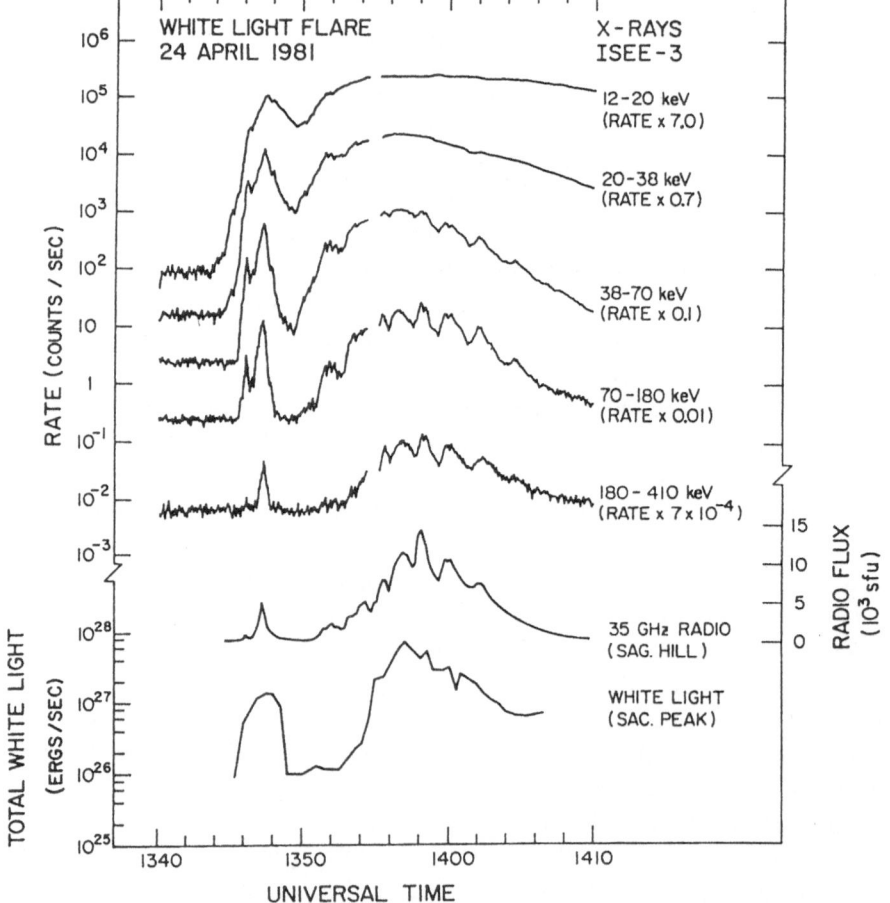

Fig. 1.   Time-correlation between hard X-ray and white-light emission (from Kane *et al.*, 1985).

2. An Approach of the WLF Problem

Simultaneous observations of hard X-ray and white-light emissions show a very good time-correlation between their impulsive phases, as shown in Figure 1. This means the mechanisms leading to those emissions are connected. It is generally believed that hard X-ray emission is produced by collisional bremsstrahlung of non-thermal electrons accelerated in the low corona and slowed down in the denser chromosphere (see Brown, 1973, for instance). The best observed time-correlations between hard X-ray and white light emission are for photons of energy greater than ≈ 70 keV, and only electrons of energy above 70 keV can reach chromosphere. We can then speculate that the chromospheric bombardment by an electron beam is, at least, somewhat responsible for white-light emission. The hard X-ray emission is fitted by a power-law distribution in energy with a spectral index γ; so, using the thick-target approximation, the electron flux is also represented by a power-law distribution with a spectral index δ which is related to γ the following way (Brown, 1972):

$$\delta = \gamma + 1 .$$

Then, in the weakly ionized low chromosphere and upper photosphere, the electrons of the beam lose their energy in collisions with neutral atoms, thus exciting and ionizing neutral hydrogen of the chromosphere.

3. Calculation

3.1. Beam excitation and ionization of the atmosphere

The rate of energy deposited by the electrons of the beam in collisional excitation and ionization of neutral hydrogen is given by (Emslie, 1978; Chambe and Hénoux, 1979)

$$\frac{dE^H}{dt} \ (\text{erg cm}^{-2}\,\text{s}^{-1}) =$$

$$= \tfrac{1}{2}(1 - x)n_{\mathrm{H}}\varDelta' \ \frac{K\mathscr{F}_1}{E_1}\left(\frac{N}{N_1}\right)^{-\delta/2}(\delta - 2)\int_0^{u_1}\frac{u^{(\delta/2)-1}}{(1-u)^{(2+\bar{\beta})/(4+\bar{\beta})}}\ du . \quad (1)$$

The electron flux is represented by a power law $\mathscr{F} \sim E^{-\delta}$ with a low-energy cut-off E_1. $\mathscr{F}_1$ is the total energy flux above E_1 and N_1 is the hydrogen column density at the deepest level electrons of energy E_1 can reach. A given column density N can only be penetrated by electrons of initial energy greater than

$$E_N = \left[\left(2 + \frac{\bar{\beta}}{2}\right)\frac{\bar{\gamma}KN}{\mu_0}\right]^{1/2} , \quad (2)$$

where μ_0 is the cosine of the angle between the initial velocity vector and the solar vertical; $K = 2\Pi e^4$; $\bar{\beta}$ and $\bar{\gamma}$ are the mean values along the electron trajectory of β and

γ, given by

$$\beta = \frac{2x\Lambda + (1 - x)(\Lambda' + \Lambda'')}{x(\Lambda - \Lambda') + \Lambda'} , \qquad \gamma = x\Lambda + (1 - x)\Lambda' , \qquad (3)$$

Λ, Λ', and Λ'' are specified in Emslie's paper. Λ represents the effects of the collisions of the electrons of the beam on the ambient free electrons and protons; Λ' and Λ'' the effects of the inelastic and elastic collisions, respectively, on neutral hydrogen atoms; x is the hydrogen ionization degree

$$u = (E_N/E)^2 ,$$

where E is the electron energy, and

$$u_1 = 1, \qquad N > N_1 ,$$
$$u_1 = (N/N_1), \quad N \leq N_1 .$$

The electrons of the beam lose their energy in elastic collisions with electrons and inelastic collisions with neutral hydrogen. The latter correspond mainly to excitation of levels $n = 2$ and $n = 3$ and to ionization of hydrogen. Therefore, the rate of energy deposit due to collisions with neutral hydrogen is directly related to the non-thermal excitation rates, to levels 2 and 3, and ionization rate:

$$\frac{dE^H}{dt} \ (eV \ s^{-1}) \simeq n_1(\chi_{12} C_{12}^B + \chi_{13} C_{13}^B + \chi_{1c} C_{1c}^B), \qquad (4)$$

where dE^H/dt is the rate of energy deposit due to collisions with hydrogen. C_{12}^B, C_{13}^B, and C_{1c}^B are, respectively, the excitation rates to levels 2 and 3 and the ionization rate due to the electron beam. χ_{12}, χ_{13}, and χ_{1c} are excitation potential of levels 2 and 3, and ionization potential, respectively. The non-thermal rates C_{ij}^B are, as an approximation, proportional to excitation and ionization cross-sections of hydrogen by 10–100 keV electrons. Mott and Massey (1965) give the relative importance of these terms: respectively, 0.53, 0.08, 0.39, leading to the following expressions of non-thermal excitation and ionization rates:

$$C_{12}^B = 2.77 \times 10^{10} \ \frac{dE^H}{dt} , \qquad C_{13}^B = 0.15 C_{12}^B , \qquad C_{1c}^B = 0.74 C_{12}^B .$$

3.2. THERMAL EXCITATION AND IONIZATION OF THE ATMOSPHERE

A plane-parallel atmosphere representation of the solar chromosphere can be adopted to describe the main features of the energy transport process during solar flares. We used semi-empirical flare model F_2 (Machado et al., 1980), which represents a bright flare. The non-flaring atmosphere has been represented by the quiet-Sun model C (Vernazza, Avrett, and Loeser, 1981). Temperature distribution of models C and F_2 is shown in Figure 2. Temperature minimum for models C and F_2, respectively, 4165 K and 4960 K.

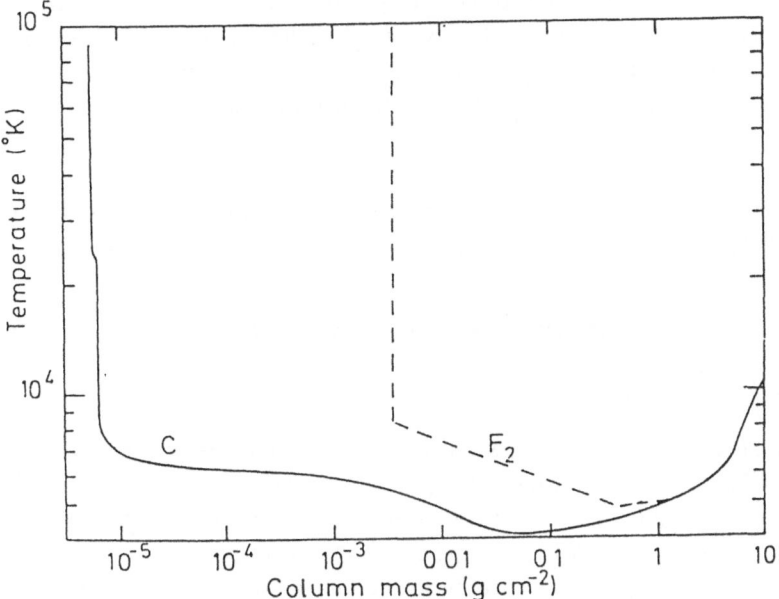

Fig. 2.   Temperature distribution for the quiet Sun (model C) and for bright flares (model F_2).

From these models, we computed the non-LTE populations of H^- and of the three first levels of hydrogen by solving the transfer equation for Balmer α, Lyman α, and Lyman β lines, and for Lyman, Balmer, Paschen, and Brackett continua, and the equation of statistical equilibrium (Aboudarham, 1986). The fourth level was supposed to be in LTE. Feautrier's (1964) method was used.

Thermal collisional excitation and ionization rates C_{ij} have been deduced from this computation. We then included non-thermal processes in statistical equilibrium equation. Their effect will be described in the next section. We also calculated radiative losses for H and H^-, the latter being the main contributor at temperature minimum region (TMR) and below (Avrett, 1980, 1985). Bound-free and free-free H^- cooling rates were computed:

$$\Phi_{H_{\overline{ff}}} = 4\Pi n_{H^-} \int \chi_{ff}(S_\nu - J_\nu) \, d\nu, \tag{5}$$

$$\Phi_{H_{\overline{bf}}} = 4\Pi n_{H^-} \int \chi_{bf} \left[n_{H^-}^* / n_{H^-} \left(\frac{2h\nu^3}{c^2} + J_\nu \right) e^{-h\nu/kt} - J_\nu \right] d\nu, \tag{5}$$

where n_{H^-} and $n_{H^-}^*$ are H^- number density in the non-LTE and LTE case, respectively. A detailed study of H^- can be found in Aboudarham and Hénoux (1986a, 1987). We can then estimate the total amount of energy deposit $\Delta\Phi$, due to both H and H^-, required to heat the concerned regions to flare temperatures. It is supposed that the energy deposit, that takes place in the average quiet Sun, does not vary from quiet Sun to flare conditions. In addition to this, we assume the energy deposit depends only on column mass m. Then the additional energy deposit $\Delta\Phi$ required, at column mass m, to increase quiet-Sun temperature to flare temperature is given by

$$\Delta\Phi_{F_2 C} = \Phi_{F_2}(m) - \Phi_C(m), \tag{6}$$

where $\Phi_{F_2}(m)$ and $\Phi_C(m)$ are, respectively, the radiative losses per unit volume inferred from model F_2 and model C.

4. Results: Beam Effect on the Atmosphere

4.1. BEAM CHARACTERISTICS AND EFFECTS ON HYDROGEN

The main features characterizing the energy deposited by the beam, as shown in Equation (1), are low-energy cut-off, E_1, initial flux above E_1, $\mathscr{F}_1$, and spectral index, δ. The standard value of 20 keV for the low-energy cut-off was adopted and the total energy flux $\mathscr{F}_1$ above 20 keV is 10^{12} ergs cm^{-2} s^{-1}. The choosen spectral index is $\delta = 4$. Those values were adopted according to the fact model F_2 represents a strong flare. Discussion on these values can be found in Aboudarham and Hénoux (1986b). In fact, only electrons of energy greater than 70 keV can reach chromosphere; this leads to an energy flux of 8×10^{10} ergs cm^{-2} s^{-1} above 70 keV.

Figure 3 shows thermal collisional excitation and ionization rates and non-thermal

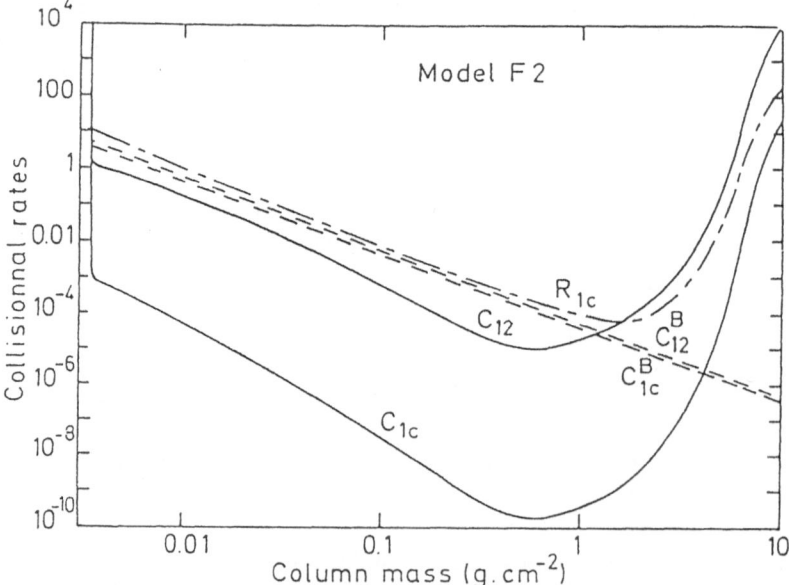

Fig. 3. Collisional excitation rate of hydrogen level 2 and collisional and radiative ionization rates of hydrogen as a function of column mass. Full lines: thermal collisional rates. Dashed lines: non-thermal collisional rates. Dotted-dashed line: photoionization rate.

ones, according to the above values characterizing the beam. We can see that non-thermal rates are much stronger than thermal ones, especially for non-thermal ionization rate which is four to six orders of magnitude higher than the thermal ionization rate in the middle chromosphere and below. This means that a detailed flaring atmosphere calculation should not neglect non-thermal processes, even for initial energy flux in the beam several orders of magnitude lower than 10^{12} ergs cm^{-2} s^{-1}.

Non-thermal ionization increases the electron number density and decreases the hydrogen level 1 population. This makes the LTE departure coefficient for levels 1, 2, 3 of hydrogen decrease. Source functions in continua decrease, leading to enhanced radiative losses, especially in the Paschen continuum, as shown in Figure 4. This enhancement of the chromospheric Paschen emission should lead to an increase in the white light emission.

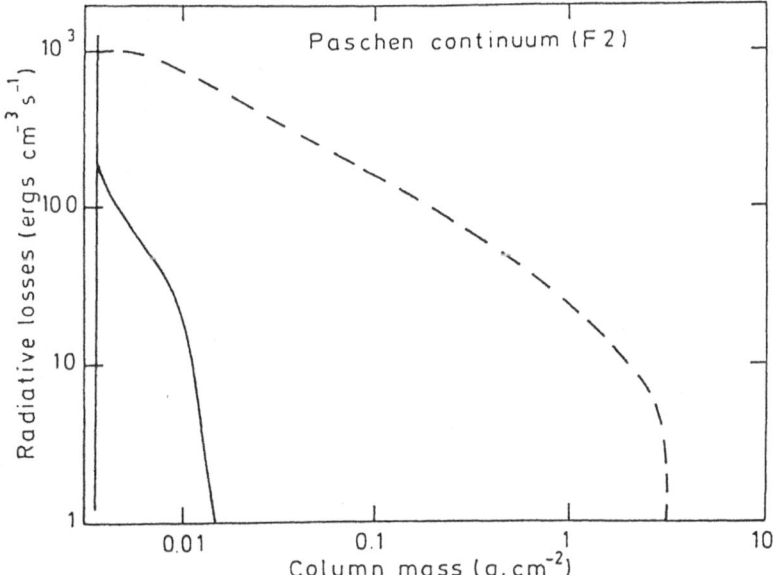

Fig. 4.   Column mass dependence of the hydrogen Paschen continuum emission. Full line: without non-thermal collisional excitation and ionization. Dashed line: non-thermal collisional excitation and ionization included.

4.2. EFFECT ON TOTAL RADIATIVE LOSSES

Figure 5 shows the column mass dependence of the difference of the total radiative losses (H + H$^-$), $\Delta\Phi_{F_2 C}$ (Equation (6)), calculated from models F_2 and C. We shall only examine the radiative losses below temperature minimum, where H$^-$ originates. For a detailed discussion on chromospheric radiative losses, see Aboudarham and Hénoux (1986b).

The increase of the electron number density leads to an increase of the H$^-$ number density. When the photoionization rate exceeds the recombination rate, H$^-$ absorbs

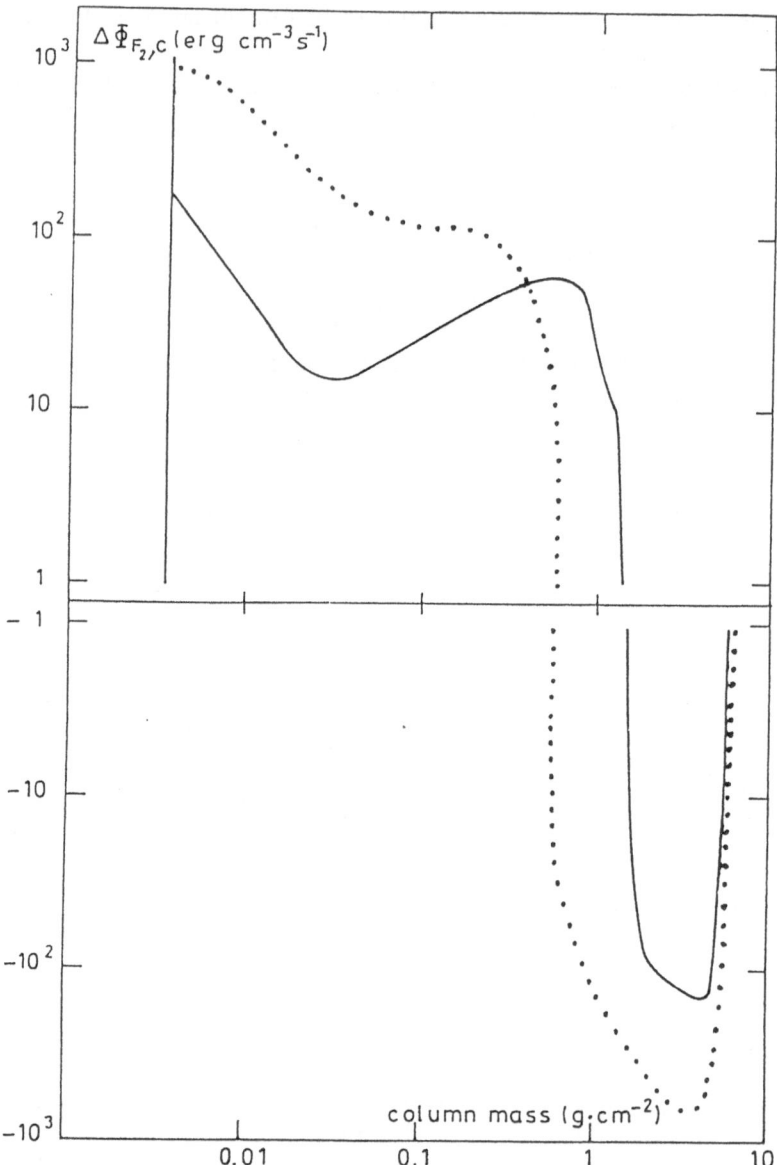

Fig. 5.   Column mass dependence of $\Delta\Phi_{F_2, C}(m)$, difference of the radiative losses (H + H$^-$) of models F_2 and C. Full line: non-thermal hydrogen excitation and ionization not included. Dotted line: non-thermal hydrogen excitation and ionization included ($\mathscr{F}_i = 10^{12}$ ergs cm^{-2} s^{-1}).

more radiation in the bound-free continuum than it radiates, and the total H$^-$ radiative losses are negative. At constant temperature, density of neutral hydrogen and radiation field, the amplitude of the bound-free cooling rate increases with electron number density. So, as we can see in Figure 5, the energy input required to heat the atmosphere from quiet Sun to flare temperatures becomes more negative at column masses between

0.6 and 6 g cm^{-2}. This means that departure from energy balance becomes stronger at this level. This implies that, when taking into account non-thermal processes due to an electron beam, the temperature of the upper photosphere at this depth must increase for the atmosphere to be in energy balance. Temperature enhancement as high as 240 K is produced at column mass $m = 2$ g cm^{-2}, as shown in Figure 6.

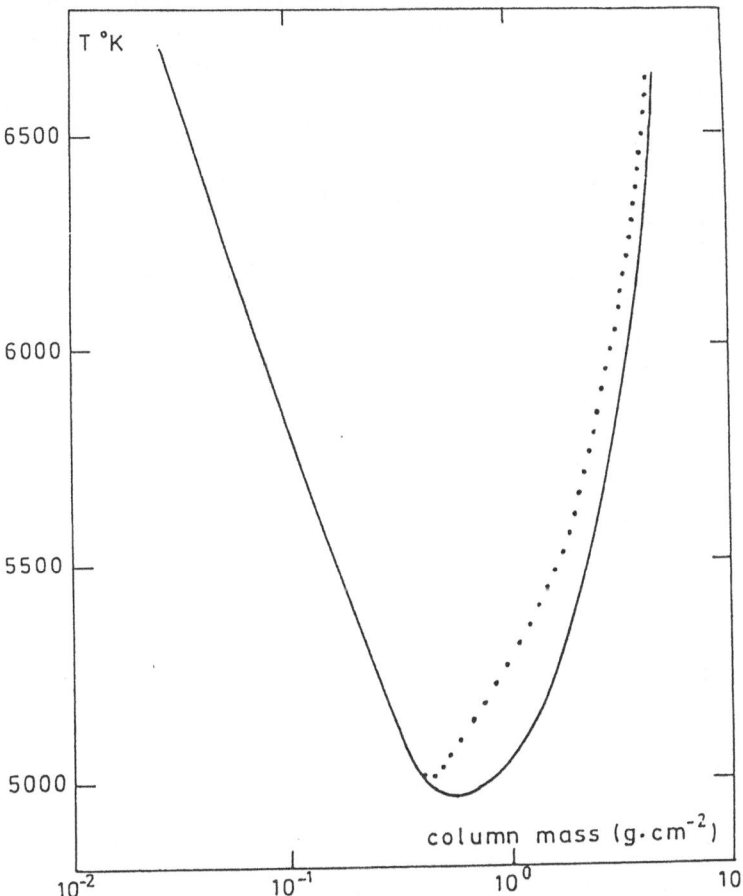

Fig. 6.   Temperature distribution illustrating the upper photospheric heating induced by non-thermal processes. Full line: semi-empirical flare model F_2. Dotted line: modified temperature distribution leading to photospheric energy balance.

4.3. EFFECT ON CONTRAST

The enhancement of both chromospheric Paschen emission and upper photospheric H$^-$ emission due to the increase of the photospheric temperature leads to an increase of the white-light continuum emission. This is quantitatively represented by the contrast C_v, defined as

$$C_v = \frac{I_v^{F_2} - I_v^C}{I_v^C} \, ,$$

where $I_\nu^{F_2}$ and I_ν^C are, respectively, the specific intensity inferred from model C and model F_2. Figure 7 shows the contrast in the visible range when photospheric radiative back-heating is not taken into account and when photospheric temperature enhancement deduced from the beam is included, compared to an observation by Neidig (1983, Figure 5, p. 293) of the 24 April, 1981 flare. Note that Neidig's observed contrast is not relative to quiet-Sun intensity, but to the mean intensity averaged over lines and continua in the wavelength region of the measurements. So, the observed contrast has been recalculated (Neidig, 1988) in quiet-Sun units, as a function of wavelength. As suggested by Neidig (1988), the spectral shape of the observed contrast might be due to atmospheric

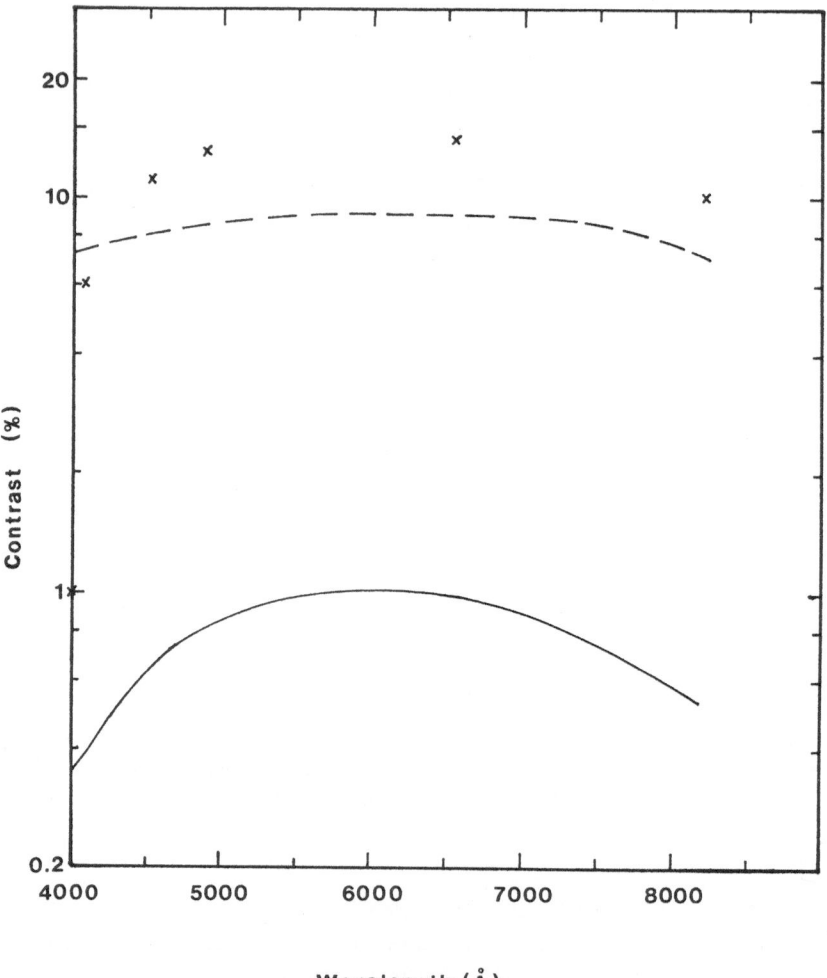

Fig. 7. Wavelength dependence of contrast C_ν between flare and quiet-Sun specific intensities, for helio-centric angle $\Theta = 53°$ ($\mu = \cos \Theta = 0.6$). Full line: non-thermal processes not included. Dashed line: non-thermal processes included. Crosses: observation by Neidig (1983) of 24 April, 1981 flare, modified data (see text).

refraction moving the flare image across the spectograph slit, so that we see a progressively fainter part of the flare at shorter wavelength. It is obvious, in Figure 7, that only the inclusion of non-thermal processes due to the electrons of the beam can explain the observed white-light emission, thus demonstrating the link between electron beams and white-light flares.

5. Discussion and Conclusion

Some observations show a very strong white-light contrast which cannot be explained with our calculations, except at large heliocentric angles. In fact, they are probably overestimated, due to two reasons: first, contrast is often measured with filters that include line contribution to the total intensity. Second, the very important broadening of chromospheric Balmer lines during flares interfere strongly with visible continuum emission.

Hudson (1972) suggested that the white-light emission observed in WLF resulted from the enhancement of the hydrogen recombination spectrum due to the increase of the hydrogen ionization by collision with non-thermal electrons. Indeed, the exact computation of the white-light hydrogen and H^- spectrum shows a significant increase of this emission due to a rise of the electron number density. The resulting increase of H^- population leads to an enhanced absorption of continuum radiation. This produces both photospheric radiative heating and an upward shift of the depth of formation of H^- radiation. These processes have opposite consequences and are time-dependent. On a short time, they may produce a decrease of the net radiation field, like negative flares observed on stars. But later on the upper photosphere radiative heating, together with the increased hydrogen recombination emission, generates a white-light flare. The main driver of the white-light emission is the hydrogen non-thermal ionization, and the dominance of the non-thermal processes over the thermal ones makes this conclusion quite independent of the assumed chromospheric temperature distribution.

An electron beam of initial energy flux $\mathscr{F}_1 = 10^{12}$ ergs cm^{-2} s^{-1} above the standard 20 keV low-energy cut-off with a spectral index $\delta = 4$ induces an average temperature enhancement in upper photosphere $(0.6 < m < 6$ g cm$^{-2})$ of 240 K, leading to a white-light contrast in good agreement with observations, demonstrating that WLF can be explained by an electron bombardment of the atmosphere during the flare, and explaining the time-correlation between hard X-ray and white-light emissions.

Acknowledgements

The authors wish to thank Dr D. F. Neidig for helpful comments and for re-calculating contrast of the 24 April, 1981 flare in quiet-Sun units in order to allow us to make comparisons with our computation, as mentioned in Section 4.3.

References

Aboudarham, J.: 1986, Ph.D. Thesis, Univ. Paris VI.

Aboudarham, J. and Hénoux, J. C.: 1986a, *Astron. Astrophys.* **156**, 73.

Aboudarham, J. and Hénoux, J. C.: 1986b, *Astron. Astrophys.* **168**, 301.

Aboudarham, J. and Hénoux, J. C.: 1987, *Astron. Astrophys.* **174**, 270.

Avrett, E. H.: 1980, in R. M. Bonnet and A. K. Dupree (eds.), *Solar Phenomena in Stars and Stellar System*, D. Reidel Publ. Co., Dordrecht, Holland.

Avrett, E. H.: 1985, in B. W. Lites (ed.), *Chromospheric Diagnostic and Modeling*, NSO, Sunspot, NM.

Brown, J. C.: 1972, *Solar Phys.* **26**, 441.

Brown, J. C.: 1973, *Solar Phys.* **28**, 151.

Chambe, G. and Hénoux, J. C.: 1979, *Astron. Astrophys.* **80**, 123.

Emslie, A. G.: 1978, *Astrophys. J.* **224**, 241.

Feautrier, P.: 1964, *Compt. Rend. Acad. Sci. Paris* **258**, 3189.

Hudson, H. S.: 1972, *Solar Phys.* **24**, 414.

Kane, S. R., Love, J. J., Neidig, D. F., and Cliver, E. W.: 1985, *Astrophys. J.* **290**, L45.

Machado, M. E., Avrett, E. H., Vernazza, J. E., and Noyes, R. W.: 1980, *Astrophys. J.* **242**, 336.

Mott, N. F. and Massey, H. S. W.: 1965, *The Theory of Atomic Collisions*, Clarendon Press, Oxford.

Neidig, D. F.: 1983, *Solar Phys.* **85**, 285.

Neidig, D. F.: 1988, private communication.

Vernazza, J. E., Avrett, E. H., and Loeser, R.: 1981, *Astrophys. J. Suppl.* **45**, 635.

SOLAR FLARE SPECTRAL DIAGNOSIS: PRESENT AND FUTURE

ESTER ANTONUCCI

Istituto di Fisica Generale dell'Università, Via P. Giuria 1, I-10125 Torino, Italy

Abstract. New perspectives in solar diagnosis have been opened in recent years with the advent of high-resolution soft X-ray spectroscopy for plasmas forming at temperatures above 10^7 K. The spectra obtained with the soft X-ray spectrometers flown during the last solar maximum on the major space missions dedicated to flares have allowed detailed studies of the hydrodynamic response of coronal loops to impulsive energy deposition and of the formation of the high-temperature plasma as a consequence of such dynamic effects. These studies are possible since high-resolution spectrometers give an accurate measure of both line intensities and profiles in important spectral regions, covering the emission of highly ionized heavy ions, which allow a direct determination of most of the crucial plasma parameters in the flare region. In response to the impulsive energy release in the flare region, while the intensity of soft X-ray lines increases, line profiles show large non-thermal broadenings and strong blue-asymmetries.

There have been important contributions in the understanding of the formation of the flare high-temperature plasma, as an effect of the hydrodynamic response of the solar atmosphere to impulsive chromospheric heating. On the other hand, the attempts to investigate the primary energy release and transport, on the basis of the soft X-ray spectral data, have not yet been entirely successful. Significant differences in the emitted spectra are expected at the very onset of flares for different energy deposition and transport processes, but the sensitivity of the present experiments is still insufficient to detect with good statistics the early stage of flares and, therefore, to allow a reliable discrimination. It is expected that future experiments with higher sensitivity will be of great importance for relating with less ambiguity the observed flare evolution in soft X-rays to the primary energy deposition in the flaring coronal loops.

1. Introduction

The flare emission in the X-ray spectral region is classically divided into two components: a non-thermal component in hard X-rays, with photon energy above 20 keV, characterizing the impulsive phase, and a thermal component in soft X-rays, with energy below approximately 20 keV, predominant in the longer-lasting gradual phase. Soft X-rays are emitted from an intense coronal source forming during the flare event at temperatures from 1 to 3×10^7 K and with an emission measure exceeding $10^{48}–10^{49}$ cm^{-3}, values well above active region ones. The hot flare plasma is confined within coronal magnetic loops. Such results were primarily obtained with proportional counters (giving temperature and emission measure) and the X-ray imaging telescopes flown during the Skylab mission, which led to the identification of a highly inhomogeneous corona, with temperature and density distributions controlled by the magnetic fields.

The major advances in the study of the flare thermal plasma during solar maximum 1980 have been achieved through the study of high-temperature lines, with the advent of high-resolution spectroscopy. The emission of the flare plasma, is in fact, rich in lines of the higher ionization stages of the most abundant ions, not observed in active regions due to their lower temperatures. Three major experiments based on high-resolution

Solar Physics **121**: 31–60, 1989.

crystal spectrometers were flown during solar cycle 21: the Solflex spectrometer on the P78/1 satellite, launched on February 24, 1979, covering the wavelength range 1.8–8.5 Å (Doschek, 1983); the Soft X-ray Polychromator on the Solar Maximum Mission, launched on February 14, 1980, with wavelength intervals 1.4–22.4 Å and 1.7–3.3 Å (Acton *et al.*, 1980); and the SOX spectrometers on the Hinotori satellite, launched on February 26, 1981, with wavelength range 1.73–1.95 Å (Tanaka *et al.*, 1982a). The most important spectral regions selected for these experiments include the H-like and He-like emission of heavy ions, such as Ca and Fe. The Fe XXV region has been observed by all three experiments. This important line group was first observed by Neupert *et al.* (1967) and spectrally resolved by Grineva *et al.* (1973). Such observations made possible the identification of the individual lines in the group. High-resolution spectroscopy of flare emission has offered the possibility to determine directly most of the thermal plasma parameters and to measure the dynamics in the flare region.

2. Diagnosis of the Thermal Flare Plasma

The study of the intensity of the hot flare lines in the spectral regions of emission of H-like and He-like ions of heavy elements gives information on the plasma by allowing determination of physical parameters such as electron temperature, differential emission measure, ionization states and, in some cases, electron density and element abundances. In addition to the line-intensity diagnosis, high-resolution spectrometers have provided, for the first time, through the measure of line profiles and thereby of Doppler temperatures and Doppler shifts, a complete picture of mass motions in the flare region. The observations have shown that the flare thermal plasma is highly dynamic at least in the early phase of the event.

2.1. DIAGNOSTIC METHODS FOR THE HE-LIKE IONS

The spectral regions of the He-like ions are rich in satellites to the He-like resonance lines, falling at relatively longer wavelengths, formed by dielectronic recombination of the He-like ions with electrons and by inner-shell excitation of Li-like ions, which become observationally important for high nuclear charge ions. Satellites give the possibility of a direct measure of both the electron temperature and the departure from ionization equilibrium of the flaring plasma. The electron temperature is measured as a function of the relative satellite to resonance line intensity in a manner which is independent of ionization balance, since the lines are coming from the same ionization stage. A similar method based on the ratio of the satellite to resonance line intensity can also be used for H-like ion spectra (Dubau *et al.*, 1981; Parmar *et al.*, 1981). The ionization equilibrium is tested by measuring, for instance, the relative intensity of the He-like to the Li-like spectrum, which is a function of the relative populations of adjacent ionization stages. The importance of dielectronic recombination satellites for directly measuring plasma parameters has been pointed out and the theory has been developed by Gabriel (1972), Bhalla, Gabriel, and Presnyakov (1975) and later works of Gabriel and co-workers. The ratio of the forbidden to intercombination lines of

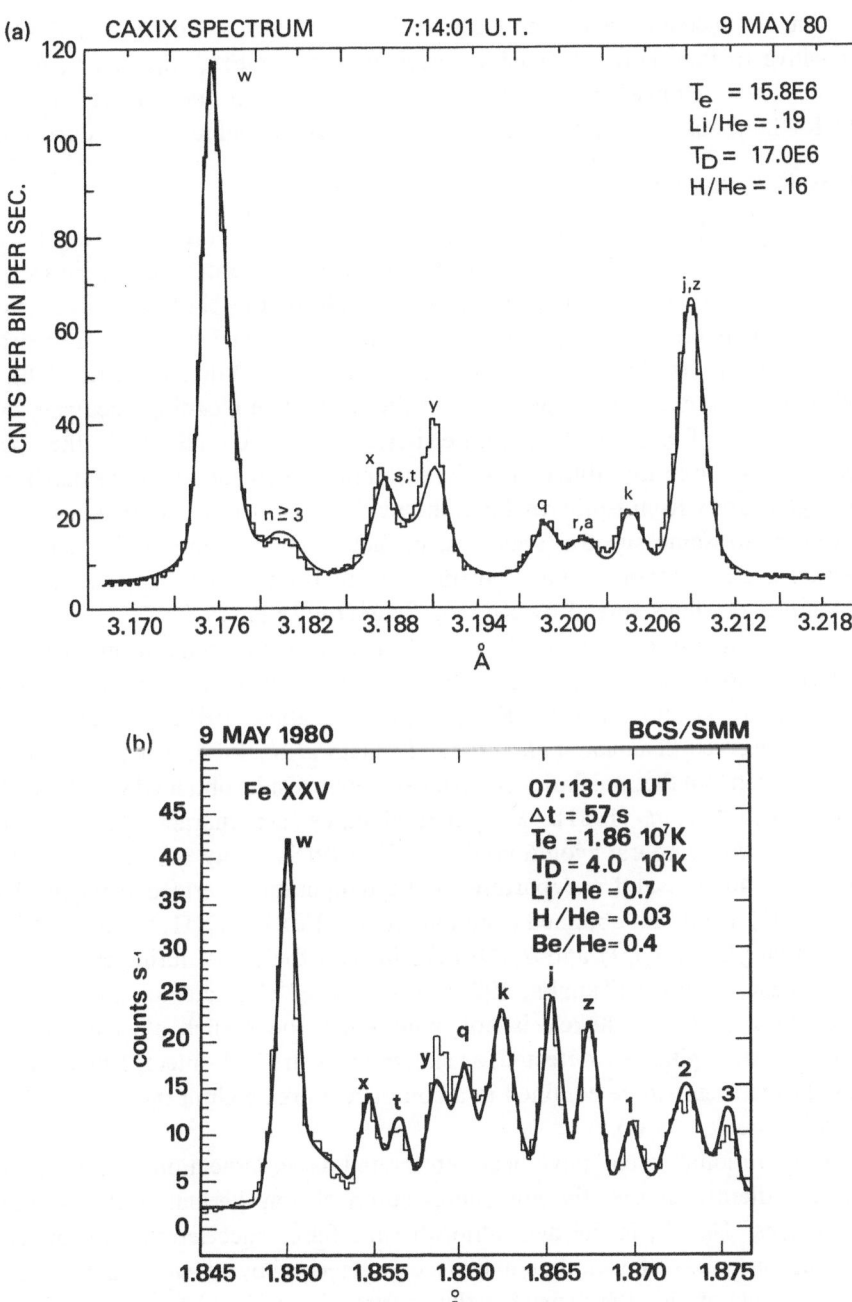

Fig. 1.   CaXIX (a) and FeXXV (b) spectra observed with the Soft X-Ray Polychromator on the Solar Maximum Mission. The continuous curves represent spectra best fitting the observations.

He-like ions are instead sensitive to electron density (Gabriel and Jordan, 1969). The ions sensitive to flare coronal densities, such as O VII, Ne IX, Mg XI, are, however, formed at lower temperatures (with emissivity peaking in the interval from 2 to 6×10^6 K) than those characterizing the hotter soft X-ray source (10^7 K) during flares.

2.2. TEMPERATURE MEASUREMENTS

The He-like and H-like ion emissions more extensively observed during flares in the course of the 1980 solar maximum are relative to Ca XIX, Ca XX, Fe XXV, and Fe XXVI. Figures 1(a) and 1(b) present an example of the He-like Ca and Fe ion emission, respectively. Plasma parameters are usually derived by fitting a theoretical spectrum to the observed one (Antonucci *et al.*, 1982; Akita, 1985; Tanaka, 1986). This method is used to take into proper account the heavy line blending occurring in this zone. The values of electron temperature derived from the emission in the spectral regions of different ions may differ when the temperature distribution in the flare plasma departs significantly from isothermal conditions. The temperature derived from a line ratio even in the same spectral region is, in fact, a mean value of the temperature distribution averaged with the line contribution functions, which in the case of high-temperature X-ray lines spread over a large temperature interval.

Temperature measurements with the SOX/Hinotori observations in Fe XXV and Fe XXVI have evidenced the existence of a 'hot', or 'superhot', dense thermal component, with temperature T_e above 3×10^7 K and density n_e above 10^{11} cm^{-3}, in half of the large flares observed (Tanaka *et al.*, 1982a; Tsuneta *et al.*, 1984; Tanaka, 1986). The discovery of such 'hot thermal flares' confirms previous results obtained with hard X-ray continuum by Lin *et al.* (1981). 'Hot thermal flares' are characterized by a close association of the temporal evolutions in the Fe XXVI intensity and hard X-ray continuum below 40 keV, and a departure of the temperature derived from the H-like Fe XXVI, T_e(H), from that obtained from the He-like Fe XXV, T_e(He), around the peak of the gradual phase. T_e(H) and T_e(He) are instead the same during the rising and decaying phase, Figure 2 (Tanaka, 1986). The gradual character of the hard X-ray emission observed below 40 keV is not maintained above such a value. At higher energies impulsive spikes are detected, as shown in Figure 3 (Tsuneta, 1987). The 'hot' thermal plasma seems to be confined in a compact source high in the loop (Takakura *et al.*, 1983; Tanaka, 1986).

Such observational results have been interpreted as an indication that, in the high-density 'hot' thermal source, the low energy cutoff of non-thermal electrons shifts to higher energies. That is, in the dense 'hot-thermal flares', acceleration of low-energy electrons becomes less efficient than in flares characterized by a lower density. This has led to the suggestion that the density in the energy release site plays an important role in partitioning the released energy into particle acceleration and local heating (Tanaka, 1987; Tsuneta, 1987; Watanabe, 1987).

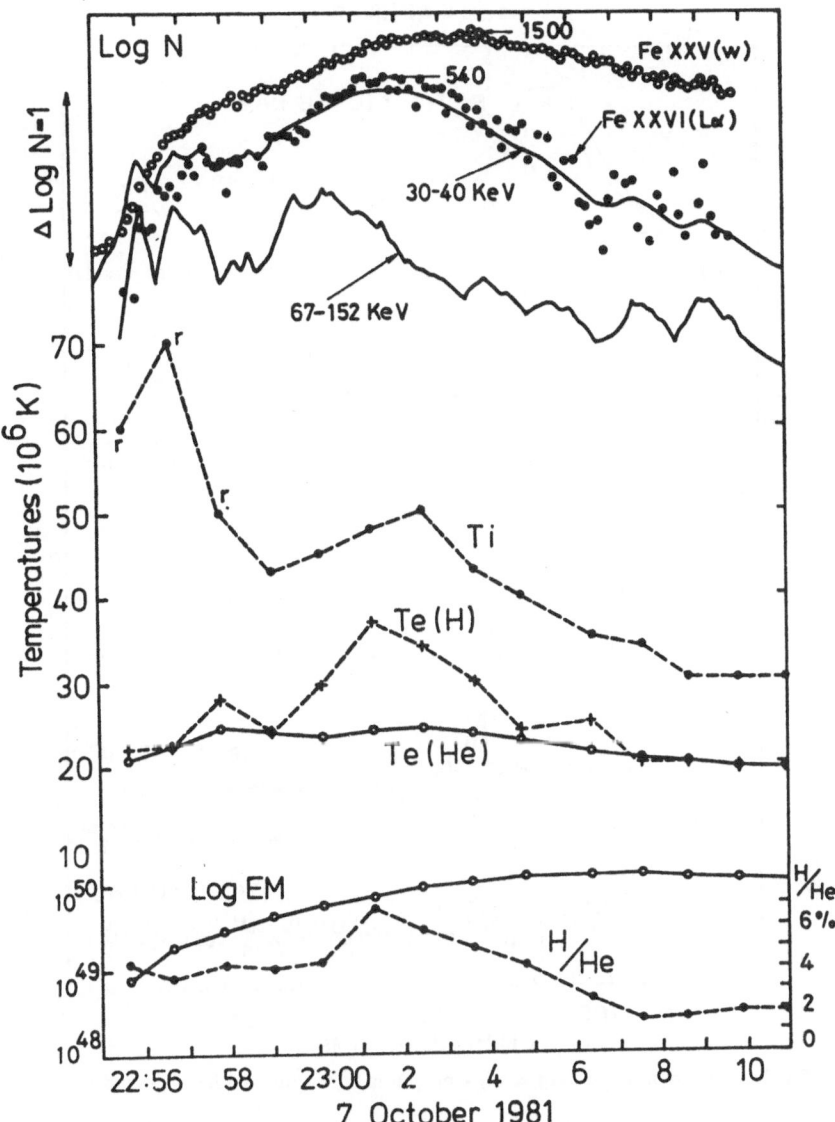

Fig. 2.   Evolution of the physical parameters derived from the Fe XXV and Fe XXVI Hinotori observations
for a 'hot thermal flare' (Tanaka, 1986).

2.3. DENSITY MEASUREMENTS

Flare densities have been measured from the He-like spectra of O VII, Ne IX, and Mg XI.
McKenzie *et al.* (1980) and Doschek *et al.* (1981) derived from the O VII emission values
of $n_e = 3 \times 10^{12}$ cm^{-3} for the April 8, 1980 flare. From Ne IX spectra, Wolfson *et al.*
(1983) derived values ranging from 3×10^{11} cm^{-3} to 1.5×10^{12} cm^{-3} in the
November 5, 1980 flare. In Mg XI spectra, Linford and Wolfson (1989) and Linford,
Lemen, and Strong (1989) measured values of about 3×10^{12} cm^{-3} for a sample of

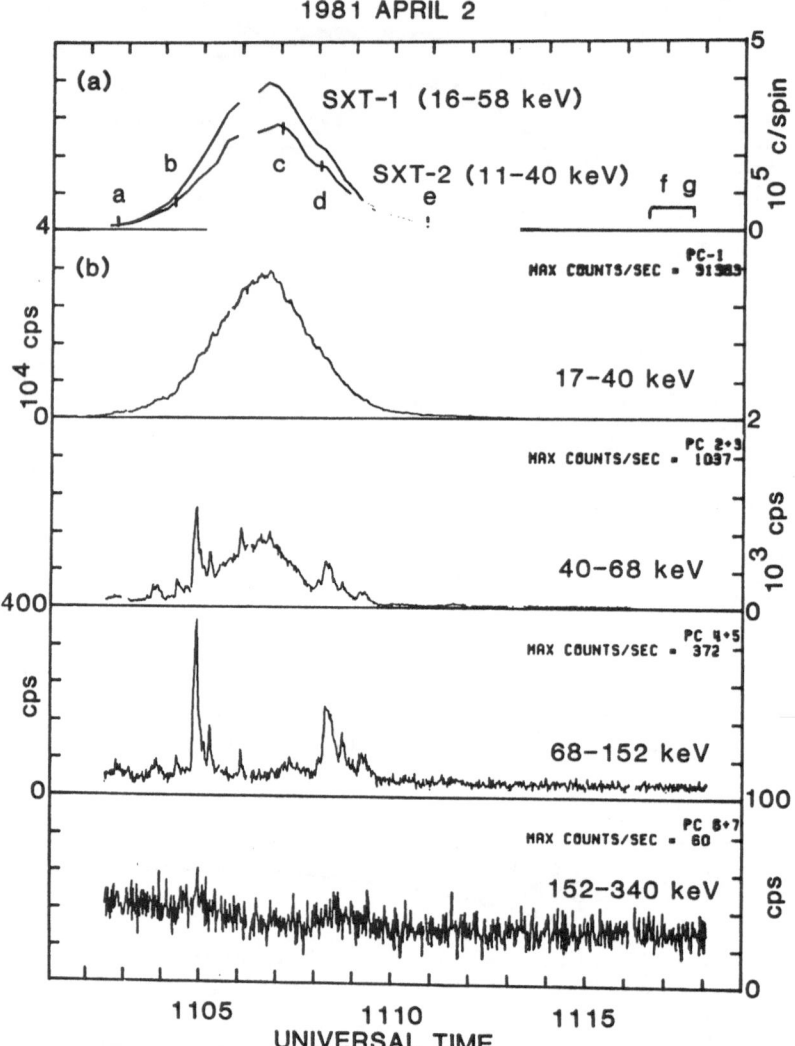

Fig. 3.   Evolution of the hard X-ray emission in a 'hot thermal flare' (Tsuneta, 1987).

flares including the May 21, 1985 event. Such direct measurements give densities which are in general higher than those inferred from the emission measure combined with information on the geometry of the flaring region. This discrepancy may be due to the fact that the ions sensitive to flare densities are emitted from a lower temperature soft X-ray source; while emission measure and geometry are usually given for a plasma at higher temperature. Such a discrepancy can, however, also suggest that the flare plasma is confined in very thin loops (e.g., Linford and Wolfson, 1989) and, therefore, the flare region is characterized by filamentary unresolved elements. The filling factor derived for the flares studied in density-sensitive emission lines ranges from 3×10^{-4} (Linford and Wolfson, 1989) to 10^{-2} (Wolfson *et al.*, 1983). These values are consistent with filling factors derived for soft X-ray loops in non-flaring conditions on the basis of scaling law considerations (Martens, van den Oord, Hoyng, 1985) and for active regions in the phase immediately preceding the flare onset by studying the flow velocities in the reconnection sites (Antonucci, Rosner, and Tsinganos, 1986).

3. Dynamics in the Flare Region

The soft X-ray emission, which is evidence for the formation of the thermal plasma at temperatures above 10^7 K, traditionally defines the flare gradual or thermal phase; while the hard X-ray emission is associated with the impulsive phase. The high-resolution observations of line profiles have, however, shown that soft X-rays also have impulsive phase signatures. The soft X-ray line profiles are in fact extremely sensitive in monitoring, although indirectly, the onset and evolution of the flare impulsive energy release. The two distinctive features of soft X-ray lines during the impulsive phase are large non-thermal broadenings, enhanced above the active region values, and a significant blue-shifted emission (Antonucci et al., 1982). Figure 4 shows the temporal evolution of the soft and hard X-ray emission and compares typical impulsive and gradual phase Ca XIX spectra. Gradual phase profiles are generally thermal, or quasi-thermal; that is, line broadenings are consistent with the average electron temperature derived from line ratios in the same spectral region. These observations of soft X-ray line profiles clearly show that the formation of the flare thermal plasma, at $1-3 \times 10^7$ K, is associated with the presence of high-velocity mass motions.

While non-thermal broadenings are interpreted as evidence for isotropic flows, or random motions, in the flare region, blue-shifts are evidence for the hydrodynamic response of the chromosphere to the impulsive energy release. In flare conditions, the chromosphere heats up to high temperatures either by thermal flux from a hot overlying coronal source or by a flux of non-thermal particles accelerated in the energy release site. The energy deposited at chromospheric level causes an expansion of the plasma, as a result of heating, which gives rise to high velocity upflows into the lower density corona: that is, to chromospheric evaporation. Their onset is roughly simultaneous with the onset of the hard X-ray emission indicating impulsive energy release. In most of the events with a detectable soft X-ray emission before the impulsive phase, non-thermal broadenings in soft X-ray lines are observed to precede blue shifts (Figure 5).

3.1. NON-THERMAL LINE BROADENINGS

Most of the interpretations relate the origin of the enhanced non-thermal line broadenings observed during the impulsive phase to the presence of convective flows, associated with chromospheric evaporation. Non-thermal broadenings have been explained as a result of a superposition of emissions from convective flows observed either in an arcade of flaring loops with different projection angles (Doschek et al., 1986), or in an asymmetrically heated loop, as in the model of Cheng, Karpen, and Doschek (1984). They have also been interpreted as a result of a velocity distribution in the evaporating plasma which is not entirely accounted for by analysing soft X-ray spectra simply in terms of

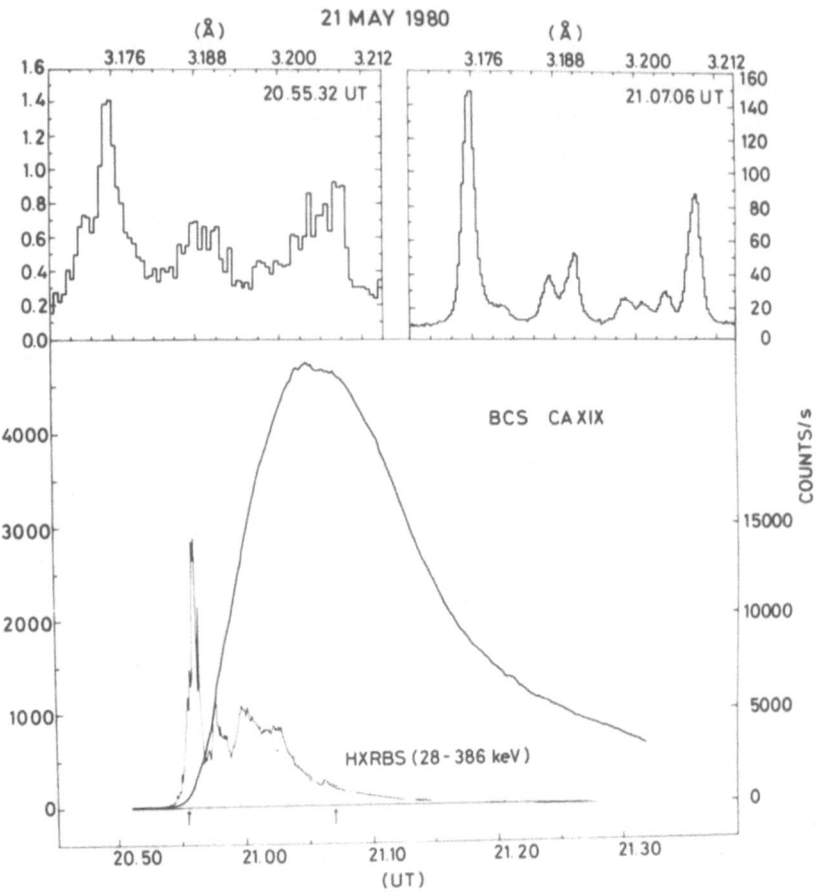

Fig. 4. Ca XIX spectra during the impulsive and gradual phase of the May 21, 1980 flare observed with the Soft X-ray Polychromator on the Solar Maximum Mission satellite, at the top. At the bottom, light curves of the soft and hard X-ray emission during the same event.

two components (Emslie and Alexander, 1987). Other interpretations, which relate enhanced broadenings to the evaporation process in a more indirect way, have also been proposed. Bornmann (1987) proposes an interpretation in terms of fluid turbulence, in which the kinetic energy of upflows transforms into thermal energy. De Jager (1985) interprets line broadenings as evidence for a temperature difference at the sites where evaporation occurs.

Although all these interpretations are plausible, they cannot predict the observed enhancement in line widths preceding the onset of blue-shifts. The non-thermal excess in soft X-ray line profiles increases significantly above the active region values already a few minutes before the onsets of the impulsive hard X-ray burst, that is of the impulsive energy release, and of evaporation (which are approximately coincident). In addition, no strong further enhancement in line broadening is observed when convective flows are first observed. As an example, if the degree of non-thermal broadening is expressed

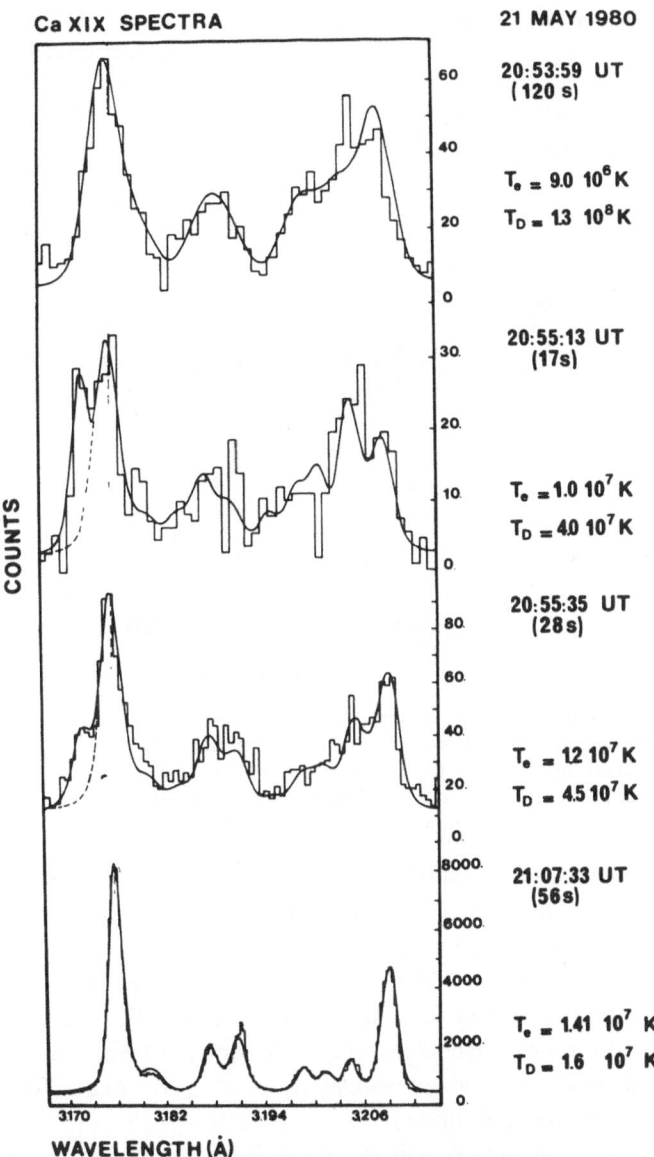

Fig. 5. Temporal sequence for the onset of enhanced non-thermal line broadenings and blue-shifts in soft X-rays during the May 21, 1980 flare. In the Ca XIX spectrum detected at 20:53:59 UT, the difference in Doppler temperature T_D and electron temperature T_e is consistent with a turbulent velocity parameter $v_t = 220$ km s^{-1}. An intense blue-shifted emission is first detected at 20:55:13 UT. The spectrum becomes thermal at the end of the impulsive phase, approximately in coincidence with the peak of the gradual phase (Antonucci *et al.*, 1985).

in terms of a turbulent velocity parameter, $v_t = [2k(T_D - T_e)/m]^{1/2}$ (where T_D is the Doppler temperature and T_e is the electron temperature obtained from line ratio), no evident variation in v_t is associated with the time T, indicating the onset of blue shifts in Ca XIX during the large April 24, 1984 flare (Figure 6). Therefore, although both

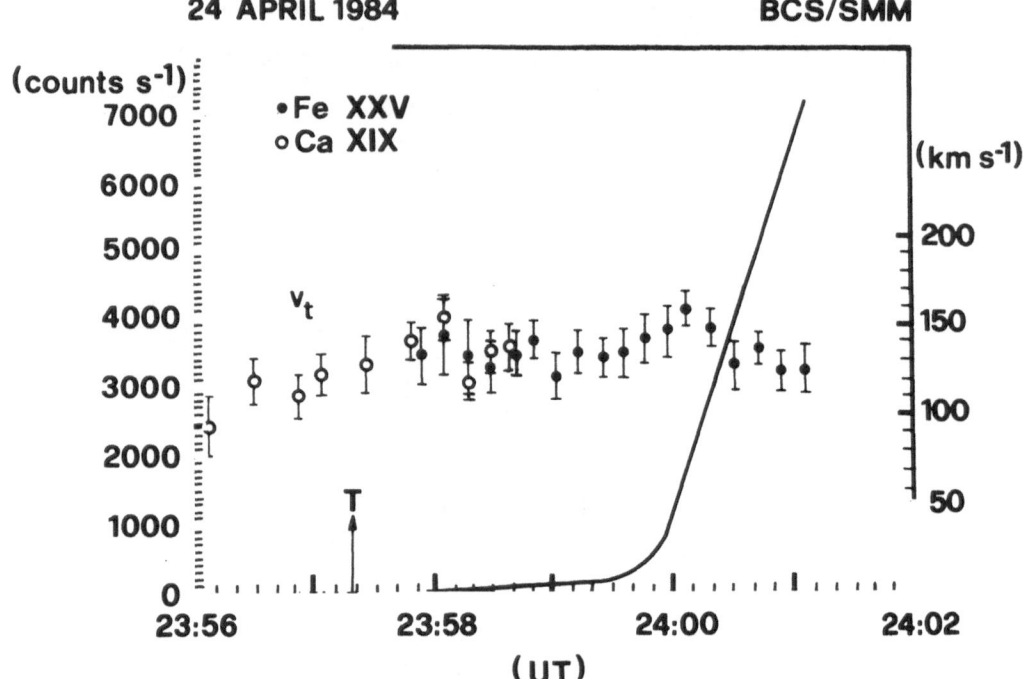

Fig. 6.   Evolution of turbulent velocity, v_t, during the early phase of the April 24 1984 flare (class X13, importance 3B), observed with the Soft X-ray Polychromator on SMM. Light symbols represent the values derived from the Ca XIX line profiles, while dark symbols refer to Fe XXV data. The continuous line represents the Ca XIX light curve. The time T represents the onset of upflows, evidence for chromospheric evaporation (Antonucci and Dodero, 1986).

non-thermal broadenings and upflows are present throughout the impulsive phase, and vanish or decrease to a minimum level at the end of it, the onset and degree of broadening are not significantly correlated with convective upflows.

Furthermore, if non-thermal widths were a result of a superposition of convective flows, they should be larger in disk flares, since vertical flows are predominant in evaporation. However, no such center-to-limb variation has been found with the present accuracy in the measure of turbulent velocity ($v_t = 100 \pm 40$ km s^{-1} for disk flares and $v_t = 120 \pm 30$ km s^{-1} for limb flares).

The consistency of turbulent velocities derived from Ca XIX and Fe XXV data, shown in Figure 6, confirms that the difference in Doppler and electron temperatures does not represent a real ion-electron temperature difference, which cannot be maintained for time intervals of minutes in a plasma at coronal flare densities.

The implication is, therefore, that enhanced non-thermal broadenings are evidence for random or 'turbulent' mass motions. Furthermore, such motions seem to be related to the onset of energy release, although not yet in an impulsive form. The initial enhancement in broadenings is, in fact, temporally correlated with the appearance of low-intensity hard X-ray emission which precedes the main burst. This indicates an initial phase of particle acceleration, associated with an initial temperature increase in the soft X-ray source. Turbulent velocities are, however, already very high and do not change appreciably when energy release becomes impulsive; that is, the degree of enhancement in turbulent velocity does not appear to be directly related to the degree of energy released.

A plausible explanation can be found by interpreting random plasma flows as a dynamic effect related to the magnetic reconnection process. In particular, it is possible that reconnection takes place in localized regions scattered throughout the flaring loop. This picture is consistent with coronal loops whose magnetic field lines become twisted by photospheric motions, leading then to local reconnection, with consequent magnetic energy dissipation throughout the loop (Antonucci, Rosner, and Tsinganos, 1986; Parker, 1987). Plasma is ejected at the Alfvén speed from a localized site, where magnetic reconnection occurs (Sweet, 1969; Parker, 1983). An isotropic flow results then from the superposition of individual flows and causes symmetric broadenings in soft X-ray lines (Antonucci, Rosner, and Tsinganos, 1986, 1987). In this hypothesis, an enhancement in line broadening occurs simultaneously with the onset of magnetic energy dissipation in the flare region and persists as long as the process continues. The same process suggested to initiate a flare can also be invoked to heat the quiet corona and active regions. Concerning the latter point, interesting observations of enhanced non-thermal excess in line profiles have been obtained by Saba *et al.* (1989) along the more intense magnetic fields, in an active region.

3.2. IMPULSIVE PHASE BLUE-SHIFTS AND CHROMOSPHERIC EVAPORATION

The properties of the soft X-ray line blue-shifts derived from studies of the impulsive phase of a large sample of flares strongly suggest that blue-shifts are evidence for chromospheric evaporation (Antonucci *et al.*, 1982; Tanaka *et al.*, 1982b; Antonucci and Dennis, 1983; Antonucci, Gabriel, and Dennis, 1984; Tanaka, 1986). Flows with velocities within 100–500 km s^{-1} are consistent with motions which are predominantly ascensional, since blue-shifts are not observed in flares occurring beyond approximately 60 deg in longitude (Antonucci, Gabriel, and Dennis, 1984; Tanaka, 1986). Such upflows are associated both spatially and temporally with an impulsive heating of the chromosphere. Their onset and duration is correlated with the impulsive hard X-ray emission, which indicates that energy is released in the flare region (Figure 7). At the time upflows are first observed, the soft X-ray emission between 3.5 and 8 keV is found to be co-spatial with the hard X-ray emission in the 16–30 keV range (within the resolution of the Hard X-Ray Imaging Spectrometer on SMM), and confined to the footpoints of the flaring loops (Figure 8).

The duration of upflows is associated with the rise in the emission from the hot coronal source forming during flares, since blue-shifts persist so long as the emission measure of the stationary component of the soft X-ray source is increasing. This is to be expected if the evaporating plasma is confined in magnetic loops. In addition, blue-shifts decrease while the emission from the thermal source increases. This can be understood in terms of a progressive decrease in the pressure difference between chromosphere and corona while the loop density increases, which causes a progressive decrease in evaporation velocities (Figure 9).

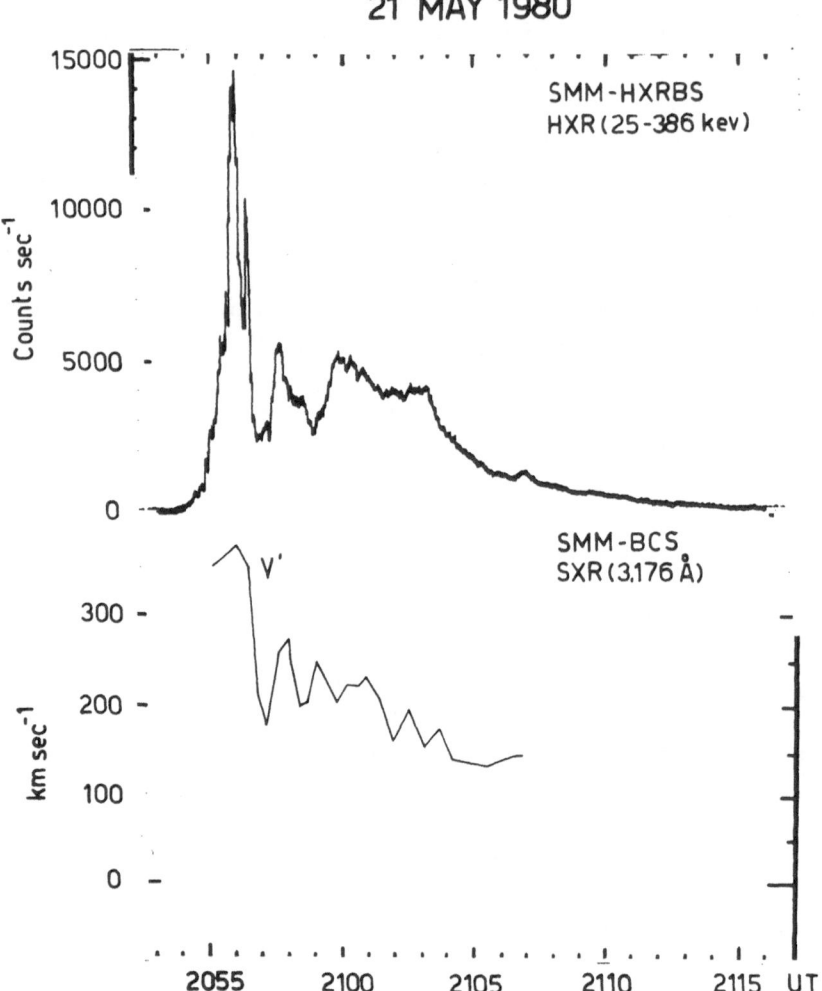

Fig. 7.   Temporal evolution of the velocities, v', derived for the upflows due to chromospheric evaporation and the impulsive emission in hard X-rays, signaling energy release, in the May 21, 1980 flare.

If the upflowing material due to chromospheric evaporation plays a dominant role in the formation of the thermal coronal plasma in flares, we expect a mass and energy balance between the evaporating material and the soft X-ray coronal source. Moreover, since density and temperature of the evaporating plasma cannot be directly measured due to the blending of the temperature diagnostic lines emitted from the evaporating and coronal plasmas, we can only assert that mass and energy balance can be ensured for plausible values of such quantities: that is, densities of the order of $1-2 \times 10^{11}$ cm^{-3}, and temperatures of $1-2 \times 10^7$ K (Antonucci *et al.*, 1982; Antonucci, Gabriel, and Dennis, 1984). Densities derived from mass balance represent lower limits, mainly because of the unresolved geometry of the flare region.

By following the same approach and requiring mass and energy balance during the

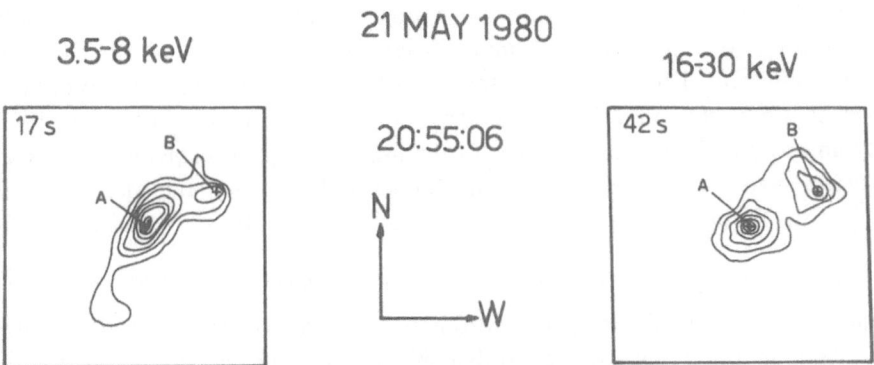

Fig. 8. Example of footpoint brightening in soft and hard X-rays at the onset of chromospheric evaporation for the May 21, 1980 flare, as observed by the SMM experiments (Antonucci *et al.*, 1985).

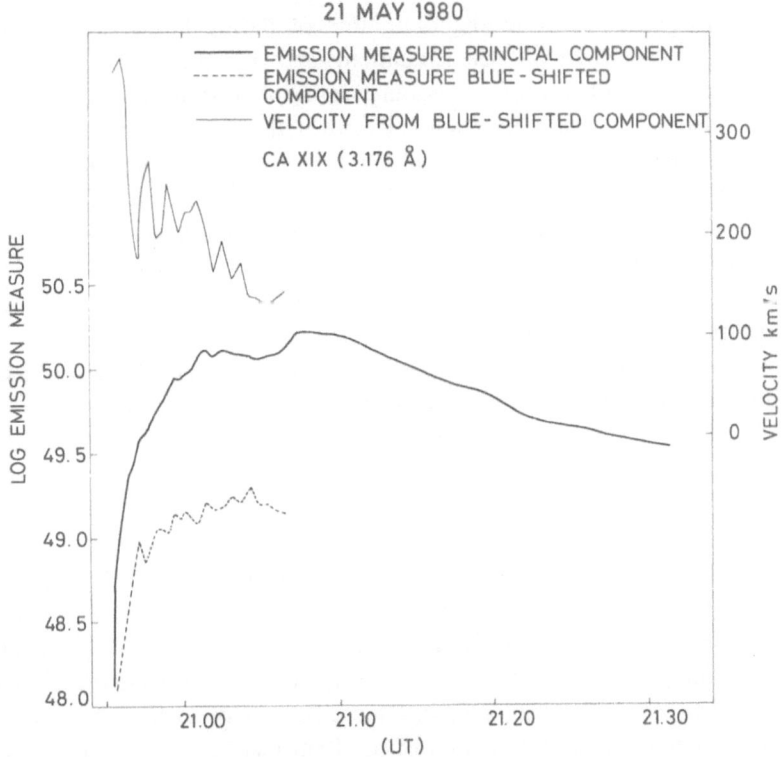

Fig. 9. Temporal evolution of the chromospheric evaporation velocity and the emission measure of the thermal soft X-ray source forming in the corona during the May 21, 1980 flare. The emission measure of the upflowing plasma is also reported as a function of time.

flare impulsive phase between upflowing and coronal plasmas, Tanaka and Zirin (1985) found the observations of the large June 6, 1982 flare consistent with the evaporation interpretation. The same procedure, however, applied to the large November 7, 1980 flare led Karpen, Doschek, and Seely (1986) to conclude that the amount of the upflowing plasma is in excess, or in other words, the upflowing plasma does not have sufficient pressure, according to the inferred density value, to expand into the coronal portion of the loop. The geometry of this flare, however, contrary to the other events studied by Antonucci, Gabriel, and Dennis (1984) and Tanaka and Zirin (1985) is totally unresolved. In general, because of the impossibility of resolving the individual sites where evaporation occurs and the difficulties in determining the effective pre-flare density in the flaring loops, we can only derive a lower limit for the density of upflows.

3.3. COMPARISON OF OBSERVATIONAL PROPERTIES AND SIMULATION RESULTS

We can investigate whether the general properties of plasma upflows are consistent with the predictions of the numerous hydrodynamical models developed in recent years to understand the hydrodynamical response of a coronal loop undergoing transient impulsive heating. For example, independently of the model assumed for energy deposition and heat input, at the onset of chromospheric evaporation the soft X-ray emission is expected to be predominant at the footpoints of flaring loops, as observed (e.g., Figure 8). This is consistent with the simulations shown in Figure 10. Ca XIX brightenings at the footpoints of a loop undergoing transient heating are predicted both for energy deposited by an electron beam injected at the top of the loop (which heats predominantly the chromosphere via Coulomb losses), or in the hypothesis of a thermal energy source located either at the apex or at the base of the loop (with consequent chromospheric heating by thermal conduction). Only for thermal energy deposition high in the corona, are the footpoint brightenings expected to be delayed with respect to the onset of flare heating, with a time lag consistent with the transit time of the thermal conduction front (Antonucci *et al.*, 1987).

The average parameters, simulated both for the evaporating plasma and for the thermal source which is forming by magnetic confinement at coronal level, shown in Figures 11 and 12, reproduce the general temporal behavior of the parameters derived from the observed soft X-ray emission (Figure 9). The relative velocity of the upflowing material with respect to the coronal soft X-ray source, v', decreases with time although the energy input flux remains constant for the entire interval considered (180 s). Both the intensity of the coronal source I_w and of the upflowing material, I'_w (Figures 11 and 12) are steadily increasing during such an interval (although I'_w decreases relative to the emission of the static soft X-ray source), as observed in Figure 9. Considering the relatively low-energy input flux, $\phi = 6\text{--}8 \times 10^9$ erg cm^{-2} s^{-1} used in these calculations, the line intensities simulated at flare onset are too weak to be observable with an instrument with sensitivity such as that of the Bent Crystal Spectrometer (Antonucci *et al.*, 1987).

The fact that model predictions in general indicate a dominant blue-shifted emission at flare onset, which is not systematically observed with the present instrumentation, has raised some questions about whether impulsive phase blue-shifts are indeed evidence for chromospheric evaporation (Doschek *et al.*, 1986; Emslie and Alexander, 1987). In fact, if the energy input flux is increased above $10^{10}\text{--}10^{11}$ erg cm^{-2} s^{-1}, the simulated blue-shifted emission is expected to dominate the coronal plasma emission, at flare

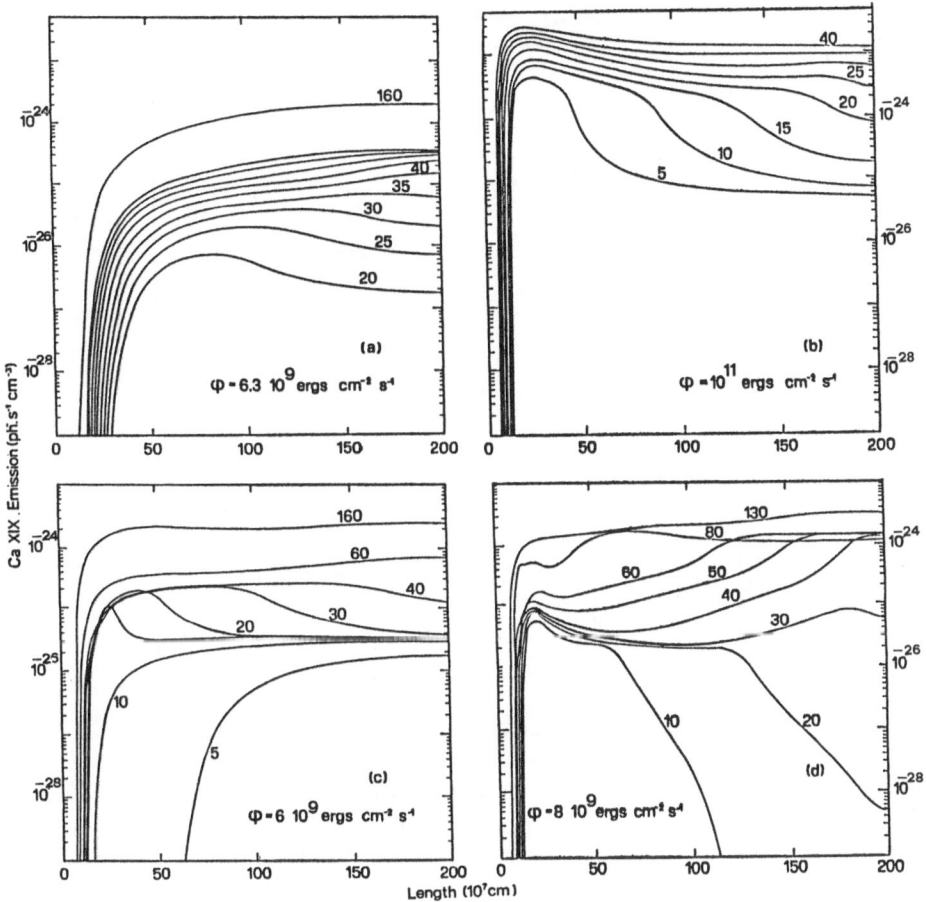

Fig. 10. Simulated curves of Ca XIX emission versus height in a flaring loop of semi-length $l/2 = 2 \times 10^9$ cm, initial density $n_e = 7 \times 10^9$ cm^{-3}, initial temperature $T_e = 3.2 \times 10^6$ K, for different energy deposition models: (a) injection of an electron beam parallel to the field lines, confining the one-dimensional fluid, with input energy flux of 6.3×10^9 erg cm^{-2} s^{-1} (with a power-law energy spectrum with lower cutoff $E_c = 10$ keV and spectral index $\delta = 8$), (b) injection of an electron beam with the same characteristics as in the previous case but with higher energy flux $\phi = 10^{11}$ erg cm^{-2} s^{-1}, (c) thermal energy deposition at the top of the loop with energy input flux $\phi = 6 \times 10^9$ erg cm^{-2} s^{-1}, (d) thermal energy deposition at the base of the loop with $\phi = 8 \times 10^9$ erg cm^{-2} s^{-1}. The simulations are performed with the unidimensional, one-fluid Palermo–Harvard hydrodynamical model (Antonucci et al., 1987).

onset, even more strongly than for the moderate energy fluxes used in the models of Figures 11 and 12. The dominant blue-shifted emission is presumably not commonly observed, since at flare onset the instrumental time resolution is degraded in order to improve the statistics of the spectra. A recent re-analysis by Tanaka (1987) of the large June 6, 1982 flare observed with the Hinotori spectrometers has shown that all the emission in Fe XXV is initially blue-shifted and the shift is consistent with velocities of the order of 300 km s^{-1} (Figure 13, frame a). A static coronal soft X-ray source,

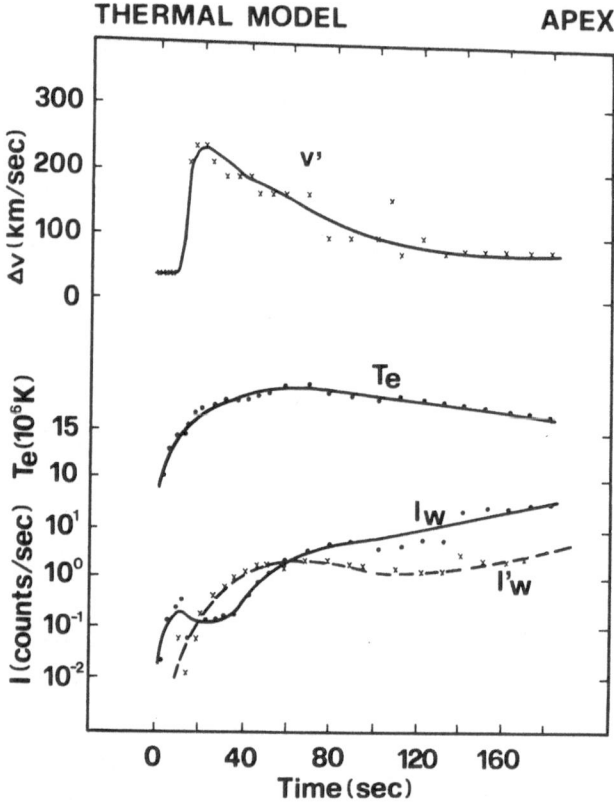

Fig. 11.   Evolution of the simulated average relative velocities of the evaporating plasma, v', the average
electron temperature in the forming coronal thermal plasma, T_e, the Ca XIX intensities, I_w and I'_w of the
coronal and upflowing plasma emission, respectively. These parameters are derived for a flaring loop with
energy deposited as in case (c) of Figure 10, with thermal heating at the top of the loop at a constant rate
for an interval of 180 s. The quantities v', T_e, I_w, and I'_w are derived from simulated Ca XIX spectra, computed
on the basis of the physical conditions predicted along the loop, in response of the flare heating. I_w and I'_w
are expressed in counts s^{-1}, expected in the Ca XIX channel of the Bent Crystal Spectrometer of SMM
(Antonucci *et al.*, 1987).

forming presumably by confinement of the evaporating plasma, appears at the rest
wavelength of Fe XXV after approximately 40 s, and continues to grow afterwards
(frames c and d of Figure 13). Only later on (frames e and f), a typical impulsive phase
spectrum with emission of the static source predominant with respect to blue emission
is observed.

Figure 14 shows an example of simulation of the evolution of Ca XIX spectra for a
model with a thermal energy source localized at the footpoints of a flaring loop. Spectral
profiles evolve showing initially the development of a dominant blue-shifted component.
After about 35 s however, the emission from the coronal source forming by confinement
of evaporating plasma in closed field lines becomes predominant. This evolution is
similar to that observed for the June 6, 1982 flare by Tanaka (1987).

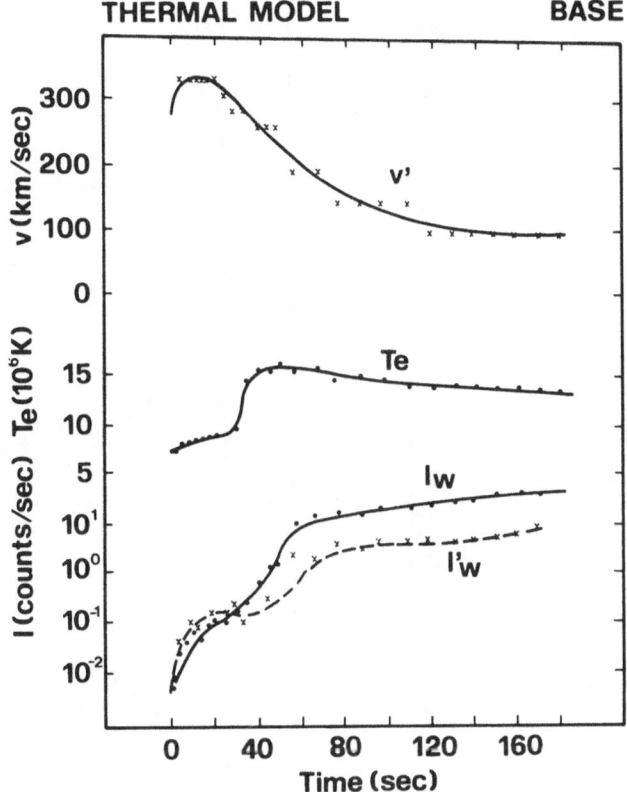

Fig. 12. Flare plasma parameters, as in Figure 11, but for thermal heating at the base of the loop (case (d) of Figure 10) (Antonucci *et al.*, 1987).

Since the simulations of Figure 14 have been derived in the case of relatively moderate energy flux input, they give initially very low counting rates, which are, for instance, below the sensitivity of an instrument such as the Bent Crystal Spectrometer (SMM), unless photons are accumulated for intervals of about 40 s (Figure 15). In this case, the typical profile of an impulsive phase spectrum is obtained. This explains how instrumental sensitivity can be responsible for observing a dominant blue emission at the very onset of chromospheric evaporation.

4. Additional Evidence for Chromospheric Evaporation

As a sufficiently strong beam of non-thermal electrons loses its energy in the chromosphere, the sudden temperature increase creates explosive chromospheric evaporation. In explosive conditions, the local pressure excess produces not only an upward

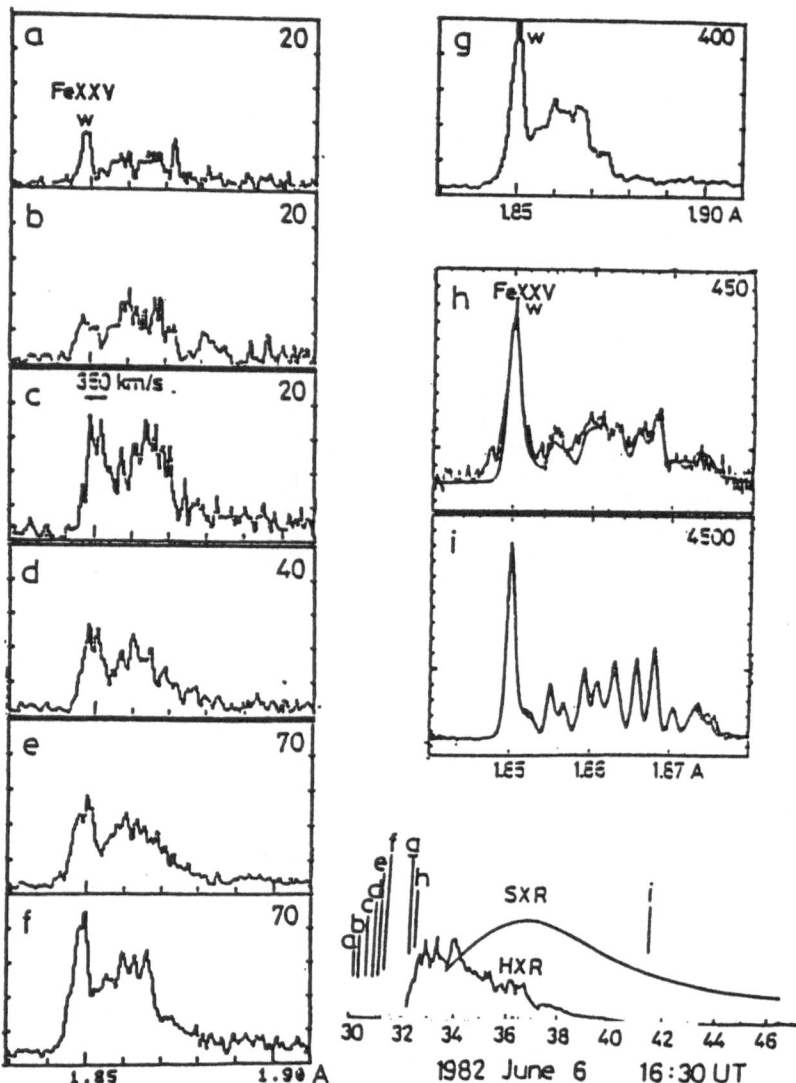

Fig. 13. Time sequence of Fe XXV spectra during the June 6, 1982 flare. The emission at 1.85 Å shows at first the development of a blue-shifted dominant resonance line (frame a, b, c, d), and the formation of the coronal source emitting an unshifted component at 1.85 Å (since frame b), which becomes predominant after about 100 s (frame e, f). The wavelengths are calibrated in reference to the Kα line at 1.94 Å (Tanaka, 1987; Watanabe, 1987, review).

expansion into the pre-flare corona, observed as blue-shifted soft X-ray emission, but it also produces a compression of the underlying unevaporated material driving a pressure wave down into the chromosphere, which can be observed as red-shifted Hα emission. At the same time, the signature expected for non-thermal electrons entering the chromosphere is the presence of broad Stark wings in Hα (Canfield, Gunkler, and

SIMULATED   SPECTRAL EMISSION

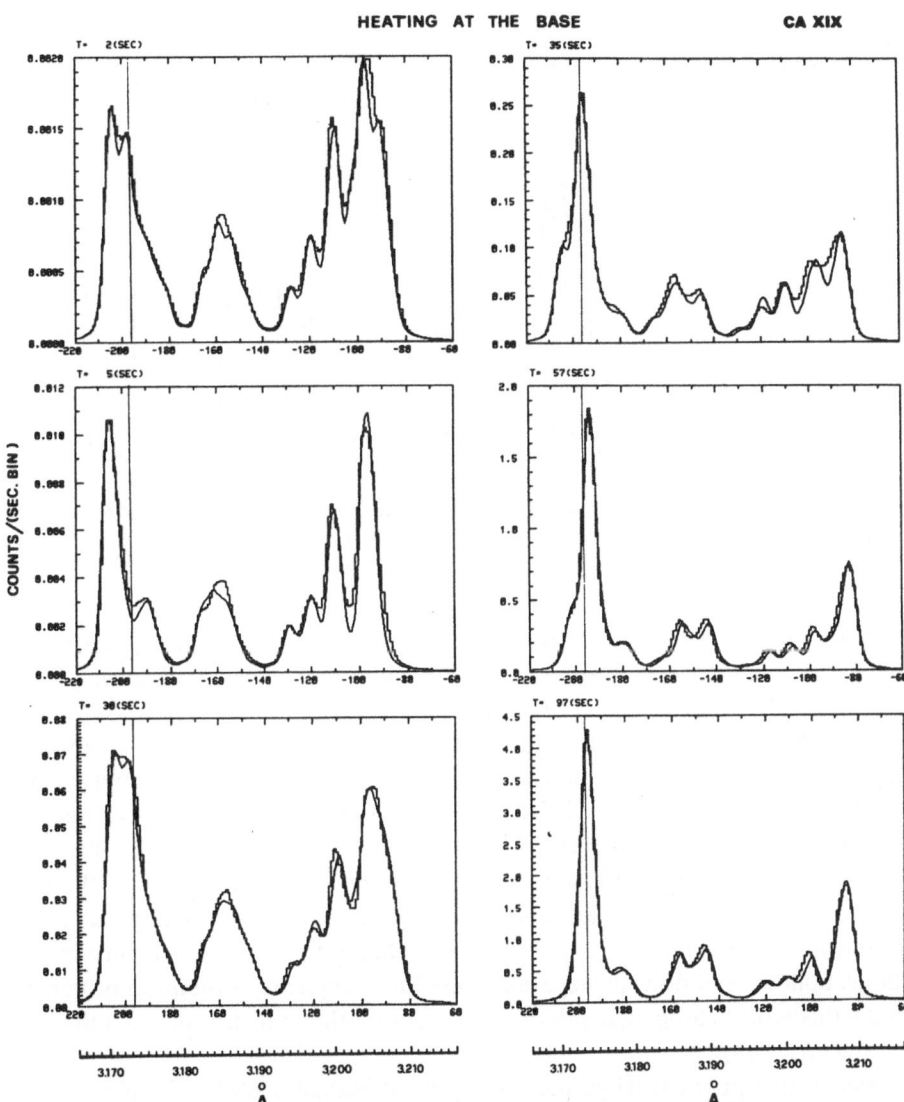

Fig. 14. Evolution of the simulated Ca XIX spectra for a thermal model of energy deposition localized at the base of the loop, as in Figure 10(d). The vertical line in each plot represents the unshifted Ca XIX resonance line at 3.176 Å. The intensity is in units of counts s^{-1} bin^{-1} of the Ca XIX channel of the Bent Crystal Spectrometer on SMM. The continuous lines represent theoretical two-component synthetic spectra fitted to the simulated spectra (stepped curves) (Antonucci *et al.*, 1987).

Ricchiazzi, 1985). Evidence for broad Stark wings due to electron beams and red-shifts in Hα (Ichimoto and Kurokawa, 1984; Canfield, 1986) have indeed been found in coincidence with the flare impulsive hard X-ray emission. The observed downward velocities, within 40–100 km s^{-1}, and lifetime of the chromospheric condensation are consistent with predictions from the simulations by Fisher, Canfield, and McClymont

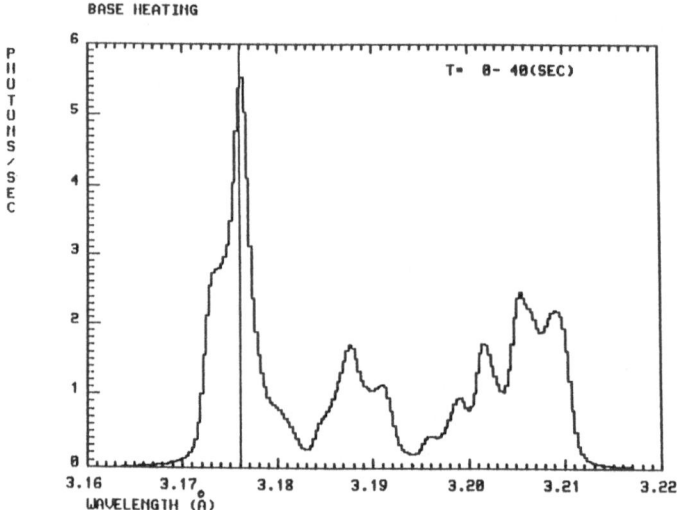

Fig. 15. Ca XIX simulated spectrum integrated over the initial 40 s for a thermal model with energy deposition localized at the base of the loop, as in Figure 10(d).

(1985). In addition, evidence for the appearance of soft X-ray emitting plasma associated with a simultaneous disappearance of Hα chromospheric material was found by Acton et al. (1982) and Gunkler et al. (1984).

While it has been possible to investigate, at least indirectly, the mass and energy balance between upflows and the forming coronal source during the impulsive phase, uniquely on the basis of soft X-ray data (Antonucci, Gabriel, and Dennis, 1984), simultaneous observations in soft X-rays and in Hα, obtained in a few flares, have recently led Canfield et al. (1989) to ascertain, within the uncertainties of the present measurements, momentum balance between the evaporating soft X-ray plasma and the downflowing Hα chromospheric plasma. The column density at which chromospheric motions, upflows and downflows, are expected to commence is found to be consistent with the theoretical stopping depth of a beam of non-thermal electrons in a thick target (Zarro et al., 1989).

Lower temperature emission lines also give additional evidence for chromospheric evaporation during the impulsive phase of flares. During the impulsive phase of three flares, an increase in the volume of the O VII source (forming at 2×10^6 K) has been observed in temporal correlation with the blue-shifts in higher temperature lines. This has been interpreted as an expansion of the source resulting from chromospheric evaporation (Bruner et al., 1988).

In a single well-resolved loop observed in Fe XV (emitted at temperature below 10^7 K) during the Skylab mission, Hiei and Widing (1979) reported a progressive brightening from the footpoint to the top of the loop. This phenomenon as well can be interpreted in terms of chromospheric evaporation.

5. Alternative Interpretative Models for Impulsive Soft X-Ray Blue-Shifts

There have been several attempts to explain soft X-ray blue-shifts during the impulsive phase of flares not simply in terms of the hydrodynamic response of coronal loops to a transient impulsive energy deposition, but by considering them as evidence for magnetohydrodynamic forces acting at the onset of flares. Karpen, Doschek, and Seely (1986) propose a magnetohydrodynamic model, where a sudden heating of twisted flux-tubes results in gas expulsion. Plasma flows away from the solar surface toward regions of weaker magnetic field at a velocity equivalent to the proton thermal velocity of a few 10^2 km s^{-1}.

In a model of reconnecting loops (Doschek *et al.*, 1986), upflows can be created within a lower density loop colliding with an emerging dense flux-tube. In this case, upflows are simply driven by the pressure difference between the two flux-tubes.

Uchida and Shibata (1988) propose a flare model based on untwisting loops. Upflows are expected in the unwinding process occurring when a twisted flux-tube emerges from the photosphere in a lower pressure region, where the plasma pressure can no longer dominate the magnetic force. In the process, plasma heated to soft X-ray temperatures is ejected predominantly upward.

Upflows in flares may be also due to coronal ions accelerated in a direction perpendicular to the magnetic field and mirrored upward in convergent flux-tubes. This effect is found to occur in a bi-dimensional model which considers the filamentation in an electron beam current (Winglee, Pritchett, and Dulk, 1988).

At the present such alternative explanations of soft X-ray blue-shifts cannot predict more than the dominant direction of the plasma motions as well as a general temporal association of the onsets of upflows and impulsive phase. They cannot, however, predict the spatial association of soft X-ray emission with footpoints at flare onset; the temporal relation of upflows to the increase in emission of the high-temperature plasma; and the decrease in upflow velocity when the density of the soft X-ray source is increasing. In addition, none of these models explains either the existence of oppositely directed flows: the upward hot evaporating material emitting soft X-rays and the downward cold chromospheric condensation observed in Hα nor the momentum balance of upflowing and downflowing plasmas, as discussed by Canfield (1986) and Canfield *et al.* (1989).

6. X-Ray Blue-Shifts as Signature of Chromospheric Heating

We have discussed the X-ray blue-shifts as evidence for chromospheric evaporation and their role in the formation of the high-temperature plasma (above 10^7 K), observed during flares. The question is whether the characteristics of the impulsive blue-shifted emission in soft X-ray lines can also be studied to investigate the primary energy release and the role of different mechanisms in heating the chromosphere to high temperature.

Several numerical models have been developed to simulate the hydrodynamic response of the plasma confined in a coronal magnetic loop to an impulsive and intense energy release. These models show that the properties of evaporation depend not only on the parameters characterizing energy release and on the process which heats the chromosphere, but also on the initial conditions and geometry of the loop. Explosive evaporation occurs for chromospheric heating by intense particle beams (e.g., Somov, Syrovatskii, and Spektor, 1981; Nagai and Emslie, 1984; MacNeice *et al.*, 1984; Fisher, Canfield, and McClymont, 1985) or by intense thermal fluxes (e.g., Smith and Harmony, 1982; Somov, Sermulina, and Spektor, 1982; MacNeice, 1986), with values above a few 10^{10} erg cm^{-2} s^{-1}. This is due to the fact that heating takes place on a time-scale shorter than the hydrodynamic expansion time-scale of the plasma. Therefore, evaporation velocity can be considered as a signature of the flux of energy released.

Evaporation velocities derived from the Ca XIX line profiles, which have been extensively studied, are typically within 100–500 km s^{-1} with average peak value around 300 km s^{-1} (Antonucci, Gabriel, and Dennis, 1984). Velocities derived from the Fe XXV line profiles show a tendency to give higher values (Feldman *et al.*, 1980; Antonucci *et al.*, 1982; Karpen, Doschek, and Seely, 1986; Tanaka, 1987). A recent study of Fe XXV line profiles detected in the high-sensitivity, high-spectral resolution channel of the Bent Crystal Spectrometer on the Solar Maximum Mission, shows that the velocity distribution in the evaporating plasma extends to much higher values. The plasma component evaporating at higher velocity is characterized by a relatively weaker emission and a higher average temperature with respect to the bulk of the evaporating plasma (Antonucci, Dodero, and Martin, 1989).

For example, the profile of the Ca XIX resonance line observed during the impulsive phase of the extremely intense April 24, 1984 flare can be reproduced by a superposition of two synthetic spectra, whose relative wavelength shift is consistent with a line-of-sight velocity of 210 km s^{-1} (Figure 16). The simultaneous profile of the Fe XXV resonance line (Figure 17(a)), observed in the high-sensitivity channel of the Bent Crystal Spectrometer (SMM), shows a weak but significant blue excess with respect to the fit consisting of two spectral components, which are Doppler-shifted in agreement with the velocity observed in Ca XIX. Such an excess in blue-emission is evidence for plasma evaporating on the average at a line-of-sight velocity of 480 km s^{-1} (Figure 17(b)).

The analysis of a set of 11 large flares, of class M and X, has shown that very high velocities are often observed in the early impulsive phase. The average velocity v_2 for the excess blue emission in Fe XXV is found to be within 500–800 km s^{-1}; while, the velocity v_1 relative to the upflows observed also in Ca XIX for the same set of flares is within 250–400 km s^{-1}. Maximum values in the velocity distribution, $v_{2, \text{max}}$, defined on the basis of the shift corresponding to $\lambda'_{2, \text{max}}$ where the intensity decreased by e^{-1} (Figure 17(b)), are within 700–1000 km s^{-1}.

The ratio of the intensities of the emission in Ca XIX and Fe XXV from the evaporating plasma moving at average velocity v_1 (first blue-shifted spectral component in Figures 16 and 17(b)) gives the possibility to estimate directly the average temperature for such a plasma. For the plasma evaporating at velocity v_2, observed only in Fe XXV, it is, instead, possible to estimate just a temperature lower limit which results from the ratio of the second blue-shifted spectral component in Fe XXV and the background intensity (at the correspondent wavelength shift) in Ca XIX. The temperature derived for the

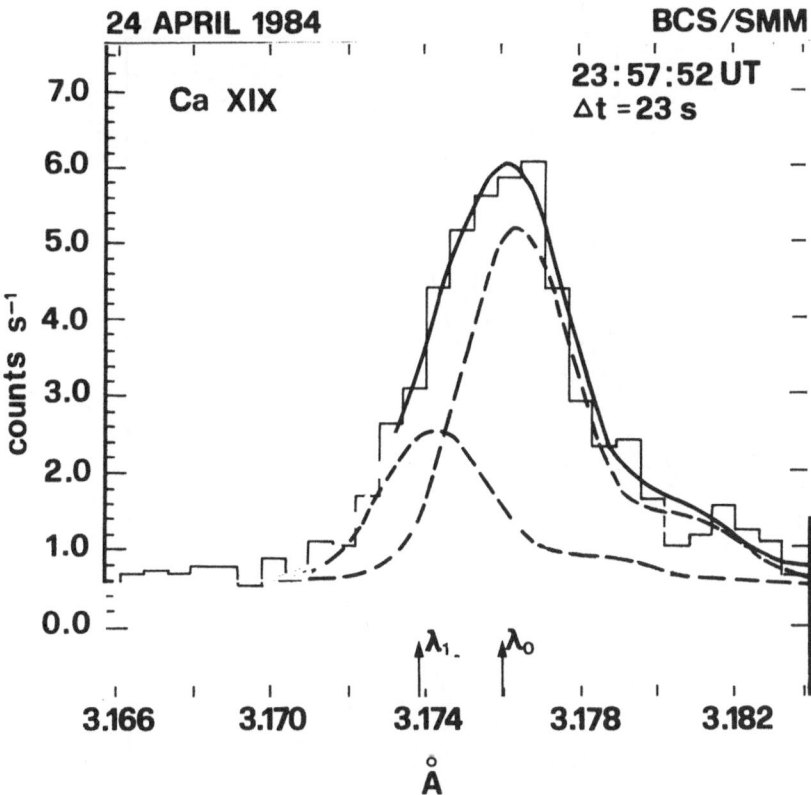

24 APRIL 1984   **BCS/SMM**

Ca XIX   23:57:52 UT
                               Δt = 23 s

Fig. 16. Ca XIX line profile (stepped line), observed during the April 24, 1984 flare, fitted with the sum (continuous line) of two spectral components (dashed lines), with relative shift consistent with a line-of-sight velocity of 210 km s⁻¹. Bins are grouped to improve the statistics (Antonucci, Dodero, and Martin, 1989).

plasma evaporating at v_1 varies within 1.3–1.8×10^7 K; while the lower limit derived for the higher velocity plasma varies within 1.8–3.2×10^7 K. In conclusion, the evaporation velocity distribution during the impulsive phase of a number of large flares extends up to velocities of about 10^3 km s⁻¹, with a high-velocity component much hotter than the more slowly evaporating one.

The observed evaporation velocities have been compared by Fisher, Canfield, and McClymont (1984) with those obtained from hydrodynamic models in an attempt to relate the observations to the heating mechanisms operating at chromospheric level and in turn to the primary energy release. These authors have expressed the maximum values obtained in the simulated velocity distributions in units of a limiting velocity, v_l, derived with a simplified method only on the basis of a reasonable chromosphere-corona pre-flare density difference (which is assumed to be about 2.35 times the sound velocity). In Figure 18, analogous to that reported by Fisher, Canfield, and McClymont (1984), dark symbols refer to results for particle-beam models in the thick-target approximation; while, light symbols refer to thermal model results. At an energy flux of a few

54

ESTER ANTONUCCI

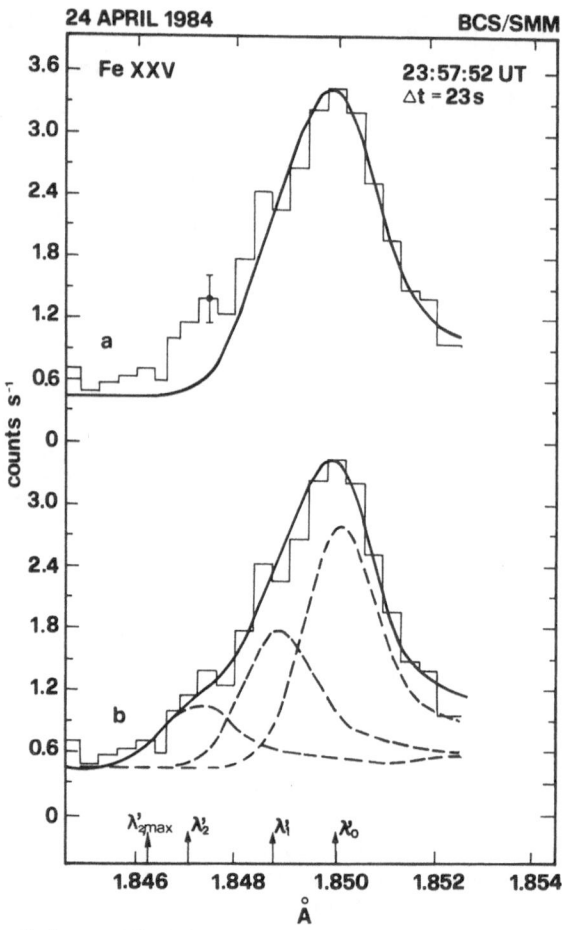

Fig. 17. Fe xxv line profile (stepped line) detected, during the April 24, 1984 flare, in the high-sensitivity Bent Crystal Spectrometer channel. Fe xxv line fitted with: (a) the sum (continuous line) of two spectral components with relative shift consistent with a line-of-sight velocity of 210 km s^{-1}; (b) the superposition of an additional third spectral component with a shift consistent with a line-of-sight velocity of 480 km s^{-1} (Antonucci, Dodero, and Martin, 1989).

10^{10} erg cm^{-2} s^{-1}, there is the transition from 'gentle' to explosive evaporation. The velocities observed for the plasma evaporating more slowly, derived from Ca xix spectra analysed by Antonucci, Gabriel, and Dennis (1984), cluster around the value $v_1/v_l = 0.2$, or, more correctly, if we consider $v_{1, max}/v_l$, maximum velocity of the slower component, around the value 0.3. On the basis of the Ca xix data (v_1/v_l), Fisher, Canfield, and McClymont (1984) have reached the conclusion that the observed values are consistent with 'gentle' chromospheric evaporation. The Fe xxv data, however, have revealed the existence of much higher evaporation velocities, at higher plasma temperatures, which expressed as v_{max}/v_l fall in the interval 0.3–0.5 (dashed region in Figure 18). This leads us to reconsider the comparison of observed and simulated results and conclude that during large flares the observed velocities are more consistent with explosive evapo-

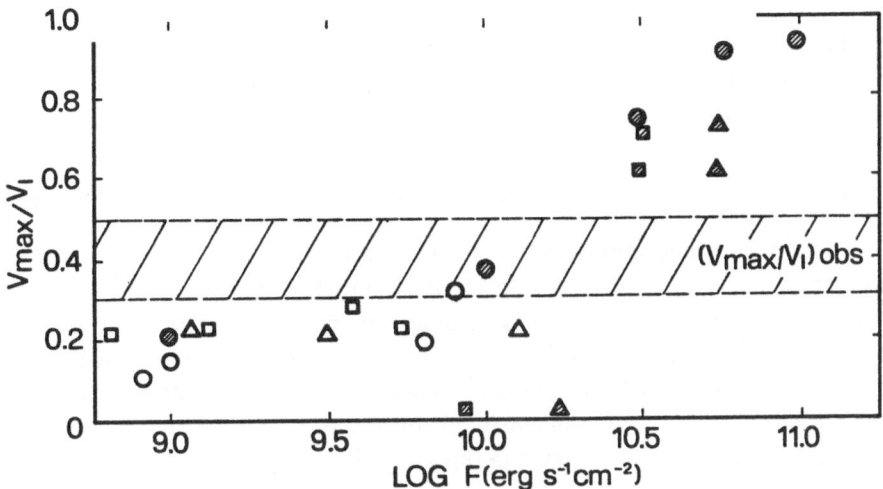

Fig. 18.   Comparison of observed and simulated maximum values v_{max} in the velocity distribution of the evaporating plasma. The velocity values are expressed in units of the limiting velocity v_l assumed equal to 235 times the sound velocity in the evaporating plasma. The dark symbols refer to thick-target particle-beam models, while light symbols refer to thermal models. The dashed region represents the interval where the observed velocities obtained with the Fe XXV spectra are found. References for flare models used in this figure are reported by Fisher, Canfield, and McClymont (1984).

ration. That is, they are consistent with velocities expected for energy fluxes above 10^{10} erg cm^{-2} s^{-1}. Energy fluxes into the chromosphere of this order are obtained for large flares from the observed hard X-ray emission, in the case that the source geometry is resolved (although there remain large uncertainties both in the definition of the parameters of the hard X-ray source and of the chromospheric areas interested by heating).

From the same comparison of numerical results and Ca XIX observations, Fisher, Canfield, and McClymont (1984) conclude that, since none of the observed velocities exceeds the velocity limit v_l, it is not necessary to invoke additional non-hydrodynamic forces to explain the impulsive phase upflows observed in soft X-rays. This conclusion remains valid even considering the higher velocity values resulting from the Fe XXV spectra (Antonucci, Dodero, and Martin, 1989). We prefer, however, to compare observations with simulated velocities independently of the velocity limit v_l, since the simulations performed for high-energy input fluxes (10 s triangular pulses with peak energy flux of 10^{11} ergs cm^{-2} s^{-1}) by Somov, Syrovatskii, and Spektor (1981) and Somov, Sermulina, and Spektor (1982), for both thermal and non-thermal cases, lead to much higher values. Explosive evaporation at velocity exceeding v_l is, in fact, reached in both cases for a loop length above 2×10^9 cm. Such authors follow the evolution of the evaporating plasma for longer periods than others and their models indicate that higher evaporation velocities can be reached for more extended loops. Therefore, the observation of high velocities, consistent with those expected for explosive evaporation (above 10^{10} ergs cm^{-2} s^{-1} in Figure 18), cannot be used to discriminate between evaporation driven by a thermal flux from a high-temperature source or by an electron beam, as it would be suggested by Figure 18, where results for thermal models are clustered in the range of energy fluxes leading to 'gentle' evaporation.

7. Dependence of Simulation Results on the Initial Plasma Conditions

The discussion on evaporation velocities has shown that numerical models of flaring loops can be extremely valuable for interpretating the observational results. Simulation results, however, depend not only on the heating function assumed for the loop but also on the geometry and initial conditions in the loop and, of course, on the assumptions which have been considered. Despite the simplifying assumptions used in the various numerical codes, such as one-dimensional, one-fluid, one-temperature models, the general behavior of the flare plasmas is reproduced.

As to the dependence on initial loop conditions, this becomes important when comparing simulated and observational results in individual flares. At the present the main difficulty for such a comparison concerns observations, which do not allow us to fully resolve the geometry and to determine the initial conditions in the pre-flare coronal loops. It is, however, crucial to understand the pre-flare state of loops, since there is evidence for large variations in the predicted flare physical parameters when varying initial conditions. For instance, in a model with energy deposition by an electron beam with fixed parameters, the evaporation varies significantly with the pre-flare density in a coronal loop of given geometry. Evaporation velocities drop from initial values above 700 km s^{-1} for an initial loop density of 2×10^9 cm^{-3}, to peak values of about 400 km s^{-1}, when this parameter increases to 3×10^{10} cm^{-3} (Figure 19). This can be understood in terms both of a variation in the heating function, resulting from a variation with density in the energy losses of the energetic electrons along the loop, and of a large variation in the corona-chromosphere pressure difference. For different initial densities, the same energy input results in different temperature conditions within the loop, as shown in Figure 20. For the denser loop at 3×10^{10} cm^{-3} the temperature peaks at the onset of energy deposition, while, for more tenuous loops temperature increases slowly to a maximum value, reached about 20 s after onset. The initial high temperature (2.5×10^7 K) in a dense loop causes a brightening in Ca XIX at the top of the loop, as shown in the simulations of the Ca XIX emission along the loop, in Figure 21(a). It is interesting to note that an initial brightening at the top of the loop has been commonly considered as evidence for heating at the top of the loop. The same effect can, however, be the result of energy losses of non-thermal electrons with a relatively soft energy spectrum, in a dense environment. For the same energy input, a tenuous loop brightens initially at the footpoints (Figure 21(b)). The above results are examples from a work by Antonucci et al. (1989).

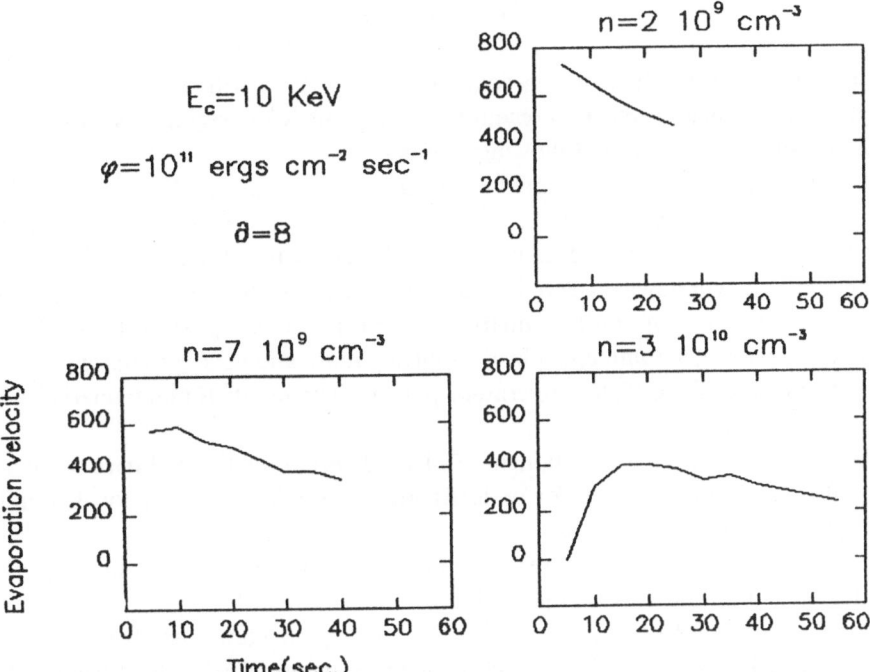

Fig. 19. Simulated averaged evaporation velocity for a non-thermal model with an electron beam represented by an energy power spectrum characterized by a cutoff E_c = 10 keV, and a spectral index δ = 8. The energy flux input is constant, $\phi = 10^{11}$ erg cm^{-2} s^{-1}, over the total interval considered. The velocity is plotted for three different initial density in the coronal loop: $n_e = 2 \times 10^9$, 7×10^9, 3×10^{10} cm^{-3} (Antonucci et al., 1989).

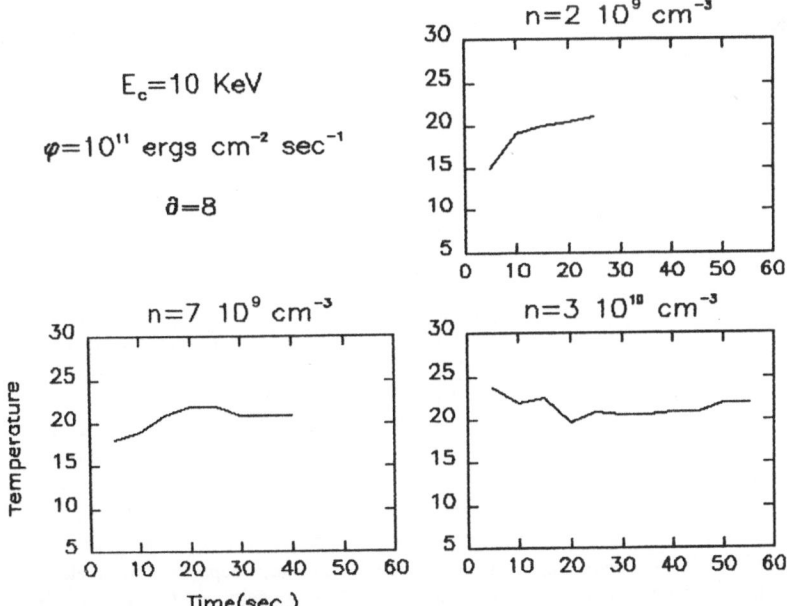

Fig. 20. Simulated average temperature (in units 10^6 K) of the flare thermal plasma for the same energy input and initial loop conditions as in Figure 19 (Antonucci et al., 1989).

8. Conclusions

The major results reached during solar maximum 1980, with the advent of crystal spectroscopy at high-wavelength resolution in the soft X-ray region from 1 to 8 Å, can be briefly summarized in the following points.

The enhancement in non-thermal broadening observed in soft X-ray lines early in a flare can be related directly to the primary energy release process, since this effect appears to be evidence for random motions in the sites where the energy release occurs.

The convective upflows observed during the flare impulsive phase can be associated with chromospheric evaporation, and from their properties it is possible to conclude that this effect is playing a dominant role in supplying mass and energy to flaring loops and in the formation of the high-temperature plasma (above 10^7 K) characteristic of the thermal or gradual phase.

The observed 'superhot' component ($3-4 \times 10^7$ K) in the thermal flare plasma may turn out to be quite important for understanding the influence of the local conditions on the acceleration mechanisms.

The flare region is probably organized in a highly filamentary structure, formed by arcades of thin 'elementary' magnetic loops.

The discussion in the previous sections has focused on a few crucial areas which will be investigated in more detail and will be better understood with the advent of future soft X-ray spectrometers with higher sensitivity, which allow us to fully exploit the high temporal resolution of the spectrometers. High sensitivity and high temporal resolution

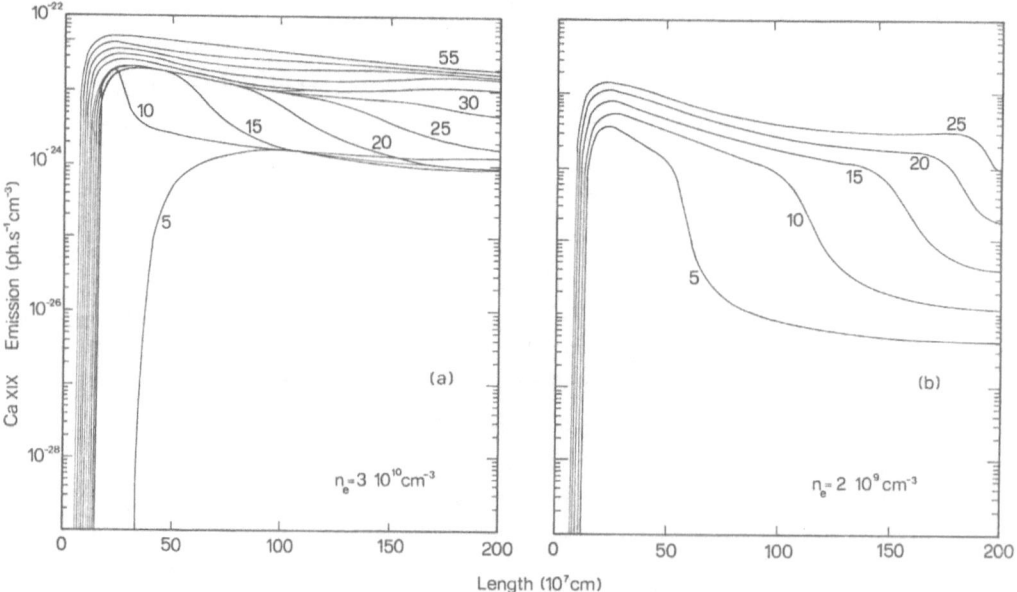

Fig. 21.   Simulated Ca XIX emission along a flaring loop, with energy deposition by an electron beam with the same characteristics as in Figure 19. The initial loop density is: (a) $n_e = 3 \times 10^{10}$ cm^{-3}; (b) $n_e = 2 \times 10^9$ cm^{-3} (Antonucci *et al.*, 1989).

are essential in the pre-onset/onset phase of flares. In fact, X-ray spectra are often well-observed when the plasma has been already heated above 10^7 K, and in most cases they become statistically significant, at flare onset, only when accumulated over 20–40 s. The study of flare onset is crucial in attempting: to determine the nature of line-broadenings, which in turn can give an insight into the primary energy release; to distinguish between chromospheric heating mechanisms, since the evolution of spectral profiles, as shown by simulations, depends more strongly on the energy deposition process in the initial 20–30 s of flares; and to confirm the role of the chromospheric evaporation in producing the soft X-ray thermal source during flares.

Another crucial requirement for future observations will be the capability of combining spectral and imaging observations at high spatial resolution in order to infer the initial geometry in the flaring region (possibly to confirm the filamentary nature of magnetic loops in active regions); to determine with more accuracy the initial sites of soft and hard X-ray emission during flares (this is important to understand both energy deposition and chromospheric heating mechanisms); and to attempt a more reliable measure of initial conditions in the flaring loop in order to improve the understanding of the response of the loop plasma to energy deposition.

References

Acton, L. W., Culhane, J. L., Gabriel, A. H., and 21 co-authors: 1980, *Solar Phys.* **65**, 53.

Acton, L. W., Canfield, R. C., Gunkler, T. A., Hudson, H. S., Kiplinger, A. L., and Leibacher, J. W.: 1982, *Astrophys. J.* **263**, 409.

Akita, K.: 1985, Ph.D. Thesis, University of Tokyo.

Antonucci, E. and Dennis, B. R.: 1983, *Solar Phys.* **86**, 67.

Antonucci, E. and Dodero, M. A.: 1986, in D. F. Neidig (ed.), *The Lower Atmosphere of Solar Flares*, NSO/Sacramento Peak, p. 363.

Antonucci, E., Gabriel, A. H., and Dennis, B. R.: 1984, *Astrophys. J.* **287**, 917.

Antonucci, E., Rosner, R., and Tsinganos, K.: 1986, *Astrophys. J.* **301**, 975.

Antonucci, E., Rosner, R., and Tsinganos, K.: 1987, in V. E. Stepanov and V. N. Obridko (eds.), *Solar Maximum Analysis*, Science Press, p. 446.

Antonucci, E., Dodero, M. A., and Martin, R.: 1989, *Astrophys. J. Suppl.* (submitted).

Antonucci, E., Gabriel, A. H., and 7 co-authors: 1982, *Solar Phys.* **78**, 107.

Antonucci, E., Dennis, B. R., Gabriel, A. H., and Simnett, G. M.: 1985, *Solar Phys.* **96**, 129.

Antonucci, E., Dodero, M. A., Peres, G., Serio, S., and Rosner, R.: 1987, *Astrophys. J.* **322**, 522.

Antonucci, E., Dodero, M. A., Peres, G., Reale, F., and Serio, S.: 1989 (in preparation).

Bhalla, C. P., Gabriel, A. H., and Presnyakov, L. P.: 1975, *Monthly Notices Roy. Astron. Soc.* **172**, 359.

Bornmann, P. L.: 1987, *Astrophys. J.* **313**, 449.

Bruner, M. E., Crannell, C. J., Goetz, F., Magun, A., and McKenzie, D. L.: 1988, NASA Technical Memorandum 87815.

Canfield, R. C.: 1986, *Adv. Space Res.* **6**, 167.

Canfield, R. C., Gunkler, T. A., and Ricchiazzi, P. J.: 1985, *Astrophys. J.* **282**, 296.

Canfield, R. C., Zarro, D. M., Metcalf, T. R., and Lemen, J. R.: 1989, *Astrophys. J.* (submitted).

Cheng, C.-C., Karpen, J. T., and Doschek, G. A.: 1984, *Astrophys. J.* **286**, 787.

de Jager, C.: 1985, *Solar Phys.* **98**, 267.

Doschek, G. A.: 1983, *Solar Phys.* **86**, 9.

Doschek, G. A., Feldman, U., Landecker, P. B., and McKenzie, D. L.: 1981, *Astrophys. J.* **249**, 372.

Doschek, G. A. *et al.*: 1986, NASA Conference Publ. 2439, 4-1/4-42.

Dubau, J., Gabriel, A. H., Loulergue, M., Steenman-Clark, L., and Volonté, S.: 1981, *Monthly Notices Roy. Astron. Soc.* **195**, 705.

Emslie, A. G. and Alexander, D.: 1987, *Solar Phys.* **110**, 295.

Feldman, V., Doschek, G. A., Kreplin, R. W., and Mariska, J. T.: 1980, *Astrophys. J.* **241**, 1175.

Fisher, G. H., Canfield, R. C., and McClymont, A. N.: 1984, *Astrophys. J.* **281**, L79.

Fisher, G. H., Canfield, R. C., and McClymont, A. N.: 1985, *Astrophys. J.* **289**, 434.

Gabriel, A. H.: 1972, *Monthly Notices Roy. Astron. Soc.* **160**, 99.

Gabriel, A. H. and Jordan, C.: 1969, *Monthly Notices Roy. Astron. Soc.* **145**, 241.

Grineva, Y. I., Karev, V. I., Korneev, V. V., Krutov, V. V., Mandelstam, S. L., Vainstein, L. A., Vasilyev, B. N., and Zhitnik, I. A.: 1973, *Solar Phys.* **29**, 441.

Gunkler, T. A., Acton, L. W., Canfield, R. C., and Kiplinger, A. L.: 1984, *Astrophys. J.* **285**, 835.

Hiei, E. and Widing, K. G.: 1979, *Solar Phys.* **61**, 407.

Ichimoto, K. and Kurokawa, K.: 1984, *Solar Phys.* **93**, 105.

Karpen, J. T., Doschek, G. A., and Seely, J. F.: 1986, *Astrophys. J.* **306**, 327.

Lin, R. P., Schwartz, R. A., Pelling, R. M., and Hurley, K. C.: 1981, *Astrophys. J.* **251**, L109.

Linford, G. A. and Wolfson, C. J.: 1989, *Astrophys. J.* (in press).

Linford, G. A., Lemen, J. R., and Strong, K. T.: 1989, *Adv. Space Res.* (in press).

Martens, P. C. H., Van den Oord, G. H. J., and Hoyng, P.: 1985, *Solar Phys.* **96**, 253.

McKenzie, D. L., Broussard, R. M., Landecker, P. B., Rugge, H. R., Young, R. M., Doschek, G. A., and Feldman, U.: 1980, *Astrophys. J.* **238**, L43.

MacNeice, P.: 1986, *Solar Phys.* **103**, 47.

MacNeice, P., McWhirter, R. W. P., Spicer, D. S., and Burgess, A.: 1984, *Solar Phys.* **90**, 357.

Nagai, F. and Emslie, A. G.: 1984, *Astrophys. J.* **279**, 896.

Neupert, W. M., Gates, W., Swartz, M., and Young, R.: 1967, *Astrophys. J.* **149**, L79.

Parker, E. N.: 1983, *Geophys. Astrophys. Fluid Dyn.* **24**, 79.

Parker, E. N.: 1987, *Solar Phys.* **111**, 297.

Parmar, A. N., Culhane, J. L., Rapley, C. G., Antonucci, E., Gabriel, A. H., and Loulergue, M.: 1981, *Monthly Notices Roy. Astron. Soc.* **197**, 29.

Saba, J. L. R. *et al.*: 1989 (in preparation).

Smith, D. F. and Harmony, D. W.: 1982, *Astrophys. J.* **252**, 800.

Somov, B. V., Syrovatskii, S. I., and Spektor, A. R.: 1981, *Solar Phys.* **73**, 145.

Somov, B. V., Sermulina, B. J., and Spektor, A. R.: 1982, *Solar Phys.* **81**, 281.

Sweet, P. A.: 1969, *Ann. Rev. Astron. Astrophys.* **7**, 149.

Takakura, T. *et al.*: 1983, *Astrophys. J.* **270**, L83.

Tanaka, K.: 1986, *Publ. Astron. Soc. Japan* **38**, 225.

Tanaka, K.: 1987, *Publ. Astron. Soc. Japan* **39**, 1.

Tanaka, K. and Zirin, H.: 1985, *Astrophys. J.* **299**, 1036.

Tanaka, K., Watanabe, T., Nishi, K., and Akita, K.: 1982a, *Astrophys. J.* **254**, L59.

Tanaka, K., Akita, K., Watanabe, T., and Nishi, K.: 1982b, *Hinotori Symposium on Solar Flares, Tokyo*, p. 43.

Tsuneta, S.: 1987, *Solar Phys.* **113**, 35.

Tsuneta, S., Nitta, N., Ohki, K., Takakura, T., Tanaka, K., and 4 co-authors: 1984, *Astrophys. J.* **284**, 827.

Uchida, Y. and Shibata, K.: 1988, *Solar Phys.* **116**, 291.

Watanabe, T.: 1987, *Solar Phys.* **113**, 107.

Winglee, R. M., Pritchett, P. L., and Dulk, G. A.: 1989, *Astrophys. J.* (in press).

Wolfson, C. J., Doyle, J. G., Leibacher, J. W., and Phillips, K. J. H.: 1983, *Astrophys. J.* **269**, 319.

Zarro, D. M., Strong, K. T., Canfield, R. C., Metcalf, T., and Saba, J. L. R.: 1989, *Astrophys. J.* (submitted).

MULTI-WAVELENGTH OBSERVATIONS OF STELLAR FLARES

P. B. BYRNE

Armagh Observatory, Armagh BT61 9DG, N. Ireland

Abstract. We present observational data on stellar flares from a range of wavelength regimes, many of which were obtained simultaneously. Physical parameters of these flares are derived and discussed in the framework of the general solar flare model. It is found that flares on dMe stars are solar-like, except in mean energy. The parameters of flares on RS CVn stars are more extreme, however, and may require new models for their interpretation.

1. Introduction

Since, in the study of stellar flares, we cannot resolve the phenomenon spatially, simultaneous observations over as many wavelength regions as possible make a vital contribution to our understanding of the event. As a result of improvements in instrumental design these observations can now cover the electromagnetic spectrum from microwaves to X-rays.

I would like to stress at the outset the amount of effort required in acquiring simultaneous data with these instruments. Obtaining good coverage of a star over the entire range of accessible wavelengths requires worldwide collaborations, especially in respect of ground-based data. Extensive collaborations have been built up over the last decade and I would like to take this opportunity to pay tribute to many observers who have taken time out from other work to assist in obtaining the best possible coverage. In spite of these efforts many of the datasets I shall describe are incomplete in their wavelength coverage.

By its very nature it is difficult to organise a review of this type in a way which will be acceptable to all. So the following scheme is a personal one. I have chosen to highlight one or other of the wavelength regions covered in each set of observations, in order of increasing wavelength, so that I can summarize the main conclusions to be derived from that particular energy domain. As a result, the same dataset will sometimes be discussed from different points of view and in different places in the review.

2. X-Ray Flares

One of the primary characteristics of a solar flare is its X-ray signature, while that of the stellar flare has been its broadband optical continuum emission. For this reason a great deal of effort has been expended to establish the relationship between the optical and X-ray emission in stellar flares. A recent, excellent example of this type of observation has been the joint observation of a soft X-ray flare on the star BY Dra by the EXOSAT LE detector and simultaneous *UBVRI* photometry (de Jager *et al.*, 1988). Figure 1 is taken from this work. The following points should be noted. First, the peak of the broadband optical emission precedes that of the soft X-rays by between 4 and

Solar Physics **121**: 61–74, 1989.

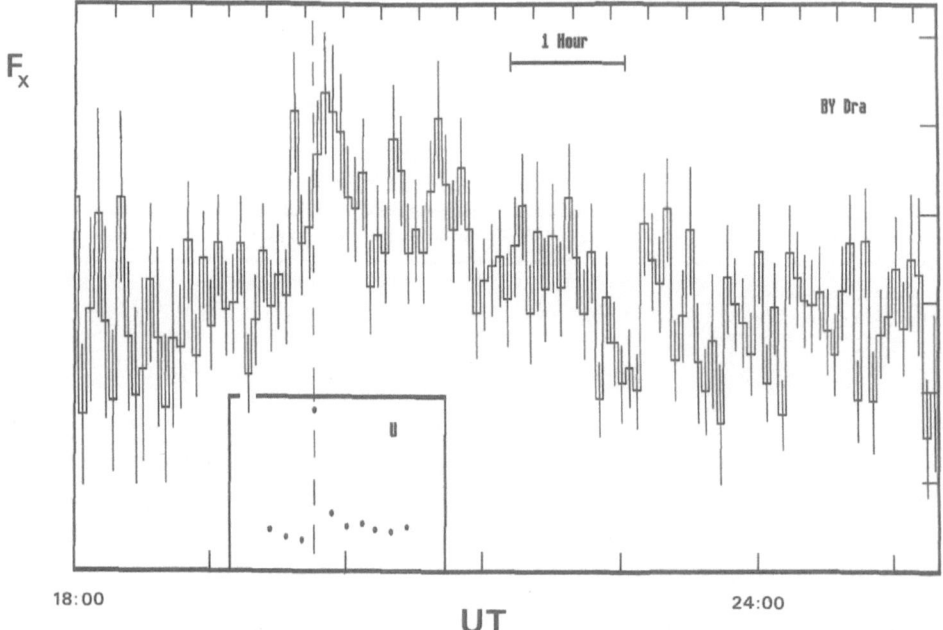

Fig. 1. A simultaneous optical and soft X-ray observation of a flare on the dMe star, BY Dra (de Jager *et al.*, 1988). The histogram shows the EXOSAT soft X-ray data, while the dots indicate the optical *U*-band light curve. The vertical dashed line is at the time of *U*-band maximum.

8 min (the resolution of the X-ray data is ≈ 4 min). Second, the time-scale of the optical flare ($\tau_{\mathrm{rise}} \approx 1$ min, $\tau_{\mathrm{fall}} \approx 5$ min) is very much shorter than that of the soft X-rays ($\tau_{\mathrm{rise}} \approx 10$ min, $\tau_{\mathrm{fall}} \approx 1$ hour). Some of the characteristics of this flare are given in Table I. Note that $E_{\mathrm{opt}} \approx E_X$.

Doyle *et al.* (1988a) observed another X-ray flare with the EXOSAT satellite on the star Gliese 644 (Wolf 630) which was recorded in both the soft (LE) and medium (ME) energy detectors (Figure 2). Here we see that the soft X-rays (LE) lag those of medium energy (ME) by ≈ 80 s. A thermal brehmsstrahlung fit to the ME data indicates that this reflects a cooling of the flare from $\approx 5 \times 10^7$ K to $\approx 2.5 \times 10^7$ K during the decay. This event was also observed in Hα, which exhibited broadening (see below). The ratio of the total energy radiated in soft X-rays to that in Hα is ≈ 25, not very different from that observed in solar flares (Neidig, 1989). Using the approximate relationships for stellar flares that $E_{\mathrm{Balmer}} \approx 3E_{\mathrm{H}\alpha}$ and $E_{\mathrm{opt}} \approx 10E_{\mathrm{Balmer}}$ yields $E_{\mathrm{opt}} \approx 1.2E_X$.

In contrast to these results on the ratio of E_{opt}/E_X, EINSTEIN X-ray and simultaneous optical observations of a flare on YZ CMi by Kahler *et al.* (1982) indicate $E_{\mathrm{opt}} \approx 0.7E_X$. An optical flare on YZ CMi observed by Doyle *et al.* (1988b) was initially reported to have no accompanying soft X-ray flare. A re-analysis of this data (Butler, Rodono, and Foing, 1989) has detected the X-ray event and indicates $E_{\mathrm{opt}}/E_X \approx 0.1$.

It would appear, therefore, that there is no unique value for E_{opt}/E_X but that it varies from flare to flare. Nevertheless, values of between ≈ 0.1 and ≈ 1 are representative.

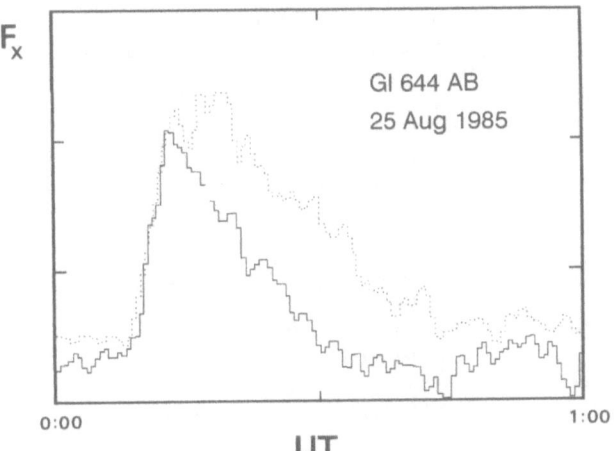

Fig. 2. EXOSAT light curves of a flare on Gliese 644 (Wolf 630) adapted from Doyle *et al.* (1988a). The dotted curve is the soft X-ray ($\approx$ 0.02–4 keV) light curve, while the solid curve is that for the medium energy X-rays ($\approx$ 2–8 keV).

Thus optical continuum emission is a very important part of the energy balance in stellar flares, being of the same order of importance as the soft X-ray emission.

Table I summarizes some of the main X-ray characteristics of representative stellar flares. Plasma temperatures of between 1×10^7 and 5×10^7 K are typical, not very different from large solar flares (Tanaka *et al.*, 1983). In general, temperatures are higher during the rise phase or at peak than later during the decay. Soft X-ray emission measures range from 10^{51}–10^{53} cm^{-3}, the lower end of which is comparable to those found in the largest solar two-ribbon flares but extending to larger values.

Electron densities, N_e, can be derived from the X-ray light curve decay times, τ_{decay}, if a simple form of cooling the flare plasma is assumed. Most authors assume either pure radiative cooling or equal contributions from cooling and conduction. Simultaneous observations of a solar flare by EINSTEIN and SMM (Schmitt, Lemen, and Zarro, 1989) indicate that the latter was a good approximation in that instance. In Table I we give electron densities based on the assumption of radiative cooling using the expression (Doyle *et al.*, 1988a)

$$N_e = \frac{3kT}{\Lambda(T)\tau_r},$$

where $\Lambda(T)$ is the plasma emissivity and τ_r is the e^{-1} decay time. Using the values in Doyle *et al.* this expression reduces to

$$N_e = \frac{7.5 \times 10^{14}}{\tau_r}.$$

Values derived in this way are typically 10^{11}–10^{12} cm^{-3}, similar to those obtained for the Sun.

Table I also contains entries for two flares observed on RS CVn stars. All of the characteristics of these flares, with the exception of the electron densities, are larger than

TABLE I

Summary of X-ray flares on dMe and RS CVn stars. E_X and E_{opt} are integrated flare energies in the soft X-ray (≈ 0.1–2 keV) and broad-band optical (Johnson UBV), respectively

Star	Source	Temperature (10^7 K)	E_X (10^{32} erg)	E_{opt} (10^{32} erg)	EM (10^{52} erg)	N_e $(10^{11} \text{ cm}^{-3})$
		dMe flares observed with EXOSAT				
BY Dra	de Jager *et al.*, 1988	1	20	40	1.2	
EQ Peg	Haisch *et al.*, 1987	2.6 (rise) 1.4 (decay)	5	(0.16) (Mg II)	1.5	2
	Kundu *et al.*	4.4 (peak) 2.7 (decay)	100	–	20	4
Wolf 630	Doyle *et al.* 1988a	5 (rise) 2.5 (decay)	12	(0.6) (Hα) [20]	3	8
		dMe flares observed with EINSTEIN				
YZ CMi	Kahler *et al.*, 1982	2 (peak)	0.36	0.12 (*U*) [≈0.3 (*UBV*)]	0.5	1
Prox Cen	Haisch *et al.*, 1981	1.7 (rise) 1.2 (decay)	0.1	–	0.1	1.4
	Haisch *et al.*, 1983	2.7 (peak)	0.35	–	0.1	5
		RS CVn flares observed with EXOSAT				
Algol	van den Oord *et al.*, 1986	5.8 (peak)	250	–	94	2.6
σ Cor Bor	van den Oord *et al.*, 1988	9.5 (rise) 6.4 (peak)	10^4	–	57 (peak) 23 (rise)	9

in the dMes discussed so far. It is clear that flares on these binary subgiant stars are about two orders of magnitude larger than in the dwarf M stars. Clearly, while it may be possible to apply the solar model to the dMe X-ray flares in a relatively straightforward way, more caution must be exercised when dealing with the RS CVns.

3. UV Line Emission

Early attempts to record the ultraviolet spectra of stellar flares were frustrated by a lack of sensitivity. With the advent of the International Ultraviolet Explorer (IUE) it became possible to record dMe and RS CVn ultraviolet spectra with sufficiently high sensitivity and spectral resolution to begin studying stellar flares in a systematic way.

The mid-UV signature of a stellar flare is well illustrated by a flare recorded by IUE on FK Aqr (Gliese 867A) by Butler *et al.* (1981) and shown in Figure 3. The figure shows the spectrum of the star in quiescence and during a 1 hour exposure during which the flare occurred. The dramatic enhancement of all of the chromospheric and transition region lines is apparent. This spectrum also illustrates the limitations of IUE for the study of stellar flares. In order to get an adequately exposed spectrum of the non-flaring star, exposures of the order 20–60 min are necessary. Furthermore, the dead-time to the

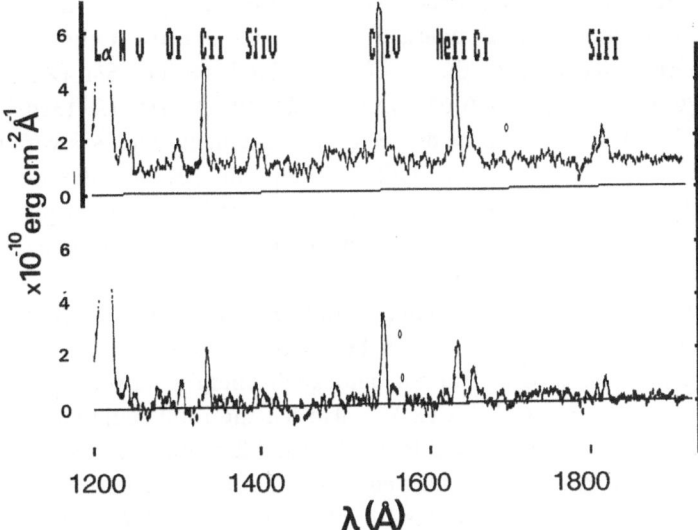

Fig. 3. An IUE short-wavelength spectrum of a flare on the dMe star, FK Aqr (a) adapted from Butler *et al.* (1981). (b) shows the star in its non-flaring state. Some of the most prominent emission lines are identified. Note the presence of a strong flare continuum.

next spectrum is of order 20 min. Since the duration of a typical UV flare appears to be much shorter than these times, no information on the time evolution of the transition region flare is possible. In spite of these restrictions, however, a great deal has been uncovered on the nature of dMe flares in the UV. Table II summarizes some of the important parameters of those stellar flares observed in the UV.

TABLE II

Summary of IUE flares

Star	Source	$E_{\rm CIV}$ (10^{30} erg)	$EM_{\rm CIV}$ (10^{52} cm^{-3})	N_e (10^{11} cm^{-3})	E_{TR} (10^{32} erg)	$E_{\rm L\alpha}$ (10^{30} erg)
		dMe stars				
FK Aqr	Butler *et al.*, 1981	34	(>0.05)	–	25	–
Prox Cen	Haisch *et al.*, 1983	0.5	(>0.001)	–	0.4	2.4
AT Mic	Bromage *et al.*, 1986	20	(>0.04)	≈10	15	40
EQ Peg	Baliunas *et al.*, 1984a	6	(>0.01)	–	4	
AD Leo	Byrne *et al.*, 1989	0.24	(>0.001)	–	0.2	4
		2.5	(0.002)	–	2	
YZ CMi	Rodono *et al.*, 1984	4	(>0.01)	–	3	
		RS CVn stars				
V711 Tau	Linsky *et al.*, 1988	14000	2	1	20000	1.3×10^5
λ And	Baliunas *et al.*, 1984b	>8000	1	$1 \leq N_e \leq 10$	>11000	55000
II Peg	Doyle *et al.*, 1989	7000	0.15	0.5	2500	–
IM Peg	Buzasi *et al.*, 1987	1×10^5	190	–	74000	–
UX Ari	Simon *et al.*, 1980	>500	–	0.7	–	–

The resonance lines of C IV ($\lambda\lambda1548/51$ Å) are the most prominent lines in the IUE short wavelength range in both quiescent and flaring spectra. Other lines seen to be enhanced during flares are those of C I ($\lambda\lambda1657-8$ Å), Si II ($\lambda\lambda1808/16/17$ Å), C II ($\lambda\lambda1335/6$ Å), Al III ($\lambda\lambda1856/63$ Å), He II ($\lambda1640$ Å), Si IV ($\lambda\lambda1394/1402$ Å), and N V ($\lambda\lambda1239/43$ Å). In general, the degree of enhancement increases with the temperature of formation of the relevant ion, with the exception of He II. All of this is consistent with a model based on solar flares wherein the transition region is pushed to higher densities during the flare with a consequent increase in the emission measure and a steepening of its temperature dependence.

Table II lists the time integrated energy emitted by each flare in the C IV resonance lines and in other important lines, where these were observed simultaneously. Energies of $\approx 10^{31}$ erg and higher are common and such energies are comparable to those observed in soft X-rays (cf. Table I). We also list the total radiative losses from the transition region as a whole, defined by the limits, $4.2 \leq \log T_e \leq 5.4$. These have been calculated by assuming a standard flare emission measure based on a mean of a number of observed flares and the radiative loss function of Raymond, Cox, and Smith (1976). Examination of Table II shows that the total transition region emission line losses, E_{TR}, are in the region $10^{32}-10^{33}$ erg. Where simultaneous X-ray observations were made (Haisch *et al.*, 1983), the transition region losses approximately equalled those in the soft X-ray region. Similarly, where simultaneous observations of the optical line emission (Baliunas and Raymond, 1984) or Lα emission (Bromage *et al.*, 1986; Byrne *et al.*, 1989) are available, these are also comparable to the transition region emission.

Strong UV continua have been seen in a number of dMe flares. Butler *et al.* (1981) observed a flat continuum over the entire IUE SWP range (1150–1950 Å) during a flare on FK Aqr, while Bromage *et al.* (1986) observed a similar continuum near the peak of a flare on AT Mic which reddened later in the flare (see Figure 4). The energy contained in these continua is considerable and may be as much as any of the other loss mechanims discussed above (see Table II). The energy in the ultraviolet continuum of the AT Mic flare was $\approx 10^5$ times that of a major solar flare.

Flares have also been observed with IUE on RS CVn stars. The most striking feature of these flares compared to those on dMes is their duration. Repeat exposures have shown C IV enhancements of about 3 over the *global* quiescent flux lasting at least 5 hours on λ And (Baliunas, Guinau, and Dupree, 1984), 7 hours on V711 Tau (Linsky *et al.*, 1988) and 6 hours on II Peg (Doyle, Byrne, and van den Oord, 1989). As a result of these long durations the energy emitted in C IV is orders of magnitude larger than in the dMes, typically $10^{34}-10^{35}$ erg. Correspondingly, the total transition region losses may be as high as $\approx 10^{37}$ erg.

It has been possible to detect electron density sensitive intersystem lines in RS CVn flare spectra, in particular C III] $\lambda1176$ Å, C III] $\lambda1909$ Å, and Si III] $\lambda1893$ Å. Combining these derived mid-transition region electron densities with the emission measures of lines such as Si IV ($\lambda1393/1402$ Å) and C IV ($\lambda1548/51$ Å) has enabled some authors to derive the volumes of the flaring plasma (see Byrne *et al.*, 1987, for a description of the technique). Volumes of order $10^{29}-10^{30}$ cm^3 and electron densities of order

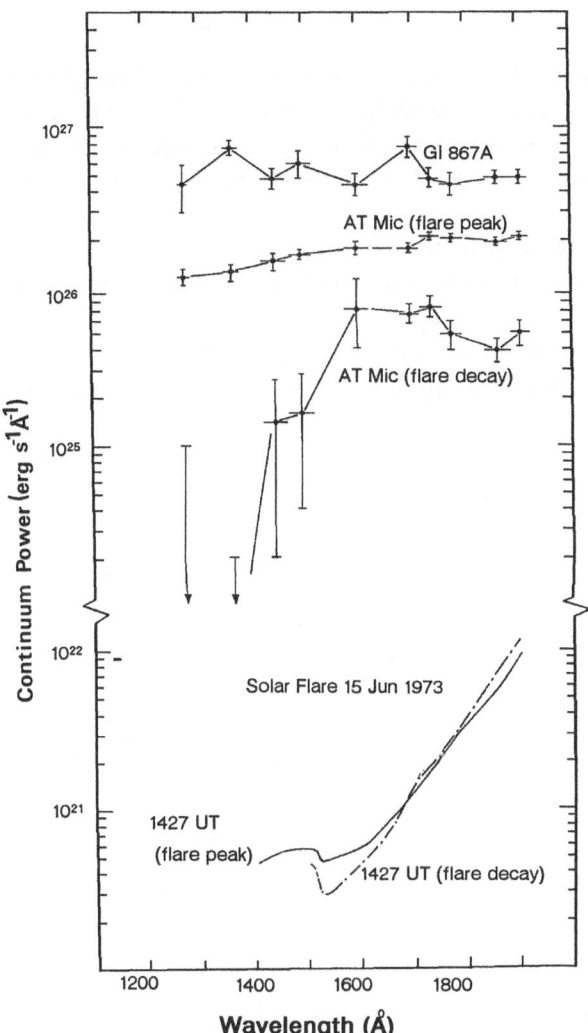

Fig. 4. A comparison of ultraviolet flare continua during two stellar flares on Gl 867A (FK Aqr) and AT Mic observed in the IUE short-wavelength region and the continuum recorded in a large solar flare (Bromage *et al.*, 1986).

$5 \times 10^{10}-10^{11}$ cm^{-3} are typical, not very much different from those of large solar two-ribbon flares. So it would appear that, in the transition region, RS CVn flares are more long-lived than solar or dMe flares but are otherwise fairly similar.

Lα and MgII h & k are important sources of cooling for the upper chromospheres of both RS CVn and dMe flares. In the case of the λ And and V711 Tau flares mentioned above the Lα and MgII losses each exceeded the total transition region losses.

4. Mass Motions

There is abundant evidence of mass motions within solar flares, ranging from Hα surges and filament disruption to blue-shifted components in Ca XIX and Fe XXV during the impulsive phase. Seeking similar evidence in stellar flares is complicated by the lack of spatial and spectral resolution. Nevertheless some significant progress has been made in this area.

Recent studies of the behaviour of the Balmer line profiles with time during a flare have produced firm evidence of broadening and asymmetry. Figure 5 shows the profiles of the Hδ and Hζ lines during a flare on the dMe star, YZ CMi, taken from Doyle *et al.* (1988b). Although taken from the same exposures the two lines shown significantly different time evolution. The profiles taken near the peak of the flare (1) both show

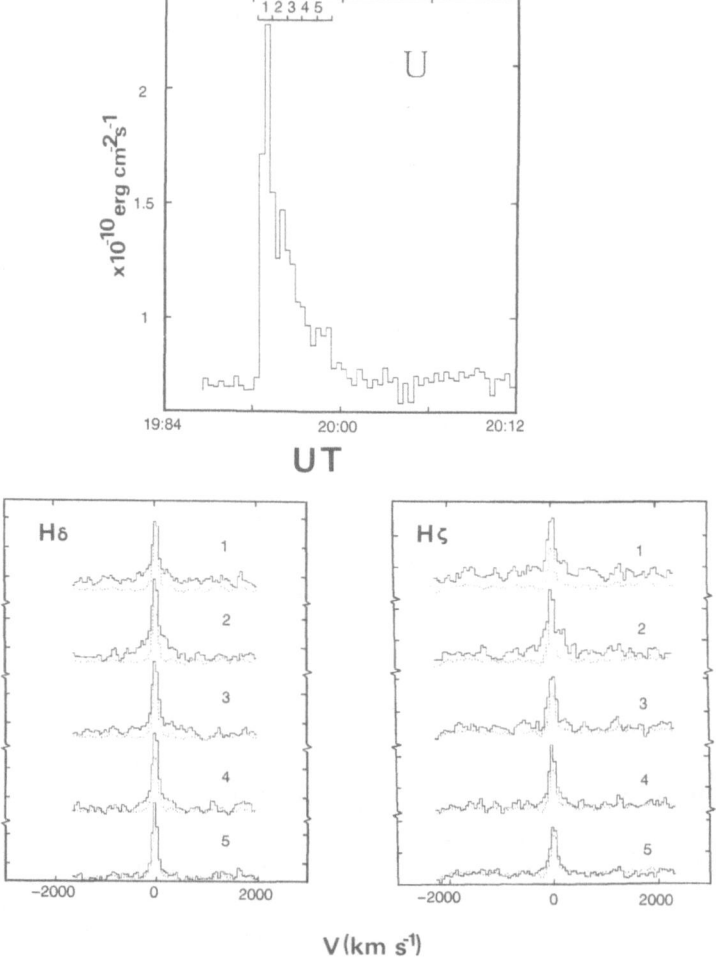

Fig. 5.   Optical *U*-band light curve (a) and corresponding Balmer Hδ (b) and Hζ (c) line profiles from a flare on ZY CMi from Doyle *et al.* (1988b). The dotted profiles in the lower panels are from a mean of several preflare spectra. The numbers refer to the time intervals during which the spectra were taken.

broadening of the core of the line compared with the mean of preflare profiles. The Hδ line, however, also shows the development of significant wings. By the time of a secondary peak in the optical light curve (2) both lines have developed extended wings. The Hδ wings have a pronounced redward asymmetry and extend out to ≈ 800 km s^{-1}. Later in the decay phase (3) these wings have weakened but they recover somewhat (4) before finally becoming unmeasurable. The excess broadening of the cores of the lines amounts to ≈ 100 km s^{-1}.

Although their time resolution was not as good as that of Doyle *et al.*, Phillips *et al.* (1988) found very similar results for Balmer line profiles in a flare on UV Cet. Rodono *et al.* (1989) also recorded extremely broad wings in Hγ and Hδ, with a pronounced red asymmetry, in a flare on AD Leo. All authors agree that the Ca II H & K lines show a relatively weak flare response and no detectable wings or wavelength shifts.

Spectroscopy of flares on RS CVn stars has taken place mainly in the ultraviolet and, again, wavelength shifts are seen. Because of detector sensitivity limitations only the Mg II h & k lines have been recorded at sufficiently high resolution and signal-to-noise to discuss these effects. Linsky *et al.* (1988) observed a large flare on V711 Tau with IUE and recorded the profiles of the h & k lines with a resolution of ≈ 19 km s^{-1}. By comparing the flare line profiles with a detailed model based on preflare data they were able to locate the flare on the K component of the binary and to derive flare-only line-profile parameters, as given in Table III. Their data were consistent with a red shift

TABLE III

Summary of wavelength shifts in dMe and RS CVn flares

Star	Source	Lines (Å)	v_{peak} (km s^{-1})	v (FWHM) Core	v (FWHM) Wings
		dMe stars			
YZ CMi	Doyle *et al.*, 1988b	Hδ, Hζ (4101, 3889 Å)	+ 50	50	200
Wolf 630	Doyle *et al.*, 1988a	Hα (6562 Å)	–	50	140
AD Leo	Rodono *et al.*, 1989	Hγ, Hδ, Hζ, Hη (4340, 4101, 3889, 3835 Å)	+ 30	250	750
UV Cet	Phillips *et al.*, 1988	Hβ, Hγ (4861, 4340 Å)	+ 100	–	–
		RS CVn stars			
II Peg	Doyle *et al.*, 1989	Mg II (2796/2803 Å)	+ 25	– 30	–
V711 Tau	Linsky *et al.*, 1988	Mg II, C IV (2796/2803, 1548/51 Å)	+ 90	–	–
λ And	Baliunas *et al.*, 1984b	Mg II (2796/2803 Å)	+ 25	–	–
UX Ari	Simon *et al.*, 1980	Mg II, Fe II (2796/2803, ≈ 2600 Å)	(– 45) (+ 475)	–	–

70   P. B. BYRNE

of ≈ 90 km s^{-1}. Earlier IUE observations of RS CVn flares on UX Ari by Simon,
Linsky, and Schiffer (1980) and on λ And by Baliunas, Guinan, and Dupree (1984) also
recorded redshifted flare light in the Mg II h & k profiles. The red shifts in the UX Ari
flare extended out to ≈ 475 km s^{-1}!

Doyle, Byrne, and van den Oord (1989) also observed a flare in the Mg II h & k lines
on the RS CVn star, II Peg. In this case a preflare spectrum, taken a short time earlier
and at almost the same orbital phase as the flare, was available and this was differenced
with the flare spectrum to derive the Mg II profile of the flare itself. The flare was of
sufficiently long duration that this was possible for two consecutive spectra of the flare.
One of these profiles is given in Figure 6 with a multi-gaussian fit superimposed. There
are two absorption features in the spectrum. One is also visible in the non-flare profile
and is due to interstellar absorption in the line of sight. The second is intrinsic to the

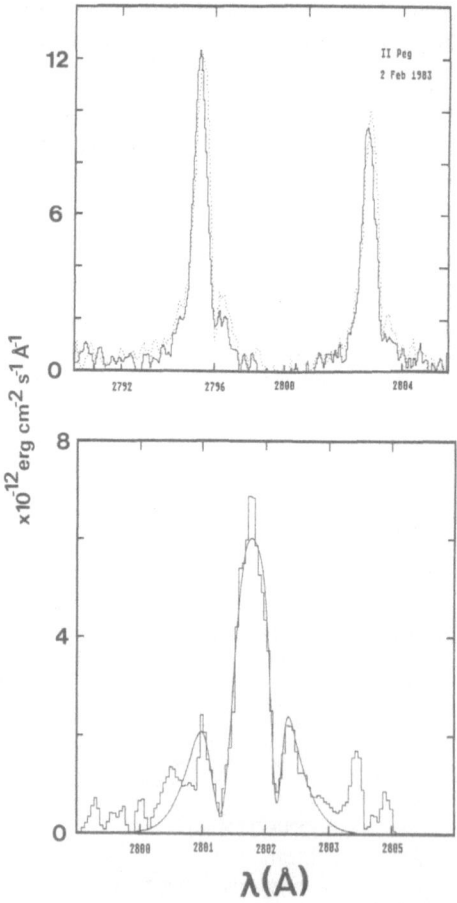

Fig. 6.   The upper panel shows the Mg II h and k line profiles immediately before (solid curve) and during
(broken curve) a flare on II Peg (adapted from Doyle, Byrne, and van den Oord, 1989). Note the apparent
redshift of the line peak. The lower panel shows the Mg II h profile of the flare light only, after subtraction
of the preflare profile and a multi-gaussian fit to this subtracted profile.

flare and apparently represents absorption by cool material overlying the flare itself. By assuming all of these components to be gaussian, the original emission profile can be reconstructed. This suggests that the Mg II emitting material in the flare was blueshifted with respect to the star by ≈ 50 km s^{-1} and the absorbing material by ≈ 80 km s^{-1}. However, because of the effect of the absorbing material on the emission profile, the original, unsubtracted flare Mg II profile appeared slightly *red*shifted with respect to the profile of the quiescent star.

Thus, while redshifts are observed generally in both dMe and RS CVn flares, these latter high resolution observations of II Peg would indicate caution in always accepting the results of low-resolution studies of the shifts in flares.

5. Microwaves

A classic example of a multi-wavelength observation of a dMe flare, including micro-waves, is the flare on AD Leo (Rodono *et al.*, 1989) shown in Figure 7. The impulsive rise in the optical *U*-band light is accompanied by a simultaneous rise in the 2 cm emission. The 6 cm emission, however, shows a much slower evolution to maximum, ≈ 10 min after the peak of the optical light. The decay time of the microwaves at both wavelengths is then comparable to the 'slow' component of the *U*-band light curve ($\tau_{decay} \approx 30$ min). This 'slow' component of the *U*-band light is known to be dominated by chromospheric emission lines, specifically the Ca II H & K and the higher H I Balmer lines. At this stage of the optical flare there is a negligible contribution from the continuum.

This pattern has been repeated in many simultaneous microwave and optical observations of flares, such as the YZ CMi flare of 3 February, 1983 (Rodono *et al.*, 1984) and another AD Leo flare on 2 February, 1983 (Gary, Byrne, and Butler, 1987). In contrast, however, an extensive study of AT Mic by Nelson *et al.* (1986) showed a very poor correlation between optical *U*-band flares and 6 cm flares. A comparable study has not been made at 2 cm, however, where the more impulsive nature of the microwave flares may make their detection easier.

Simultaneous microwave and optical observations of flares are summarized in Table IV. It will be noted in this table that there is no evident correlation between the peak flux density in microwaves and the total energy in optical or X-ray radiation. This would suggest that we are sampling a different radiating plasma at these different wavelengths.

6. Conclusions

Considerable progress has been made in the study of stellar flares, both on dMe and RS CVn stars, as a result of multi-wavelength studies of the phenomenon. The basic solar model of a localised injection of energy at the base of a magnetic loop structure and its subsequent cooling, at least in part, by radiation appears to serve us well for stellar flares also.

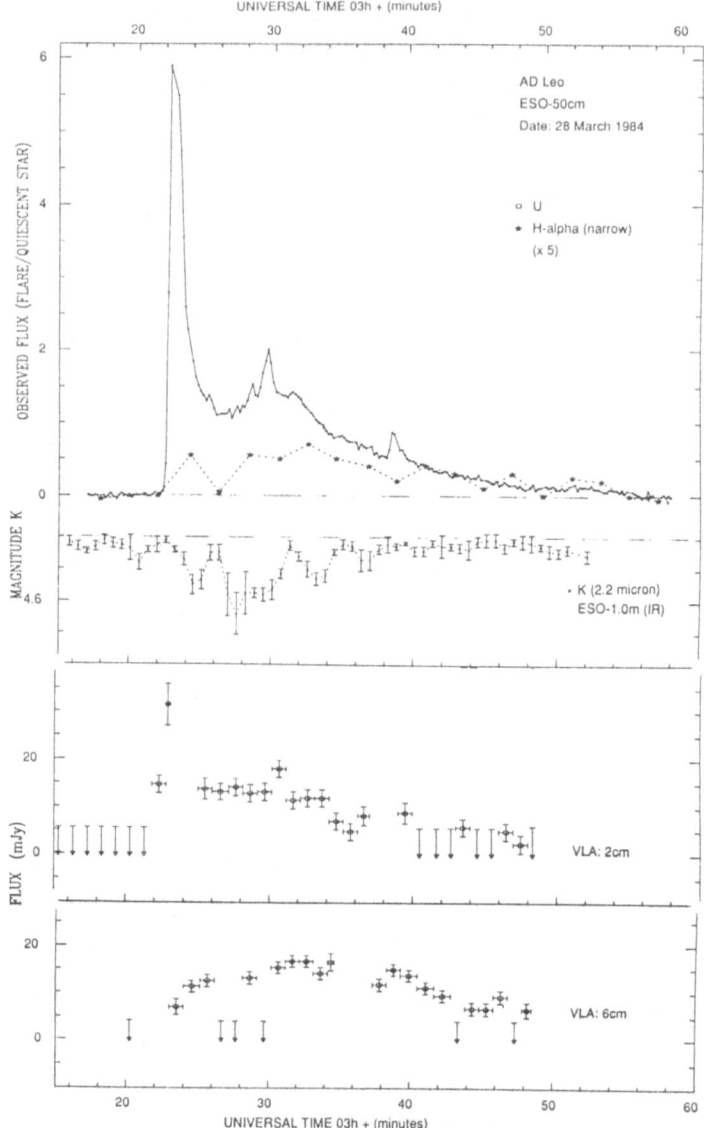

Fig. 7.   Simultaneous optical (*UBVRI*), infrared (*K*-band) and microwave (2 and 6 cm) observations of a
large flare on AD Leo from Rodono *et al.* (1989).

Table V summarizes what we consider the most reasonable average parameters of
stellar flares to be based on the observational survey made here. It will be noted from
this table that the basic parameters of the dMe star flares are not of a different order
of magnitude to solar flares except in mean energy per flare. In the case of the RS CVn
stars, however, this is not the case. Evidently the binary nature of the RS CVs, combined
with their lower gravities, is capable of boosting the energy content per flare. Whether
this occurs through a more efficient dynamo mechanism or more efficient magnetic flux

TABLE IV

Summary of microwave flares on dMe stars

Star	Source	S (2 cm) (mJy)	S (6 cm) (mJy)	S (20 cm) (mJy)	E_X (10^{30} erg)	E_{opt} (10^{30} erg)
UV Cet	Kundu et al., 1988	–	2.8	10	6	–
		–	3	35	3	–
AT Mic	Nelson et al., 1986	–	17	–	–	70
EQ Peg	Kundu et al., 1980	–	> 12	> 10	10^4	–
AD Leo	Byrne et al., 1989	–	3	–	–	15 (U)
	Kundu et al., 1988	–	4.5	–	?	–
	Rodono et al., 1984	32	16	–	–	300 (U)
YZ CMi	Rodono et al., 1989	< 2	5.5	< 2	–	200 (U)
	Kundu et al., 1988	–	2	15	20	–

TABLE V

Summary of dMe and RS CVn flare parameters

	dMe	RS CVn
Soft X-ray emission measure (cm^{-3})	10^{51}–10^{53}	10^{53}–10^{54}
X-ray temperature (K)	1–5×10^7	6–10×10^7
X-ray electron density ($\tau_{1/2}$) (cm^{-3})	1–8×10^{11}	3–9×10^{11}
Integrated energy (erg)	10^{31}–10^{34}	10^{34}–10^{36}
C IV emission measure (cm^{-3})	10^{49}–2×10^{50} (?)	10^{51}–10^{54}
Transition region electron density (cm^{-3})	$\leq 10^{12}$	5–10×10^{10}
Integrated transition region energy (erg)	10^{31}–10^{33}	10^{35}–10^{37}
Line peak velocities (km s^{-1})	$+30$–$+50$	$+25$–$+90$
Turbulent velocities (km s^{-1})	50–250	60–120

concentration or some other means of tapping the orbital energy of the binary is not yet clear. What is clear is that continued multi-wavelength studies of the flare phenomenon can contribute to a better understanding of the phenomenon.

References

Baliunas, S. L. and Raymond, J. C.: 1984a, *Astrophys. J.* **282**, 728.

Baliunas, S. L., Guinan, E. F., and Dupree, A. K.: 1984b, *Astrophys. J.* **282**, 733.

Bromage, G. E., Phillips, K. J. H., Dufton, P. L., and Kingston, A. E.: 1986, *Monthly Notices Roy. Astron. Soc.* **220**, 1021.

Butler, C. J., Rodono, M., and Foing, B. H.: 1989, *Astron. Astrophys.* (in press).

Butler, C. J., Byrne, P. B., Andrews, A. D., and Doyle, J. G.: 1981, *Monthly Notices Roy. Astron. Soc.* **197**, 815.

Buzasi, D. L., Ramsey, L. W., and Huenemoerder, D. P.: 1987, *Astrophys. J.* **322**, 353.

Byrne, P. B., Doyle, J. G., Brown, A., Linsky, J. L., and Rodono, M.: 1987, *Astron. Astrophys.* **180**, 172.

Byrne, P. B., Brown, A., Butler, C. J., Gary, D. E., Linsky, J. L., and Simon, T.: 1989, *Astron. Astrophys.* (in preparation).

Cohen, L., Feldman, U., and Doschek, G. A.: 1978, *Astrophys. J. Suppl.* **37**, 393.

De Jager, C., Heise, J., Avagoloupis, S., Cutispoto, G., Kieboom, K., Herr, R. B., Landini, M., Langerwerff,

A. F., Mavridis, L. N., Melikian, A. S., Molenaar, R., Monsignori-Fossi, B. C., Nations, H. L., Pallavicini, R., Piirola, V., Rodono, M., Seeds, M. A., van den Oord, G. H. J., Vilhu, O., and Waelkins, C.: 1988, *Astron. Astrophys.* **156**, 95.

Doyle, J. G., Byrne, P. B., and van den Oord, G. H. J.: 1989, *Astron. Astrophys.* (in press).

Doyle, J. G., Butler, C. J., Callinan, P. J., Tagliaferri, G., de la Reza, R., Torres, C. A., and Quast, G.: 1988a, *Astron. Astrophys.* **191**, 79.

Doyle, J. G., Butler, C. J., Byrne, P. B., and van den Oord, G. H. J.: 1988b, *Astron. Astrophys.* **193**, 229.

Gary, D. E., Byrne, P. B., and Butler, C. J.: 1987, in J. L. Linsky and R. E. Stencel (eds.), *Cool Stars, Stellar Systems and the Sun*, Proc. 5th Cambridge Workshop on Cool Stars, p. 106.

Haisch, B. M., Linsky, J. L., Slee, O. B., Siegman, B. C., Nikoloff, I., Candy, H., Harwood, D., Verveer, A., Quinn, P. J., Wilson, I., Page, A. A., Higson, P., and Seward, F. D.: 1981, *Astrophys. J.* **245**, 1009.

Haisch, B. M., Linsky, J. L., Bornman, P. L., Antiochos, S. K., Golub, L., and Vaiana, G. S.: 1983, *Astrophys. J.* **267**, 280.

Haisch, B. M., Butler, C. J., Doyle, J. G., and Rodono, M.: 1987, *Astron. Astrophys.* **181**, 96.

Kahler, S. *et al.*: 1982, *Astrophys. J.* **252**, 239.

Kundu, M. R., Pallavicini, R., White, S. M., and Jackson, P. D.: 1988, *Astron. Astrophys.* **195**, 159.

Linsky, J. L., Neff, J. E., Brown, A., Gross, B. D., Simon, T., Andrews, A. D., Rodono, M., and Feldman, P. A.: 1988, *Astron. Astrophys.* **211**, 173.

Neidig, D.: 1988, *Solar Phys.* **121**, 261 (this issue).

Nelson, G. J., Robinson, R. D., Slee, O. B., Ashley, M. C. B., Hyland, A. R., Touhy, I. R., Nikoloff, I., and Vaughan, A. E.: 1986, *Monthly Notices Roy. Astron. Soc.* **220**, 91.

Phillips, K. J. H., Bromage, G. E., Dufton, P. L., Keenan, F. P., and Kingston, A. E.: 1988, *Monthly Notices Roy. Astron. Soc.* **235**, 573.

Raymond, J. C., Cox, D. P., and Smith, B. W.: 1976, *Astrophys. J.* **204**, 290.

Rodono, M., Cutispoto, G., Catalano, S., Linsky, J. L., Gibson, D. M., Brown, A., Haisch, B. M., Butler, C. J., Byrne, P. B., Andrews, A. D., Doyle, J. G., Gary, D. E., Henry, G. W., Russo, G., Vittone, A., Scaltriti, F., and Foing, B.: 1984, *Proc. 4th European IUE Conf.*, ESA SP-218, p. 247.

Rodono, M., Houdebine, E., Catalano, S., Foing, B., Butler, C. J., Scaltriti, F., Cutispoto, G., Gary, D. E., and Gibson, D. M.: 1989, *Proc. IAU Colloq. 104 – Poster Papers*, Reprints Catania Obs. (in press).

Simon, T., Linsky, J. L., and Schiffer, F. H.: 1980, *Astrophys. J.* **239**, 911.

Schmitt, J. H., Lemen, J. R., and Zarro, D.: 1989, *Solar Phys.* **121**, 361 (this issue).

Tanaka, K., Nitta, N., Akita, K., and Watanabe, T.: 1983, *Solar Phys.* **86**, 91.

van den Oord, G. H. J., White, N. E., Culhane, J. L., Parmar, A. N., Kellett, B. J., Kahn, S., and Kuijpers, J.: 1986, *Astrophys. J.* **301**, 262.

van den Oord, G. H. J., Mewe, R., and Brinkman, A. C.: 1988, *Astron. Astrophys.* **205**, 181.

SOLAR FLARES: THE IMPULSIVE PHASE

BRIAN R. DENNIS

Solar Physics Branch, Laboratory for Astronomy and Solar Physics, Goddard Space Flight Center, Greenbelt, MD 20771, U.S.A.

and

RICHARD A. SCHWARTZ

STX Corporation, Lanham, MD 20706, U.S.A.

Abstract. Only during the previous solar cycle have systematic observations begun to be made with the sensitivity and time resolution, and the continuous coverage required to catch the impulsive phase and measure the rapid variations present in many wavelength ranges. Observations in X-rays, gamma-rays, UV, Hα, and radio wavelengths all reveal rapid variations during the impulsive phase and have contributed to our understanding of the different phenomena involved. Results have been obtained from several spacecraft, from rocket and balloon flights, and from ground-based observations. These are reviewed in the context of a simple single loop flare model with a view to showing what results are consistent with this model and what the major problems are in our understanding of the impulsive phase. New instrumentation planned for observations during the present Cycle 22 will provide a concerted attack on the impulsive phase as part of the Max '91 program.

1. Introduction

The terms 'impulsive' and 'gradual' were originally proposed by Covington and Harvey (1958) to describe two broad classes of microwave bursts. Kane (1969) first recognized the two components in energetic X-ray bursts. Today, the terms are used to describe the different phases of flares as observed in many wavelength ranges. Unfortunately, there is often confusion with this terminology since what appears impulsive in one wavelength range, e.g., hard X-rays, may appear gradual in another wavelength range, e.g., soft X-rays. Nevertheless, the distinction between the two phases is important since it is assumed that during the impulsive phase energy is being released 'impulsively', i.e., on time-scales of seconds or less, whereas during the gradual phase either energy is being released more gradually on time-scales of minutes to tens of minutes or no energy is being released at all.

The usual assumption is that solar flares result from the release of free energy in coronal magnetic fields either through reconnection or some other form of magnetic dissipation. The impulsive and gradual phases must then be interpreted as resulting from different energy release processes, different modes of magnetic dissipation, or different magnetic configurations. In many cases, though not all, the gradual phase is thought to result from the slow decay of the energy released during the impulsive phase with no new energy release being required. This idea that all the energy of a solar flare is released during the impulsive phase may explain why, as pointed out by Sturrock *et al.* (1984), it has often been 'implicitly assumed that to explain the impulsive phase is to explain the complete flare'. This idea may be true for many smaller flares where the impulsive phase appears as the dominant feature but for other flares, particularly the larger ones,

Solar Physics **121**: 75–94, 1989.

gradually varying hard X-ray and microwave emissions lasting as long as an hour or more indicate clearly that energy continues to be fed into the flare long after any impulsive phase is over.

Further indications that the impulsive phase is not the complete story of flare energy release are the observations of coronal mass ejections (CMEs) lifting off *before* the impulsive phase of the presumably associated flare (Simnett and Harrison, 1985; Harrison *et al.*, 1985; Harrison, 1986). Kahler *et al.* (1988) also report that filament eruptions begin before the onset of the impulsive phase and evolve smoothly through the flare. Thus, we must conclude from these observations that in those cases, the impulsive flare was a consequence of the associated filament eruption or CME rather than that the eruption or the CME was a consequence of the flare. This places a completely new light on the significance of the impulsive phase for those events. It must be remembered that the total kinetic energy associated with the mass motion of a CME can be considerably larger than the total energy released during the associated flare.

In spite of this new understanding of the relation between the different aspects of the energy release phenomena, it is still true that 'the principal theoretical flare problem is that of sufficiently rapid primary energy release' (Brown, Smith, and Spicer, 1981). Hard X-rays are observed with such intensities that, given the standard interpretation of them as collisional bremsstrahlung from high-energy electrons in the flare plasma, energy release rates as high as 10^{30} ergs s^{-1} are required to accelerate the emitting electrons during the impulsive phase. Simpler considerations of the $\gtrsim 10^{32}$ ergs released in the biggest flares in a characteristic flare duration time of $\sim 10^3$ s show that sustained energy release rates of at least 10^{29} erg s^{-1} are required. Such energy release rates are extremely challenging theoretically given the magnetic field strengths and configurations believed to exist in the corona. It is for this reason that increasing emphasis is being placed on observations of the impulsive phase in many different wavelength ranges. This paper constitutes a review of the more recent observations of impulsive phase phenomena at all wavelengths where evidence of such phenomena is found. It is hoped that this review will be valuable to theoreticians, who will see what observations are available to compare with their model predictions, and to observers, who can use it as a basis for comparing with observations from the much improved instrumentation planned for the next maximum in solar activity during Cycle 22.

In order to provide a common basis for discussing the observations in different wavelength ranges, we present in the next section a simple flare model that is consistent with many impulsive phase observations. This model also serves to show where in the solar atmosphere the different emissions may originate and how they may be related to one another and to the various phenomena of the impulsive phase. We have broken down the paper into three sections – energy release, energy transport, and energy loss – and show how the observations in different energy ranges provide information on these processes. The following sections contain discussions of observations in hard X-rays and γ-rays, microwaves and other radio waves, soft X-rays, UV and EUV wavelengths, and Hα. Finally, a brief discussion is given of what to expect from the planned Max '91 program of new observations during the current cycle of solar activity.

2. A Simple Flare Model

It has been known since Skylab observations showed the ubiquity of magnetic loops in the solar atmosphere that such loops must play a dominant role in the flare process. Consequently, it is not surprising that the possible flare model illustrated in Figure 1 (Dennis *et al.*, 1986; Gurman, 1987) is based on a magnetic loop extending into the corona. While a single loop is shown in the figure, much more complicated field geometries involving arcades of loops are usually present in all but the simplest of flares. Indeed, the most intense impulsive flares tend to occur in magnetically complex regions (Švestka, 1976).

According to this simple model, free magnetic energy in the current-carrying loop is

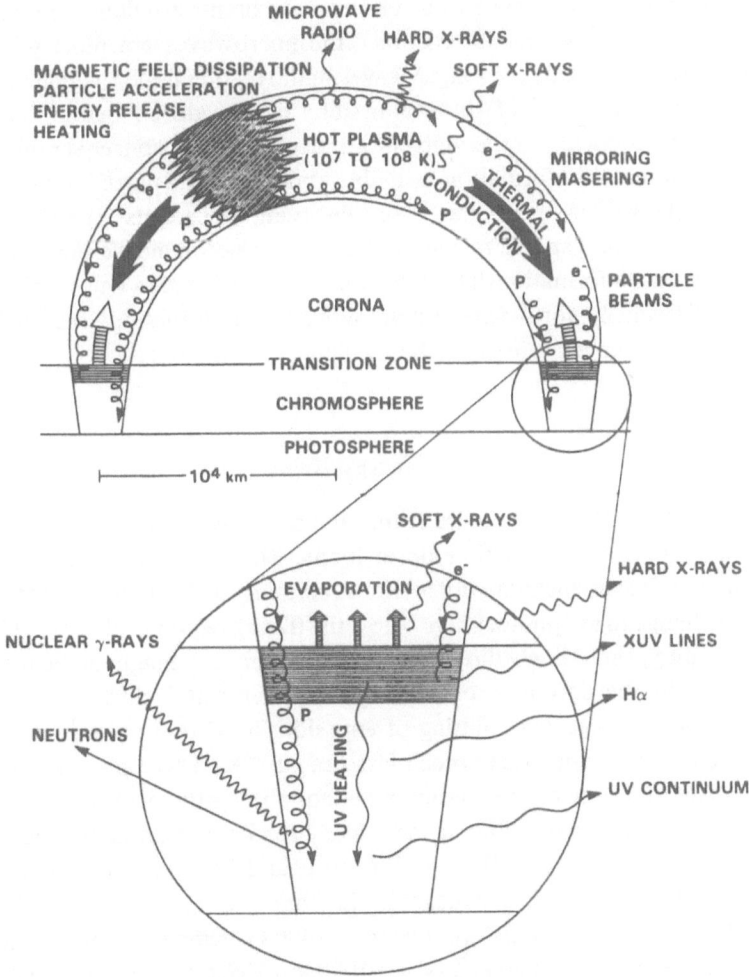

Fig. 1. Schematic diagram of a simple flare showing some of the physical processes that are believed to occur and where in the solar atmosphere different emissions are thought to originate.

dissipated by some ill-defined mechanism probably in the coronal part of the loop. As a result of this dissipation, the coronal plasma is heated, possibly to temperatures in excess of 10^8 K, and electrons and protons are impulsively accelerated to high energies. The division of the released energy between the heated plasma and the accelerated particles is still largely unknown and the subject of great controversy. The accelerated particles propagate along the magnetic field lines and interact with the ambient gas in the legs of the loop, with the most energetic ones penetrating to the loop footpoints in the chromosphere.

As indicated in the figure, a variety of observable emissions are produced from the different parts of the loop. From the point of view of understanding the particle acceleration process and, hence, the energy release process itself, the hard X-rays, γ-rays, radio waves, and neutrons are the most important emissions. The hard X-ray and γ-ray continuum is believed to be electron-ion bremsstrahlung, the γ-ray lines and neutrons result from nuclear interactions, the microwaves are most probably gryro-synchrotron emission, and the longer wavelength radio emission is plasma wave emission of various types. All of these emissions are produced before the accelerated particles lose their energy to the ambient atmosphere, and, consequently, they contain the greatest amount of information available about how the particles were accelerated. Increasing emphasis has been placed on observing these emissions with ever greater temporal, spectral, and spatial resolution (plus polarization measurements where possible) to extract this information. This push for greater resolution will continue during the maximum of Cycle 22, for we are still not close to the resolutions required in many of these wavelength ranges to extract the available information.

3. Energy Release

We assume, as illustrated in Figure 1, that the impulsive energy release takes place in the coronal part of the magnetic loop or loops. The evidence for this location of the release site is not overwhelming but it almost certainly lies above the photosphere since no sudden change in the photosphere below the flaring region is observed (Kahler *et al.*, 1980). Currently, there is no direct way to determine the magnetic field strength and configuration in the corona – only the photospheric and chromospheric fields are measured from the Zeeman splitting of emission lines from various partially-ionized atoms generally at temperatures much less than 10^6 K. These photospheric fields can be extrapolated into the corona assuming either a potential field or a 'force-free' field.

Observations show that a flare is most likely to occur in an active region where the magnetic shear is the greatest (Hagyard, Moore, and Emslie, 1984), thus implying that there is considerable twist in the magnetic field lines of the loop. Moore, Hagyard, and Davis (1987) show that a flare will occur in 1000 G fields if a critical shear angle of 80–85° is exceeded for a distance of $> 10\,000$ km along the polarity inversion line. It may well be that the flare energy is derived from the dissipation of the poloidal component of the field in the loop produced by a current along the loop. At present such theories are speculative because of the limitations of the magnetic field measurements and there is little hope in the foreseeable future of being able to directly measure the coronal magnetic fields in a flaring region with anything like the accuracy required to detect the reduction in free energy expected during a typical impulsive flare. Indeed today the uncertainty in measurements of the total magnetic energy in an active region is generally greater than or of the same order as the energy released in the flare. Thus, while definite before-to-after magnetic changes have been detected in a few flares (e.g., Moore et al., 1984), in no case has it been shown that the observed change quantitatively accounts for the flare energy (Moore, 1988).

In spite of this limitation, changes in the magnetic field can be seen in observations of filaments of chromospheric material that reside in and trace out sheared magnetic fields over magnetic inversion lines. During a flare these filaments are seen to expand and untwist indicating a decrease in the magnetic energy (e.g., Kurokawa et al., 1987; Moore, 1988). Moore (1988) has shown that for three eruptive events the decrease in magnetic energy of 10^{30}–10^{32} ergs is of the same order as the total energy released in the flare and/or coronal mass ejections. This provides strong support for the idea that the flare energy comes from the magnetic field, at least for this particular kind of flare involving a filament eruption.

There is hope that future microwave observations made with high spatial resolution at many frequencies can provide a means of determining magnetic field strengths in the corona and transition zone (Kundu and Lang, 1985; Holman, 1986). Webb et al. (1987), for example, have been, 'able to deduce or place constraints on the magnetic field strengths within, and their variations along', six coronal loops using microwave observations at 1.45 GHz (20 cm) and 4.9 GHz (6 cm) together with photospheric magnetograms and soft X-ray images and the dipole loop models of Holman and Kundu (1985). Holman (1986) has pointed out that the present determinations of coronal magnetic field strengths are ambiguous since there are two, or possibly three, different contributions to the observed microwave emission, thermal bremsstrahlung (free-free), thermal gyroresonance (cyclotron) emission, and possibly non-thermal gyrosynchrotron emission. The magnitude of each of these contributions must be determined before the observations can be unambiguously interpreted.

The contribution from free-free emission can be determined if simultaneous soft X-ray images are available to provide measures of the electron temperature and emission measure. The gyroresonance emission depends on the magnetic field strength and electron temperature but it can be present at several possible harmonics of the electron gyrofrequency. Holman and Kundu (1985) have computed the expected thermal gyroresonance emission for two-dimensional dipole loop models, the simplest non-trivial configuration that might be expected, and these predictions can be used as a basis for comparing with observations. They show that, in order for future observations to provide unambiguous magnetic field information useful in determining the preflare conditions of a loop, the microwave maps must be made with high spatial resolution and at many closely spaced frequencies. Coordinated EUV and/or soft X-ray images should be obtained to determine the temperature and emission measure distributions so that the thermal bremsstrahlung contribution can be determined and the thermal gyroresonance temperature dependence computed.

Measurements of the magnetic field topology during an impulsive flare are also difficult to obtain but for different reasons. Although one type of emission, gyro-synchrotron emission, dominates at this time, the fluctuations are much more rapid and short time-resolution observations at many different frequencies are desirable. Consequently, the 3–10 s capability of the VLA at only one or two frequencies becomes a serious limiting factor. Hoyng *et al.* (1983), Dulk and Dennis (1982), and Schmahl, Kundu, and Dennis (1985) have combined VLA snapshot maps at 2 and/or 6 cm with single-frequency flux measurements and hard X-ray observations to determine a magnetic field strength of ~ 550 G in the flaring region. No mapping of the field was possible, however, although Hoyng *et al.* were able to show a bending of the field lines by $\gtrsim 80°$ over a distance of $3''$ suggesting the top of a magnetic loop, albeit a considerably smaller loop than the one suggested by the X-ray images.

The closest we have come to observing the energy release process may be by observing the decimetric radio spikes present during some impulsive flares. Benz (1985) showed that events are observed between 100 and 1000 MHz with $\sim 10\,000$ spikes suggesting that the energy release process was fragmented with each spike resulting from an energy release of 10^{26} ergs within 0.05 s.

Significant metric and decimetric radio emission is usually observed during the impulsive phase, often in the form of type III bursts. These bursts are narrow band and drift rapidly with time generally towards lower frequencies. They are interpreted as resulting from an electron beam passing upwards through the corona with the radio emission produced at the leading front of the beam where it is unstable to the production of Langmuir waves that subsequently couple into electromagnetic waves at the local plasma frequency or its harmonic. The bursts are generally observed at frequencies of $\lesssim 500$ MHz corresponding to densities of $\lesssim 10^9$ cm^{-3} and altitudes of $\gtrsim 10^5$ km although they have now been seen at frequencies as high as 5 GHz (Benz, private communication). Comparisons between type III and hard X-ray bursts show poor correlations with isolated type IIIs – only 3 % are correlated with hard X-rays – but good correlation with groups of type IIIs (Kane, 1981) especially those with type V continuum (Stewart, 1978). The correlation becomes even better for type IIIs with high starting frequencies (Benz, Bernold, and Dennis, 1983) presumably because a beam that produces such a burst is denser and hence becomes unstable earlier along its trajectory from the acceleration site (Benz, 1987). The denser the beam, the bigger the hard X-ray flux it or its downward directed counterpart produces.

Some type III bursts are observed to extend to the vicinity of the Earth. *In situ* measurements of the source electrons have verified the electron beam origin and have established that the number of electrons required to generate a type III burst is several orders of magnitude smaller than the number required to generate a detectable flux of X-rays with current instrumentation.

4. Energy Transport

At least five possible forms have been proposed for the energy transport from the energy release site in the corona to the region of energy dissipation in the lower corona and chromosphere at the loop footpoints. They are as follows with the X-ray production mechanism shown in brackets:

– Thermal plasma with a temperature $T \geq 10^8$ K (thermal bremsstrahlung X-rays).
– Fast electrons with energies ≥ 20 keV (thick-target bremsstrahlung X-rays).
– Relativistic electrons (X-rays by inverse Compton interactions).
– Protons with energies ≤ 1 MeV (fast electrons produced at footpoints by ill-defined mechanism produce bremsstrahlung X-rays).
– Protons with energies > 1 MeV (inverse bremsstrahlung X-rays).

The first two possibilities in this list constitute the well-known thermal and thick-target models, respectively. In the thermal model (Brown, Melrose, and Spicer, 1979; Smith and Lilliequist, 1979; Batchelor et al., 1985), the energy release goes to impulsively heat the plasma near the release site to a temperature of $\geq 10^8$ K. This plasma is temporarily confined behind ion-acoustic conduction fronts that form in the loop and move at the ion sound speed (10^8–10^9 cm s^{-1}) down the legs of the loop to the footpoints taking ~ 20 s for a 30 000 km long loop if the density is 10^{10} cm^{-3}. In the thick-target model, electrons are accelerated high in a magnetic loop and propagate along the guiding field lines, producing X-ray bremsstrahlung and atmospheric heating as they proceed. The higher energy electrons lose most of their energy in the higher density regions of the lower corona and upper chromosphere. This thick-target model has been the most successful in explaining the largest fraction of the observations during the impulsive flares but still cannot be considered as proven.

The other three possibilities given above involving relativistic electrons or protons of different energies have met with limited success in explaining the observations of impulsive flares. However, there are several scientists working to better understand the implications of these models and to make predictions that can be tested against observations. It is fair to say that none of these models can, as yet, be definitively ruled out.

Models involving relativistic electrons require much less total electron energy (Brown, 1976) than the other models but are perhaps the least successful in explaining the observations. The hard X-rays are produced from the relativistic electrons as synchrotron or inverse Compton radiation but the required electron energies are very high, > 1 GeV for synchrotron and > 10 MeV for inverse Compton radiation. Electrons with such high energies and in sufficient numbers to produce the observed X-ray fluxes are not consistent with microwave burst intensities nor are they detected in interplanetary space (Brown, Smith, and Spicer, 1981).

Models involving protons as the primary accelerated particles have seen a resurgence of interest recently in attempts to explain some apparent difficulties with the thick-target bremsstrahlung model. In particular, Simnett (1986) has proposed that the bulk of the energy in the impulsive phase is initially transferred to protons with energies between

100 keV and 1 MeV. Martens (1988) has developed a model for the generation of proton beams in two-ribbon flares. In this model neutral beams are generated by direct electric field acceleration making protons with typical energies of 200 keV the main carriers of the beam energy. Henoux *et al.* (1988) have recently presented observations of Hα linear polarization in a flare as evidence for the existence of the atmospheric bombardment by deka-keV protons. Their suggestion that impact linear polarization of chromospheric lines can be used as a diagnostic of deka-keV protons opens up the possibility for detecting these enigmatic protons and determining if they do in fact play a dominant role during impulsive flares. Previously, the best hope for detecting these protons was through the predicted red-shifted component of the Lα line resulting from the decay of excited hydrogen atoms produced in charge exchange interactions between the protons and the ambient plasma (Orrall and Zirker, 1976).

Heristchi (1986) has argued that bremsstrahlung by fast protons ($E_p \gg 1$ MeV) on stationary electrons may be the origin of hard X-ray emission in flares. However, as pointed out by Emslie and Brown (1985), such a model, while energetically feasible, requires a number of fast protons which is three orders of magnitude higher than that required to produce the observed gamma-ray fluxes by nuclear reactions (e.g., Ramaty and Murphy, 1987). Heristchi (private communication) has countered this claim by pointing out that the following factors not considered by Emslie and Brown may in fact remove this discrepancy:

(i) a factor of two decrease in the number of protons required because of the lower energy loss rates, and hence greater bremsstrahlung efficiency, in the near-neutral deeper layers of the atmosphere where the protons mostly interact;

(ii) an uncertainty of a factor of ~ 8 in the extrapolation of the proton spectrum from the 60 MeV or so responsible for emission of 30 keV hard X-rays (the lower limit of the SMM HXRBS energy range) to the 20 MeV responsible for γ-ray emission;

(iii) a general confusion of up to a factor of 6 in the quantitative ratio between particle and photon energy contents in a thick-target model (cf. discrepancies between Brown (1971); Hoyng, Brown, and van Beek (1976); and Emslie, Phillips and Dennis (1986)); and

(iv) a factor that could be as large as 10 resulting from a high degree of directivity of the emitted bremsstrahlung, producing a given photon yield for a smaller flux of protons than in a calculation assuming isotropic emission.

If these factors are all taken together, *and in the same direction*, they can remove the discrepancy between hard X-ray and gamma-ray yields by reducing the number of 20 MeV protons required to a value much less than that used by Emslie and Brown (1985). However, Emslie (private communication) points out that factor (ii) could work the other way, i.e., there could equally well be a spectral steepening below 60 MeV; factor (iii) is probably not an issue any more since early errors and misprints have now been corrected (e.g., Emslie, Phillips, and Dennis, 1986); and factor (iv) depends on the position and orientation of the flaring loop on the Sun. Therefore, the consistency claimed by Heristchi seems to be extremely unlikely at best.

An equally serious problem of this model is the large numbers of pions and neutrons that would be produced in nuclear interactions of the high-energy protons required to produce the observed γ-ray continuum up to ~ 10 MeV. The electron/positron and γ-ray decay products of the pions should result in far more γ-rays in the 10 to 100 MeV range than are observed with the Gamma-Ray Spectrometer on SMM (Forrest, private communication). The neutron fluxes would also be higher than those observed. Thus, although this model is intriguing and it is being further refined by Heristchi (private communication), it seems to be beset with several problems that make it inconsistent with the γ-ray and neutron observations.

5. The Energy Loss Region

In this section we focus on the energetic particles as they enter the higher density footpoints of the lower corona and the chromosphere. Here, some small fraction of their energy goes into radiation while the rest is lost to collisions heating the cooler ambient plasma which in turn yields its own radiation signatures. In this section we examine observations not only of that immediate radiation from the energetic particles, but also that resulting from heating and collisional excitations.

First, we describe the hard X-ray and microwave radiation produced by the energetic electrons, citing observations describing its spectral, temporal, and spatial morphology. We note whether these measurements support a thick-target beam interpretation or some thermal or trap model. Then we look at the implications of the γ-rays produced by energetic ions colliding with the solar atmosphere. Next, we describe the UV, soft X-rays, and Hα emanating from the apparent footpoint products of the energy input associated with the fast electrons. While considerable qualitative support exists for the general picture of particle acceleration and transport within loops, it must be emphasized that all of the observations are subject to varying interpretation due to the uncertainties in many important parameters and unavoidable integrations over space, time, and energy.

5.1. ENERGETIC PARTICLE RADIATION SIGNATURES

The most direct way to study the energetic particle populations is from the radiation produced as they move through the solar atmosphere. Bremsstrahlung X-rays are produced by the fast electrons, most from ~ 10 to 100 keV but ranging to above 10 MeV. Despite the fact that the photon production cross-section is quite broad, simplifying assumptions allow one to relate the X-ray spectrum and the injected and instantaneous electron distributions (Brown, 1971). The overlying solar atmosphere is transparent to this radiation but detection is limited to space-borne photon-counting instruments with event energy obtained by pulse-height analysis. Imaging can only be achieved using some variant of a masking technique. Fast electrons also produce microwaves by gyro-synchrotron emission as they move along the magnetic field. Even though there is significant absorption in high field regions at lower frequency, the opacity drops well below unity at the higher frequencies. The spectrum produced depends on the magnetic field strength, harmonic number, and direction. Consequently, compared to the hard

X-ray spectrum, it is more difficult to relate it directly to the spectrum of source electrons. The great advantage of microwaves is that they can be imaged inter-ferometrically on the ground in both left and right circular polarization from < 1 GHZ to over 20 GHZ. Both line and continuum γ-ray emission are produced as the energetic protons and other ions (10–30 MeV) move through the solar atmosphere. Spectro-scopic measurements reveal the composition and spectra of these energetic ions.

5.2. HARD X-RAYS FROM FAST ELECTRONS

In the thick-target model, the energetic electrons stream through the low density corona and lose most of their energy in Coulomb collisions in the higher density plasma at the footpoints. The effective range of the electrons, computed by integrating the expression given by Trubnikov (1965) for the energy loss rate in a cool medium, is $(E/20 \text{ keV})^2 \times 7.7 \times 10^{19} \text{ cm}^{-2}$. Consequently, most deka-keV electrons are stopped by a column depth of about 10^{20} cm^{-2}, i.e., most of the loop length for coronal densities of 10^9–10^{10} cm^{-3}. Also, since the ratio of bremsstrahlung to collisional loss is about 1 in 10^5, most of the energy flux does not appear in the form of hard X-rays but instead heats the ambient medium. While the actual electron-ion bremsstrahlung X-rays are not energetically important, they provide the most direct and most easily interpretable information on the fast electrons, and these are energetically important.

5.3. EVIDENCE FOR FOOTPOINT EMISSION

The most direct evidence that the impulsive hard X-rays are produced at the footpoints of loops was provided by the Hard X-ray Imaging Spectrometer (HXIS) on SMM when it revealed widely separated bright patches in 16–30 keV images during impulsive flares on 10 April, 21 May, and 5 November, 1980 (Duijveman, Hoyng, and Machado, 1982). The Solar X-ray Telescope (SXT) on Hinotori also showed double sources during some impulsive flares (Ohki *et al.*, 1983). For other flares (Kane, 1983) stereoscopic measure-ments obtained from spacecraft off the Earth–Sun line have shown that 95% of the emission at 150 keV comes from a height of less than 2500 km above the photosphere, i.e., consistent with a footpoint source.

5.4. SPECTRA

A balloon-borne high-resolution spectrometer (Lin *et al.*, 1981; Lin and Schwartz, 1987) has revealed the details of the 15–200 keV photon spectrum for a moderately large flare (GOES class M6). Earlier measurements had shown that the spectrum was consistent with a power-law from 20–70 keV and steepened at higher energies. Such a 'broken' power-law spectrum is also consistent with the spectrum expected from a thermal distribution of electrons. Also, it had been noted that for individual spikes within a flare, the spectrum would harden to the peak and then soften again on the fall. It also seemed that the largest flares were also the hardest. The high-resolution spectra, on the other hand (Lin and Schwartz, 1987), were not consistent with an isothermal shape on either short (2 s) or long (30 s) time-scales. The spectra show an evolving double power-law form suggestive of the dc electric field acceleration seen in the lower magnetosphere during aurorae. It was also shown that the soft-hard-soft spectral evolution was most pronounced at energies above 30 keV. Furthermore, there was a clearly thermal hard X-ray component which first appeared near the peak of the event, but it was not impulsive and dominated the emission as the power-law tail diminished. Lin *et al.* (1981) named this the super-hot component, at 30–35 million K some 10–15 million K higher than the normal soft X-ray emitting plasma commonly seen during this and other flares; its emission measure was about 10% of the emission measure of the 20 million K plasma. This super-hot component has not been recognized previously because of the much poorer energy resolution of earlier hard X-ray spectrometers but it is probably present in most flares.

5.5. TIME STRUCTURES IN HARD X-RAYS

Important information is revealed about the energy release process and interaction region by the rise, decay, and delay times of hard X-ray bursts. Flare light curves consist of single bursts or a series of bursts with widths typically ranging from seconds to tens of seconds, although faster structures are sometimes superimposed on a slower profile. Hard X-ray flares also cover >4 orders of magnitude in peak flux ranging from microflares (Lin *et al.*, 1984), detected down to the limits of sensitivity, and up to giant events which can saturate all available detectors.

Commonly observed rise times of a fraction to several seconds are thought to indicate the time development of the acceleration process. Individual spikes in microflares (Lin *et al.*, 1984; Simnett and Dennis, 1987) may actually show the fundamental units of impulsive energy release (Parker, 1988). However, some extremely rapid rise times of 10's of milliseconds are thought to be characteristic of the electron propagation times (Kiplinger *et al.*, 1983; Lu and Petrosian, 1988).

For an energetic electron either precipitating into the lower corona or becoming trapped in a low-density loop, the ambient electron density, n, seen along its trajectory sets an upper limit to the burst decay time τ_ε (there may be additional loss times). Kiplinger *et al.* (1983) show that

$$\tau_\varepsilon < \left(\frac{1}{E}\frac{dE}{dt}\right)^{-1} = 2 \times 10^8 \, E^{3/2} \, n^{-1} \, \text{s} = 1.8 \left(\frac{E}{20}\right)^{3/2} n_{10}^{-1} \, \text{s} \, ,$$

where n_{10} is the density in units of 10^{10} cm^{-3}. The soft-hard-soft spectral evolution typical of most flares suggests that the decay is governed by changes in electron injection/acceleration rather than by trapping of the electrons. Trapping in a low-density loop would result in the progressive hardening of the X-ray spectrum and this is sometimes observed, especially in the large, gradual bursts. The fact that many high-energy impulsive bursts decay within a few seconds supports the idea that the electrons *are* precipitating into the higher densities of the low corona and chromosphere.

During most bursts, X-rays peak simultaneously at all energies but increasing delays have been seen in the peak times at progressively higher energy X-rays for a number of flares (Bai *et al.*, 1983a; Bai and Dennis, 1985; Ohki *et al.*, 1983; Schwartz, 1984;

or see Vlahos *et al.*, 1986). There are two competing explanations for this phenomenon. The first is that the low-density traps which could produce slow decays could also produce progressive delays for electrons injected simultaneously at all energies (Bai and Ramaty, 1979; Vilmer, Kane, and Trottet, 1982). The second is a second-step acceleration process where energetic particles above some threshold are further accelerated by an energy-dependent process (see Bai *et al.*, 1983b). This is believed related to the fact that energy-dependent hard X-ray delays have been mostly observed in flares which produced observable nuclear γ-rays and/or energetic interplanetary protons (Bai and Ramaty, 1979; Bai *et al.*, 1983a, b; Bai and Dennis, 1985; Ohki *et al.*, 1983). Also a trap model really requires a progression of delays but Schwartz (1984) showed that for good statistical data, there was an absence of any delays at low energies.

5.6. DIRECTIVITY OF HARD X-RAYS

Any directivity of the hard X-ray flux could be a key diagnostic of the energetic electron population because bremsstrahlung is emitted preferentially in the direction of the incident electron (Henoux, 1975; Langer and Petrosian, 1977). A good measure of directivity could be made using identical instruments with good sensitivity above several hundred keV (necessary because Compton backscatter is expected to wash out the effect at lower energies) placed 90° apart relative to the Sun. To date this has not been possible but two recent studies (Vestrand *et al.*, 1987; Kane *et al.*, 1988) have attempted to measure it using existing sets of less than ideal data. A set of 39 joint ISEE-3 and PVO observations, with angular separations ranging from 1 to 66 degrees, does not indicate any systematic directivity (Kane *et al.*, 1988). In contrast, Vestrand *et al.* (1987) have made a statistical study of the > 300 keV flux observed by a single instrument, GRS, where they find convincing evidence of spectral hardening for flares closer to the limb. However, there are problems with both studies – the first must be accurately inter-calibrated and has a small sample of ideal events while the second is prone to selection effects. A much clearer case for directivity can be made for very high energy electron bremsstrahlung. Above 10 MeV, GRS has observed continuum photon emission primarily from flares near the solar limb (Vestrand *et al.*, 1987). These γ-rays are believed to be from ultra-relativistic electrons emitting close to their footpoint turnaround and hence mostly moving parallel to the surface. Their bremsstrahlung is highly beamed in the direction of electron motion thus explaining these results.

The energetic electrons also emit gyro-synchrotron microwave radiation in addition to hard X-rays. Using the peak fluxes as a function of energy for a sample of ~400 events, Kosugi, Dennis, and Kai (1988) determined that for impulsive flares the best correlation was obtained between X-rays $\leq$80 keV and the 17 GHZ microwaves. Assuming then that both emissions come from the same population, they concluded that ~20 GHZ flux comes from $\leq$200 keV electrons streaming through a ~900 G field in a layer 3–10 × 10^3 km thick. Few electrons are reflected and trapped in the loop and 'the thermal model is incompatible with the observations'. However, many other studies show the microwaves originating near the top of the loop (Leach and Petrosian, 1983) even for flares with footpoint hard X-rays.

5.7. Gamma Rays from Energetic Particles

Gamma-ray lines were first detected during the great flares of August 1972 (Chupp *et al.*, 1973; Chupp, Forest, and Suri, 1975). Since the launch of SMM, GRS has detected line emission in many flares and a far greater understanding of the morphology of these events has been obtained (Chupp, 1984). Although it has not yet been possible to fly the appropriate detectors to directly image the MeV photons, we at least know that the majority of energetic ions move towards the Sun and not into space because the outgoing particle flux of events detected on interplanetary spacecraft is usually only $\leq 1\%$ of that necessary to produce the observed γ-rays (Murphy and Ramaty, 1984).

One of the more important results from the last maximum is that for some events the peaks of the γ-ray emission are coincident (± 1 s) with the hard X-ray peaks (Chupp, 1984). From earlier observations (Chupp *et al.*, 1973; Chupp, Forrest, and Suri, 1975), it had been thought that additional time, at least 10's of seconds, was required to accelerate ions up to the 10's of MeV necessary to produce γ-ray line emission. The best example of coincident peaks were found for energy bands from 40 keV to 25 MeV for the flare of 1982 February 8 (Chupp, 1984). Also, fast decay times imply that the energy loss region must be of high density precluding the trapping of the bulk of the ions high in the corona (Share *et al.*, 1983; Ramaty and Murphy, 1984; Murphy and Ramaty, 1984).

The spectroscopic analysis of the γ-ray lines is expected to yield rich results when these measurements can be obtained during the coming maximum using liquid nitrogen cooled germanium detectors. To date the only high-resolution results were serendipitous HEAO-3 measurements of 2.223 MeV photons which leaked through a thick CsI anti-coincidence shield (Prince *et al.*, 1982). This is a narrow line resulting from the capture of neutrons on hydrogen. From other flares it has been learned that this line is suppressed for limb events relative to disc flares (Murphy and Ramaty, 1984) showing that the photons must be produced in the photosphere as the result of a downward directed energetic particle flux.

Another important result is that the nuclear line component may be present in all flares. Forrest (1983) has extrapolated the 270 keV to 1 MeV bremsstrahlung spectrum to the nuclear line range (4–8 MeV) for 65 flares and attributed any excess to nuclear processes. This excess was not always seen, but the absence may be attributed to the sensitivity in the nuclear range. The hypothesis of an omnipresent high energy ion acceleration is quite controversial as it covers only a small fraction of the observed hard X-ray flares (~ 1 in 100) due, at least in part, to the more difficult task of detecting higher energy γ-rays (falling spectrum and detection cross-section). In contrast, Bai and Dennis (1985) and Bai (1986) have determined that there are distinctive characteristics to γ-ray line producing flares.

5.8. UV–Hard X-ray Comparison

Several groups (Poland *et al.*, 1982; Machado, Duijveman, and Dennis, 1982) have found evidence for the cospatiality of the O v transition zone line and impulsive X-rays to the resolution of SMM HXIS and the UV Spectrometer Polarimeter (UVSP). A comparison of hard X-ray observations with coincident, high time resolution observations of hard X-ray, UV line and continuum emissions was reported by Orwig and Woodgate (1986). They found that in one flare on 1984 May 20, the UV continuum and hard X-ray emissions were simultaneous to within 0.1 s. A detailed cross-correlation analysis of the three emissions in another flare on 1985 April 24 showed that spiky features in the UV line and UV continuum emissions were simultaneous to within 0.1 s, but both UV emissions were delayed with respect to the corresponding hard X-ray features by up to 0.3 s. Cheng *et al.* (1988) have repeated the analysis for this event and present similar results for other events. These observations place strict limitations on the energy propagation times from the corona through the transition region to the lower chromosphere during the flare impulsive phase.

The simultaneity to within 0.3 s between the hard X-ray peaks and transition zone lines can be understood qualitatively in the thick-target model. However, attempts to quantify the expected O v flux have met with limited success (e.g., Poland *et al.*, 1984; Emslie and Nagai, 1985; Mariska and Poland, 1985).

The simultaneity between the hard X-rays and continuum, which is generally believed to originate from close to the temperature minium region, is much more difficult to understand at first sight. The high-energy electrons do not make it down to this low level in the chromosphere so that direct heating is not possible. Thermal conduction would take too long and is totally ineffective at these depths in any case (Emslie, Brown, and Machado, 1981). Furthermore, the extremely large energy deposition rates required at the temperature minimum region are unlikely to be attained by any canonical energy transport mechanism like accelerated particles or EUV heating (Machado and Mauas, 1987). Instead, Machado and Mauas (1987) have proposed that the source of the emission is Si II in the temperature minimum region created by photo-ionization due to line emission (mostly from the C IV resonance line at 1549 Å) from the transition region.

5.9. Iron $K\alpha$

Transient iron $K\alpha$ radiation is another possible diagnostic of electron beaming and/or the height of the hard X-ray production region (Tanaka, Watanabe, and Nitta, 1984; Emslie, Phillips, and Dennis, 1986). The radiation is produced after the removal of the K-shell electron either by photo-ionization (fluorescense) or by electron-impact. During a flare, there may be three sources of excitation: the soft X-ray plasma spectrum above 7.1 keV, the non-thermal X-rays above 7.1 keV, or an electron beam passing through the chromosphere. The efficiency of either fluorescence response depends upon the height of the X-ray source above the photosphere because of the difference in the solid angle subtended below.

Tanaka, Watanabe, and Nitta (1984) used the Hinotori instruments to measure the X-ray spectrum from 1.5 to > 100 keV during the 1981 July 28 flare. During the rise of the impulsive phase, before the soft X-ray plasma became too intense, the power-law spectrum extended to as low as 7 keV. Furthermore, they found that the fluorescence from the soft X-ray plasma could not account for all of the iron $K\alpha$ emission at that time.

However, the excess could be explained if height of the X-ray source is close to zero, or higher if there is an electron beam penetrating the chromosphere. Either of those cases is consistent with a footpoint source for the hard X-rays.

5.10. SOFT X-RAYS

When the energy released in the corona reaches the cooler plasma near the footpoints, heating occurs resulting in a thermal plasma with a bulk temperature of 15–20 million K. The main increase in this hot plasma appears during the impulsive phase although it is usually detectable, albeit at a low level, for at least a minute before the beginning of the hard X-ray burst (see Priest *et al.*, 1986, for discussion). This may be a thermal precursor to the main energy release or the hard X-ray flux may be below the instrumental threshold level.

There are two principal observational requirements in soft X-rays which must be met by a successful model for the impulsive phase. The first is that the dynamic input of energy during the impulsive phase must be accounted for by a corresponding increase in the energy content of soft X-ray plasma. One of the early results (Neupert, 1968) of spaceborne instrumentation was that the smooth rise of the soft X-ray light curve resembled the integral of the impulsive hard X-ray light curve as is expected in the thick-target model. It is also expected to some degree in thermal conduction and joule heating models. The second point of agreement should be between the plasma dynamics revealed by spectroscopic line measurements and the consequences of a rapid energy input into a cooler stable region.

Although both hard X-rays and soft X-rays have been observed for thousands of flares, determining both the energy content of the initial fast electrons and the resultant soft X-ray emitting plasma is a process plagued with observational difficulties (Wu *et al.*, 1986). To obtain the fast electron spectrum, a power-law is normally fit to the hard X-ray spectrum and then the electron distribution is obtained from the fit parameters (Brown, 1971; Lin and Hudson, 1976). In a typical case, most of the electron energy is within ~ 10 keV of the low-energy cutoff, usually assumed to be ~ 20 keV, in a spectral region unresolvable by most hard X-ray detectors. The energy content of the soft X-ray plasma is also difficult to determine, since most measurements give only the temperature and emission measure while the filled volume (or density) must be otherwise estimated or guessed.

If the injection of energy into the transition region at the base of the loop exceeds an energy flux threshold defined by the peak of the radiative loss function and the pre-flare density (Fisher (1987) estimates a threshold $\gtrsim 10^{10}$ ergs cm^{-2} s^{-1}), the resultant heat cannot be radiated away fast enough leading to a rapid increase in temperature and volume. This process results in what is known as explosive chromospheric evaporation (Antonucci *et al.*, 1982; Antonucci and Dennis, 1983) and it may explain the blue-shifted emission lines observed in 80% of M and X disk flares detected with BCS. The blue-shifted component is absent in flares past 60 degrees in longitude as expected for upwardly directed flows.

5.11. Hα OBSERVATIONS

Of all the flare emissions, only in the optical Hα line does the current instrumentation permit high resolution images and spectra with a rapid sampling time (Acton *et al.*, 1982). The radiation comes from the lower chromosphere below the region where the UV and EUV are formed. While the path length to the optical source is almost the same as for the UV source, the column depth is increasing rapidly and this quickly attenuates an incident electron beam. With arc sec resolution, Hα images locate particle precipitation and energy loss with respect to other data such as vector magnetograms. Additionally, the radiative loss rate can reflect the rapid input and dissipation of energy just as seen in the EUV continuum.

The new generation of spectroheliogrammetry has spawned a concomitant effort to predict the spectra as a function of the energy release process. In particular, Canfield, Gunkler, and Ricchiazi (1984) have made predictions of the flare atmosphere under the condition of impulsive energy release to predict the Hα profile, which typically shows a central reversal. Additionally, only for high values of the energetic electron flux ($\gtrsim 10^{10}$ ergs cm^{-2} s^{-1} above 20 keV) are there broad non-gaussian wings in the Hα profile. Canfield and Gunkler (1985) have found evidence for such enhanced-wing signatures well correlated in space and time with hard X-ray emission during the compact solar flare of 7 May, 1980.

These observations provide strong support for the thick-target model although it is difficult to determine quantitatively the electron flux and spectrum required to produce the measured enhanced wings. As mentioned earlier, these high values of the energy flux are considered necessary to produce explosive chromospheric evaporation which gives rise to the upflowing plasma evidenced by the soft X-ray blue-shifts. Zarro *et al.* (1988) have seen redshifts in the Hα spectra indicating the corresponding plasma downflows into the cooler chromosphere. They have shown momentum balance to within the factor of two uncertainties of the measurements in both spectral ranges. Importantly, these spectral features are characteristic only of the impulsive phase with qualitatively different features observed during the following gradual phase of flares.

6. Conclusion

It is clear that as a result of the explosion of new data on impulsive phase phenomena obtained over the previous solar maximum, we now have a much clearer picture of a solar flare. Nevertheless, the fundamental longstanding problems remain unanswered. The thick-target loop model emphasized in this paper is consistent with much of the new data but it is clearly a great oversimplification. It does, however, serve to focus our thinking on the outstanding issues that elude our comprehension. These include the following principal flare problems:

(1) The rapid rate of energy release – as high as 10^{30} erg s^{-1} in some flares.

(2) The rapid acceleration of protons and electrons to relativistic energies in seconds or less.

(3) The number of particles accelerated exceeds the number initially in the coronal

loop. How is a sufficient number of particles transported from the higher density regions into the accelerator on the required time-scales?

(4) The electrical current associated with the electron beam would produce an unreasonably large magnetic field of $\gtrsim 10^9$ G unless it is divided into $\gtrsim 10^6$ filaments (Spicer and Sudan, 1984; Holman, 1985).

(5) The total energy in fast electrons appears to be larger than the thermal energy of the soft X-ray emitting plasma, especially for flares where the hard X-ray spectrum is measured to be a power law down to energies as low or lower than 10 keV (Kahler and Kreplin, 1971; Tanaka, Watanabe, and Nitta, 1984).

The new instrumentation planned for the current cycle of solar activity will have exciting new capabilities to address these and other issues in ways never before possible. As stated in the Max '91 report (Dennis and Canfield, 1988), "*The diagnostic power of this new instrumentation is qualitatively different from what was available during the previous solar cycle. It allows us to go deeper than the question of what a flare is; it allows us to gather spectra and images that are relevant to the question of what causes a flare to happen in the first place. We can seriously address not only the impulsive energy release with its attendant heating and particle acceleration, but also the magnetic and thermal environment that leads to it.*"

Currently planned instruments include the core space missions – the Japanese Solar-A spacecraft and the Gamma Ray Observatory (GRO) – and advanced instruments on high altitude balloons and rockets and at groundbased observatories. With this comprehensive Max '91 program of observations we can look forward to obtaining the following observations:

– Hard X-ray images with arc sec angular resolution that will, for the first time, fully resolve flaring magnetic loops and trace the evolution of the electron spectrum along the loops.

– Hard X-ray and γ-ray spectra with keV energy resolution to resolve, for the first time, γ-ray lines and determine their widths and shapes and also clearly separate the thermal and non-thermal components of the hard X-ray spectrum.

– Vector magnetograms with better spatial resolution and stability than previously possible to quantitatively measure, for the first time, the energy content of active regions before, during, and after a flare.

– Microwave imaging spectroscopy with arc sec angular resolution, subsecond time resolution, and better than 10% spectral resolution to exploit the microwave diagnostics of coronal magnetic fields, energetic electrons, and pre- and post-flare plasmas.

The Max '91 program is being developed to coordinate these and other observations in order to optimize the scientific return. Solar activity is rising at an unprecedented rate toward what may be an unexpectedly early and intense maximum in 1990. All indications are that the new observations will yield a prolific scientific bonanza; we stand expectantly and impatiently waiting for the show to begin.

Acknowledgements

We gratefully acknowledge the support given to us by all participants in the SMM project and in the Max '91 program and for all the people around the world who have so freely given of their time, energy, and data in the ongoing effort to educate us about solar flares. The SMM project was funded by NASA and one of us (RAS) was supported under NASA contract NAS-5-28752.

References

Acton, L. W., Canfield, R. C., Gunkler, T. A., Hudson, H. S., Kiplinger, A. L., and Leibacher, J. W.: 1982, *Astrophys. J.* **263**, 409.
Antonucci, E. and Dennis, B. R.: 1983, *Solar Phys.* **86**, 67.
Antonucci, E., Gabriel, A. H., Acton, L. W., Culhane, J. L., Doyle, J. G., Leibacher, J. W., Machado, M. E., and Orwig, L. E.: 1982, *Solar Phys.* **78**, 107.
Bai, T.: 1986, *Astrophys. J.* **308**, 912.
Bai, T. and Dennis, B. R.: 1985, *Astrophys. J.* **292**, 699.
Bai, T. and Ramaty, R.: 1979, *Astrophys. J.* **227**, 1072.
Bai, T., Dennis, B. R., Kiplinger, A. L., Orwig, L. E., and Frost, K. J.: 1983a, *Solar Phys.* **86**, 409.
Bai, T., Hudson, H. S., Pelling, R. M., Lin, R. P., Schwartz, R. A., and von Rosenvinge, T. T.: 1983b, *Astrophys. J.* **267**, 433.
Batchelor, D. A., Crannell, C. J., Wiehl, H. J., and Magun, A.: 1985, *Astrophys. J.* **295**, 258.
Benz, A. O.: 1985, *Solar Phys.* **96**, 357.
Benz, A. O.: 1987, *Proc. 21st ESLAB Symposium,* ESA SP-275, Bolkesjo, Norway, pp. 105–108.
Benz, A. O., Bernold, T. E. X., and Dennis, B. R.: 1983, *Astrophys. J.* **271**, 355.
Brown, J. C.: 1971, *Solar Phys.* **18**, 489.
Brown, J. C.: 1976, *Phil. Trans. Roy. Soc. London* **281**, 473.
Brown, J. C., Melrose, D. B., and Spicer, D. S.: 1979, *Astrophys. J.* **228**, 592.
Brown, J. C., Smith, D. F., and Spicer, D. S.: 1981, in Stuart Jordan (ed.), *The Sun as a Star, Monograph Series on Nonthermal Phenomena in Stellar Atmospheres,* NASA SP-450, Washington, DC, p. 181.
Canfield, R. C. and Gunkler, T. A.: 1985, *Astrophys. J.* **288**, 353.
Canfield, R. C., Gunkler, T. A., and Ricchiazzi, P. J.: 1984, *Astrophys. J.* **282**, 296.
Cheng, C.-C., Vanderveen, L., Orwig, L. E., and Tandberg-Hanssen, E.: 1988, *Astrophys. J.* **330**, 480.
Chupp, E. L.: 1984, *Ann. Rev. Astron. Astrophys.* **22**, 359.
Chupp, E. L., Forest, D. J., and Suri, A. N.: 1975, in S. Kane (ed.), 'Solar Gamma, X and EUV Radiations', *IAU Symp.* **68**, 341.
Chupp, E. L., Forrest, D. J., Higbie, P. R., Suri, A. N., Tsai, C., and Dunphy, R. P.: 1973, *Nature* **241**, 333.
Covington, A. E. and Harvey, G. A.: 1958, *J. Roy. Astron. Soc. Can.* **52**, 2962.
Dennis, B. R. and Canfield, R. C. (eds.): 1988, *Max '91, Flare Research at the Next Solar Maximum,* NASA, Greenbelt, MD.
Dennis, B. R., Chupp, E. L., Crannell, C. J., and 40 co-authors: 1986, *An Advanced Payload for the Exploration of High Energy Processes on the Active Sun,* Report of the Max '91 Science Study Committee, NASA, Greenbelt, MD.
Duijveman, A., Hoyng, P., and Machado, M. E.: 1982, *Solar Phys.* **81**, 137.
Dulk, G. A. and Dennis, B. R.: 1982, *Astrophys. J.* **260**, 844.
Emslie, A. G. and Brown, J. C.: 1985, *Astrophys. J.* **295**, 648.
Emslie, A. G. and Nagai, F.: 1985, *Astrophys. J.* **288**, 779.
Emslie, A. G., Brown, J. C., and Machado, M. E.: 1981, *Astrophys. J.* **246**, 337.
Emslie, A. G., Phillips, K. J. H., and Dennis, B. R.: 1986, *Solar Phys.* **103**, 89.
Fisher, G. H.: 1987, *Astrophys. J.* **317**, 502.
Forrest, D. J.: 1983, in M. L. Burns, A. K. Harding, and R. Ramaty (eds.), *Positron-Electron Pairs in Astrophysics,* AIP, New York, p. 3.
Gurman, J. B.: 1987, *NASA's Solar Maximum Mission: A Look at a New Sun,* NASA, Greenbelt, MD.

Hagyard, M. J., Moore, R. L., and Emslie, A. G.: 1984, *Adv. Space Res.* **4**, 71.

Harrison, R. A.: 1986, *Astron. Astrophys.* **162**, 283.

Harrison, R. A., Waggett, P. W., Bentley, R. D., Phillips, K. J. H., Bruner, M., Dryer, M., and Simnett, G. M.: 1985, *Solar Phys.* **97**, 387.

Henoux, J. C.: 1975, *Solar Phys.* **42**, 219.

Henoux, J.-C., Chambe, G., Feautrier, N., and Sahal, S.: 1988, *Nature* (submitted).

Heristchi, D.: 1986, *Astrophys. J.* **311**, 474.

Holman, G. D.: 1985, *Astrophys. J.* **293**, 584.

Holman, G. D.: 1986, in A. I. Poland (ed.), *Coronal and Prominence Plasmas*, NASA CP-2442, Greenbelt, MD, p. 297.

Holman, G. D. and Kundu, M. R.: 1985, *Astrophys. J.* **292**, 291.

Hoyng, P., Brown, J. C., and van Beek, H. F.: 1976, *Solar Phys.* **48**, 197.

Hoyng, P., Marsh, K., Zirin, H., and Dennis, B. R.: 1983, *Astrophys. J.* **268**, 865.

Kahler, S. W. and Kreplin, R. W.: 1971, *Astrophys. J.* **168**, 531.

Kahler, S., Spicer, D., Uchida, Y., and Zirin, H.: 1980, in P. Sturrock (ed.), *Solar Flares, A Monograph from the Skylab Solar Workshop II*, University Press, Boulder, CO, pp. 83–114.

Kahler, S. W., Moore, R. L., Kane, S. R., and Zirin, H.: 1988, *Astrophys. J.* **328**, 824.

Kane, S. R.: 1969, *Astrophys. J.* **157**, L139.

Kane, S. R.: 1981, *Astrophys. J.* **247**, 1113.

Kane, S. R.: 1983, *Solar Phys.* **86**, 355.

Kane, S. R., Fenimore, E. E., Klebesadel, R. W., and Laros, J. G.: 1988, *Astrophys. J.* **326**, 1017.

Kiplinger, A. L., Dennis, B. R., Frost, K. J., and Orwig, L. E.: 1983, *Astrophys. J.* **273**, 783.

Kosugi, T., Dennis, B. R., and Kai, K: 1988, *Astrophys. J.* **324**, 1118.

Kundu, M. R. and Lang, K. R.: 1985, *Science* **228**, 9.

Kurokawa, H., Hanaoka, Y., Shibata, K., and Uchida, Y.: 1987, *Solar Phys.* **108**, 251.

Langer, S. H. and Petrosian, V.: 1977, *Astrophys. J.* **215**, 666.

Leach, J. and Petrosian, V.: 1983, *Astrophys. J.* **269**, 715.

Lin R. P. and Hudson, H. S.: 1976, *Solar Phys.* **50**, 153.

Lin, R. P. and Schwartz, R. A.: 1987, *Astrophys. J.* **312**, 462.

Lin, R. P., Schwartz, R. A., Pelling, R. M., and Hurley, K. C.: 1981, *Astrophys. J.* **251**, L109.

Lin, R. P., Schwartz, R. A., Kane, S. R., Pelling, R. M., and Hurley, K. C.: 1984, *Astrophys. J.* **283**, 421.

Lu, E. T. and Petrosian, V.: 1988, *Astrophys. J.* **327**, 405.

Machado, M. E. and Mauas, P. J.: 1987, in B. R. Dennis, L. E. Orwig, and A. L. Kiplinger (eds.), *Rapid Fluctuations in Solar Flares*, NASA CP-2449, Greenbelt, MD, p. 271.

Machado, M. E., Duijveman, A., and Dennis, B. R.: 1982, *Solar Phys.* **79**, 85.

Mariska, J. T. and Poland, A. I.: 1985, *Solar Phys.* **96**, 317.

Martens, P. C. H.: 1988, *Astrophys. J.* **330**, L131.

Moore, R. L.: 1988, *Astrophys. J.* **324**, 1132.

Moore, R. L., Hagyard, M. J., and Davis, J. M.: 1987, *Solar Phys.* **113**, 347.

Moore, R. L., Hurford, G. J., Jones, J. P., and Kane, S. R.: 1984, *Astrophys. J.* **276**, 379.

Murphy, R. J. and Ramaty, R.: 1984, *Adv. Space Res.* **4**, 127.

Neupert, W. M.: 1968, *Astrophys. J.* **153**, L59.

Ohki, K., Takakura, T., Tsuneta, S., and Nitta, N.: 1983, *Solar Phys.* **86**, 301.

Orrall, F. Q. and Zirker, J. B.: 1976, *Astrophys. J.* **208**, 618.

Orwig, L. E. and Woodgate, B. E.: 1986, in D. F. Neidig (ed.), *The Lower Atmosphere in Solar Flares*, Proc. NSO/SMM Symposium, Sunspot, NM, pp. 306–317.

Parker, E. N.: 1988, *Astrophys. J.* **330**, 474.

Poland, A. I., Machado, M. E., Wolfson, C. J., Frost, K. J., Woodgate, B. E., Shine, R. A., Kenny, P. J., and Cheng, C.-C.: 1982, *Solar Phys.* **78**, 201.

Poland, A. I., Orwig, L. E., Mariska, J. T., Nakatsuka, R. S., and Auer, L. H.: 1984, *Astrophys. J.* **280**, 457.

Priest, E. R., Gaizauskas, V., Hagyard, M. J., Schmahl, E. J., and Webb, D. F.: 1986, in M. R.Kundu and B. E. Woodgate (eds.), *Energetic Phenomena on the Sun*, NASA CP-2439, Greenbelt, MD, pp. 1-1.

Prince, T. A., Ling, J. C., Mahoney, W. A., Riegler, G. R., and Jacobson, A. S.: 1982, *Astrophys. J.* **255**, L81.

Ramaty, R. and Murphy, R. J.: 1984, in S. Woosley (ed.), *High Energy Transients in Astrophysics*, AIP, New York, p. 628.

Ramaty, R. and Murphy, R. J.: 1987, *Space Sci. Rev.* **45**, 213.

Schmahl, E. J., Kundu, M. R., and Dennis, B. R.: 1985, *Astrophys. J.* **299**, 1017.

Schwartz, R. A.: 1984, 'High Resolution Hard X-Ray Spectra of Solar and Cosmic Sources', Ph.D. Thesis, Univ. of California, Berkeley, CA.

Share, G. H., Chupp, E. L., Forrest, D. J., and Rieger, E.: 1983, in M. L. Burns, A. K. Harding, and R. Ramatay (eds.), *Positron-Electron Pairs in Astrophysics*, AIP, New York, p. 15.

Simnett, G. M.: 1986, *Solar Phys.* **106**, 165.

Simnett, G. M. and Dennis, B. R.: 1987, in B. R. Dennis, L. E. Orwig, and A. L. Kiplinger (eds.), *Rapid Fluctuations in Solar Flares*, NASA CP-2449, Greenbelt, MD, p. 123.

Simnett, G. M. and Harrison, R. A.: 1985, *Solar Phys.* **99**, 291.

Smith, D. F. and Lilliequist, C. G.: 1979, *Astrophys. J.* **232**, 582.

Spicer, D. S. and Sudan, R. N.: 1984, *Astrophys. J.* **280**, 448.

Stewart, R. T.: 1978, *Solar Phys.* **58**, 121.

Sturrock, P. A., Kaufmann, P., Moore, R. L., and Smith, D. F.: 1984, *Solar Phys.* **94**, 341.

Švestka, Z.: 1976, *Solar Flares*, D. Reidel Publ. Co., Dordrecht, Holland.

Tanaka, K., Watanabe, T., and Nitta, N.: 1984, *Astrophys. J.* **282**, 793.

Trubnikov, B. A.: 1965, *Particle Interactions in a Fully Ionized Plasma*, Consultants Bureau, New York, Vol. 1.

Vestrand, W. T., Forrest, D. J., Chupp, E. L., Rieger, E., and Share, G. H.: 1987, *Astrophys. J.* **322**, 1010.

Vilmer, N., Kane, S. R., and Trottet, G.: 1982, *Astron. Astrophys.* **108**, 306.

Vlahos, L., Machado, M. E., Ramaty, R., Murphy, R. J., and 23 co-authors: 1986, in M. R. Kundu and B. E. Woodgate (eds.), *Energetic Phenomena on the Sun*, NASA CP-2439, Greenbelt, MD, pp 2-1.

Webb, D. F., Holman, G. D., Davis, J. M., Kundu, M. R., and Shevgaonkar, R. K.: 1987, *Astrophys. J.* **315**, 716.

Wu, S. T., de Jager, C., Dennis, B. R., Hudson, H. S., Simnett, G. M., Strong, K. T., Bentley, R. D., and Bornmann, P. L.: 1986, in M. R. Kundu and B. E. Woodgate (eds.), *Energetic Phenomena on the Sun, The SMM Flare Workshop Proceedings*, NASA CP-2439, Greenbelt, MD.

Zarro, D. M., Canfield, R. C., Strong, K. T., and Metcalf, T. R.: 1988, *Astrophys. J.* **324**, 582.

ELECTRON ACCELERATION IN SOLAR FLARES

WOLFGANG DRÖGE* and PETER MEYER

Enrico Fermi Institute, University of Chicago, 933 East 56th Street, Chicago, IL 60637, U.S.A.

PAUL EVENSON

Bartol Research Institute, University of Delaware, Newark, DE 19716, U.S.A.

and

DAN MOSES

American Science and Engineering Inc., Fort Washington, Cambridge, MA 02139, U.S.A.

Abstract. For the period September 1978 to December 1982 we have identified 55 solar flare particle events for which our instruments on board the ISEE-3 (ICE) spacecraft detected electrons above 10 MeV. Combining our data with those from the ULEWAT spectrometer (MPI Garching and University of Maryland) electron spectra in the range from 0.1 to 100 MeV were obtained. The observed spectral shapes can be divided into two classes. The spectra of the one class can be fit by a single power law in rigidity over the entire observed range. The spectra of the other class deviate from a power law, instead exhibiting a steepening at low rigidities and a flattening at high rigidities. Events with power-law spectra are associated with impulsive (< 1 hr duration) soft X-ray emission, whereas events with hardening spectra are associated with long-duration (> 1 hr) soft X-ray emission. The characteristics of long-duration events are consistent with diffusive shock acceleration taking place high in the corona. Electron spectra of short-duration flares are well reproduced by the distribution functions derived from a model assuming simultaneous second-order Fermi acceleration and Coulomb losses operating in closed flare loops.

1. Introduction

Energetic particle spectra observed in interplanetary space following solar flares are one measure of particle spectra at their solar acceleration sites. The study of these particles offers a unique opportunity to understand high-energy acceleration processes and to obtain information on the physical conditions at the flare site. In large solar flares, electrons are accelerated to energies of more than 100 MeV and ions to energies of up to several GeV.

In this paper we present an analysis of a recent study (Moses *et al.*, 1989) of the shapes of solar flare electron spectra in the energy range 0.1–100 MeV. Previous surveys of electron flare spectra have been conducted by Datlowe (1971), Lin, Mewaldt, and van Hollebeke (1982), Evenson *et al.* (1984), and Cane, McGuire, and von Rosenvinge (1986). Compared to the earlier studies, our recent work has the advantage of a nearly continuous coverage over a four year period of high solar activity and a superior energy range and resolution.

It is found that solar flare electrons exhibit a variety of spectral shapes in the range from mildly relativistic to ultrarelativistic energies. When expressed as a function of rigidity, the spectra can be divided into two broad groups. The spectra of one group are

---

* Present address: Institut für Kernphysik, Kiel University, Olshausenstr. 40, D-2300 Kiel, F.R.G.

Solar Physics **121** (1989) 95–103.
© 1989 by Kluwer Academic Publishers.

consistent with power laws in rigidity, whereas the spectra of the other group sometimes become excessively hard above 5 MeV, an effect not noticed in previous observations. We find that our identification of two classes of spectral shapes correlates well with the classification scheme of interplanetary particle events based upon the 1–8 Å soft X-ray duration of the associated flare, as was suggested by Kahler *et al.* (1984) and Cane, McGuire, and von Rosenvinge (1986).

We have modeled the observed electron spectra by diffusive shock acceleration (Axford, Leer, and Skadron, 1977) and stochastic acceleration models including Coulomb losses (Steinacker, Dröge, and Schlickeiser, 1988). We find that for any given event, the stochastic acceleration model gives good fits. Diffusive shock acceleration is consistent with most of the long-duration soft X-ray events, but fails to explain short-duration spectra.

2. Data Analysis

The observations were made with two instruments onboard the *International Sun–Earth Explorer 3* (ISEE-3/ICE) spacecraft: the ULEWAT spectrometer (MPI/University of Maryland) which measured the electron flux in the energy range 0.075–1.3 MeV, and the University of Chicago MEH spectrometer which measured the electron flux in the energy range 5–100 MeV. The time interval of the survey extends from launch of ISEE-3 in August 1978 through December 1982. The ISEE-3 spacecraft was positioned at the Earth–Sun Lagrangian point well outside of the Earth's geomagnetic field.

The electron events for the survey were chosen in the following manner. All events which had a significant ($> 3\sigma$) flux enhancement at energies above 10 MeV and which could be associated with a flare were considered. To minimize propagation effects only events with a well-defined flux rise time were selected; events which were disturbed by an interplanetary shock were removed from the survey. A total of 55 events were found to satisfy these criteria.

For each event the electron spectrum was constructed by taking the maximum flux in each energy interval. It was shown by Lin, Mewaldt, and van Hollebeke (1982) that time of maximum spectra have the same shape and slope as the original injection spectra, if the propagation through the interplanetary medium can be described by standard diffusion treatment and if the observations are at energies sufficient to neglect the effects of convection and adiabatic deceleration.

A full description of the instruments, the analysis method used as well as a detailed table with the properties of each event will be presented elsewhere (Moses *et al.*, 1989).

3. Results

Figure 1 shows the electron spectrum for the 19 December, 1982 event. In order to allow a direct comparison of the observed spectra with spectra predicted by shock and stochastic acceleration models we have plotted the electron flux density in energy space $J(E)$ which is proportional to the electron number density in momentum space $N(p)$,

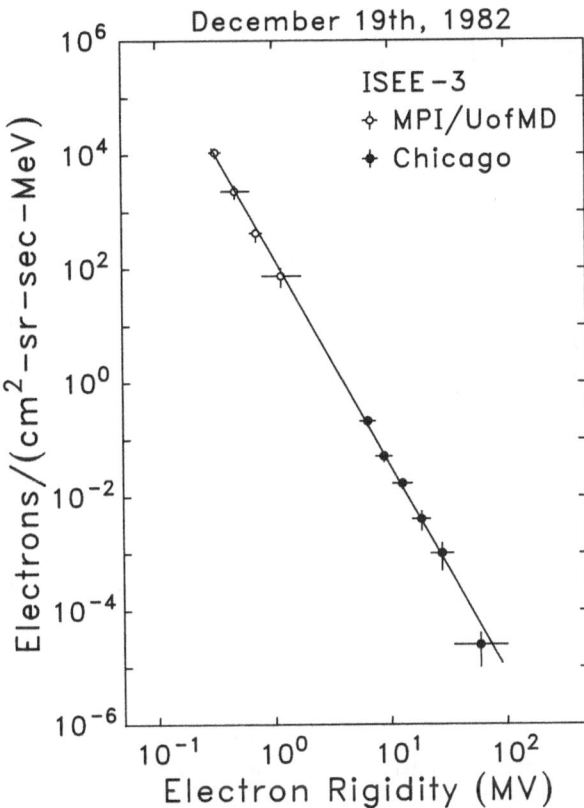

Fig. 1.   Electron spectrum of the long-duration flare from 19 December, 1982. The spectrum is fitted by solution (4) with $p_0 = 0.3$ MeV c^{-1}, $p_c = 0.1$ MeV c^{-1}, and $\alpha T = 0.2$. A shock acceleration model with a compression ratio $r = 2.1$ can reproduce the spectrum as well.

as a function of rigidity. As can be seen from Figure 1, the spectrum of this event can be well fitted by a single power law in rigidity (respectively momentum). Figure 2 shows the event of 7 June, 1980 (3:12 UT). The spectral shape of this event is quite different. A power-law spectrum is observed at rigidities above ~ 5 MeV, but the spectrum becomes much steeper towards lower rigidities.

We find that, with regard to their spectral shapes, the electron events of our survey can be divided into two broad groups. The spectra of one group are reasonably well fitted by single power laws in rigidity, whereas the spectra of the other group deviate from single power laws: they are flatter at high rigidities and steeper at low rigidities. In order to find a classification scheme we have made separate power-law fits to the low- and high-rigidity part of each spectrum and we have determined the duration at 10% of the peak flux of the 1–8 Å soft X-ray emission of the parent flare. The ratio of the low- to the high-rigidity spectral index as a function of the soft X-ray duration is shown in Figure 3. Events associated with long-duration soft X-ray emission (LDE) have spectral index ratios close to unity. This indicates that their spectra can be modeled

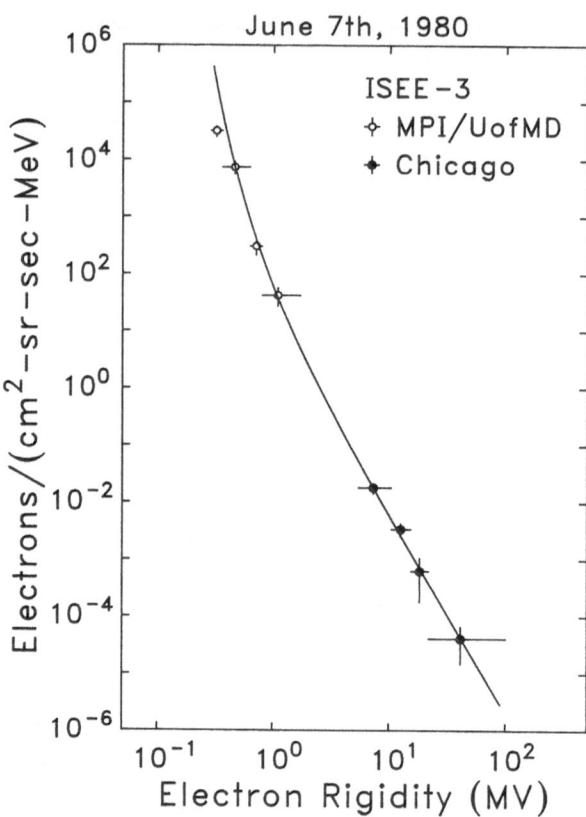

Fig. 2. Electron spectrum of the short-duration flare from 7 June, 1982 (3:12 UT). The spectrum is fitted by solution (4) with $p_0 = 0.3$ MeV c^{-1}, $p_c = 0.93$ MeV c^{-1}, and $\alpha T = 0.3$.

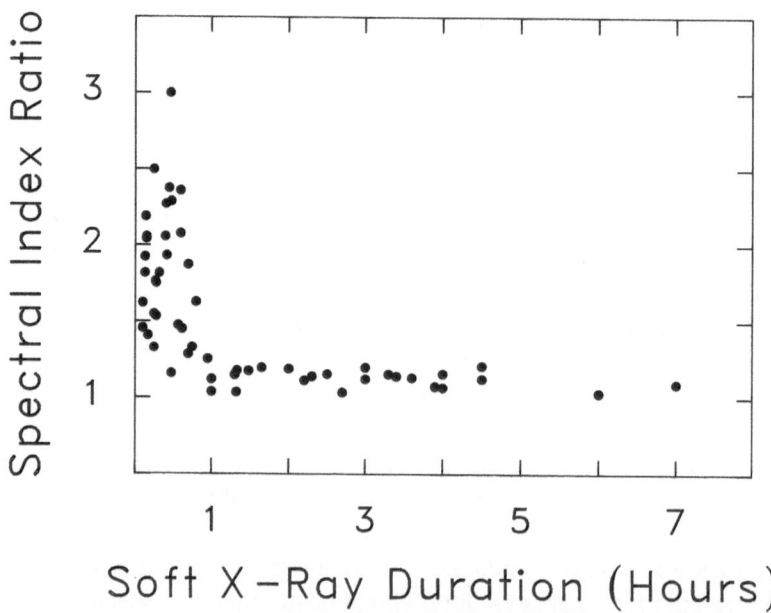

Fig. 3. Low-rigidity to high-rigidity spectral slope ratios plotted as a function of the soft X-ray duration. Events with a ratio ≈ 1 can be fitted with a single power law.

by one single power law over the entire range in rigidity; although some LDEs might exhibit a slight steepening at low rigidities. Events associated with short-duration soft X-ray emission (SDE) have spectral index ratios well above unity, spectra of SDEs can, therefore, not be modeled by single power laws in rigidity. As can be seen from Figure 3, the events group into two distinct populations rather than exhibiting a continuous correlation between the spectral index ratio and the soft X-ray duration. The separation of the two populations occurs at roughly 1 hour. These results suggest that non-thermal (particle acceleration) and thermal (soft X-ray emission) features of a flare are related.

The classification of solar flare electron spectra is congruent with the classification of other flare characteristics with respect to their soft X-ray emission reported in earlier studies. The fact that soft X-ray events could be divided into two classes was first recognized by Pallavicini, Serio, and Vaiana (1977). They found that flares with short duration (≤ 1 hour, Class I events) have smaller volumes, occur at lower heights ($\leq 10^4$ km) and have higher energy densities. In contrast, long duration (Class II) flares have larger volumes and lower energy densities and occur at greater heights ($\approx 5 \times 10^4$ km). Long-duration flares also are often associated with coronal mass ejections (Kahler et al., 1984) and interplanetary shocks (Cane, 1985), which in general is not the case for short-duration flares.

Evenson et al. (1984), studying electron events observed during 1980 which form a subset of the present survey, identified a class of events which correlate with gamma-ray flares and which have an unusually high abundance of electrons relative to protons. Figure 4 shows the electron to proton ratio as a function of the soft X-ray duration for the full data set. The electron flux has been evaluated at 14.5 MeV from the power-law fits to the high-rigidity part of the spectrum and is compared with the proton flux in the 9–23 MeV range for all events where proton data are available (from Cane, McGuire, and von Rosenvinge, 1986). Although there is a large variability in the data, it can be seen from Figure 4 that the typical electron to proton ratio of SDEs events is approximately two orders of magnitude higher than that of LDEs. The higher electron to proton ratio for SDEs is due to the fact that they have much weaker proton fluxes, both classes have comparable electron fluxes. Again, the separation between the two classes occurs at a duration of roughly 1 hour.

4. Discussion

The fact that the spectra of long-duration events can be modeled by a single power law in rigidity indicates that one acceleration mechanism is operating over the whole observed energy range from 0.1 to 100 MeV. Stochastic acceleration in turbulent plasmas (cf. Ramaty, 1979) reproduces the observed spectra and is a mechanism commonly considered as candidate for the acceleration of relativistic solar flare electrons. On the other hand, a model involving acceleration at a shock wave passing through the corona can explain the spectra as well. Because LDEs are often associated with interplanetary shocks, a shock might be involved in particle acceleration in this class of flares. The observed spectral indices for LDEs are in the range 3.1–4.7,

WOLFGANG DRÖGE ET AL.

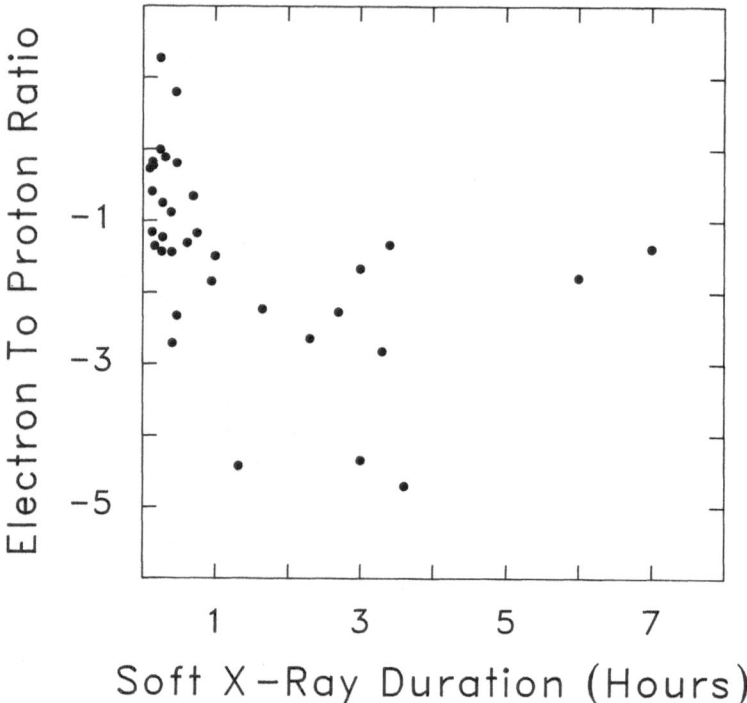

Fig. 4.   Electron to proton flux ratios as a function of the soft X-ray duration.

corresponding to compression ratios in the range 1.8–2.4 in a diffusive shock accelera-tion model. The electron spectra alone cannot distinguish between these options.

There are several possible explanations for the spectral shape of short-duration events. A superposition of two components – a steep spectrum at low energies related to acceleration in the impulsive phase of the event, and a harder spectrum of particles accelerated in a second step/second phase – is one possibility. We note that simple diffusive shock acceleration cannot model the spectra of SDEs. However, if the back-reaction process on shocks due to their acceleration of particles (Drury, 1984) is considered, the resulting spectrum can be of the form of SDE spectra. Back-reaction effects might be important for shocks in an environment with high energy densities, as can be expected for impulsive flares, but a detailed model involving an application of this effect on acceleration in solar flares is not yet available.

The model we have applied to fit the spectra of short-duration events is the one by Steinacker, Dröge, and Schlickeiser (1988), who demonstrated that stochastic accelera-tion taking into account ionization energy losses can produce the required steepening of the spectra at low rigidities. In this model it is assumed that electron acceleration in SDEs takes place in compact magnetic loops low in the corona where the plasma density is high and ionization losses cannot be neglected.

As was shown by Dröge and Schlickeiser (1986), the effects of stochastic acceleration, additional energy gain and loss processes and particle escape can be incorporated in a transport equation in momentum space

$$\frac{\partial f}{\partial t} - \frac{1}{p^2} \frac{\partial}{\partial p} \left(p^2 D(p) \frac{\partial f}{\partial p} \right) + \frac{1}{p^2} \frac{\partial}{\partial p} (p^2 (\dot{p}_G + \dot{p}_L) f) + \frac{f}{T(p)} = Q(p), \quad (1)$$

where p is the particle momentum, $N(p) = 4\pi p^2 f(p, t)$ the number of particles per unit momentum and volume, and $\dot{p}_G$ and $\dot{p}_L$ are the systematic gain and loss rates of the particles. The effects of particle escape out of the acceleration region have been combined in the escape time T, and $Q(p, t)$ represents sources and sinks of the particles. The momentum diffusion coefficient $D(p)$ depends on the nature of the stochastic process and is of the form

$$D(p) = \frac{p^2 \alpha}{3\beta l}, \quad (2)$$

where $\alpha = V^2/(lc)$, $\beta = v/c$, V is the velocity of the fluid elements or the Alfvén velocity, v is the particle velocity, l is an effective mean free path against particle scattering off the plasma waves. The loss rate due to Coulomb collisions of electrons in a fully-ionized hydrogen plasma has been considered by Bai (1982). It was found that for reasonable flare parameters ($T \approx 10^7$ K, $n_e = n_p = 10^{10}$ cm^{-3}) the loss rate can be written as

$$-\dot{p}_{\text{Coul}} = \frac{mc}{\tau_L \beta^2} \quad (3)$$

($\tau_L = 0.146 \times 10^{13}(n \text{ cm}^{-3})^{-1}$ s) at all momenta of interest.

In order to obtain asymptotic power laws at high rigidities we have to make the assumption $\alpha T = $ const. In this case, the steady-state solution of Equation (1) for monoenergetic injection $Q(p) \sim \delta(p - p_0)$ is of the form

$$N(p > p_0) \sim p^{1/2 - \mu} M \left(\mu - \tfrac{1}{2}, 1 + 2\mu, \frac{4p_c}{p} \right), \quad (4)$$

where M is the Kummer function or confluent hypergeometric function (Abramowitz and Stegun, 1965), $\mu = \sqrt{\tfrac{9}{4} + 3/\alpha T}$, and $p_c = 3mc/(4\alpha\tau_L)$ is a characteristic momentum where the gain rate due to stochastic acceleration and the loss rate are equal.

We have made fits to all electron spectra with both a single power law in rigidity and the Kummer function solution (4) and applied a χ^2-test for the goodness of the fit. For the long-duration events, both spectral forms provide good fits to the data (cf. Figure 1). For short-duration events, single power laws generally give poor fits. Here the Kummer function solution improves the quality of the fits drastically, giving fits with values for χ^2 per degree of freedom similar to the ones obtained for long-duration events (cf. Figure 2).

Events which are modeled by a single power law require one fit parameter, i.e., the compression ratio in a diffusive shock acceleration model or the acceleration parameter

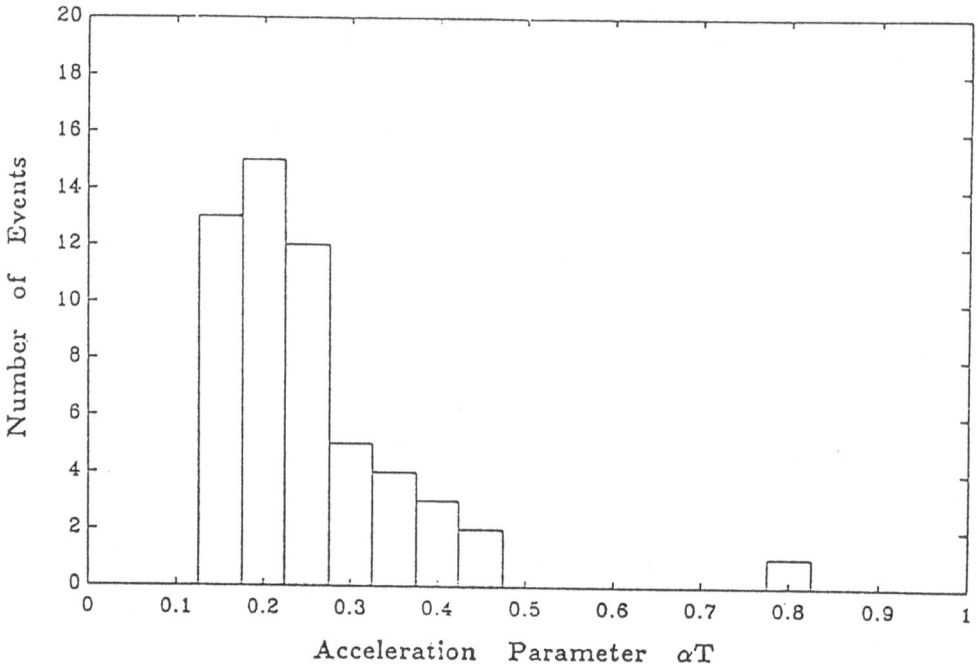

Fig. 5.   Distribution function of the stochastic acceleration parameter αT. The event with $\alpha T = 0.8$ is the flare of 29 June, 1980 which had an exceptionally hard spectrum at high rigidities.

αT for stochastic acceleration without losses. In a stochastic acceleration model which includes ionization losses, the additional fit parameter is the characteristic momentum p_c. This parameter describes where the spectrum begins to deviate from the asymptotic power law at high rigidity and turns into a steepening towards lower rigidity.

In Figure 5 we show the distribution of the fit parameter αT for all events of our survey. Values for αT are in the range 0.12 to 0.8. As mentioned above, at high rigidities the spectra of short-duration events are harder than the spectra of long-duration events. The average value of αT is 0.19 for SDEs, whereas it is 0.3 for LDEs. The values of the second fit parameter range from $p_c \approx 3$ MeV c^{-1} required to model events where the spectral steepening is most pronounced, to $p_c \rightarrow 0$ in which case the Kummer function solution becomes a power law over the entire observed range to fit LDE spectra. We are in the process of studying possible correlations between the two fit parameters and between the fit parameters and observables of the parent flares (Dröge et al., 1989). We note that the values of αT we obtain for the spectra of our survey are on average one order of magnitude larger than values of αT required to fit proton spectra with stochastic acceleration (cf. McGuire and von Rosenvinge, 1984). This might indicate that protons interact with different plasma waves or have different escape times.

We believe that stochastic acceleration including ionization losses is a promising model for the acceleration of energetic electrons in impulsive solar flares, although other explanations are possible. This model also provides good fits for long-duration events, but a further study of the properties of each class of flares is required to decide whether the same acceleration mechanism operating under different boundary conditions or rather two distinct acceleration mechanisms are more likely to accelerate electrons in impulsive and long-duration flares.

Acknowledgements

We are grateful to Dr Dietrich Hovestadt for making available the electron data from the ULEWAT experiment. We wish to thank Leo Krawczyk for invaluable assistance in the data analysis. This work was supported in part by NASA contracts NAS5–28500 and NAS5–25496. W.D. gratefully acknowledges support as a Lynen Fellow of the Alexander von Humboldt-Stiftung.

References

Abramowitz, M. and Stegun, I. A.: 1965, *Handbook of Mathematical Functions*, Dover, New York.
Axford, W. I., Leer, E., and Skadron, G.: 1977, *Proc. 15th Int. Cosmic Ray Conf. (Plovdiv)* **11**, 132.
Bai, T.: 1982, *Astrophys. J.* **259**, 342.
Cane, H. V.: 1985, *J. Geophys. Res.* **90**, 191.
Cane, H. V., McGuire, R. E., and von Rosenvinge, T. T.: 1986, *Astrophys. J.* **301**, 448.
Datlowe, D.: 1971, *Solar Phys.* **17**, 436.
Dröge, W. and Schlickeiser, R.: 1986, *Astrophys. J.* **305**, 909.
Dröge, W. *et al.*: 1989 (in preparation).
Drury, L. O'C.: 1984, *Adv. Space Res.* **4**, 185.
Evenson, P., Meyer, P., Yanagita, S., and Forrest, D.: 1984, *Astrophys. J.* **283**, 439.
Kahler, S. W., Sheeley, N. R., Howard, R. A., Koomen, R. J., Michels, D. J., McGuire, R. E., von Rosenvinge, T. T., and Reames, D. V.: 1984, *J. Geophys. Res.* **89**, 9683.
Lin, R. P., Mewaldt, R. A., and van Hollebeke, M. A. I.: 1982, *Astrophys. J.* **253**, 949.
McGuire, R. E. and von Rosenvinge, T. T.: 1984, *Adv. Space Res.* **4**, 117.
Moses, D., Dröge, W., Meyer, P., and Evenson, P.: 1989, *Astrophys. J.* (submitted).
Pallavicini, R., Serio, S., and Vaiana, G.: 1977, *Astrophys. J.* **216**, 108.
Ramaty, R.: 1979, in J. Arons, C. Max, and C. McKee (eds.), *Particle Acceleration Mechanisms in Astrophysics*, AIP, No. 56, AIP, New York, p. 135.
Steinacker, J., Dröge, W., and Schlickeiser, R.: 1988, *Solar Phys.* **115**, 313.

MODELS OF FLARING LOOPS

A. GORDON EMSLIE

Department of Physics, UAH, Huntsville, AL 35899, U.S.A.

Abstract. We review the somewhat questionable concept of an isolated flare loop and the various physical mechanisms believed to be responsible, to some degree, for energy transport within the loop structure. Observational evidence suggests a predominant role for high-energy electrons as an energy transport mechanism, and we explore the consequences of such a scenario in some detail, focusing on radiation signatures in the soft X-ray, hard X-ray, and EUV wavebands, as observed by recent satellite observatories. We find that the predictions of flare loop models are in fact in excellent agreement with these observations, reinforcing both the notion of the loop as a fundamental component of solar flares and the belief that electron acceleration is an integral part of the flare energy release process.

1. The Concept of a Flare Loop

Although the magnetic field topology and geometry in regions of flaring activity may be (of necessity?) complicated, there has nevertheless been considerable enthusiasm in recent years devoted to the modeling of the behavior of an isolated looplike structure in response to flare energy dissipation. Such flare 'loops' are defined by a bundle of magnetic field lines, considered to be rigid, and usually the region outside of this bundle is considered to be field-free. The material in the loop is treated, in most cases, as a single fluid free to move in one dimension only, i.e., along the field lines within the bundle. Energy (whether in the form of accelerated particles, heat, hydrodynamical kinetic energy, or waves) also flows along this single dimension, is converted into heat and mass motions, and is ultimately radiated away.

Convenient as this construct may be for theoretical modelers, there is substantial cause to question the assumptions inherent within it. First, the bundle of field lines under consideration is in all probability part of a more global field structure, and is highlighted against the background field only because of the preferential concentration of radiating material along the field lines of the 'loop'. This would not be a serious problem if the field lines were infinitely rigid, since energy flow across field lines (and, therefore, into or out of the bundle) is strongly suppressed, except in the case of radiative transport (see, e.g., the discussion of cross-field energy transport by maser-produced microwaves; Melrose and Dulk, 1982, 1984). However, such an assumption of rigid field lines neglects both the magnetohydrodynamic response of the atmosphere and the possibility of dynamic evolution of the field through further reconnection with adjacent field lines as the flare proceeds. Gas pressures developed as a result of flare heating can in fact reach values comparable to the pressure of the ambient magnetic field (Emslie, 1981a). We see, therefore, that the assumption of an isolated, rigid, unyielding, non-evolving field structure is a highly questionable one.

Second, a single-fluid treatment of the energized gas is probably inadequate. Many heating mechanisms (e.g., Joule heating, electron bombardment) heat primarily elec-

Solar Physics **121**: 105–115, 1989.

trons, while some (e.g., ion-cyclotron waves) preferentially heat the ions. Electron-ion temperature equilibration times can be a substantial fraction of the duration of the impulsive phase of a flare, so that at least a two-fluid treatment is called for at such early times in the event. Shoub (1983) has shown that, even in the quiet Sun, there exist regions of steep temperature gradient (such as the chromosphere/corona transition region) in which the momentum distribution function for the particles is probably far from Maxwellian, demanding a kinetic treatment in order to correctly deduce radiative loss rates, thermal conductive fluxes, etc. The even steeper temperature gradients found in flares make this result even more pertinent (Karpen and DeVore, 1987).

Even accepting the assumptions of the model, the modeling of the flare loops in still in its infancy. To begin with, one must satisfactorily deal with the complicated interplay between hydrodynamics, plasma physics, radiative transfer, and atomic and nuclear physics necessary to model the response of the atmosphere and the associated radiative signatures. The formidable complexity of this problem rules out analytic treatments (although with some *drastic* simplifications a crude analytic understanding of the qualitative physics taking place is attainable; Brown and Emslie, 1989; see Section 3). We are thus driven to the use of numerical computer codes. To date only approximate numerical treatments exist, and even these involve the use of sophisticated computational techniques and require many hours of CPU time on large machines. Even then, results from different codes can apparently yield substantially different results, even when attempting to solve the same 'benchmark' problem. These discrepancies can sometime be resolved through a realization that different numerical schemes (e.g., Lagrangian vs Eulerian) must of necessity handle a given problem in substantially different ways; however, there still remain differences that cannot readily be accounted for (see Kopp *et al.*, 1986, for a discussion). At this stage, therefore, results from even single-fluid codes must be treated with some caution.

A further problem is that numerical codes, in order to represent a physical model of an actual flare, require the specification of not only initial and/or boundary conditions (see Section 3), but also the appropriate source terms in the momentum (Brown and Craig, 1984; McClymont and Canfield, 1984, Tamres, Canfield, and McClymont, 1986) and energy equations. However, the form of the flare energy input itself is not well known at this time, awaiting an understanding of the deep mysteries by which magnetic free energy is converted into accelerated particles and/or heat. Indeed, one of the major goals of flare loop modeling is to constrain the primary energy release mechanism through comparison of the predictions of candidate energy transport models with observation. A secondary, although equally challenging, goal is to understand the processes of energy transport themselves, using the flare as our 'laboratory' to analyze processes that are impossible to study in controlled terrestrial experiments.

Despite these somewhat disparaging remarks, however, a great deal of success has been achieved in explaining the observed characteristics of solar flares in terms of energization of a loop or series of loops. In the next section we shall examine the candidate energy transport mechanisms themselves, before discussing their application to flare loop modeling in subsequent sections.

2. Energy Transport Mechanisms

Numerous mechanisms have been invoked to account for the spread of energy, generated by magnetic field dissipation in the primary energy release site, throughout the flare volume. They can be generally categorized into transport by particles, waves (including heat conduction), mass motions, and radiation.

2.1. PARTICLE BOMBARDMENT

Prompt acceleration of both electrons and protons during the impulsive phase of a solar flare appears to be a common characteristic (see, e.g., Ramaty, 1986; Vlahos *et al.*, 1986). Emslie (1983) has shown that both electrons and protons can contribute significantly to atmospheric heating, with protons characteristically heating at great depths (lower chromosphere and temperature minimum) and electrons predominating in the higher levels (corona and upper chromosphere). Due to the much lower energy per particle involved, electron heating involves a much greater particle flux and, hence, beam-associated electrical current. Thus, in determining the dynamics of the particles and the dissipation of their energy, we must consider not only 'single particle' processes, such as Coulomb collisions (Brown, 1972; Emslie, 1978) and magnetic field variation (Leach and Petrosian, 1981; Chandrashekar and Emslie, 1986), but also 'collective' processes, such as the driving of a beam-neutralizing return current (Knight and Sturrock, 1977; Emslie, 1980) and the possible generation of Langmuir and electron-cyclotron waves (Vlahos and Papadopoulos, 1979; Emslie and Smith, 1984; McClements, 1987a, b). By contrast, in the case of proton bombardment only single-particle processes need be considered, since the number of particles, and the associated electrical current, are relatively small. Note that in general energy transport by high energy particles is extremely rapid (see Section 4).

2.2. MHD WAVES

Because of the large ($\gtrsim 10^8$ cm s^{-1}) Alfvén velocities appropriate to flaring coronae, propagation of MHD waves represents another rapid a method of transporting energy throughout the flare volume. However, such waves are only significantly damped in regions of high electrical resistivity, and as such probably only play a significant role in the heating of the cool, partially ionized, lower layers of the atmosphere, around the temperature minimum region (Emslie and Sturrock, 1982). However, MHD waves propagating *perpendicular* to field lines may also represent an effective way of removing energy from the primary energy release region (see Spicer, Mariska, and Boris, 1986).

2.3. HEAT CONDUCTION

Due to its strong dependence on temperature, heat conduction is a very effective mechanism of energy transport in the flaring corona (Emslie, 1985). By the same token, it is totally ineffective in the cool chromospheric layers, where radiation (Section 2.5) instead dominates (Emslie, Brown, and Machado, 1981). In the presence of a strong magnetic field, the heat flow is strongly anisotropic, transport parallel to the magnetic field being orders of magnitude more important than transport in the perpendicular direction. Although traditionally treated in the classical, diffusive, approach (Spitzer, 1962), the true processes of heat conduction in solar flares can be extremely complicated and varied. In regions of steep temperature gradient, the local temperature gradient is no longer, as in diffusive treatments, sufficient to determine the local heat flux; instead more global considerations, requiring a consideration of the variation of temperature over several mean free paths on either side of the region of interest, apply (e.g., Karpen and DeVore, 1987). The large heat fluxes associated with flare energization may approach the saturated heat flux limit (the maximum rate at which electrons with mean thermal speed v_e can transport energy), so that the (free-streaming) flux $F \approx n v_e^3$ (v_e is the thermal velocity) becomes functionally dependent on the local density and temperature, rather than on the temperature and temperature gradient. In addition, the Onsager relations (e.g., Spitzer, 1962) imply that a heat flux has an associated electrical current. If driven to a suitably high level, the neutralizing return current (Knight and Sturrock, 1977) will excite plasma waves which in turn inhibit the conductive flux (see, e.g., Brown, Melrose, and Spicer, 1979; Smith and Lilliequist, 1979). Therefore, a full and proper treatment of conductive heat transfer in flares is a complex issue, requiring sophisticated numerical schemes to yield results which approach the true physical behavior of the plasma.

2.4. MASS MOTIONS

The steep pressure gradients established during the impulsive heating of preflare material are responsible for driving energetically significant bulk flows, as first pointed out by Craig and McClymont (1976). Indeed, all present-day models of flare loops (e.g., Pallavicini *et al.*, 1983; Nagai and Emslie, 1984; Fisher, Canfield, and McClymont, 1985; MacNeice *et al.*, 1984) have as their core a system of hydrodynamic equations. Driven sufficiently, such a system can develop hydrodynamic shocks, whose viscous dissipation can also be energetically significant. We shall return to a more detailed discussion of mass motions in flare loops in Sections 3 and 5.

2.5. RADIATION

Long recognized as a valuable *diagnostic* of flare energetics, it is now becoming increasingly apparent that the radiation field itself may contribute significantly to the energy budget of the flare, particularly in the optically thick lower layers of the atmosphere. Radiation 'backwarming' by soft X-ray radiation is an effective mechanism for heating the upper chromosphere (Somov, 1985; Machado, 1978), while EUV radiation produced in the transition region can be responsible for depositing significant energy around the temperature minimum (Emslie and Machado, 1979), as well as inducing charges in opacity that lead to enhanced absorption of photospheric radiation (Machado, Emslie, and Mauas, 1986). The complicated interplay between lines and continua of the plethora of atomic species to be found throughout a flare loop (from Fe XXVI in the corona to neutral hydrogen and metals in the lower chromosphere) makes a through treatment of radiative energy balance exceedingly complicated, even in the quiet Sun (Vernazza,

Avrett, and Loeser, 1981). Recently, however, preliminary models of radiative energy balance in flares, both empirical (Machado *et al.*, 1980; Avrett, Machado, and Kurucz, 1986) and theoretical (McClymont and Canfield, 1983; Canfield, Fisher, and McClymont, 1983) have been put forward.

3. Hydrodynamic Modeling of Flare Loops – Some General Comments

The basic physics underlying the hydrodynamic response of the solar atmosphere to flare energy input is relatively straightforward (Brown and Emslie, 1989). The deposited energy first causes a temperature rise in the ambient material. Pressure gradients set up by this (generally nonuniform) heating then establish mass motions, as a result of which the density of the material changes. The subtle interplay between the energy input and output terms (which as we have seen in the preceding section, are functions of ambient parameters such as temperature, temperature gradient, and density) and the ambient conditions themselves renders a thorough understanding of the hydrodynamic response less than straightforward. The nature of the hydrodynamic equations themselves (a set of nonlinear coupled partial differential equations with integro-differential source terms) further adds to the complexity of the problem. Furthermore, the assumed initial and boundary conditions used in the definition of the problem in fact play a critical role in the evolution of the system. To see this, we note that the set of hydrodynamic equations generally possesses three sets of characteristics. One of these is stationary and corresponds to the local increase in temperature due to flare energy deposition. The other two are moving (one upward, one downward) and correspond to propagation of information, typically at a speed around the sound speed. Consider the downward-propagating characteristics. At a given depth z_0, there is a corresponding time (t_0, say) at which the first downward-moving characteristic from the top of the atmosphere arrives there. For times $t < t_0$, the specification of initial conditions for all z, plus information on the local heating, are clearly sufficient to prescribe the properties on all characteristics passing through $z = z_0$ and, hence, to determine the behavior of the gas there. However, for $t > t_0$, initialized information from the top boundary ($z = 0$) has already passed $z = z_0$, and the downward-propagating characteristic through z_0 thus originates at the top of the atmosphere (the $z = 0$ plane) at a finite value of t (t^*, say). Therefore, prescription of the characteristics through z_0 past time $t = t_0$ requires knowledge of the *boundary* conditions on $z = 0$ up to time t^*. (For details of this argument, see Brown and Emslie, 1989.)

Most hydrodynamic simulations of flare loops provide this necessary boundary condition by considering one-half of a symmetrical loop, thereby effectively imposing the symmetry condition that the velocity vanish at the loop apex. However, such perfect symmetry is hardly likely to be realized in practice, and it would be a worthwhile exercise to consider in detail the effect that more realistic boundary conditions have on the solutions.

4. The Transition Region as a Diagnostic of Energy Transport Mechanisms

Radiation from the chromosphere/corona region ($10^4 \lesssim T \lesssim 10^6$ K) is dominated by optically thin UV lines. The intensity of such lines is given by the formula

$$I = A \int n^2 \frac{dz}{dT} \, dT,$$

where A is a constant depending on the abundance of the responsible element relative to hydrogen and on atomic parameters relevant to the line in question. The quality $n^2(dz/dT)$ is known as the Differential Emission Measure (DEM) and depends only on the structure of the atmosphere. Typically, as flare heating within a loop proceeds, the temperatures at which a given line is produced are realized at greater and greater depths (the transition region 'moves' downward), resulting in an enhancement of the n^2 factor. However, if the coronal part of the loop is also appreciably heated, then the larger conductive heat losses through the transition region necessitate a higher temperature gradient (and so lower dz/dT) at a given temperature. Thus the behavior of the DEM is a result of two competing factors. In a model in which the primary energy transport mechanism is thermal conduction from a hot corona, we would expect DEM and, therefore, the intensity in transition region UV lines, to *decrease* in response to flare energy input (Emslie and Nagai, 1985). Observations, however, show that UV line intensities are in fact well correlated with hard X-ray burst time profiles (Poland *et al.*, 1984; Orwig and Woodgate, 1986). Since hard X-ray bursts are a good diagnostic of the energy input rate to the flare (e.g., Brown, 1971), we see that the assumption of conductively-dominated energetics is incompatible with the observations. On the other hand, Emslie and Nagai (1985) and Mariska and Poland (1985) have shown that the more spatially distributed heating corresponding to collisional degradation of non-thermal electron results is more modest decreases in (dz/dT) coupled with substantial increases in n^2, so that the behavior of the DEM in such an energy transport model is indeed in accord with the hard X-ray/UV correlation observations. This success of the electron-heated hypothesis (or, more significantly, the failure of the thermal conduction hypothesis) leads us to consider electron-heated loop models in more detail in the next section.

5. Soft X-Ray Diagnostics of Electron-Heated Loops

One of the most powerful diagnostics of energy transport in flare loops is the shift and broadening of soft X-ray line profiles resulting from the flows set up by the flare-associated heating. Antonucci *et al.* (1982) have reported observations of broadening and blue asymmetry in the profile of the Ca XIX resonance line at 3.177 Å ('w', in the notation of Gabriel, 1972). They model this by a double gaussian fit to the line profile, the components of which are typically separated by a few mÅ, corresponding to a few hundred km s^{-1} of upflow velocity. Emslie and Alexander (1987) used the electron-heated hydrodynamic simulation results of Nagai and Emslie (1984) to synthesize

Ca XIX '*w*' profiles and found general agreement of both the absolute intensity and shape of the line with observation. (By contrast, Cheng, Karpen, and Doschek (1984) have modeled the response of a loop to a heat input with a gaussian spatial distribution. This heating function is *ad hoc*, is not based on any of the energy transport mechanisms discussed in Section 2, and in fact results in Ca XIX '*w*' profiles which disagree significantly with the observational data.) However, Emslie and Alexander found no evidence for a stationary component, as reported by Antonucci *et al.* (1982), and suggested that this may have been due to an error in the absolute calibration of the Solar Maximum Mission (SMM) Bent Crystal Spectrometer (BCS) instrument used. More recent work by McClements and Alexander (1989) has shown that there is no significant center-to-limb variation in the wavelength of the principal component, implying that this component is indeed stationary, contrary to the model results of Emslie and Alexander (1987). Mariska, Emslie, and Li (1989) and Li, Emslie, and Mariska (1989) note that Emslie and Alexander's loop length was sufficiently large that no significant accumulation of upwelling material at the apex of the loop occurred over the duration of the simulation, and suggested that the use of shorter loops would more naturally give rise to a stationary component at the loop apex. They consequently carried out a series of numerical models describing the hydrodynamic response of a loop to nonthermal electron input, varying the flux and spectrum of the injected electrons and computing the Ca XIX '*w*' line profiles for each model. Their results show clearly that, while blueshifted principal components *do* persist during the first few tens of seconds of energy input (in agreement with the modeling of Emslie and Alexander), a strong stationary component develops after about 30 s a time of order L/c_s, where L is the half-length of the loop ($\approx 10^9$ cm) and c_s the sound speed (roughly the velocity of the upward moving material). Averaged over the 30–50 s typical of the observations (Antonucci *et al.*, 1982), the overall Ca XIX '*w*' profile, therefore, shows a principal component that corresponds to the strong emission late in the impulsive phase (and is therefore stationary), with the earlier blueshifts manifesting themselves only in a blue 'shoulder' on the averaged line profile. These time-integrated profiles agree very well with the observations, providing considerable support for the electron-heated model. In addition, the intensity of the blueshifted component of the Ca XIX '*w*' line correlates well (but *nonlinearly*) with the electron flux; together with hard X-ray bremsstrahlung observations, for which the intensity varies *linearly* with the total number of injected electrons, we can deduce the area over which the electrons are injected, a quantity not only of considerable interest to theoreticians, but also directly comparable with spatial images of the flare in, for example, the Hα line.

6. Other Results Supporting Simple Loop Models

In this section we briefly summarize results from other areas which also lend support to the simple electron-heated loop model concept.

6.1. THE EUV TO HARD X-RAY FLUX RATIO

McClymont and Canfield (1986) have shown that, within a reasonable scatter, a simple (power-law) relationship exists between the fluxes in hard X-rays and broadband EUV (10–1030 Å) from event to event. Such a relationship is in fact simply explained by a model in which all events correspond to the injection of nonthermal electrons into loops of the same area (of order 10^{16} cm^2). In such a scenario, a doubling (say) of the electron flux produces a doubling of the hard X-ray bremsstrahlung yield; it also overheats more of the chromosphere to coronal temperatures, so that a smaller fraction of the electron energy is deposited in cooler EUV-emitting layers. The observed EUV/hard X-ray flux relationship is consistent with heating by an electron beam with a power-law energy spectrum $E^{-\delta}$, with $\delta \approx 5$ (McClymont and Canfield, 1986). Furthermore, LaRosa and Emslie (1988) have shown that the large chromospheric beam fluxes implied by the model for large events require some form of a beam focusing (e.g., a convergence of the uiding magnetic field lines) to be present, so as to avoid embarrassingly high fluxes at the injection point, which would excite collective plasma instabilities and so saturate the bremsstrahlung yield at a level below that observed (see Emslie, 1980).

6.2. MOMENTUM BALANCE

Zarro *et al.* (1988; see also Canfield *et al.*, 1988) have shown that the upward momentum of the Ca XIX-emitting material (see Section 5) is in general balanced, to order of magnitude, by the downward momentum associated with the well-known redshift in Hα profiles (e.g., Švestka, 1976). This confirms that the blueshifted component of the Ca XIX profile is produced by gas energized by heating of material in the upper atmosphere, such as by electron bombardment at the footpoints of a loop. (If the blueshifted material was instead associated with a mechanism which did not involve impulsive energy input, such as, for example, the rising of prominence material, then the momentum recoil would be absorbed by the whole Sun and would not have a clear observational signature.)

6.3. EVOLUTION OF HARD X-RAY SPATIAL STRUCTURE

In a simple electron-heated loop model, the hard X-rays are initially emitted predominantly from the dense chromospheric footpoints. Later, as a result of two factors, both of which are direct consequences of the collisional heating affected by the electron beam: (i) the evaporation of chromospheric material and (ii) the creation, through collisional heating, of a hot coronal source of thermal hard X-rays, the emission should become more diffusely distributed throughout the entire loop volume (Emslie, 1981b; Brown and Emslie, 1987). Results from the Hard X-Ray Imaging Spectrometer (HXIS) on SMM (e.g., Hoyng *et al.*, 1981) show that this simple behavior indeed occurs; however, spatial resolution on the order of a few arc sec, together with temporal resolution of a few seconds, is necessary to trace the evolution of the structure from the initial double footpoint structure into the later diffuse source, and so test the relative roles of the factors above. Such instrumentation, using Fourier-transform spectrometers, is a major thrust of the Max '91 program for flare study at the next solar maximum (Dennis *et al.*, 1988).

6.4. Hα PROFILES AND MORPHOLOGY

Canfield, Gunkler, and Ricchiazzi (1984) report that the Hα profiles during the impulsive phase show broad wings which can only be satisfactorily explained by the Stark broadening associated with collisional electron heating of dense chromospheric layers to Hα-emitting temperatures (see, however, Doschek *et al.*, 1986). The fluxes inferred from modeling of the line profile are consistent with those required to produce the hard X-ray burst (Canfield and Gayley, 1987). In addition, there is ample observational evidence showing that the brightenings occur at the feet of magnetic structures (e.g., Švestka, 1976; Tandberg-Hanssen and Emslie, 1988).

6.5. IRON Kα EMISSION

Tanaka, Watanabe, and Nitta (1984) and Emslie, Phillips, and Dennis (1985) have shown that impulsive enhancements in the iron Kα line (produced by inner-shell ionization of a near-neutral iron atom, follow by $2p \rightarrow 1s$ recombination into the inner-shell vacancy) occur simultaneously with hard X-ray bursts and that the intensity of the line is consistent with that expected from collisional ionization due to beamed non-thermal electrons in a simple loop geometry.

7. Conclusions

We have shown that, despite the obviously heuristic foundations on which the concept of a flare 'loop' is based, the modeling of such loops in fact yield results which are consistent with a large variety of observational data. This establishes on a more secure foundation two notions: namely (i) that magnetic loops are a fundamental constituent of solar flares, and (ii) that the primary energy release mechanism is an efficient particle accelerator. Evidently models that specifically invoke loop geometries (e.g., Spicer, 1977a, b; Colgate, 1978; Tajima, Brunel, and Sakai, 1982) should be more thoroughly pursued. We here also note recent observational evidence suggestive of energy *release* in loop-like structures. A detailed and systematic analysis of hard X-ray imaging compared with vector magnetograms demonstrated that, while many flares are triggered by the interaction of distinct flux systems, most of the energy released occurred within the loop of greatest shear (Machado *et al.*, 1988). However, it must be re-emphasized that the models to date are clearly oversimplistic (e.g., assumed topology and geometry of the 'loop', treatment of the interrelationship between hydrodynamics and radiative transfer, etc.). Furthermore, the theoretical problem of how the required large flux of nonthermal electrons can be impulsively accelerated remains a formidable one (Heyvaerts, 1981; Vlahos *et al.*, 1986). It, therefore, remains to be seen whether these loop models will survive the more intense scrutiny that will be provided by the state-of-the-art instrumentation now planned for the next solar maximum.

Acknowledgements

I thank T. N. LaRosa for discussion and comments and J. C. Brown for clarifying the discussion of characteristics in Section 3. This work was supported by NASA Grants NAGW-294 and NAG5–500, and NSF Grants ATM-8715195 and AST-8351058.

References

Antonucci, E., Gabriel, A. H., Acton, L. W., Culhane, J. L., Doyle, J. G., Leibacher, J. W., Machado, M. E., Orwig, L. E., and Rapley, C. G.: 1987, *Solar Phys.* **78**, 107.
Avrett, E. H., Machado, M. E., and Kurucz, R. L.: 1986, in D. F. Neidig (ed.), *The Lower Atmosphere of Solar Flares*, National Solar Observatory, p. 216.
Brown, D. G. and Emslie, A. G.: 1987, *Solar Phys.* **110**, 305.
Brown, J. C.: 1971, *Solar Phys.* **18**, 489
Brown, J. C.: 1972, *Solar Phys.* **26**, 441.
Brown, J. C. and Craig, I. J. D.: 1984, *Astron. Astrophys.* **130**, L5.
Brown, J. C. and Emslie, A. G.: 1989, *Astrophys. J.* (in press).
Brown, J. C., Melrose, D. B., and Spicer, D. S.: 1979, *Astrophys. J.* **228**, 592.
Canfield, R. C. and Gayley, K. G.: 1987, *Astrophys. J.* **322**, 999.
Canfield, R. C., Fisher, G. H., and McClymont, A. N.: 1983, *Astrophys. J.* **265**, 507.
Canfield, R. C., Gunkler, T. A., and Ricchiazzi, P. J.: 1984, *Astrophys. J.* **282**, 296.
Canfield, R. C., Zarro, D. M., Metcalf, T. R., and Lemen, J. R.: 1988, *Bull. Am. Astron. Soc.* **20**, 688.
Chandrashekar, S. and Emslie, A. G.: 1986, *Solar Phys.* **107**, 83.
Cheng, C.-C., Karpen, J. T., and Doschek, G. A.: 1984, *Astrophys. J.* **286**, 787.
Colgate, S. A.: 1978, *Astrophys. J.* **221**, 1068.
Craig, I. J. D. and McClymont, A. N.: 1976, *Solar Phys.* **50**, 133.
Dennis, B. R. *et al.*: 1988, *Max '91 – Flare Research at the Next Solar Maximum*, NASA.
Doschek, G. A. *et al.*: 1986, *Energetic Phenomena on the Sun*, NASA CP-2439, pp. 4–39.
Emslie, A. G.: 1978, *Astrophys. J.* **224**, 241.
Emslie, A. G.: 1980, *Astrophys. J.* **235**, 1055.
Emslie, A. G.: 1981a, *Astrophys. Letters* **22**, 171.
Emslie, A. G.: 1981b, *Astrophys. J.* **245**, 711.
Emslie, A. G.: 1983, *Solar Phys.* **84**, 263.
Emslie, A. G.: 1985, *Solar Phys.* **98**, 281.
Emslie, A. G. and Alexander, D.: 1987, *Solar Phys.* **110**, 295.
Emslie, A. G. and Nagai, F.: 1985, *Astrophys. J.* **288**, 779.
Emslie, A. G. and Smith, D. F.: 1984, *Astrophys. J.* **279**, 882.
Emslie, A. G. and Sturrock, P. A.: 1982, *Solar Phys.* **80**, 99.
Emslie, A. G., Brown, J. C., and Machado, M. E.: 1981, *Astrophys. J.* **246**, 337.
Emslie, A. G., Phillips, K. J. H., and Dennis, B. R.: 1985, *Solar Phys.* **103**, 89.
Fisher, G. H., Canfield, R. C., and McClymont, A. N.: 1985, *Astrophys. J.* **289**, 414.
Gabriel, A. H.: 1972, *Monthly Notices Roy. Astron. Soc.* **160**, 99.
Heyvaerts, J.: 1981, in E. R. Priest (ed.), *Solar Flare Magnetohydrodynamics*, Gordon and Breach Publ. Co., New York.
Hoyng, P. *et al.*: 1981, *Astrophys. J.* **246**, L155.
Karpen, J. T. and DeVore, C. R.: 1987, *Astrophys. J.* **320**, 904.
Knight, J. W. and Sturrock, P. A.: 1977, *Astrophys. J.* **218**, 306.
Kopp, R. A., Fisher, G. H., MacNeice, P., McWhirter, R. W. P., and Peres, G.: 1986, *Energetic Phenomena on the Sun*, NASA CP-2439, Chapter 7.
LaRosa, T. N. and Emslie, A. G.: 1988, *Astrophys. J.* **326**, 997.
Leach, J. and Petrosian, V.: 1981, *Astrophys. J.* **251**, 781.
Li, P., Emslie, A. G., and Mariska, J. T.: 1989, *Astrophys. J.* (in press).
Machado, M. E.: 1978, *Solar Phys.* **60**, 341.

Machado, M. E., Emslie, A. G., and Mauas, P. J.: 1986, *Astron. Astrophys.* **159**, 33.

Machado, M. E., Avrett, E. H., Vernazza, J. E., and Noyes, R. W.: 1980, *Astrophys. J.* **242**, 336.

Machado, M. E., Moore, R. L., Hernandez, A. M., Rovira, M. G., Hagyard, M. J., and Smith, J. B., Jr.: 1988, *Astrophys. J.* **326**, 425.

MacNeice, P., McWhirter, R. W. P., Spicer, D. S., and Burgess, A.: 1984, *Solar Phys.* **90**, 357.

Mariska, J. T. and Poland, A. I.: 1985, *Solar Phys.* **96**, 317.

Mariska, J. T., Emslie, A. G., and Li, P.: 1989, *Astrophys. J.* (in press).

McClements, K. G.: 1987a, *Astron. Astrophys.* **175**, 255.

McClements, K. G.: 1987b, *Solar Phys.* **109**, 355.

McClements, K. G. and Alexander, D.: 1989, *Solar Phys.* (submitted).

McClymont, A. N. and Canfield, R. C.: 1983, *Astrophys. J.* **265**, 483.

McClymont, A. N. and Canfield, R. C.: 1984, *Astron. Astrophys.* **136**, L1.

McClymont, A. N. and Canfield, R. C.: 1986, *Astrophys. J.* **305**, 936.

Melrose, D. B. and Dulk, G. A.: 1982, *Astrophys. J.* **259**, 844.

Melrose, D. B. and Dulk, G. A.: 1984, *Astrophys. J.* **282**, 308.

Nagai, F. and Emslie, A. G.: 1984, *Astrophys. J.* **279**, 896.

Orwig, L. E. and Woodgate, B. E.: 1986, in D. F. Neidig (ed.), *The Lower Atmosphere of Solar Flares*, National Solar Observatory, p. 306.

Pallavicini, R., Peres, G., Serio, S., Vaiana, G., Acton, L., Leibacher, J., and Rosner, R.: 1983, *Astrophys. J.* **270**, 270.

Poland, A. I., Orwig, L. E., Mariska, J. T., Nakatsuka, R., and Auer, L. H.: 1984, *Astrophys. J.* **280**, 457.

Ramaty, R.: 1986, in P. A. Sturrock (ed.), *Physics of the Sun*, Vol. II, D. Reidel Publ. Co., Dordrecht, Holland, p. 291.

Shoub, E. C.: 1983, *Astrophys. J.* **266**, 339.

Smith, D. F. and Lilliequist, C. G.: 1979, *Astrophys. J.* **232**, 582.

Somov, B. V.: 1975, *Solar Phys.* **42**, 235.

Spicer, D. S.: 1977a, NRL Report 8036.

Spicer, D. S.: 1977b, *Solar Phys.* **53**, 305.

Spicer, D. S., Mariska, J. T., and Boris, J. P.: 1986, in P. A. Sturrock (ed.), *Physics of the Sun*, Vol. II, D. Reidel Publ. Co., Dordrecht, Holland, p. 181.

Spitzer, L. W., Jr.: 1962, *Physics of Fully Ionized Gases*, Intergraph.

Švestka, Z.: 1976, *Solar Flares*, D. Reidel Publ. Co., Dordrecht, Holland.

Tajima, T., Brunel, F., and Sakai, J.: 1982, *Astrophys. J.* **258**, L45.

Tamres, D. H., Canfield, R. C., and McClymont, A. N.: 1986, *Astrophys. J.* **309**, 409.

Tanaka, K., Watanabe, T., and Nitta, N.: 1984, *Astrophys. J.* **282**, 793.

Tandberg-Hanssen, E. and Emslie, A. G.: 1988, *The Physics of Solar Flares*, Cambridge University Press, Cambridge.

Vernazza, J. E., Avrett, E. H., and Loeser, R.: 1981, *Astrophys. J. Suppl.* **45**, 635.

Vlahos, L. and Papadopoulos, K.: 1979, *Astrophys. J.* **233**, 717.

Vlahos, L. *et al.*: 1986, in M. R. Kundu and B. E. Woodgate (eds.), *Energetic Phenomena on the Sun*, NASA CP-2439.

Zarro, D., Canfield, R. C., Strong, K. T., and Metcalf, T. R.: 1988, *Astrophys. J.* **324**, 582.

STELLAR FLARE SPECTRAL DIAGNOTICS:
PRESENT AND FUTURE*

BERNARD H. FOING[†]

Institut d'Astrophysique Spatiale, IAS/LPSP BP 10, 91371 Verrières-le-Buisson, France

Abstract. Stellar spectral diagnostics are of utmost importance to test fundamental concepts of flare physics such as particle beam versus suprathermal heating, atmospheric response, mass motions, microflaring, statistics and recurrence of flares, flare activity and stellar interior. We review some of these diagnostics (from photometry, optical, and ultraviolet spectroscopy at medium- and high-spectral resolution, X-ray, and radio observations). Specific diagnostics from line and continuum fluxes, density sensitive lines, broadening and velocity field effects and the comparison with semi-empirical models are also described.

Some results on stellar flares obtained from previous multi-wavelength observing campaigns are presented. Future satellite missions and ground-based observatories, with new techniques for obtaining high spectral and temporal resolution, are discussed in light of their possible contribution to our understanding of solar and stellar flares.

1. Fundamental Scientific Issues

The advent of new satellites and large ground-based telescopes with increasing sensitivities, spectral resolution and time resolution will allow us to obtain spectral data of solar quality on stellar flares. These new stellar flare observations, in addition to the data gathered over the last 10 years, will provide a deeper understanding in several fundamental areas.

1.1. ENERGY TRANSPORT MECHANISM: PARTICLE BEAM VS HEAT CONDUCTION

Two types of theories have been proposed to explain how the flare energy, which is released in the upper corona by magnetic field reconnection, is transported down to the lower layers; these theories assume either beams of energetic particles or thermal conduction. Different predictions come from these theories: Brown (1971) gives the absorption coefficient of an electron beam according to the energy per electron. Therefore, depending on the penetration length, electrons of a given energy (e.g., 20 keV) can reach the chromosphere, transporting significant energy and producing a detectable signature, for example in Balmer line wings. If the electrons are stopped in the corona, thermal conduction is then required to transport most of the energy downwards. These analyses have been refined in order to take into account non-classical Coulomb effects and the role of currents and wave-particle interaction. Emslie and Nagai (1985) predict a different temporal evolution of transition region lines by the two mechanisms, which can be checked observationally using far UV and extreme UV spectroscopy.

Canfield and Gunkler (1985) have calculated Hα enhanced wings as a signature of

---

* Based partially on observations obtained at European Southern Observatory, Canada-France Hawaii Telescope, and with IUE and EXOSAT satellites.

† Present address: ESA/ESTEC Space Science Dept., P.O. Box 299, 2200 AG Noordwijk, The Netherlands.

Solar Physics **121**: 117–133, 1989.
© 1989 *Kluwer Academic Publishers.*

nonthermal electron beams during the chromospheric evaporation phase. These predictions require a test with high-resolution spectra and sufficient time resolution.

Heritschi *et al.* (1989) have discussed how proton beams seem better able than electron beams to explain solar flare observations in a more physically consistent way. Hénoux (1989) showed that the observed polarisation in the Hα line tends to also favour the role of protons. All these aspects require observational checks with high time resolution during the stellar flare impulsive phase.

1.2. ATMOSPHERIC RESPONSE TO FLARES

In order to constrain our knowledge of flare physical quantities such as the overall energy budget, density and volumes, it is necessary to observe the atmospheric response to stellar flares. The spectral energy distribution and decay time-scales reflect the atmospheric parameters. In this respect, solar white-light flares are relevant, but cooler stars provide a low photospheric background against which the flare signature can be seen with high contrast. In addition, rapid rotation and deep convection zones in some stars lead to efficient magnetic field generation and to frequent and energetic flare activity. Stellar flare studies require high speed photometric and spectroscopic observations from the ground, and simultaneous multi-wavelength coverage from radio observations and satellites at higher energy (X-ray, UV).

On the Sun, flare diagnostics from accelerated particles observed at 1 AU, γ-ray and hard X-ray radiation contain very valuable information. The energy carried away by ejected material and accelerated particles may in fact dominate in the solar flare energy budget.

In the stellar flare case, particles, γ-rays, and hard X-rays are unobservable with available instruments. For hard X-rays, a collecting area equivalent to a soccer field was quoted (Schmitt, Lemen, and Zarro, 1989) for detecting solar-like flares on nearby stars. However, as hard X-rays arise mainly for bremsstrahlung emission by high-energy electrons, gyrosynchrotron radio emission may provide a proxy for the characterisation of these particles.

Soft X-ray flare measurements have been obtained using the Einstein and Exosat satellites; they have permitted us to measure the X-ray luminosity, temperatures, time-scales, and overall X-ray losses, which can constrain loop models to provide lengths and electronic densities (Haisch, 1983).

Ultraviolet observations obtained with IUE permit emission measure analysis of the transition region plasma; also line ratios can provide density diagnostics, and UV continua provide information on the chromospheric temperature structure during flares.

Optical observations show the response of different continua, hydrogen Balmer and Paschen series lines, and lines formed at higher excitation temperatures or by different processes (collisional, photo-ionisation, photo-excitation).

Infrared measurements, such as those obtained by Rodonò *et al.* (1984) showing the first detection of a 'negative infrared flare', can provide constraints on theories predicting changes in the H^- opacity, or inverse Compton effect of relativistic electrons on IR photons.

Radio observations, once we have identified the emission/absorption mechanisms (gyrosynchrotron, self-absorption, ...) will give information on magnetic field intensity and geometry, and on the electron beam parameters, especially with the advent of dynamic radio spectra observations (Bastian and Bookbinder, 1987).

Different solar flare models have been developed by Machado *et al.* (1980), Damé and Cram (1983), Aboudarham and Hénoux (1986), and Avrett, Machado, and Kurucz (1986). They allow us to estimate where in the atmosphere the radiation at a given wavelength comes from and to quantify some open issues. What are the heating processes? How can we derive a complete energy budget from the available X-ray, transition region line, and optical studies? How can we derive densities, volumes, and temperature regimes of the flaring plasma? What do the time-scales (rise and decay at different wavelength) tell us about the dynamics of flares and atmospheric properties? All these questions require simultaneous multi-wavelength observations.

1.3. MASS MOTIONS, EJECTED COMPONENTS, AND MOMENTUM BALANCE

Solar observations from SMM (Antonucci *et al.*, 1982, 1984) show the presence of Ca XIX blueshifts of the hot plasma during the impulsive phase, suggesting a chromospheric evaporation scenario. Simultaneous Hα redshifts as studied by Zarro and Canfield (1989) confirm this interpretation and indicate a momentum balance between the evaporated material ejected at a velocity estimated from these blueshifts and the compressed chromospheric Hα material. There is no chance to measure such XUV line blueshifts on stars, but stellar Hα redshifts should be observable. An indirect diagnostic of the 'well' area where impulsive explosions occur is possible from photometry (de Jager *et al.*, 1986, 1989). From these observations, one can test whether this scenario of chromospheric evaporation and impulsive phase explosion and corresponding momentum balance applies.

Evidence of filament eruption as obscuring material has been reported by Haisch *et al.* (1983) to fit X-ray spectral data during a flare on Proxima Centauri. Rodonò *et al.* (1979) have reported negative preflare dips from optical light curves, indicative of ejected material obscuring the disk. Hénoux (1988) studied this effect, calling it a black and white flare in analogy with the effects of Scotch whisky. Collier-Cameron and Robinson (1988, 1989) observed in the star AB Dor that such ejected material can be traced from its velocity-varying absorption in Hα when these clouds transit over the visible disk, and they developed a method of cloud imaging from time-resolved high-resolution spectroscopy.

Mass motions can be studied from their Doppler effect in line profiles, such as the Mg II line wing broadening reported during a flare on the star UX Ari (Simon, Linsky, and Schiffer, 1980). Linsky *et al.* (1989) and Neff, Brown, and Linsky (1989) reported IUE observations of a flare on V711 Tau showing observable shifts and broadening in the Mg II and C IV lines. A blue-shifted emitting Hα component suggesting a rising material during a flare was reported on UV Cet. Also, Doyle *et al.* (1989) reported Balmer line broadenings, which may be interpreted as turbulence and possible multiple velocity components rather than the Stark effect.

Solar observations from the ground and from satellites show discrete ejections in the form of coronal mass ejections, eruptive filaments, sprays, and surges. Observations with the HRTS instrument (Brueckner and Bartoe, 1983; Brueckner *et al.*, 1986) show jets, coronal bullets, and turbulent flows at small scales.

An important question is to learn what part of a stellar wind comes from coronal mass ejections, coronal transients, or other aspects of flares. In order to learn the role of flares in the mass and momentum balance, it is necessary to measure mass motions with high spectral resolution, and to probe the ejected material quantities by different techniques (photometric absorption in X-ray, UV, optical, and spectral imaging).

1.4. MICROFLARING, FLARING, AND CORONAL HEATING

The origin of the coronal heating is still unknown. One hypothesis is that that solar-like coronae are heated by flaring and microflaring (Butler *et al.*, 1986). As has been studied in the laboratory and in the magnetosphere, preflare plasma is believed to convert magnetic energy into heat by reconnection and the dissipation of magnetic waves or electric currents. There is indeed an indication of this conversion at small scales. On the Sun hard X-ray microflares have been reported by Lin *et al.* (1984) and also UV line brightenings by Withbroe, Habbal, and Ronan (1985). Also, X-ray brightenings of active region loops observed with SMM confirm the existence of underlying microflare activity (Haisch *et al.*, 1989). Parker (1988) has invoked microflares or even nanoflares as a contribution to coronal heating.

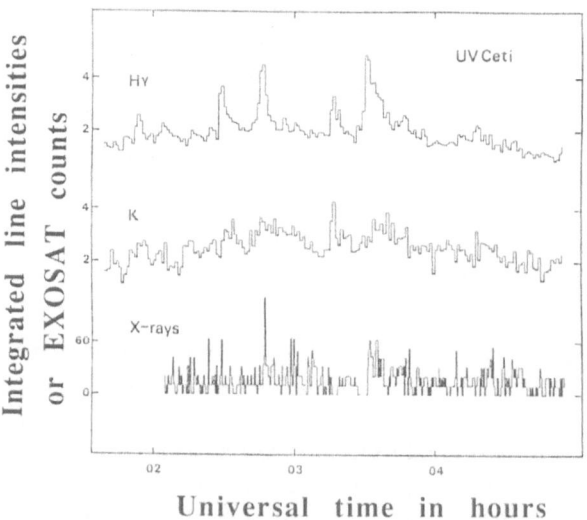

Fig. 1.   The soft X-ray flux detected with the Exosat CMA from UV Ceti on 6 December, 1984 is compared with the Hγ and Ca II K fluxes measured from the ESO 3.6 m telescope. Correlated events in the three diagnostics confirm the existence of small flares and imply that much of the low-level X-ray flux previously considered to be quiescent probably originates from small flare events (Butler *et al.*, 1986). One can notice also the different time behaviour of the line fluxes, with an impulsive component for the Hγ line, while the Ca II K is more gradual (courtesy, Butler, Rodonò, and Foing, 1988).

On flare stars, a correlation was found between the optical flare power and the average X-ray luminosity independently by Doyle and Butler (1985) and Skumanich (1985, 1986). From X-ray measurements the general presence of a hot component at $T > 10^7$ K consistent with continuous flaring is found in a wide range of cool stars. Einstein, and especially the Exosat satellite with its long eccentric orbit have permitted studies of the short-term X-ray variability and the detection of low-energy flares.

Making use of simultaneous measurements from Exosat and ESO 3.6 m spectroscopy, Butler et al. (1986) have reported a temporal correlation between Hγ flux and X-rays in some dMe stars. Sporadic X-ray variations (cf. Figure 1) which are a signature of solar-like compact flares, contribute significantly to the average X-ray flux, indicating that the concept of a steady and quiescent corona may be of little value for these stars. By extrapolating the X-ray vs Hγ correlation to smaller energy, Butler et al. (1986) have argued that microvariations in the Balmer lines (observations made possible with the high S/N allowed by large ground-based telescopes) may correspond to X-ray microvariations below the detection threshold of Exosat and other X-ray satellites. Future statistical studies with more sensitive instruments and correcting for the detection threshold and for the merging of several small events are required for progress in this area.

1.5. STATISTICS AND RECURRENCE OF FLARES

The basic question of what powers the flares has received a promising answer from solar observations and theory; the twisting of magnetic loop footpoints by turbulent-convective photospheric and sub-photospheric flows may be the magnetic energy storage process. Other changes of field configurations, such as magnetic field cancellation and association with the magnetic inversion line have been observed on solar flares. Van Ballegooijen (1985) suggested that the distribution of flare frequency $v(E)$ versus energy E corresponds to the distribution of magnetic field energy $B^2/8\pi$ for a given scale L. This hypothesis may be checked from statistical studies of solar Hα flares, hard X-rays flares, and microflares.

From statistics of stellar optical flares, Gershberg and Shakhovskaya (1983), and Pettersen, Coleman, and Evans (1984) derive a power-law histogram $v(E) \sim E^{-\beta}$ with $\beta \sim 0.5$–1.2. However, the statistical treatment of low-energy events must be refined. Studies can be extended to include small flares and microflares by spectroscopic monitoring of the Hα or Balmer lines for events with no detectable optical continuum counterpart.

It is important to have continuous coverage to have complete statistics and to assess the flare-integrated energy content. The classification and physical understanding of different flare phenomena may be useful in determining different key parameters and processes. The studies of flare recurrence either on short time-scales (homologous or sympathetic flares, destabilisation of nearby loops, etc.) or on longer time-scales (clustering periodicities, flaring active longitudes), may provide some clues on the small-scale and large-scale distribution of magnetic field on the stellar surface.

1.6. FLARE ACTIVITY AS A DIAGNOSTIC OF MAGNETIC AND INTERIOR PROPERTIES

We have seen that X-ray observations provide information on magnetic loop scale lengths and that radio gyrosynchrotron emission from stellar flares allows us to estimate the magnetic field intensity. Several empirically-derived parameters are needed to establish a link between flare activity and the global properties of a star: (1) an estimate of the bolometric flux of flare emission from the observed optical, UV and X-ray radiation, (2) a measurement of mass motions as an indicator of matter expansion or chromospheric evaporation during the flares, and (3) a diagnostic of conduction transport. The overall energy content may be compared with that inferred from magnetic field distribution measurements through the Zeeman line-broadening technique (Saar, Linsky, and Beckers, 1986), the degree of spottedness (cf. Rodono *et al.*, 1986), or the distribution of chromospheric active regions (Foing *et al.*, 1988). Additionally, from inverse imaging techniques such as rotational modulation (Char and Foing, 1989), Doppler imaging (Vogt and Penrod, 1983; Vogt, Penrod, and Hatzes, 1987; Jankov and Foing, 1987), Zeeman–Doppler (Donati, Semel, and Praderie, 1988), or eclipse imaging, one may crudely reconstruct the distribution of stellar active structures. These techniques will provide critical data with which to test our ideas about flare build-up in the stellar context. Studies of flare frequencies and energies also provide a tool for probing the magnetic field distribution and sub-photospheric dynamics.

The subsurface source of flare energy can be studied by separating the dependence on spectral type (convection zone depth), rotation, age, and evolution (cf. Skumanich, 1986). Different kinds of flares on other stars should be studied especially in the extreme cases of very thin convection zones (minimal expected dynamo) and of magnetic field covering nearly all of the stellar surface. Also, the effect of interconnecting magnetic fields in close binary systems should be considered.

2. Spectral Diagnostics

2.1. PHOTOMETRIC DIAGNOSTICS

Since the bulk of photospheric radiation from late-type stars is emitted in the optical and infrared range, sufficient contrast is required to detect and quantify flares against this background. Whereas solar optical white-light flares are rare events, broad-band optical enhancements are observed regularly in M-dwarf flare stars, due to the faint background, especially in the U band. Thus, U-band enhancements up to 5 magnitudes are observed while I-band enhancements are only a few hundredths of magnitude.

During the impulsive phase of solar white-light flares, the optical continuum shows a strong Balmer discontinuity, and is generally flat during the decay phase. Rust (1986) attributes the impulsive spectrum to heating in the lower chromosphere producing H free-bound and free-free emission, while the decay phase emission is produced by the H^- continuum. The mechanism by which the strong soft X-rays could penetrate the upper photosphere, ionize H and produce sufficient H^- emission could be tested by detailed simultaneous optical and X-ray observations.

2.2. OPTICAL SPECTROSCOPY

Optical spectroscopy at low resolution permits a more accurate investigation of the free-free, and the Paschen and Balmer bound-free continua. Flare spectra at medium resolution such as the one displayed in Figure 2 provide measurements of line fluxes.

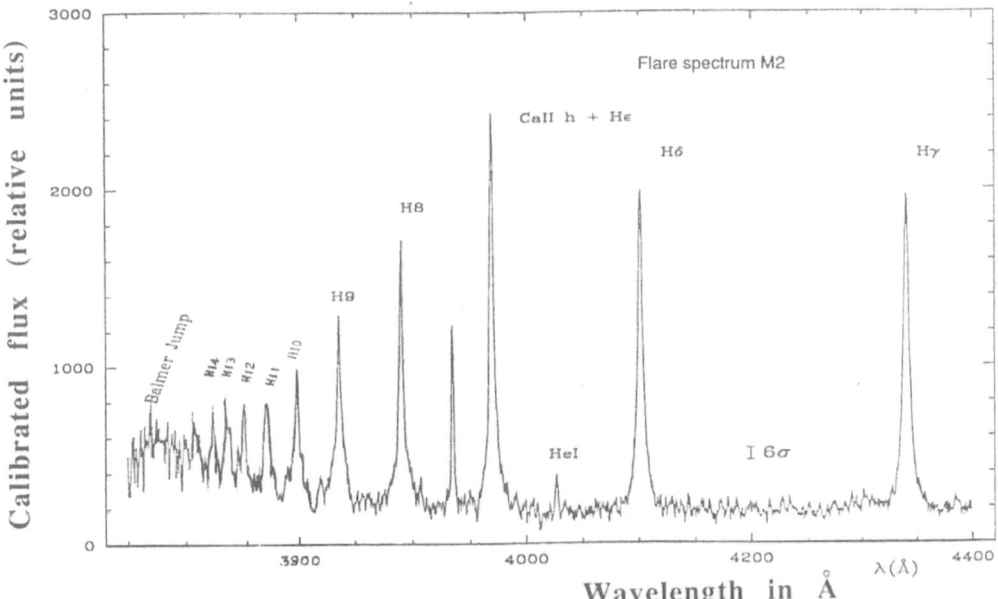

Fig. 2. Calibrated AD Leo flare differential spectrum (difference between flare and preflare spectra) obtained at the ESO 3.6 m telescope with the IDS spectrograph from 3600–4400 Å on 28 March, 1984. The spectrum is the difference between raw flare spectrum (at time indicated by tag M2 on upper Figure 3(b)) and a mean preflare spectrum. One can observe the increase of Balmer emission and wing broadening, the appearance of higher members of the Balmer series, and the emission of He I line at 4026 Å (courtesy, Rodonò *et al.*, 1989).

Balmer decrements may be used to diagnose typical temperatures and densities during flares (Butler *et al.*, 1987). The time behaviour of different line fluxes can be followed as in Figure 3. Chromospheric line fluxes strengthen during flares and decay more slowly than the optical continuum. For a given line, the dependence of line emission efficiency on temperature may be calculated, assuming for instance 'partial LTE' between higher levels of Rydberg series (for which Saha and Boltzmann equations are assumed valid for the corresponding populations), and the line flux evolution would reflect the change of the electron temperature during the decay phase (Houdebine *et al.*, 1989). Some constraints on the temperature and density structure can be obtained from line fluxes, especially in the optical range from Balmer lines, He and higher excitation lines. The role of photoionisation and photoexcitation of these lines by soft X-ray and XUV radiation should be estimated. The broadening and merging of higher Balmer lines

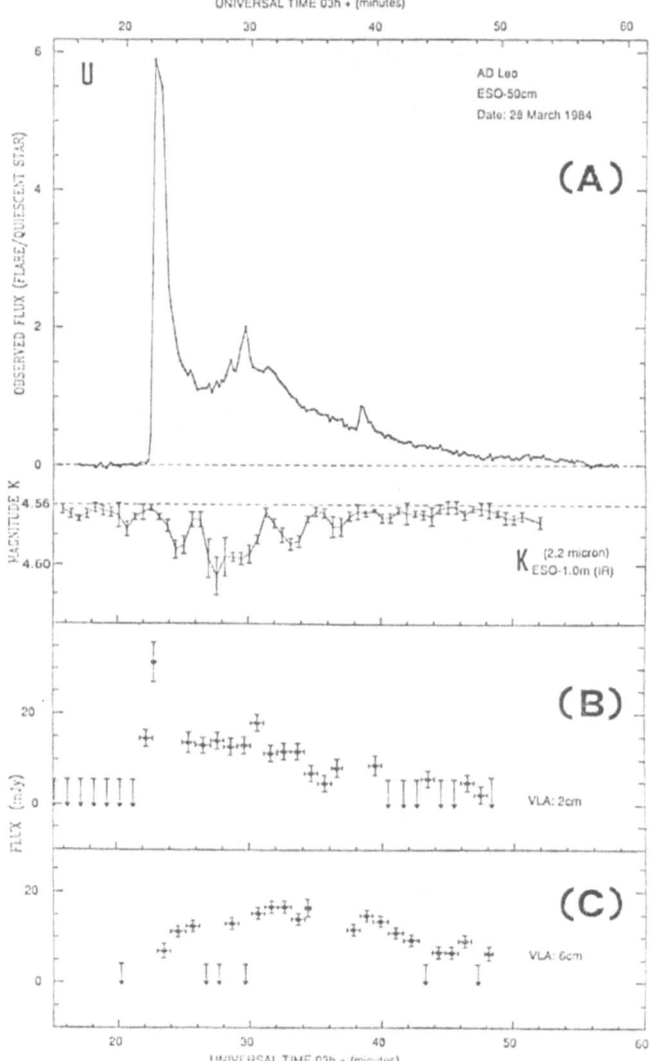

Fig. 3a.   Time evolution of broad-band flux measurements during the AD Leo flare: (A) showing distinct impulsive and gradual phases in the U band and the first detected 'infrared negative flare' in K band at 2.2 micron; (B) the impulsive behaviour in the VLA 2 cm radio flux; (C) the gradual increase and decay of the VLA 6 cm band.

dominated by the Stark effect can be used to estimate electron densities in the chromosphere (Donati-Falchi, Falciani, and Smaldone, 1985). Measurement of the broadening of the lower Balmer lines, which are less affected by the Stark effect, together with line shifts, provide information on the large-scale flows during flares.

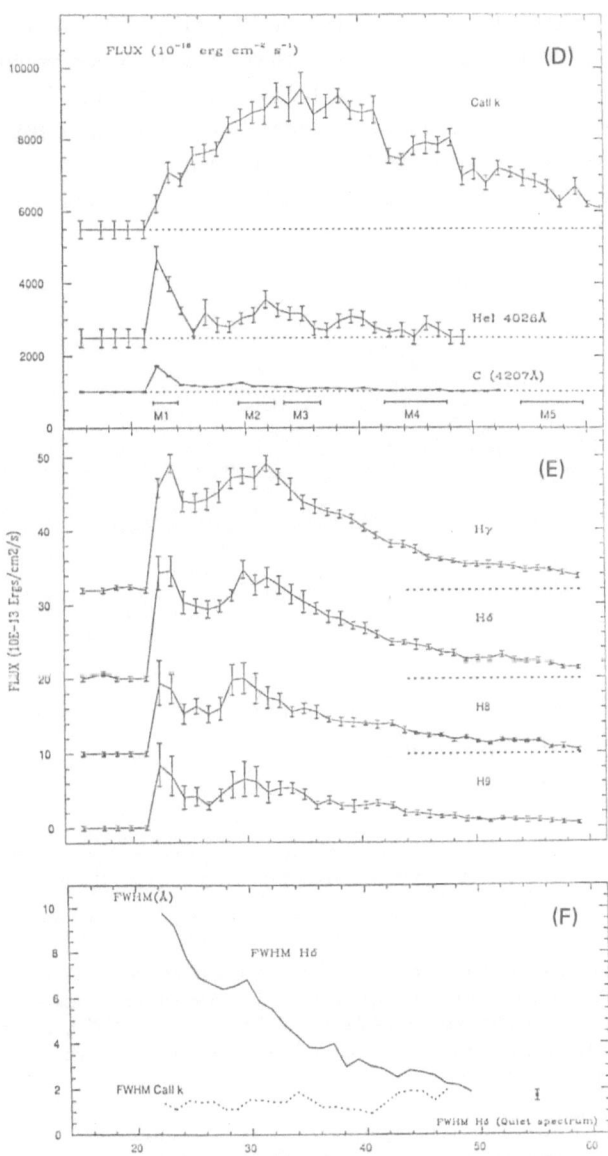

Fig. 3b.   Variation of line fluxes from ESO 3.6 m spectroscopy of the AD Leo flare: (D) Ca II K line flux evolution showing only a gradual phase; impulsive variation of He I 4026 Å line similar to the variation of the continuum band at 4207 Å; (E) evolution of Balmer line fluxes with different delays and time profiles; (F) variation of the Balmer Hδ line width during the flare from 10 Å to its preflare value of 2 Å FWHM, while no variation of Ca II width is detected (courtesy, Rodonò et al., 1989a).

2.3. Optical diagnostics at high spectral resolution

At high spectral resolution (20 000–100 000) and reasonable signal to noise (> 50) optical flare spectroscopy allows the study of Stark broadening in the Balmer lines, He I and Ca II lines; velocity fields associated with thermal, turbulent, or directed motion; and the time and velocity-dependent signature of ejected components (cf. Figures 4 and 5) in emission or in absorption. Also high spectral resolution is required to separate blends by photospheric lines or molecular bands which are important for cool stars, and allows us to measure the core filling and flare signature in photospheric lines. The observed chromospheric profiles such as those shown in Figures 4 and 5 can be com-

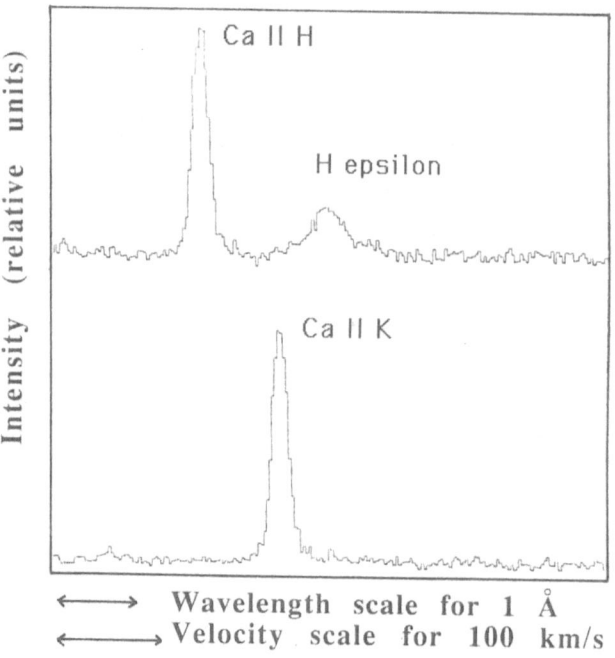

Fig. 4.   AD Leo composite spectrum (of 4 spectra taken out of noticeable flare equivalent to a 2.5 hour exposure) obtained at the CFH 3.6 m telescope with the Coudé spectrometer and reticon detector at 50 000 spectral resolution. Note the difference in the wings of Ca II H & K and Hε lines, and the corresponding broadening scale. The line profiles can be compared with semi-empirical chromospheric calculations.

pared with semi-empirical calculations (cf. Figure 6) using atmospheric models with different temperature stratifications. Those models consistent with the observations provide a vertical scale and information on the depth of formation of each spectral line, and also indicate the likely excitation and ionisation mechanisms. The presence of velocity gradients can be diagnosed from line shifts, asymmetries and detailed line profiles but this diagnostic is crude because of the interplay between the flare geometry in 3 dimensions, the dynamics, the NLTE radiative transfer effects and redistribution effects. However, it is hoped that a new generation of flare models which include the coupling between dynamics and radiation self-consistently will shed light on the principal mechanisms at work during the atmospheric response to flares.

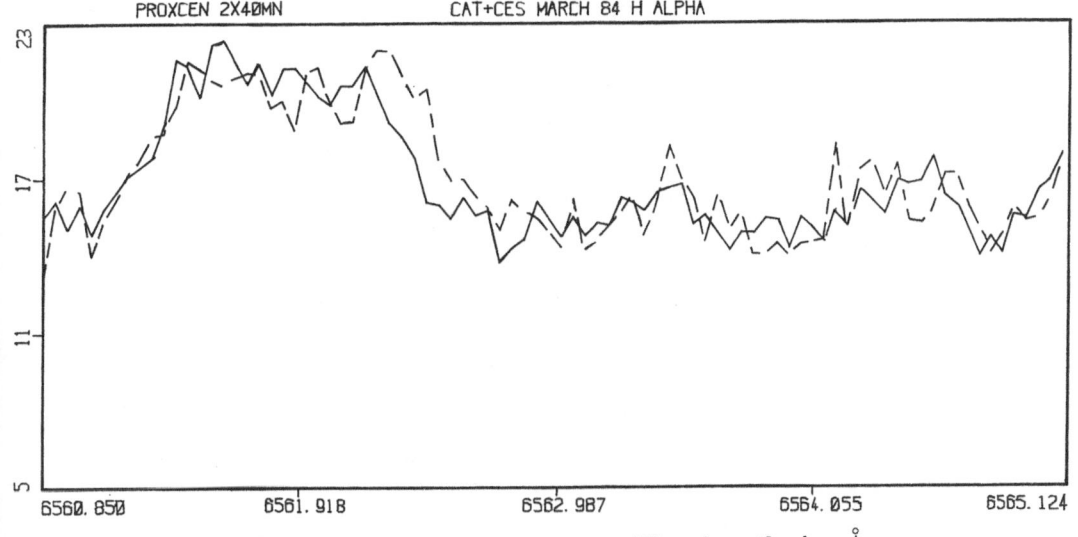

Fig. 5.   Comparison between two 40 min exposure spectra of Prox Cen in Hα observed with the CAT 1.4 m telescope + CES spectrometer and reticon at ESO. The superposition shows an asymmetric broadening of the line, possibly due to a Doppler-shifted component. Provided the S/N and time resolution are sufficient, line asymmetries and velocity components can be monitored as flare motion diagnostics, as suggested in these Hα studies.

2.4. ULTRAVIOLET SPECTRAL DIAGNOSTICS

There is an enormous literature on solar flare optical + UV + X-ray flare diagnostics (cf. Feldman, 1981; Dere and Mason, 1981; and Dupree, 1978, for these aspects).

For stars, the 1200–3200 Å ultraviolet range covered by the IUE spectrograph is rich in spectral emission lines of species such as C I, O I, Si II, Fe II formed at 4000–6000 K at the base of the chromosphere, the Lα line and the C II 1335 Å at top of the chromosphere, and lines of Si III, C III, Si IV, O IV, C IV, and N V formed in the transition region at 30 000–150 000 K. For collisionally excited resonance lines, the surface flux is related to the emission measure $EM = \int_{AT} N_e^2 \, dh$ over the temperature range ΔT of line emission. For intersystem lines which are collisionally excited but depopulated by line radiation and collisions (proportional to the electron density), the observed line ratios provide a measure of electron densities and thus the emitting volume at the temperature of line formation. The presence in the UV spectrum of lines formed over a wide range of temperature and excited by different processes permits us to infer the distribution of emission measure with temperature and to constrain assumptions about the geometry, temperature distribution, and electronic density stratification.

Optically thick chromospheric resonance lines and other transitions have been computed using various non-LTE codes in order to characterise the temperature and density distribution in active stars and flares. Rapid increases in UV line fluxes and continua

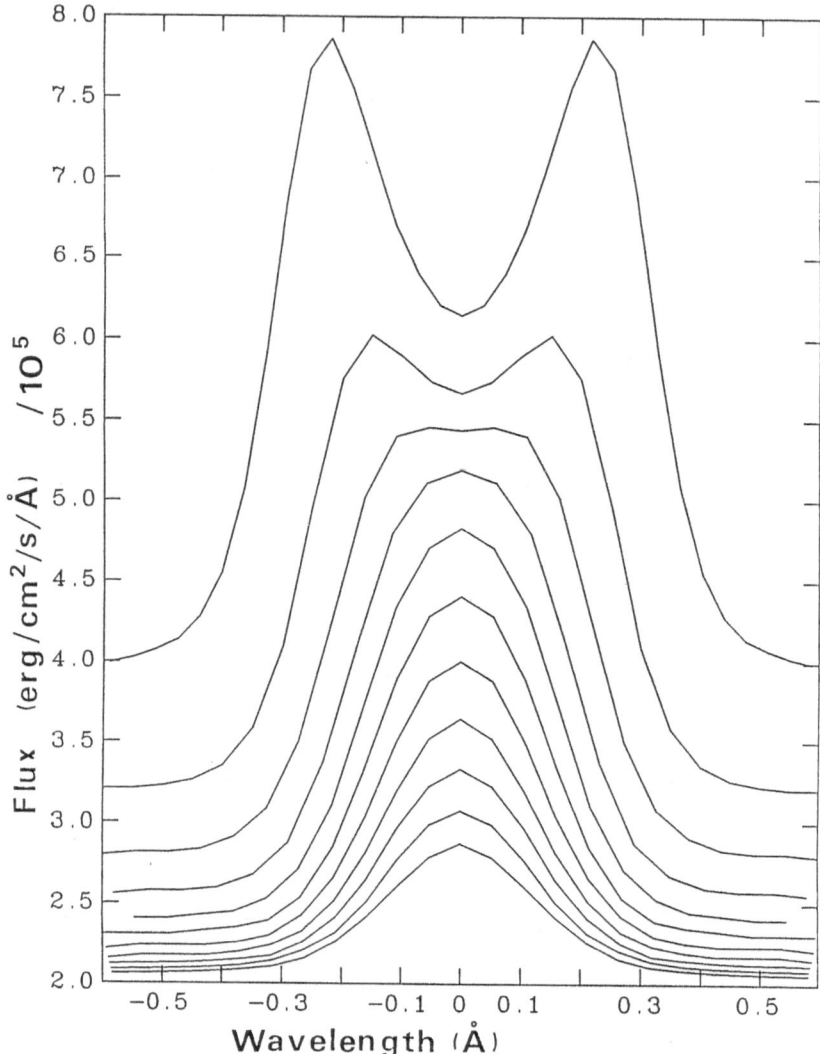

Fig. 6. Calculation of Balmer line profiles for a dMe atmospheric model using an adaptation of the M. Carlsson code with 16 levels and 12 transitions considered for the H atom. The diagram shows the lines from Hγ down to H14 (courtesy, Houdebine and Panagi, 1989).

have been detected during flares on dMe, RS CVn, and other flare stars. In particular, UV continua as observed on the Sun at 1600 and 2200 Å allow us to probe the temperature minimum region and middle photosphere, which play important roles in the flare-energy balance (cf. Foing, Bonnet, and Bruner, 1986; Foing *et al.*, 1986).

Because of its limited sensitivity and operational constraints, IUE cannot observe stellar flares with time resolution less than 10 minutes. The Goddard high-resolution spectrograph and the faint object spectrograph on space telescope will improve the sensitivity, time resolution and spectral resolution significantly over that of IUE but with a decreased spectral range and scheduling flexibility. Also the large oversubscription will make it difficult to monitor flare stars and to coordinate multi-wavelength observations. Other instruments such as the ST imaging spectrograph, or the Lyman mission will extend the spectral range. With its 900–1200 Å prime spectral region and 30 000 resolution, Lyman will observe the H-Ly series, the important O VI, S VI, C III lines at temperatures between 2×10^5 and 10^6 K, and include lines of the ions N I–III, P II–V, S III–VI, C I–IV. Useful density ratios are available for several of these ions. Also the coverage of the 100–912 Å range will extend the temperature range up to 2×10^7 K by observation of ionisation stages of Fe II–XXIV, and of the He I and He II Lyman series. Simultaneous coverage of the complete 1200–3200 Å range is also required in a future successor to IUE. USSR instruments such as SUVT-170 planned for a 1995 launch with spectrometers covering the 1100–1900 Å and 1900–3500 Å ranges with resolution modes of 0.1, 3, and 30 Å, or the EUVITA instrument on Spectrum X planned for 1993 should be useful for flare studies.

These future satellites with their anticipated UV coverage, spectral resolution, and temporal resolution of 10–100 s, should permit studies of thermal conduction, chromospheric evaporation, mass ejections, flare expansion, radiation, and dynamics on a number of flare stars.

2.5. X-RAY SPECTRAL DIAGNOSTICS

The use of X-ray photometry and spectroscopy was described previously by Schmitt (1989) in this conference as a way to study the 10^6–10^8 K plasmas in stellar flares. Calculations of the emergent spectrum show the dominant role of emission lines for $T < 2 \times 10^6$ K and of the bremsstrahlung continuum at higher temperatures. Low-resolution spectroscopy ($E/\Delta E = 10$–30), as was achieved with the Einstein solid state spectrometer, allows one to match an observed spectrum with a theoretical spectrum from a 2-temperature plasma folded with the instrumental response. This technique can also be applied to the analysis of low-resolution spectra to be obtained with the JET-X instrument onboard Spectrum-X, and later with higher throughput and time resolution by instruments on AXAF and by XMM/Focal plane imager. Moderate resolution spectroscopy ($E/\Delta E = 100$–300) such as with the Objective Grating Spectrometer (OGS) on Einstein and Transmission Grating Spectrometer (TGS) on Exosat allowed to resolve spectral lines and to infer EM(T) distribution for the coronae on Capella and σ^2 CrB (Mewe et al., 1982). Also density-sensitive ratios from He, Be, and C iso-electronic sequences allow us to infer densities and thus emitting volumes of the hot flaring plasma. With the enormous throughput of grating spectrometers on AXAF and XMM, moderate resolution with 100 s time resolution of bright flares on dMe and RS CVn system should be feasible (Linsky, 1987).

2.6. RADIO OBSERVATIONS

The usefulness of the radio spectral region for studying the hot thermal plasma and nonthermal electrons in coronal flare plasmas is described in Kuijpers (1989) and

Kundu, White, and Schmahl (1989). The spectral, temporal, and polarisation properties of different emission mechanisms, both coherent or incoherent are described in reviews by Kuijpers (1985), Dulk (1985), and Melrose (1987).

Gyrosynchroton emission from mildly relativistic electrons in magnetic loops was invoked by Linsky and Gary (1983) for dMe stars and also by Mutel *et al.* (1985) to explain VLBI observations of RS CVn systems. However, radio emission during flares with brightness temperature $> 10^{13}$ K (confirmed from spike rise times faster than 0.2 s in AD Leo, Lang *et al.*, 1985) and 100% circular polarisation is explained as a coherent process such as electron-cyclotron maser or plasma radiation. Recently, Bastian and Bookbinder (1987) obtained the first dynamic spectra of flares on UV Ceti analogous to radio bursts observed on the Sun, thus inaugurating a new tool for radio flare studies.

3. Multi-Wavelength Diagnostics

For solving the scientific questions raised earlier, diagnostics are available in different wavelength ranges and at various spectral resolutions. These tools help to describe some of the existing processes at work during flares and to quantify some physical properties of flare plasmas. In order to give a more complete description at different heights and temperature regimes in the flaring atmosphere, these diagnosis should be used simultaneously (Foing *et al.*, 1988; Linsky, 1988).

As an example, we show the results of a multi-wavelength flare campaign (cf. Rodonò *et al.*, 1984, 1989a, b) on AD Leo, including ultraviolet spectroscopy with IUE, *U*-band photometry, optical spectroscopy in the range 3600–4400 Å, infrared photometry from ESO and radio observations with the VLA. The *U*-band photometry (Figure 3(a)) shows a rapid increase followed by several spikes. Simultaneous with the *U* spikes and with a similar temporal light curve, the 2 cm radio emission showed an increase during the flare. The 6 cm emission, on the other hand, gradually reached a peak flux some 15 min later than the impulsive phase. A very exciting result came from the infrared photometry with the 2.2 μm *K*-band showing a decrease or 'infrared negative flare' simultaneous with the optical flare but with characteristics different from those predicted by the inverse Compton effect or H^- opacity explanations.

The Ca II K line fluxes (cf. Figures 2 and 3(b)) show a much more slowly increasing excitation during the gradual phase of the flare than the Balmer lines. The He I 4026 Å line exhibits an impulsive behaviour in the flare spectra. The flux in the Balmer lines began to increase coincident with the continuum flare, but with a delay in the peak emission and a double-peaked light curve. Most of the Balmer emission occurs in the cooling plasma during the flare decay phase. Different time behaviour of the Balmer line fluxes can occur (Houdebine *et al.*, 1989) as the result of different flare temperatures as a function of time which in turn determine the 'emission efficiency' which is different for each line. A difference in the Balmer line broadening was also observed (e.g., a width of 10 Å in the Hδ Balmer line during the flare but only 2 Å in the quiet spectrum), whereas the Ca II K width appears to remain constant within our measurement uncertainties.

The temporal correlation of soft X-rays and Balmer Hγ line emission reported by Butler *et al.* (1986) from coordinated Exosat observations and ESO optical spectroscopy is an example of coordinated observations that provide insights into physical processes and the energy balance during flares. From observations of flares in soft X-rays and Balmer lines on different stars (such as UV Ceti, YZ CMi, Gl 644, and the Sun), Butler, Rodonò, and Foing (1988) propose that there is an almost equal amount of energy in the time-averaged soft X-ray flare emission and in the flux from the Balmer and Lyman lines. This puzzling result, which appears to be valid over four decades of flare energy, should be tested systematically by observing Lyman and Balmer H lines together with lines at higher excitation and simultaneously with the soft X-ray and EUV Lyman continuum below the 912 Å limit, that may control the ionization and excitation of H and other important transitions.

Coordinated multi-wavelength campaigns (Foing *et al.*, 1988) involving future X-ray and UV satellites and ground-based instrumentation with good simultaneity and continuous coverage are needed to extend our understanding of stellar flares. Such campaigns will require special efforts for organisation, observations, calibration, data analysis, and theoretical interpretation.

4. Conclusions: Requirements for Future Stellar Flare Observations

For the diagnostics of flares, we have seen that multi-band photometry over the whole range (radio, infrared, optical, ultraviolet, X-ray) with time resolution of 0.1–1 s is necessary for the timing of the flare impulsive phase. Medium-resolution spectroscopy in the optical and ultraviolet with a time resolution of 10–30 s is required to study line fluxes and continua diagnostics, for emission measure analysis, and to evaluate density-sensitive line ratios. High-resolution spectroscopy of line profiles provides insight into the plasma dynamics, and the radiative and excitation mechanisms.

All of these tools covering the full electromagnetic range must be used simultaneously with enough time resolution to study the flare plasma over its complete temperature range and including its nonthermal high-energy component. The development of new instruments and satellite missions must take into account this multi-wavelength coverage requirement. The need for simultaneous and continuous coverage requires the organisation of coordinated multi-wavelength observing campaigns (Foing *et al.*, 1988), and the use of networks of photometers and spectrometers around the globe (Catala and Foing, 1988). An important ingredient is the knowledge of the magnetic field distribution and large-scale motions for studying build-up conditions prior to stellar flares. Finally, a strong interaction between solar and stellar flare physics, together with atomic, plasma, and radiative transfer physics should extend the success of this IAU Colloquium on *Solar and Stellar Flares*. Those communities should jointly develop diagnostic methods, interpretative tools and theories for understanding the new observational results made possible with the next generation of space and ground-based instruments.

Acknowledgements

For the collaborative programmes described in this article, I acknowledge support for observation facilities and for the exchange of scientists from ESA, NASA and SERC, ESO, CFH, CNRS, and EEC funding for European Network of Research Laboratories on Stellar Activity.

I wish to thank my collaborators involved in the different scientific aspects that I reviewed here, especially M. Rodonò, C. J. Butler, E. Houdebine for providing some material and figures discussed here, and Drs B. M. Haisch, J. E. Neff, J. L. Linsky, A. Brown, P. B. Byrne, J. G. Byrne, J. G. Doyle, P. Panagi, S. Catalano, for enlightening discussions. I thank J. L. Linsky for a detailed revision of the manuscript.

References

Aboudarham, J. and Hénoux, J. C.: 1986, *Astron. Astrophys.* **156**, 73.
Antonucci, E., Gabriel, A. H., and Dennis, B. R.: 1984, *Astrophys. J.* **287**, 917.
Antonucci, E. *et al.*: 1982, *Solar Phys.* **78**, 107.
Avrett, E. M., Machado, M. E., and Kurucz, R. L.: 1986, in D. Neidig (ed.), *The Lower Atmosphere of Solar Flares*, National Solar Observatory/SPO.
Bastian, T. S. and Bookbinder, J.: 1987, *Nature* **326**, 678.
Brown, J. C.: 1971, *Solar Phys.* **18**, 489.
Brueckner, G. E. and Bartoe, J. D.: 1983, *Astrophys. J.* **272**, 329.
Brueckner, G. E. *et al.*: 1986, *Adv. Space Res.* **6**, No. 8, 263.
Butler, C. J., Rodonò, M., and Foing, B. H.: 1988, *Astron. Astrophys.* **206**, L1.
Butler, C. J., Rodonò, M., Foing, B. H., and Haisch, B. M.: 1986, *Nature* **321**, 679.
Butler, C. J., Doyle, J. G., Foing, B. H., and Rodonò, M.: 1988, in O. Havnes and B. R. Pettersen (eds.), *Tromso Midnight Sun Conference on Activity in Cool Star Envelopes*.
Canfield, R. D. and Gunkler, T. A.: 1985, *Astrophys. J.* **288**, 353.
Catala, C. and Foing, B. H. (eds.): 1988, *1st MUSICOS Workshop on MUlti SIte COntinuous Spectroscopy*, Observatoire de Paris-Meudon.
Char, S. and Foing, B. H.: 1989, in *Modeling the Stellar Environment: How and Why?*, Editions Frontières, Gif/Yvette.
Collier-Cameron, A. and Robinson, R. D.: 1988, in A. K. Dupree and M. T. Lago (eds.), *Formation and Evolution of Low Mass Stars*, D. Reidel Publ. Co., Dordrecht, Holland.
Collier-Cameron, A. and Robinson, R. D.: 1989, *Monthly Notices Roy. Astron. Soc.* **236**, 57.
Damé, L. and Cram, L.: 1983, *Solar Phys.* **87**, 329.
De Jager, C. *et al.*: 1986, *Astron. Astrophys.* **156**, 95.
De Jager, C. *et al.*: 1989, *Astron. Astrophys.* **211**, 157.
Dere, K. P. and Mason, E.: 1981, in F. Q. Orrall (ed.), *Solar Active Regions*.
Donati, J. F., Semel, M., and Praderie, F.: 1988, in C. Catala and B. H. Foing (eds.), *1st MUSICOS Workshop on MUlti SIte COntinuous Spectroscopy*, Observatoire de Paris-Meudon, p. 37.
Donati-Falchi, A., Falciani, R., and Smaldone, L. A.: 1985, *Astron. Astrophys.* **152**, 165.
Doyle, J. G. and Butler, C. J.: 1985, *Nature* **313**, 378.
Doyle, J. G. *et al.*: 1989, *Astron. Astrophys.* (in press).
Dulk, G. A.: 1985, *Ann. Rev. Astron. Astrophys.* **23**, 169.
Dupree, A. K.: 1978, *Adv. Atomic Molecular Phys.* **14**, 393.
Emslie, G. and Nagai, F.: 1985, *Astrophys. J.* **288**, 779.
Feldman, U.: 1981, *Phys. Scripta* **24**, 681.
Foing, B. H.: 1989, *Irish Astron. J.* (in press).
Foing, B. H., Bonnet, R. M., and Bruner, M.: 1986, *Astron. Astrophys.* **162**, 292.
Foing, B. H. *et al.*: 1986, in D. Neidig (ed.), *The Lower Atmosphere of Solar Flares*, NSO/SPO, p. 319.
Foing, B. H., Butler, C. J., Haisch, B. M., Linsky, J. L., and Rodonò, M.: 1988, in C. Jaschek and C. Sterken (eds.), *Coordination of Observational Projects*, Cambridge University Press, Cambridge, p. 197.

Gershberg, M. R. E. and Shakhovskaya, N. I.: 1983, *Astrophys. Space Sci.* **95**, 235.

Haisch, B.: 1983, in P. B. Byrne and M. Rodonò (eds.), *Activity in Red Dwarf Stars*, D. Reidel Publ. Co., Dordrecht, Holland, p. 255.

Haisch, B. M. *et al.*: 1983, *Astrophys. J.* **267**, 280.

Haisch, B. M., Strong, K. T., Harrison, R. A., and Gary, G. A.: 1989, *Astrophys. J. Suppl.* **68**, 371.

Hénoux, J. C.: 1989, unpublished poster at *IAU Colloq. 104 on Solar and Stellar Flares*.

Hénoux, J. C. and Aboudarham, J.: 1988, in C. Catala and B. H. Foing (eds.), *1st MUSICOS Workshop MUlti SIte COntinuous Spectroscopy*, Observatoire de Paris-Meudon, p. 89.

Heritschi, D., Raadu, M. A., Vial, J. C., and Malherbe, J. M.: 1989, in B. M. Haisch and M. Rodonò (eds.), *IAU Colloq.* **104**, 'Solar and Stellar Flares', Poster Papers, Publ. Catania Astrophys. Obs., Special Volume, p. 321.

Houdebine, E. R. and Panagi, P.: 1989, *Astron. Astrophys.* (submitted).

Houdebine, E. R., Butler, C. J., Rodonò, M., Panagi, P., and Foing, B. H.: 1989, in B. M. Haisch and M. Rodonò (eds.), *IAU Colloq.* **104**, 'Solar and Stellar Flares', Poster Papers, Publ. Catania Astrophys. Obs., Special Volume, p. 59.

Jankov, S. and Foing, B. H.: 1987, in J. L. Linsky and R. E. Stencel (eds.), *Cool Stars, Stellar Systems, and the Sun*, Springer-Verlag, Berlin.

Kuijpers, J.: 1985, in R. M. Hjellming and D. M. Gibson (eds.), *Radio Stars*, D. Reidel Publ. Co., Dordrecht, p. 185.

Kuijpers, J.: 1989, in *IAU Colloq. 104 on Solar and Stellar Flares* (this volume).

Kundu, M. R., White, S. M., and Schmahl, E. J.: 1989, in *IAU Colloq. 104 on Solar and Stellar Flares* (this volume).

Lang, K. *et al.*: 1985, in M. Zeilik and D. M. Gibson (eds.), *Cool Stars, Stellar Systems and the Sun*, Springer-Verlag, Berlin.

Lin, R. P., Schwartz, R. A., Kane, S. R., Pelling, R. M., and Hurley, K. C.: 1984, *Astrophys. J.* **283**, 421.

Linsky, J. L.: 1987, *Astrophys. Letters and Comm.* **26**, 21.

Linsky, J. L.: 1988, in F. Cordova (ed.), *Multiwavelength Astrophysics*, Cambridge University Press, Cambridge.

Linsky, J. L. and Gary, D. E.: 1983, *Astrophys. J.* **274**, 776.

Linsky, J. L. *et al.*: 1989, *Astron. Astrophys.* **211**, 173.

Neff, J. E., Brown, A., and Linsky, J. L.: 1989, in B. M. Haisch and M. Rodonò (eds.), *IAU Colloq.* **104**, 'Solar and Stellar Flares', Poster Papers, Publ. Catania Astrophys. Obs., Special Volume, p. 111.

Parker, E. N.: 1988, *Astrophys. J.* **330**, 474.

Pettersen, B. R., Coleman, L. A., and Evans, D. S.: 1984, *Astrophys. J. Suppl.* **54**, 375.

Rodonò, M., Pucillo, M., Sedmak, G., and de Biase, G. A.: 1979, *Astron. Astrophys.* **76**, 242.

Rodonò, M. *et al.*: 1984, in *Proc. 4th IUE Conference*, ESA SP-218, p. 247.

Rodonò, M., Foing, B. H., Linsky, J. L., Butler, J. C., Haisch, B. M., Gary, D. E., and Gibson, D. M.: 1985, *ESO Messenger* **39**, 9.

Rodonò, M. *et al.*: 1986, *Astron. Astrophys.* **165**, 135.

Rodonò, M. *et al.*: 1989a, in B. M. Haisch and M. Rodonò (eds.), *IAU Colloq.* **104**, 'Solar and Stellar Flares', Poster Papers, Publ. Catania Astrophys. Obs., Special Volume, p. 53.

Rodonò, M. *et al.*: 1989b (in preparation).

Rust, D.: 1986, in D. Neidig (ed.), *The Lower Atmosphere of Solar Flares*, NSO/SPO.

Saar, S. H., Linsky, J. L., and Beckers, J. M.: 1986, *Astrophys. J.* **302**, 777.

Schmitt, J. H., Lemen, J. R., and Zarro, D.: 1989, in *IAU Colloq. 104 on Solar and Stellar Flares* (this volume).

Simon, T., Linsky, J. L., and Schiffer, F. H.: 1980, *Astrophys. J.* **239**, 911.

Skumanich, A.: 1985, *Australian J. Phys.* **38**, No. 6.

Skumanich, A.: 1986, *Astrophys. J.* **309**, 858.

van Ballegooijen, A. A.: 1985, *Astrophys. J.* **298**, 421.

Vogt, S. S. and Penrod, G. D.: 1983, *Publ. Astron. Soc. Pacific* **95**, 565.

Vogt, S. S., Penrod, G. D., and Hatzes, A. P.: 1987, *Astrophys. J.* **321**, 496.

Withbroe, G. L., Habbal, S. R., and Ronan, R.: 1985, *Solar Phys.* **95**, 297.

Zarro, D. M. and Canfield, R. C.: 1989, in B. M. Haisch and M. Rodonò (eds.), *IAU Colloq.* **104**, 'Solar and Stellar Flares', Poster Papers, Publ. Catania Astrophys. Obs., Special Volume, p. 53.

PREFLARE ACTIVITY

V. GAIZAUSKAS

Herzberg Institute of Astrophysics, National Research Council of Canada, Ottawa, Canada K1A OR6

Abstract. Magnetic reconnection at current sheets or in current-bearing arches in the solar atmosphere is generally accepted as the mechanism responsible for the sudden energy release in solar flares. Attempts have so far been unsuccessful to isolate from the observations some unique preconditions which would be necessary and sufficient to ensure rapid conversion of energy by this process. Here we survey recent multi-wavelength observations which illustrate the variety of preflare activity. Multiple structures are now believed to participate in the energy release. Dynamic global coupling of the magnetic fields between a flaring site and the rest of an activity complex is seen from the data to be an important aspect of preflare activity.

1. Introduction

It is generally accepted that flares draw their power from the free energy stored in stressed magnetic fields (Švestka, 1976). The rapid transformation of magnetic energy into plasma heating and particle acceleration is believed to occur through magnetic reconnection when the stresses exceed a critical threshold. A possible alternative is the formation of double layers (see Kuijpers, these proceedings). Reconnection mechanisms may involve magnetic tearing of single coronal loops (Spicer, 1976; Van Hoven, 1976) or the merging of current sheets (see Priest, 1985, for an extended review). Other plausible mechanisms incorporated into flare models have been categorized according to their drivers by Spicer and Brown (1981). By studying the preflare state we hope to constrain the choice of driving mechanisms, always bearing in mind that different regimes of plasma physics may apply as the flare evolves from the preflare to the impulsive phase (Van Hoven and Hurford, 1986). The most important questions to be settled from preflare observations concern the geometry of the magnetic field, the stresses applied to them from photospheric to coronal heights, and the changing physical properties of the plasma trapped in those fields.

Because the transition is not perfectly abrupt, there is always some arbitrariness about the state, preflare or flaring, to which a particular phenomenon belongs. The difficulty is aggravated by the fact that phenomena assigned to the preflare phase do occur even in the absence of any flare. The problem is illustrated by the idealized flux curve in Figure 1 which is adapted from actual cm-wavelength observations of a solar flare. Without spatial resolution, e.g., observing a stellar flare, there is no certainty that the small transient labelled 'precursor' has anything to do with the flare itself. This type of discrete transient will be discussed in Section 2 along with phenomena, sometimes offered as evidence for preheating, which might produce a small plateau as the emission begins to rise (Figure 1).

In Figure 1 the weak sporadic emission labelled 'preflash', part of the onset of the flare, signals the initial release of high-energy particles. It is manifested as decimetric radio pulsations, metric type III bursts, and hard as well as soft X-rays (Benz *et al.*,

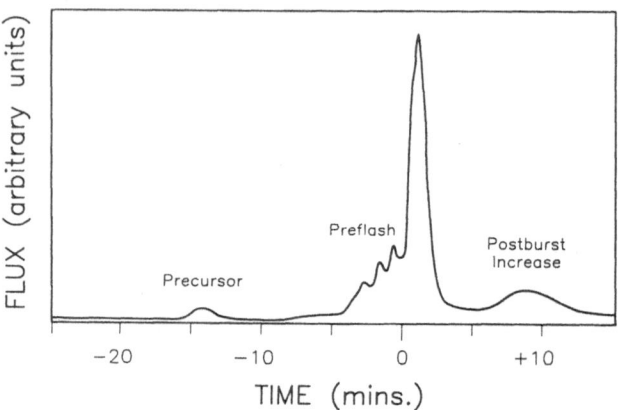

Fig. 1.   Idealized flux profile of a flare detected in an energy band which senses a distinct impulsive phase
at $t = 0$.

1983; Raoult *et al.*, 1985; Machado, Orwig, and Antonucci, 1986; Kundu *et al.*, 1987). These spatially-resolved observations suggest that a morphological transition, involving different plasmas, occurs between the onset and impulsive phases of a flare. A comprehensive study of X-ray emission in flares by Machado *et al.* (1988) is explicit on this point. The basic structure of a flare consists of an initiating bipole impacted against one or more adjacent bipoles. During flare onset the X-ray emission is concentrated in the initiating bipole, but during the impulsive phase most of the energy is released inside the initiating bipole and/or inside one of the adjacent bipoles rather than at the point of impact between them. The timing and nature of these phenomena implies a continuous process of acceleration that starts before the flash phase and continues beyond the impulsive phase of the flare (Vlahos *et al.*, 1986).

Even though they precede the impulsive phase by as much as a few minutes, the intimate relation of preflash phenomena to the main energy release of flare energy sets them outside the scope of this paper. Emphasis will be given instead in Section 3 to evolutionary processes on the scale of hours and even days which are linked to the storage of excess energy at the site of a flare. Because more detailed evidence has been recently acquired on the most distinctive type of preflare event, the erupting filament, all of Section 4 is devoted to a discussion of the changes preceding three eruptions.

Although this review deals exclusively with examples of solar flares, global interactions are stressed wherever possible as having possible application to other stars.

2. Precursors and Preheating

A precursor is here defined as a transient event preceding the impulsive phase, possibly even before the onset and not necessarily at the site of the flare itself (Priest *et al.*, 1986, Section 1.4). Under this broad definition, earlier flares can qualify as precursors because their frequency of occurrence may sometimes reach a crescendo just before a major flare (Dodson and Hedeman, 1976). We include two special cases: earlier flares in the same location and with similar emission patterns (homologous flares); earlier flares in different locations but erupting in near synchronism (sympathetic flares).

2.1. HOMOLOGOUS FLARES

Homology is commonly found during periods of frequent flare activity (Woodgate, 1982; Zirin, 1983). We can infer that the same flare-producing stresses act on magnetic field configurations for hours, sometimes days. The rate of repetition of flares belonging to the same homologous series can be as rapid as a few per hour or as infrequent as one in several days (Martres *et al.*, 1984). But no consistent relationship has been found during SMY between the repetition rate of homologous flares and the brightness of their X-ray, optical, or microwave emissions (Gaizauskas, 1982; Woodgate, 1982). The stresses within an active region need not, therefore, be uniformly applied with time. Indeed, Machado (1985) finds that homology is clearly correlated with the development of magnetic shear in an active region.

The comparison of several series of recurrent events should isolate essential preflare factors, but that hope is only partially realized. Woodgate *et al.* (1984) find that flow patterns and magnetic shear qualify for an essential role in some but not every homologous set which they investigated. Strict homology, in terms of identical size, shape and location of flare kernels, is never realized because of the incessant evolution of magnetic fields on a small scale (Gaizauskas, 1982; Zirin, 1983; Machado, 1985).

2.2. SYMPATHETIC FLARES

Flare-related linkages between adjacent and even remote active regions ought to be a natural consequence of the interwoven loop structure of the corona as validated by Skylab. A single excitation could in principle be transmitted along a multiplicity of linkages and produce flare-like responses at several remote sites. Different statistical tests of large samples of flares have however led to opposite conclusions: no significant increase in the excess beyond the random rate of flares coinciding to within 20 min for pairs of active regions more than 30° apart (Fritzová-Švestková, Chase, and Švestka, 1976); a significant correlation between the maximum and onset phases of flares in solar regions spaced from 19° up to 104° apart (Ogir, 1981).

Case studies of flares using modern mapping methods in microwaves and in X-rays now show energy being transferred between sites in adjacent active regions. Kundu, Rust, and Bobrowsky (1983) estimate a velocity ≥ 6000 km s^{-1} for 6-cm emission advancing along a loop $\sim 10^5$ km long joining faint flare-like Hα brightenings. The simultaneous microwave and X-ray observations of Kundu *et al.* (1984) find an X-ray brightening advancing along a loop 1.3×10^5 km long joining two regions in the same complex of activity. The speed of the X-ray front (~ 100 km s^{-1}) and its orientation are consistent with a disturbance propagating between the two subflares which occurred with a delay of ~ 25 min near opposite footpoints of the interconnecting loop. Double microwave events at 1.8 cm wavelength, selected for the similarity of the time profiles of both components, have been studied by Nakajima *et al.* (1985). Their spatially-resolved data show that a secondary microwave burst can occur at a distance of

10^5–10^6 km from the primary flare site. For two of these events a new X-ray source was observed adjacent to the secondary source and was associated with faint Hα brightenings. The velocity deduced from the measurements ($\geqslant 10^4$ km s^{-1}) implies that the exciting agent must be high-energy electrons produced at the primary flare site. This is unequivocal evidence that one flare can trigger another, albeit a weak one, at a remote site.

2.3. SOFT X-RAY PRECURSORS

Transient enhancements in soft X-rays commonly appear in data from Skylab (Kahler, 1979; Webb, 1985) and from SMM (Section 1.4 of Priest et al., 1986) as loops or kernels close to, but not necessarily at, flare sites for many minutes preceding the impulsive phase. A soft X-ray precursor may coincide in place and shape with features belonging to the impulsive phase as in the case of a limb flare observed by de Jager et al. (1983). Spatial coincidence between x-ray precursor and flare is usually partial, a situation which is consistent with flare energy being released within two or more interacting magnetic loops (Machado et al., 1988).

Flare-related coronal mass ejections (CMEs) support the view that the interaction between magnetic loops is a basic ingredient of the flare process. Weak soft X-ray bursts have been observed at a time coincident with the projected onset of a CME associated with a flare; precursor bursts occur some tens of minutes prior to the impulsive phase and are located in one foot of a pre-existing large coronal arch (Harrison et al., 1985; Simnett and Harrison, 1985; Harrison, 1986). The rest of the arch brightens weakly within minutes. Outward motion of plasma emitting X-rays during the precursor burst indicates that a CME has been launched (Harrison et al., 1985; Harrison, 1986). Because the associated flares sit to one side of the centreline of the ejecta, Harrison proposes that (i) a small magnetic feature interacts with one foot of a large pre-existing arch; (ii) a CME is launched without an explosive release of energy; (iii) a flare is subsequently triggered at the site of the precursor by prevailing post-launch conditions. It is worth noting that neither radio emission nor hard X-rays ≥ 15 keV have been associated with these soft X-ray precursors (Simnett and Harrison, 1985).

2.4. MICROWAVE PRECURSORS

Radio precursors were discovered long ago to consist of changes in intensity and/or polarization of the microwaves emitted from an active region some tens of minutes before the onset of a flare (see Kundu, 1965, and references therein). Recent samplings of many flares indicate, however, that radio precursors do not occur in a majority of cases. For bursts at 17 GHz with a peak flux density > 10 s.f.u., the probability of weaker preceding activity is only 25% (Kai, Nakajima, and Kosugi, 1983). The probability increases to $\sim 50\%$ if much stronger bursts (> 500 s.f.u. at 17 GHz) are selected. The average delay between precursor and main bursts is 25 min. In samples of major flares observed interferometrically at 10.7 GHz, Hurford and Zirin (1982) find distinctive changes to occur in only 11% of their cases a few minutes to tens of minutes before the beginning of the impulsive phase. The most common signature is a step-like increase in amplitude accompanied by a decrease or a reversal in the degree of polarization. Out of eight bursts observed with supersynthesis arrays at 2, 6, and 20 cm wavelength by Willson (1983) only one exhibits detectable preburst heating. Taken together, these results with different instruments suggest that precursors are not a general feature of the burst process at centimeter wavelengths. But this conclusion should be treated with caution because, as Hurford, Read, and Zirin (1984) show, pre-impulsive bursts may occur in such narrow spectral bands that they can easily be missed by instruments responding to a single frequency.

Those radio precursors which are observed at high spatial resolution provide important insights to the initiation of a flare. Morphological transitions of the emitting structures, continuing for minutes to tens of minutes through the preflare and impulsive phases, are a sign of interacting, pre-existing loops (Kundu *et al.*, 1982; Willson and Lang, 1984; Kundu and Shevgaonkar, 1985; Lang and Willson, 1986). Dramatic changes in polarization are observed at centimeter wavelengths to begin from minutes up to an hour before a flare and to continue through the impulsive phase (Lang, 1979; Kundu *et al.*, 1982; Kundu, Schmahl, and Velusamy, 1982; Willson and Lang, 1984; Kundu, 1986). The preflare reorganization of magnetic field implied by these results can be interpreted in different ways: as propagation effects in a highly structured and highly dynamic magnetoionic plasma (Willson, 1983; Van Hoven and Hurford, 1986); as the appearance of new magnetic structures low in the corona; or in some cases simply as sources switching on and off at different times and at different locations within a magnetically complex region. It is tempting to explain the polarization changes in terms of new magnetic flux emerging from beneath the solar surface (Kundu, 1986), but concomitant observations of the lower atmosphere on a fine scale are rare. Indeed in the one case where observations of changing polarization at centimeter wavelengths are reinforced by optical observations at high spatial and temporal resolution, none of the usual signatures of emerging flux are detected (see below, Section 4.3).

2.5. ULTRAVIOLET PRECURSORS

Transient brightenings of UV lines over a broad range of amplitudes are common on a small spatial scale above active regions (Van Hoven *et al.*, 1980; Cheng *et al.*, 1981; Cheng, Tandberg-Hanssen, and Orwig, 1984; Porter, Toomre, and Gebbie, 1984; Cheng and Tandberg-Hanssen, 1986; Priest *et al.*, 1986, Section 1.4.4). Some of the UV kernels with intense preflare activity become flares. But others do not join in a later flare which may be concentrated nearby in entirely new kernels (Cheng *et al.*, 1981; Cheng, Tandberg-Hanssen, and Orwig, 1984). A change in the frequency of small-scale events before and after a flare has been sought with inconclusive results owing to inadequate statistics (Porter, Toomre, and Gebbie, 1984).

In those few cases of good spatial and temporal coverage obtained simultaneously at multiple wavelengths, transient UV preflare activity occurs in very inhomogeneous surroundings. The limb flare of 30 April, 1980 erupted at the junction of a small loop rising into an overlying structure (Woodgate *et al.*, 1981; de Jager *et al.*, 1983). Strong UV and X-ray brightenings at the footpoint of the rising loop precede that flare.

Repeated transient UV brightenings are associated with the upwelling, twisting, and disruption of an active-region filament in the 20 min leading up to a 2-ribbon flare (Kundu et al., 1985). Differences between preflare UV structures from one flare to the next impede our ability to designate a reliable precursor against the varying UV background.

2.6. PREFLARE HEATING

Current sheet models (e.g., Heyvaerts, Priest, and Rust, 1977) and the unstable arch model (Spicer, 1976) of flares predict a pre-heating phase. Analysis of spatially-unresolved soft X-ray emission from many flares observed by an early satellite indicates a tendency for X-ray bursts to begin, on the average, about two minutes earlier than their associated Hα flares (Thomas and Teske, 1971). Recent case studies of spatially-resolved events do show pre-heated structures many minutes before they flare at microwave frequencies (Kundu et al., 1982; Lang and Willson, 1984; Willson, 1984; Kundu and Shevgaonkar, 1985) and in the EUV (Cheng et al., 1982, 1983).

Surveys of many Skylab soft X-ray images do not, however, indicate a requirement for coronal preflare heating lasting longer than 2 min in small flares consisting of just one or two loops (Kahler, 1979); they do reveal cases where structures displaced from the flare site brighten tens of minutes before onset (Webb, 1985). In a complementary study of a randomly selected set of moderate-sized flares observed with ~2.5 arc min resolution by the Mapping X-ray Heliometer on OSO-8, Wolfson (1982) finds intensity variations in soft X-rays before a flare to be no different from the variations in the same active region when no flare occurs. Furthermore, slowly varying emission in harder X-rays (3.5–8 keV) occurs even in the absence of flares and is located predominantly over polarity inversion lines where volumes of hot plasma have temperatures exceeding 10^7 K (Schadee, de Jager, and Švestka, 1983). These facts support Kahler's (1979) position that high temperatures and /or densities in the preflare region are not generally pre-requisites for a flare.

2.7. SURGING ARCHES

In a sample of 58 flares, Mouradian, Martres, and Soru-Escaut (1983) find that just over half are preceded by a surging arch – a transient absorbing feature visible in solar regions at wavelengths displaced from the central core of Hα. Red- and blue-shifted components are visible simultaneously from the first appearance of the structure; they are not cospatial. Initially linear in shape, the structure arches expands rapidly and becomes a complex assortment of multiple strands about the time the associated flare erupts in the same active region. The average delay between the beginning of the surging arch and the flare is 11 min in this sample. An early example of this phenomenon was noted by Athay and Moreton (1961) in conjuction with a blast wave produced by a flare (the same event is illustrated in more detail by Moreton, 1961). The relationship of the surging arch, presumably a rising magnetic loop, to the short-term evolution of the local magnetic field and, hence, to the flare itself is still obscure.

The rich variety of precursors surveyed above underscores the complexity of the preflare state. A typical signature of that state cannot yet be isolated from existing observations of precursors.

3. Evolution of Magnetic Fields

Flare activity is intimately connected with evolving magnetic fields (see reviews and references therein by Martin (1980) and by Priest *et al.* (1986)). Changing fields can create a current sheet (e.g., Heyvaerts, Priest, and Rust, 1977) and current networks (e.g., Hénoux and Somov, 1987). The destabilization of these current-forming processes can be initiated by a sudden injection of a new magnetic bipole. But it can also result from steady evolutionary trends which often get obscured by numerous short-lived changes in magnetic structures. We examine below the emergence of magnetic flux and the creation of magnetic shear, two factors commonly associated with flares, within the context of magnetic evolution on the Sun.

3.1. EMERGING FLUX AND MAGNETIC COMPLEXITY

Numerous flares erupt early in the development of an Emerging Flux Region (EFR), but they are minor in the case of a new region growing in isolation (Bruzek, 1967). The large flares with the interesting physics erupt where old and new magnetic flux interact. Interactions are frequent because active regions do not form at random on the Sun. They tend to appear in clusters called 'complexes' or 'nests' of activity which survive for at least several rotations (Gaizauskas *et al.*, 1983; Castenmiller, Zwaan, and van der Zalm, 1986). During its active lifetime, an activity complex will be refreshed by injections of new magnetic flux in the form of bipolar active regions at a rate equivalent to their disappearance; the mean lifetime of a single bipole is about 2 weeks. Thus the detection on a remote star of a large magnetic structure need not be interpreted as one immense starspot; it may be instead an agglomeration of many spots.

Complexes are illustrated in the magnetograms of Figure 2 at the same solar longitude in the southern hemisphere on successive rotations of the Sun in 1980. The elongation of the activity complex in the June 22 magnetogram (right) is created by the emergence and expansion of 17 small- to medium-sized bipolar regions along the inclined polarity inversion line where previously there had been just three large regions. During the May rotation (left) only one new region forms in the complex during its entire disk passage. The day following its appearance is marked by intense flare activity centered on the small new region. During the June rotation two-thirds of the flaring sites are located in or on the boundaries of the 17 new regions (Martin *et al.*, 1982). Yet the level of flare activity measured as counts of events or as total peak emission in soft X-rays is comparable for both rotations (Gaizauskas, 1982). If emerging flux in itself always dominates the flare process, the level of flaring during the second rotation should greatly exceed that of the first. Since this does not happen, the individual flux emergences must fulfill other conditions in order to be flare-effective.

The single small region that emerged on the first rotation contributed a disproportionate share of the strong flares from the activity complex. Within the single penumbra comprising that small region, pores of opposite magnetic polarity were arranged in a complex pattern and were moving rapidly (> 0.1 km s^{-1}) relative to each other (Nagy, 1983). A similar situation arose on a much greater scale in McMath 15403,

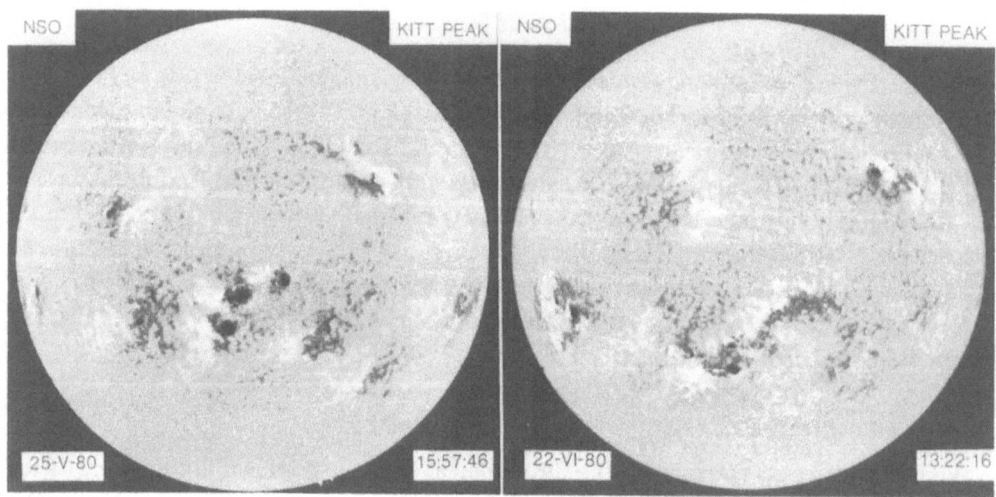

Fig. 2. A large complex of activity (sunspot 'nest') crossing the central meridian on successive solar rotations in May and June 1980. Orientation: north is at the top, and west is to the right. Whole-disk magnetograms from the National Solar Observatory supplied courtesy of J. W. Harvey.

a region which by itself produced 10% of the 324 flares emitting soft X-rays stronger than class M1 in a two-year period studied by Gaizauskas and McIntosh (1986b). A large new sunspot group formed inside an existing large spot on 10 July, 1978 (Sattarov, 1983). The expansion of the complex magnetic pattern resulting from this superposition in McMath 15403 (Dezsö et al., 1980) was accompanied by one of the strongest flaring episodes of Cycle 21.

Extreme magnetic complexity which has long been recognized as advantageous for producing great flares (Giovanelli, 1939) is a consequence of the Sun's tendency to keep pushing magnetic flux up within long-lived nests of activity. Observations show that the complexity can be prefabricated before any spots emerge (Zirin and Tanaka, 1973), or it can be created after originally unpaired bipoles emerge and push together as they grow (Tang, 1983; Zirin and Liggett, 1987). Flare activity stays enhanced in a nest of activity during its initial one or two disk passages but diminishes thereafter (Gaizauskas and McIntosh, 1986b).

Bai (1988) has examined recurrent flare activity using data from three solar cycles. He finds that superactive regions with major flares appear more frequently in certain areas on the Sun which he calls 'hot spots'. The hot spots are active intermittently and can be traced from one sunspot cycle to the next. Bai attributes hot spots to long-lived subsurface activity which rotates rigidly but at slightly different rates in the northern and southern hemispheres.

Proof of existence of a long-lived global pattern of active sources would eventually simplify flare prediction. Clues to a large-scale pattern can be found in the discoveries by Soru-Escaut, Martres, and Mouradian (1985) and by Mouradian et al. (1987) of a relation between long-lived Hα filaments with anomalous rates of rotation, the emergence of active centers, and the flare productivity in those centers. McIntosh and Wilson

(1985) present additional observational evidence which suggests that large-scale solar magnetic fields are organized as cellular patterns with sunspot regions forming on long-lived, updrafting boundaries of the cells. They claim that the trajectories of Hα filaments which last for many solar rotations can be used to define areas of enhanced shear and vorticity in the local flow which are later identified with the location of large flare-active regions.

In summary, the appearance and growth of new magnetic flux is a necessary precondition for flares. Small flares are common during the early phase of an EFR; large flares are often rooted in new, rapidly growing and complex patterns of magnetic flux. But the vast bulk of magnetic flux appears at the surface without producing flares as strong as an M1 event in X-rays – only 8% of the active regions met this condition in the two-year period studied by Gaizauskas and McIntosh (1986b). The proximity, orientation, and motion of the new flux relative to existing flux in an activity complex have, therefore, to meet stringent conditions in order to stimulate powerful flares.

3.2. MAGNETIC SHEAR AND ELECTRIC CURRENTS

The growth of an EFR inside an activity complex comprising several bipolar regions deforms the composite magnetic field of the system. The work done during the reconfiguration of the magnetic field into a non-potential form becomes available as free energy to power flares wherever and whenever a fast and efficient dissipation mechanism can be activated. Processes for transforming a quasi-stable magnetic configuration from a passive to a highly dynamic state are reviewed in Priest (1981) and by Spicer (1982). Here we draw attention to some observations of the quasi-static preflare state which are relevant to setting initial conditions in models of these processes.

From the alignment of chromospheric fibrils adjacent to magnetic polarity inversions in active regions, it is commonly inferred that the magnetic field must be sheared along and above the inversion lines. Direct confirmation of magnetic shear at the photospheric level comes from measurements with vector magnetographs of the transverse component of the magnetic field. The azimuthal difference between the observed field and the calculated potential field is a measure of the degree of magnetic shear. Shear so-defined attains maximum values at the sites of flare onset located along the magnetic polarity inversion line in the active region studied by Hagyard et al. (1984). The frequency and magnitude of flares increases in accord with the growth and relaxation of the shear in this same active region (Krall et al., 1982). Rapid spot motions and high inferred velocity shear also coincide with increased flare activity (Krall et al., 1982; Gesztelyi and Kalman, 1986; Kovacs and Dezsö, 1986). Analysis of several active regions shows, however, that sites of strong magnetic shear can be created and persist along their magnetic inversion lines without becoming sites of flare kernels (Athay, Jones, and Zirin, 1986; Hagyard and Rabin, 1986). The concept of a critical shear as a preflare threshold seems not to be generally valid. The formation of flare-related shear has been observationally associated with flux emergence (Zirin, 1983), flux submergence (Rabin, Moore, and Hagyard, 1985), flux cancellation (Martin, Livi, and Wang, 1985), and vortical motions in EFR (Martres et al., 1982). But a flux emergence leading to the direct and long-enduring collision of two large sunspot umbrae of opposite polarity was not flare-effective until the interconnecting fibrils indicated a sudden transition from a current-free to a sheared magnetic configuration (Gaizauskas and Harvey, 1986a).

The peak electric current along the line-of-sight derived from vector magnetograms coincides exactly with the sites of flare initiation in a five-kernel flare observed by Lin and Gaizauskas (1987). Using photospheric and chromospheric data for the same event, Hagyard (1988) was able to model the current-bearing structures at the site of flare onset: two arcades of loops with a radius of only 45 km, each carrying currents of 15×10^{10} A, and oriented about 60° to the inversion line. Interacting loops on this scale cannot account for all of the observed kernels or their distribution. For this same active region, longer loops interconnecting other flare sites have been inferred from X-ray and chromospheric data (Machado *et al.*, 1983; Ding *et al.*, 1987).

Multi-kernel flares imply a global network of currents coupling different parts of activity complexes. The formation of multiple current sheets in the corona by systematic photospheric motions has been modelled for the case of a quadrupolar field (Baum and Bratenahl, 1980; Hénoux and Somov, 1987; Low and Wolfson, 1988). An example is shown in Figure 3 of the current system formed when two bipoles are coaligned

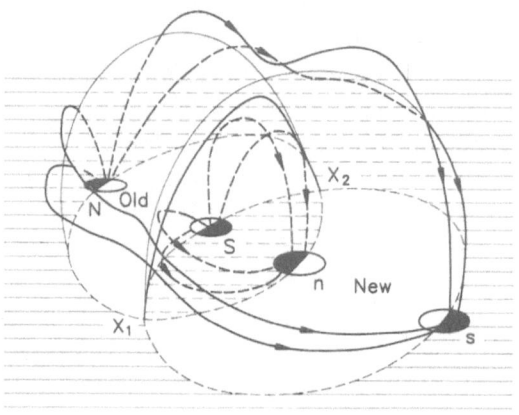

Fig. 3.   Three-dimensional representation of coronal currents and magnetic field separatrix surfaces for a quadrupole. The magnetic bipoles depict a new active region forming in line with an old one. The heavy arc joining X_1 and X_2 is the separator between two cells of magnetic flux. Adapted from Hénoux and Somov (1987).

end-to-end. The coronal field is partitioned into flux cells by special surface distributions of magnetic field lines called separatrices. The ovoid-shaped separatrices intersect along separator lines, regions of interaction between current systems generated by footpoint motions in the separatrix cells. Reconnection at the separator may release energy not in a confined region but over entire separatrix surfaces, thereby affecting many magnetic field loops. The energy stored in both the separator current and the individual loops when released forms multi-kernel flares in the process. The significant point for flare build-up is that the current along the separator can be increased beyond an instability threshold by an excess relative to a potential field of magnetic flux in one of the separatrix cells and a deficiency in another, or by excessive kinetic energy (i.e., footpoint motions) in any cell. In this formulation of the problem, boundary conditions at remote parts of an activity complex cannot be ignored relative to the local conditions at the site of a flare.

4. Erupting Filaments

Enhanced mass motions in active-region filaments are one of the best preflare indicators. The association of erupting filaments with two-ribbon flares receives considerable attention because the geometry of the situation lends itself to theoretical modelling with predictive capabilities. It must be emphasized, however, that not all large flares are accompanied by erupting filaments, nor do all flare-related filament eruptions have a preflare phase. The reader is referred to Martin (1980) for a review of the observational aspects of filament activation and to Low (1982) and Priest (1982) for theoretical discussions of the stability of these structures.

A specific mechanism which is often invoked to destabilize a filament is reconnection at a current sheet between newly emerging or evolving flux and an overlying filament (Heyvaerts, Priest, and Rust, 1977). A quantitative verification of this model is difficult because changes in magnetic flux on a small scale are constantly happening against a backdrop of slowly-evolving magnetic patterns in any well-developed activity complex. Three specific examples of flare-related eruptions are discussed below to illustrate the scope of the problem and to point out alternative mechanisms.

4.1. FLARE OF 21 MAY, 1980

This major flare began at 20:50 UT and was well-observed by the SMM spacecraft (de Jager and Švestka, 1985). Hoyng et al. (1981) attribute the destabilization of a long filament over an extended polarity-inversion line to the emergence nearby of a bipolar region containing a new pore. Subsequent analysis of magnetograms by Harvey (1982) suggests that the pore formed not by emergence but by the compression of existing flux at the surface. New flux did appear nearby as patches of polarity opposite to their unipolar surroundings. But the net flux directly beneath the activated filament actually decreased. The effect can be seen in the vicinity of the arrow on the magnetogram of 20:15 in Figure 4. The sharp alteration of the polarity-inversion line at this location is reflected in the broadening of the neutral-line filament and the appearance of additional fine strands within it; compare the filtergrams at 15:56 and 19:37 UT in Figure 4. The parting of the dark filament is as likely due to the substantial reduction of negative-polarity (black) flux as to the injection of new flux; compare the widening gap in the magnetograms of 20:15 and 21:47 UT in Figure 4 at the location of the arrow. The cancellation of magnetic flux erodes the continuity of the filamentary structure. The filament did not erupt or disappear; instead, Hα-emitting material was ejected from one end of the filament in the form of a spray (McCabe et al., 1986).

This flare is a paradigm for the empirical rule first expressed by Martres et al. (1968): evolving magnetic features of one polarity involved in a flare are increasing at the same time as features of the opposite polarity are decreasing.

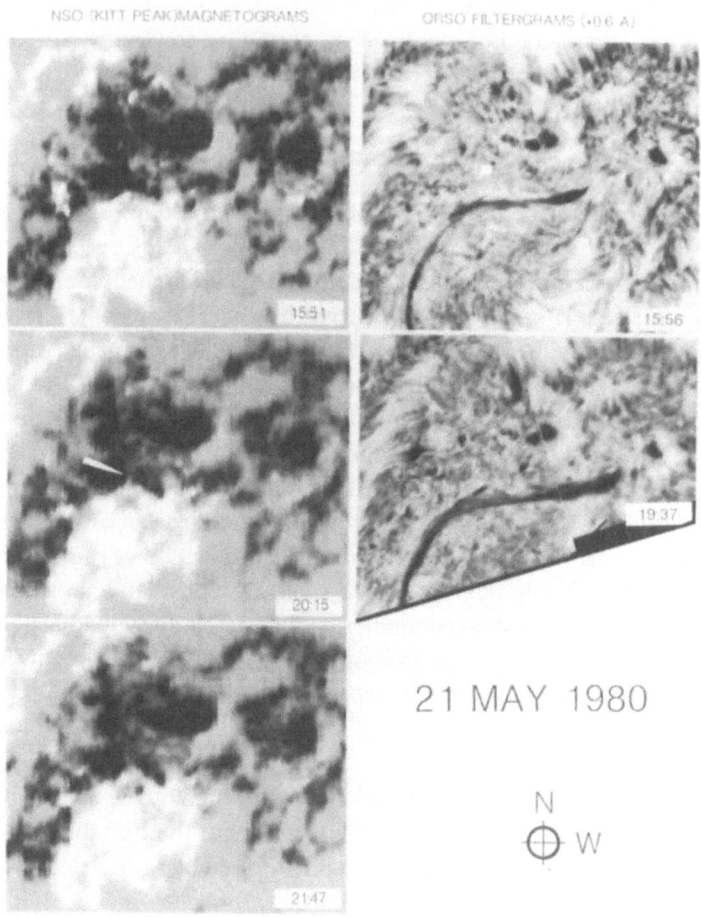

Fig. 4.   Changes associated with the major flare beginning at 20:50 UT on 21 May, 1980. Left column: longitudinal component of the magnetic field. Right column: chromospheric fine structure. The arrow on the magnetogram at 20:15 UT points to a widening gap in negative (black) polarity flux. Field of view = 200″ × 165″. NSO magnetograms supplied by courtesy of J. W. Harvey.

4.2. FLARE OF 22 JUNE, 1980

The activation of a filament in Boulder Region 2517 (Hale 16918) prior to the eruption of a flare (Imp. 2) at 13:05 UT on 22 June, 1980 has been attributed by Simon *et al.* (1984) to magnetic reconnection between the filament and new emerging flux. The essential circumstances are depicted schematically in Figure 5(b) where pores O_8 and O_9 are inferred by Simon *et al.* to be a new EFR straddling the filament at the polarity inversion line of a large complex of activity. The observed convergence of these pores is, however, contrary to the spreading motion normally associated with an EFR. The chromospheric velocity feature associated with these pores by Raadu *et al.* (1988) is much weaker than the usual loop flows in an EFR. The designation of O_8–O_9 as an

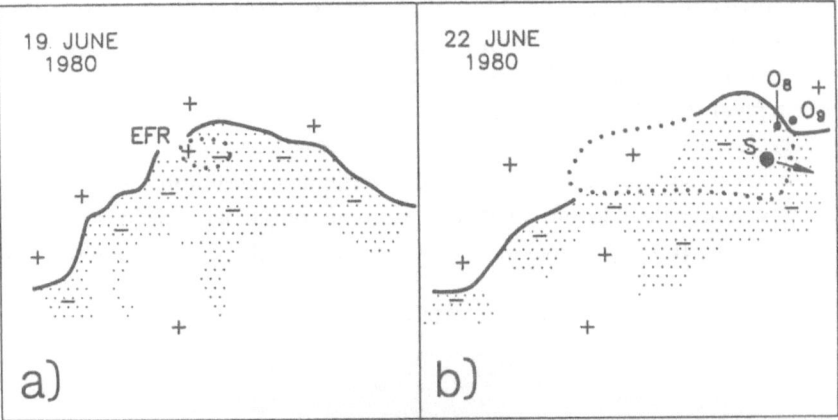

Fig. 5.   Schematic view of new bipolar magnetic flux expanding inside an existing complex fo activity. (a) A new EFR (oval outline of heavy dots) appears next to a 'neutral line' filament (full line). (b) After 4 days the new region (outline of heavy dots) has expanded and reshaped the polarity inversion line. Fresh magnetic flux surrounding the westward-moving leader spot S has pushed older flux against the reshaped 'neutral line' near pores O_8 and O_9. The stippled area roughly corresponds to the outline of negative magnetic flux in Hale 16918 (Martin *et al.*, 1982). The locations of the filaments on the indicated dates are adapted from Martin *et al.* (1984).

EFR, or any magnetic connection between these pores is, therefore, debatable. The most significant preflare phenomenon reported by Simon *et al.* is the very rapid motion of bright Hα emission surrounding O_8 towards the filament just minutes before the filament erupts and the first flare kernels appear. The moving bright feature may be an early signature of magnetic reconnection but, since the orientation of interacting magnetic fields remains problematical, the circumstances do not lead themselves to a definitive test of the Emerging Flux Model. The small-scale changes noted above are embedded in a much larger evolutionary change (Martin *et al.*, 1984). An EFR forms on 19 June adjacent to the polarity inversion line (Figure 5(a)). As the original EFR expands, another EFR emerges within it. In four days the expanded bipolar region compresses pre-existing magnetic flux at the polarity inversion line and distorts the shape of the original trajectory of the filament (Figure 5(b)). The relative effect on the filament's stability of the small-scale stresses imposed by motion of O_8 vs large-scale evolutionary stresses driven by emerging flux has yet to be assessed.

4.3. Flare of 25 June, 1980

This two-ribbon flare (Imp. 1B) with impulsive phase at 15:51 UT was jointly observed by the VLA, the SMM spacecraft, and ground-based observatories. It was inferred from the detection at 6 cm wavelength of changing polarized structures 15 min before the flare that emerging flux might be acting as a trigger (Kundu *et al.*, 1982). Subsequent analysis of photospheric and chromospheric data ruled out emerging flux beneath the location of the polarized microwave-emitting features. It was shown instead that a filament at this location began, more than two hours before the flare, to rise with uniform accelera-tion and then, at 20 min before the flare, to execute a more complex series of motions

including untwisting (Kundu *et al.*, 1985; Gaizauskas, 1984). The preflare untwisting of
the filament overlapped in space and time with the changing polarized microwave
structures and may be physically related to them.

The prolonged rise of the filament followed by its untwisting and disruption form a
single progression with the eventual onset of the flare. Did a particular disturbance
initiate the irreversible outward motion of the filament beginning of 13:40 UT? In
Figure 6 we examine the chromospheric environment of the filament (enclosed between
two arrows) near the beginning of its ascent. The field of view includes only the

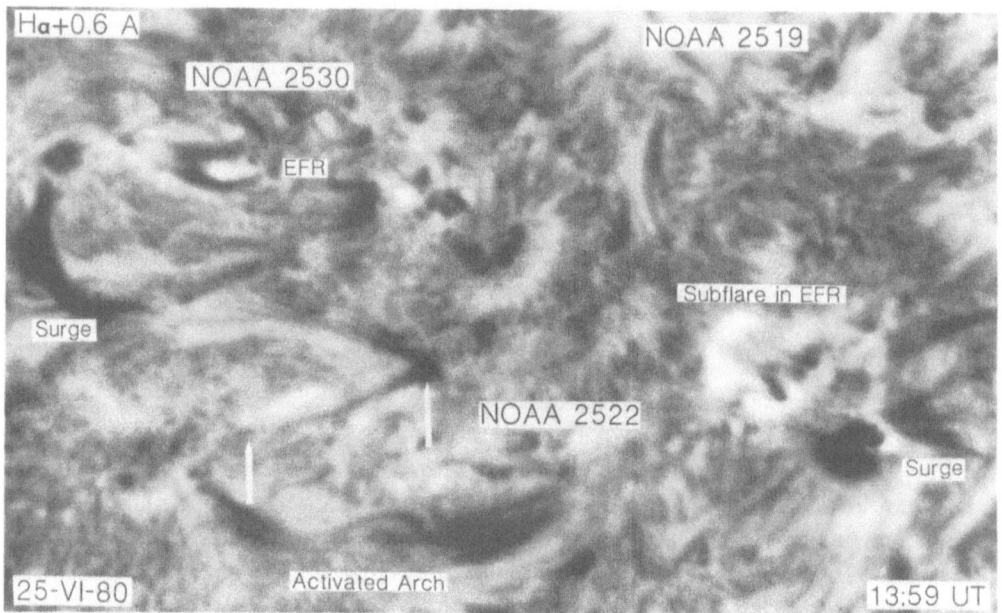

Fig. 6. A close-up view of the chromosphere in part of an activity complex two hours before the small
filament enclosed by the arrows erupts immediately prior to a class 1B flare. Dynamic features are labelled
and described in the text. *Top*: north; *right*: west. Off-band Hα filtergram from the Ottawa River Solar
Observatory. Field of view = 290" × 175".

southeastern corner of the elongated complex of activity which straddles the central
meridian in the magnetogram of 22 June, 1980 (Figure 2). The portion of the complex
depicted in Figure 6 consists of three numbered bipolar regions in different stages of
growth. At the time shown, region 2519 is decaying while 2522 and 2530 each contain
vigorously growing EFR. We note the following:

– A flare is erupting inside the EFR of region 2522.

– Surges are in progress in the leading spot of 2522 and the trailing spot of 25530.
Surges recur at these locations throughout the day.

– Just below the label 'NOAA 2522' there is a large intrusion in a unipolar area of
flux of opposite magnetic polarity (Schmahl, 1982). This isolated pole connects from
time to time via an 'activated arch' to the spot which anchors the eastern end of the
marked filament. The arch is called 'activated' here because its visibility is strongly

dependent on wavelenth shifts of the Hα line, implying velocity flows in the structure. The activation in progress in Figure 6 begins at 13:40 UT coincident with the gradual rise in the marked filament but subsides within a half-hour.

– The large spots near either end of the marked filament are moving in opposite directions with the high relative velocity of 0.2 km s^{-1} (Schmahl, 1982).

Any one of the above coincidental phenomena cannot be singled out as 'causing' the initial ascent of the filament. Together, however, they imply a global dynamism in the complex of activity which steadily alters the equilibrium of the filament. The filament tries to readjust by gradually rising until it reaches a state where further equilibrium becomes impossible and a catastrophe results. The 25 June, 1980 filament eruption is an ideal candidate for the application of nonlinear dynamics and the study of chaotic behaviour (Martens, 1984).

On the face of it, emerging flux can be suspected of playing a role during each of the three flare-associated filament eruptions discussed above. But when we probe more deeply, quantitative data which would permit a definitive test of the Emerging Flux Model may elude us, or the flux may emerge remotely to influence the equilibrium of the filament indirectly rather than directly through a current sheet.

5. Concluding Remarks

The most challenging aspects of the flare problem deal with the sudden release of energy in a compact volume. A tendency has developed to seek the answers from conditions existing at the affected volume or in a single loop which contains that volume. But observational evidence reviewed here shows that multiple structures are involved in the energy release and that they cannot be considered as uncoupled from a global background. Now, theoretical tools are being developed to cope with a much more complicated and dynamic, hence realistic, magnetic environment than an arcade of loops.

The absence of a singular class of precursor that could reliably foretell a flare, even though observers have sought them assiduously for years, and the disparity among flares (excluding homologues) observed in different active regions, means that preflare transient behaviour does not always arise from the same mechanism. If, as now seems likely, multiple structures are involved in a flare, preflare transient emission may represent a momentarily successful bid by one of several structures to counter a loss of equilibrium. The exact nature of the process depends upon the coupling between structures adjacent to the flare and the rest of the activity complex. This widespread dependence increases the scope for variety among preflare phenomena.

The complexity of the preflare state will not be resolved without many more comprehensive, multi-wavelength analyses of preflare periods observed jointly by spacecraft and ground-based instruments. In those analyses, closer attention should be paid to the way developing large-scale patterns of magnetic fields affect current structures on a local scale.

References

Athay, R. G. and Moreton, G. E.: 1961, *Astrophys. J.* **133**, 935.
Athay, R. G., Jones, H. P., and Zirin, H.: 1986, *Astrophys. J.* **303**, 877.
Bai, T.: 1988, *Astrophys. J.* **328**, 860.
Baum, P. J. and Bratenahl, A.: 1980, *Solar Phys.* **67**, 245.
Benz, A. O., Barrow, C. H., Dennis, B. R., Pick, M., Raoult, A., and Simnett, G.: 1983, *Solar Phys.* **83**, 267.
Bruzek, A.: 1967, *Solar Phys.* **2**, 451.
Castenmiller, M. J. M., Zwaan, C., and van der Zalm, E. B. J.: 1986, *Solar Phys.* **105**, 237.
Cheng, C.-C. and Tandberg-Hanssen, E.: 1986, *Astrophys. J.* **309**, 421.
Cheng, C.-C., Tandberg-Hanssen, E., and Orwig, L. E.: 1984, *Astrophys. J.* **278**, 853.
Cheng, C.-C., Tandberg-Hanssen, E., Bruner, E. C., Orwig, L., Frost, K. J., Kenny, P. J., Woodgate, B. E., and Shine, R. A.: 1981, *Astrophys. J.* **248**, L39.
Cheng, C.-C., Bruner, E. C., Tandberg-Hanssen, E., Woodgate, B. E., Shine, R. A., Kenny, P. J., Henze, W., and Poletto, G.: 1982, *Astrophys. J.* **253**, 353.
Cheng, C.-C., Pallavicini, R., Acton, L. W., and Tandberg-Hanssen, E.: 1985, *Astrophys. J.* **298**, 887.
De Jager, C. and Švestka, Z.: 1985, *Solar Phys.* **100**, 435.
De Jager, C., Machado, M. E., Schadee, A., Strong, K. T., Švestka, Z., Woodgate, B.E., and van Tend, W.: 1983, *Solar Phys.* **84**, 204.
Dezsö, L., Gesztelyi, L., Kondas, L., Kovacs, A., and Rostas, S.: 1980, *Solar Phys.* **67**, 317.
Ding, Y. J., Hagyard, M. J., Deloach, A. E., Hong, Q. R., and Liu, X. P.: 1987, *Solar Phys.* **109**, 307.
Dodson, H. W. and Hedeman, E. R.: 1976, *Solar Phys.* **47**, 267.
Fritzová-Švestková, L., Chase, R. C., and Švestka, Z.: 1976, *Solar Phys.* **48**, 275.
Gaizauskas, V.: 1982, *Adv. Space Res.* **2**, No. 11, 11.
Gaizauskas, V.: 1985, in C. de Jager and B. Chen (eds.), *Proc. Kunming Workshop on Solar Physics and Interplanetary Travelling Phenomena*, Science Press, Beijing, 710.
Gaizauskas, V. and Harvey, K. L.: 1986a, *Adv. Space Res.* **6**, No. 6, 17.
Gaizauskas, V. and McIntosh, P. S.: 1986b, in P. A. Simon, G. Heckman, and M. A. Shea (eds.), *Solar-Terrestrial Predictions*, Proc. Workshop Meudon, France, NOAA/AFGL, p. 126.
Gaizauskas, V., Harvey, K. L., Harvey, J. W., and Zwaan, C.: 1983, *Astrophys. J.* **265**, 1056.
Gesztelyi, L. and Kalman, B.: 1986, *Adv. Space Res.* **6**, No. 6, 29.
Giovanelli, R. G.: 1939, *Astrophys. J.* **89**, 555.
Hagyard, M. J.: 1988, *Solar Phys.* **115**, 107.
Hagyard, M. J. and Rabin, D. M.: 1986, *Adv. Space Res.* **6**, No. 6, 7.
Hagyard, M. J., Smith, J. R., Jr., Teuber, D., and West, E. A.: 1984, *Solar Phys.* **91**, 115.
Harrison, R. A.: 1986, *Astron. Astrophys.* **162**, 283.
Harrison, R. A., Waggett, P. W., Bentley, R. D., Phillips, K. J. H., Bruner, M., Dryer, M., and Simnett, G. M.: 1985, *Solar Phys.* **97**, 387.
Harvey, J. W.: 1982, *Adv. Space Res.* **2**, No. 11, 31.
Hénoux, J.-C. and Somov, B. V.: 1987, *Astron. Astrophys.* **185**, 306.
Heyvaerts, J., Priest, E. R., and Rust, D. M.: 1977, *Astrophys. J.* **216**, 123.
Hoyng, P. and 11 co-authors: 1981, *Astrophys. J.* **246**, L155.
Hurford, G. J. and Zirin, H.: 1982, AFGL-TR-82–0117.
Hurford, G. J., Read, J. B., and Zirin, H.: 1984, *Solar Phys.* **94**, 413.
Kahler, S.: 1979, *Solar Phys.* **62**, 347.
Kai, K., Nakajima, H., and Kosugi, T.: 1983, *Publ. Astron. Soc. Japan* **35**, 285.
Kovacs, A. and Dezsö, L.: 1986, *Adv. Space Res.* **6**, No. 6, 29.
Krall, K. R., Smith, J. B., Jr., Hagyard, M. J., West, E. A., and Cummings, N. T.: 1982, *Solar Phys.* **79**, 59.
Kundu, M. R.: 1965, *Solar Radio Astronomy*, John Wiley Interscience, New York.
Kundu, M. R.: 1986, *Adv. Space Res.* **6**, No. 6, 93.
Kundu, M. R. and Shevgaonkar, R. K.: 1985, *Astrophys. J.* **291**, 860.
Kundu, M. R., Rust, D. M., and Bobrowsky, M.: 1983, *Astrophys. J.* **265**, 1084.
Kundu, M. R., Schmahl, E. J., and Velusamy, T.: 1982, *Astrophys. J.* **253**, 963.
Kundu, M. R., Schmahl, E. J., Velusamy, T., and Vlahos, L.: 1982, *Astron. Astrophys.* **108**, 188.
Kundu, M. R., Gaizauskas, V., Woodgate, B. E., Schmahl, E. J., Shine, R., and Jones, H. P.: 1985, *Astrophys. J. Suppl.* **57**, 621.

Kundu, M. R., Gopalswamy, N., Saba, J. L. R., Schmelz, J. T., and Strong, K. T.: 1987, *Solar Phys.* **114**, 273.

Lang, K. R.: 1980, in R. F. Donnelly (ed.), *Solar-Terrestrial Predictions Proceedings III, Solar Activity Predictions*, US Dept. of Commerce, p. C-131.

Lang, K. R. and Willson, R. F.: 1986, *Adv. Space Res.* **6**, No. 6, 97.

Lin, Y. Z. and Gaizauskas, V.: 1987, *Solar Phys.* **109**, 81.

Low, B. C.: 1982, *Rev. Geophys. Space Phys.* **20**, 145.

Low, B. C. and Wolfson, R.: 1988, *Astrophys. J.* **324**, 547.

Machado, M. E.: 1985, *Solar Phys.* **99**, 159.

Machado, M. E., Orwig, L. E., and Antonucci, E.: 1986, *Adv. Space Res.* **6**, No. 6, 101.

Machado, M. E., Somov, B. V., Rovira, M. G., and de Jager, C.: 1983, *Solar Phys.* **85**, 157.

Machado, M. E., Moore, R. L., Hernandez, A. M., Rovira, M. G., Hagyard, M. J., and Smith, J. B., Jr.: 1988, *Astrophys. J.* **326**, 425.

Marsh, K. A.: 1978, *Solar Phys.* **59**, 105.

Martens, P. C. H.: 1984, *Phys. Reports* **115**, 315.

Martin, S. F.: 1980, *Solar Phys.* **68**, 217.

Martin, S. F., Livi, S. H. B., and Wang, J.: 1985, *Australian J. Phys.* **38**, 929.

Martin, S. F., Dezsö, L., Antalová, A., Kučera, A., and Harvey, K. L.: 1982, *Adv. Space Res.* **2**, No. 11, 39.

Martin, S. F. and 11 co-authors: 1984, *Adv. Space Res.* **4**, No. 7, 61.

Martres, M.-J., Michard, R., Soru-Iscovici, I., and Tsap, T.: 1968, in K. O. Kiepenheuer (ed.), 'Structure and Development of Solar Active Regions', *IAU Symp.* **35**, 311.

Martres, M.-J., Rayrole, J., Semel, M., Soru-Escaut, I., Tanaka, K., Makita, M., and Moriyama, F.: 1982, *Publ. Astron. Soc. Japan* **34**, 299.

Martres, M.-J., Woodgate, B. E., Mein, N., Mouradian, Z., Rayrole, J., Schmieder, B., Simon, G., and Soru-Escaut, I.: 1984, *Adv. Space Res.* **4**, No. 7, 5.

McCabe, M. K., Švestka, Z., Howard, R. A., Jackson, B. V., and Sheeley, N. R.: 1986, *Solar Phys.* **103**, 399.

McIntosh, P. S. and Wilson, P. R.: 1985, *Solar Phys.* **97**, 59.

Moreton, G. E.: 1961, *Sky Telesc.* **21**, 145.

Mouradian, Z., Martres, M.-J., and Soru-Escaut, I.: 1983, *Solar Phys.* **87**, 309.

Mouradian, Z., Martres, M.-J., Soru-Escaut, I., and Gesztelyi, L.: 1987, *Astron. Astrophys.* **183**, 129.

Nagy, I.: 1983, *Publ. Debrecen Obs.* **5**, 107.

Nakajima, H., Dennis, B. R., Hoyng, P., Nelson, G., Kosugi, T., and Kai, K.: 1985, *Astrophys. J.* **288**, 806.

Ogir, M. B.: 1981, *Izv. Krymsk. Astrofiz. Obs.* **64**, 118.

Porter, J. G., Toomre, J., and Gebbie, K. B.: 1984, *Astrophys. J.* **283**, 879.

Porter, J. G., Moore, R. L., Reichmann, E. J., Engvold, O., and Harvey, K. L.: 1987, *Astrophys. J.* **323**, 380.

Priest, E. R. (ed.): 1981, *Solar Flare Magnetohydrodynamics*, Gordon and Breach, London.

Priest, E. R.: 1982, *Solar Magnetohydrodynamics*, D. Reidel Publ. Co., Dordrecht, Holland.

Priest, E. R.: 1985, *Rep. Prog. Phys.* **48**, 955.

Priest, E. R., Gaizauskas, V., Hagyard, M. J., Schmahl, E. J., and Webb, D. F.: 1986, in M. R. Kundu and B. E. Woodgate (eds.), *Energetic Phenomena on the Sun*, NASA CP-2439, Ch. 1.

Raadu, M. A., Schmieder, B., Mein, N., and Gesztelyi, L.: 1988, *Astron. Astrophys.* **197**, 289.

Rabin, D. M., Moore, R. L., and Hagyard, M. J.: 1984, *Astrophys. J.* **287**, 404.

Raoult, A., Pick, M., Dennis, B. R., and Kane, S. R.: 1985, *Astrophys. J.* **299**, 1027.

Rust, D. M. and Roy, J.-R.: 1975, AFCRL-TR-0437, 61.

Rust, D. M., Nakagawa, Y., and Neupert, W. M.: 1975, *Solar Phys.* **41**, 392.

Sattarov, I.: 1983, *Astron. Zh.* **60**, 350.

Schadee, A., de Jager, C., and Švestka, Z.: 1983, *Solar Phys.* **89**, 287.

Schmahl, E. J.: 1982, *Adv. Space Res.* **2**, No. 11, 73.

Simnett, G. M. and Harrison, R. A.: 1985, *Solar Phys.* **99**, 291.

Simon, G., Mein, N., Mein, P., and Gesztelyi, L.: 1984, *Solar Phys.* **93**, 325.

Soru-Escaut, I., Martres, M.-J., and Mouradian, Z.: 1985, *Astron. Astrophys.* **145**, 19.

Spicer, D. S.: 1976, *An Unstable Arch Model of a Flare*, NRL report 8036.

Spicer, D. S.: 1982, *Space Sci. Rev.* **31**, 351.

Spicer, D. S. and Brown, J. C.: 1981, in S. Jordan (ed.), *The Sun as a Star*, NASA SP-450.

Švestka, Z.: 1976, *Solar Flares*, D. Reidel Publ. Co., Dordrecht, Holland.

Tang, F.: 1983, *Solar Phys.* **89**, 43.

Thomas, R. J. and Teske, R. G.: 1971, *Solar Phys.* **16**, 431.

Van Hoven, G.: 1976, *Solar Phys.* **49**, 95.

Van Hoven, G. and Hurford, G. J.: 1986, *Adv. Space Res.* **6**, No. 6, 83.

Van Hoven, G. and 18 co-authors: 1980, in P. Sturrock (ed.), *Solar Flares*, Colorado Associated University Press, Boulder, Ch. 1.

Vlahos, L., Machado, M. E., Ramaty, R., and Murphy, R. J.: 1986, in M. R. Kundu and B. E. Woodgate (eds.), *Energetic Phenomena on the Sun*, NASA CP-2439, Ch. 2.

Webb, D. F.: 1985, *Solar Phys.* **97**, 321.

Willson, R. F.: 1983, *Solar Phys.* **83**, 285.

Willson, R. F.: 1984, *Solar Phys.* **92**, 189.

Willson, R. F. and Lang, K.R.: 1984, *Astrophys. J.* **279**, 427.

Wolfson, C. J.: 1982, *Solar Phys.* **76**, 377.

Woodgate, B. E.: 1982, *Adv. Space Res.* **2**, No. 11, 61.

Woodgate, B. E. and 15 co-authors: 1981, *Astrophys. J.* **244**, L133.

Woodgate, B. E. and 8 co-authors: 1984, *Adv. Space Res.* **4**, No. 7, 5.

Zirin, H.: 1983, *Astrophys. J.* **274**, 900.

Zirin, H. and Liggett, M. A.: 1987, *Solar Phys.* **113**, 267.

Zirin, H. and Tanaka, K.: 1973, *Solar Phys.* **32**, 173.

Zwaan, C.: 1987, *Ann. Rev. Astron. Astrophys.* **25**, 83.

SIMULTANEOUS MULTI-FREQUENCY IMAGING
OBSERVATIONS OF SOLAR MICROWAVE BURSTS

M. R. KUNDU, S. M. WHITE, and E. J. SCHMAHL

Astronomy Program, Univ. of Maryland, College Park, MD 20742, U.S.A.

Abstract. We review the results of simultaneous two-frequency imaging observations of solar microwave bursts with the Very Large Array. Simultaneous 2 and 6 cm observations have been made of bursts which are optically thin at both frequencies, or optically thick at the lower frequency. In the latter case the source structure may differ at the two frequencies, but the two sources usually seem to be related. However, this is not always true of simultaneous 6 and 20 cm observations. The results have implications for the analysis of non-imaging radio data of solar and stellar flares.

1. Introduction

The analysis of the energetics of both stellar and most solar flares relies on the interpretation of non-imaging radio and/or X-ray observations. The type of analysis is often of the following type (e.g., Batchelor *et al.*, 1985): assuming that the radio emission at a frequency below the peak of the spectrum is optically thick thermal emission (flux $\sim Af^2T$), one uses the flux at a given frequency f together with a temperature T from X-ray data to calculate an area A. Departure from the f^2 law gives the variation of the optically-thick area with frequency, and time dependence is used to deduce the rate at which the source grows in the impulsive phase. There is an implied assumption that the structure of the source is simple, and that, in some sense, the sources one is seeing at the different frequencies are the same.

It is important to check this non-imaging technique against imaging observations. Using two sub-arrays of about 13 antennae each, the VLA* can be used to image solar microwave bursts simultaneously at two frequencies. Here we review the results of such observations from the last solar maximum. There are several different types of burst found. The simplest case is that in which the time profiles of the two frequencies are identical and the maps show a simple source at the same location at both frequencies. This situation implies that one is seeing the same electrons at both frequencies, and one can then accurately calculate the magnetic field, particle density and distribution in the source. Naturally, this is rare, but there is one good example (October 1, 1980).

A second type of burst shows similar time profiles but different locations for the two frequencies: when the high frequency emission comes from two small sources and the lower frequency from a large featureless source, this can be interpreted as footpoint plus loop-top emission, from one population of non-thermal electrons in the loop. There are other cases in which the different time profiles and different spatial locations of the

---

* The Very Large Array is a facility of the National Radio Astronomy Observatory, which is operated by Associated Universities, Inc., under contract with the National Science Foundation.

Solar Physics **121** (1989) 153–161.
© 1989 *by Kluwer Academic Publishers.*

sources at the two frequencies imply that one is seeing two populations of electrons, usually one thermal component and another non-thermal. Particularly at 20 cm (1.4 GHz) one rarely seems to see the same population of electrons which radiates at higher frequencies. In some events the spatial structure of the source region is so complex that no simple conclusions can be drawn. In the space available here we discuss the observations briefly, and summarize the implications for analysis of non-imaging data.

2. Simple Optically-Thin Non-Thermal Burst

Figure 1(a) shows the time profile of the flux from a burst on October 1, 1980 (Kundu, Velusamy, and White, 1987), as observed in the VLA's shortest spacings. The time profiles at the two wavelengths, 6 cm (5 GHz) and 2 cm (15 GHz), are essentially indentical. In Figure 1(b), we plot the time evolution of the peak brightness temperature in the map during the impulsive phase of the flare, together with the degree of polarization at both wavelengths. This confirms the close resemblance of the two time profiles.

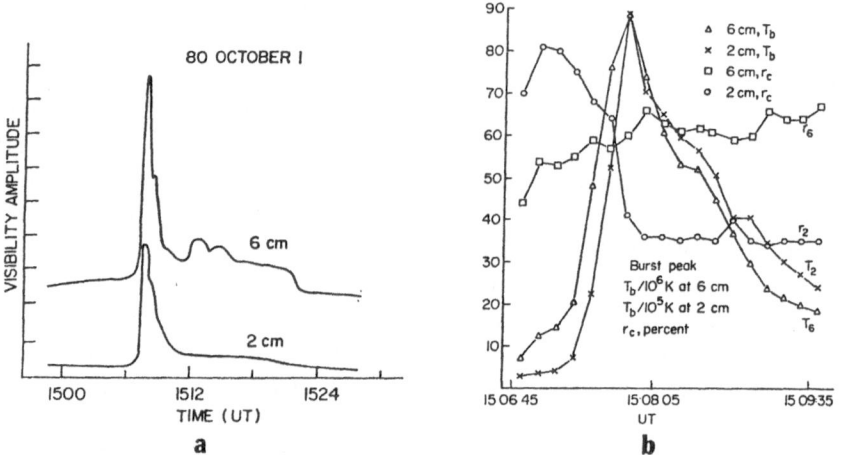

Fig. 1. (a) The time profile of the flux in the shortest spacings at the VLA at 2 and 6 cm wavelengths. (b) The time profiles of the maximum brightness temperatures in the 2 and 6 cm maps during the peak of the burst in (a), together with the variation of the degree of circular polarization at both wavelengths.

Figure 2 shows the structure of the pre-event region at 6 cm (Figure 2(a)), the 6 and 2 cm burst sources in I (Figure 2(b, c)), and the location of the burst source superposed on a Kitt Peak magnetogram of the region (Figure 2(d)). This indicates that the burst occurred over the neutral line between a predominantly positive magnetic field region and an intruding negative region, and strongly suggests that reconnection between the regions of opposite polarity played a role in the onset of the burst.

The spatial locations of the 6 and 2 cm burst sources in Figure 2, together with the similar time evolutions, suggest that we are seeing the same electrons in the same location at both wavelengths. The brightness temperature of the 6 cm emission indicates

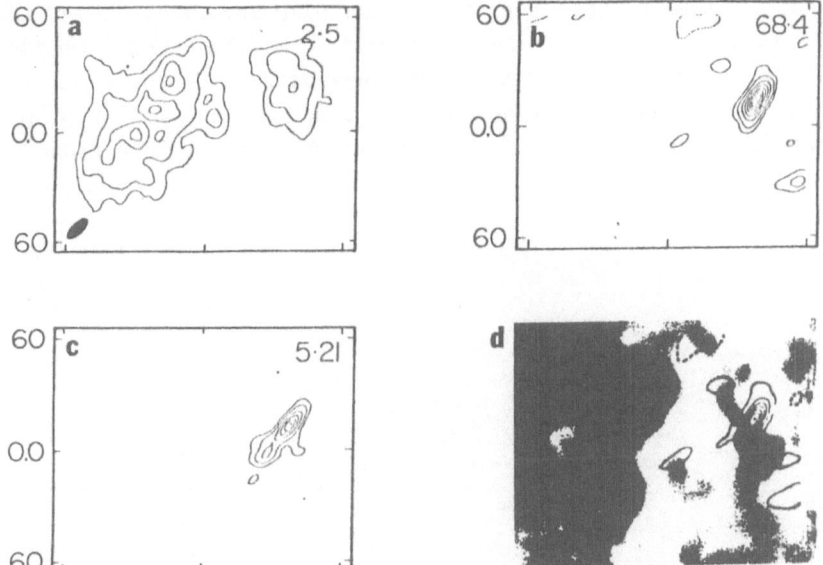

Fig. 2.   (a) The pre-flare map of the active region at 6 cm. (b, c) The I maps of the burst peak at 6 and 2 cm, respectively. (d) The Kitt Peak magnetogram of the active region, together with the contours of the burst source from (b) superposed.

non-thermal gyrosynchrotron as the likely emission mechanism. The high degree of polarization at both wavelengths, together with the spectrum of the burst obtained from patrol telescopes which gradually decreases from 2.8 to 20 GHz, suggests that both 2 and 6 cm are in the optically thin range. With this information the full parameters in the source region can be derived, assuming a homogeneous source: we find a surprisingly flat spectral index of 1.44 for the energy distribution of the electrons, an electron number density of 9×10^4 cm^{-3} (for the non-thermal radiating electrons above 10 keV), a total energy of about 8×10^{23} ergs in the electrons, and a magnetic field of 390 G. We assume that the electrons are trapped in a compact low-lying loop, and that this explains the relatively high magnetic field for a 6 cm non-thermal source. In order for the assumption of a homogeneous source to be valid, it may be necessary to assume either that the electrons are trapped at the top of the loop due to a highly anisotropic pitch angle distribution, or that the magnetic field strength does not change greatly along the loop (which can be achieved if the loop cross-section is roughly constant). In practice, we believe that it is most likely that there is a range of magnetic fields in the loop, and that the value 390 G is most representative of this range.

3. Weak Burst Optically Thick at 6 cm and Optically Thin at 2 cm

In the first burst discussed the 6 cm emission was optically thin, since it was highly polarized and the burst spectrum from patrol telescopes indicated a spectrum decreasing from about 15 cm (to shorter wavelengths). Usually, however, solar microwave bursts have a spectral peak at around 4 cm (8 GHz), and 6 cm emission is then optically thick while 2 cm emission is optically thin. Dulk, Bastian, and Kane (1986) observed such a burst at 6 and 2 cm and in hard X-rays on November 22, 1981. Figure 3(a) shows the time evolution of the flux from the burst in each wavelength range, and Figure 3(b) shows the relative locations of the 2 cm (solid lines) and 6 cm (dashed lines) sources.

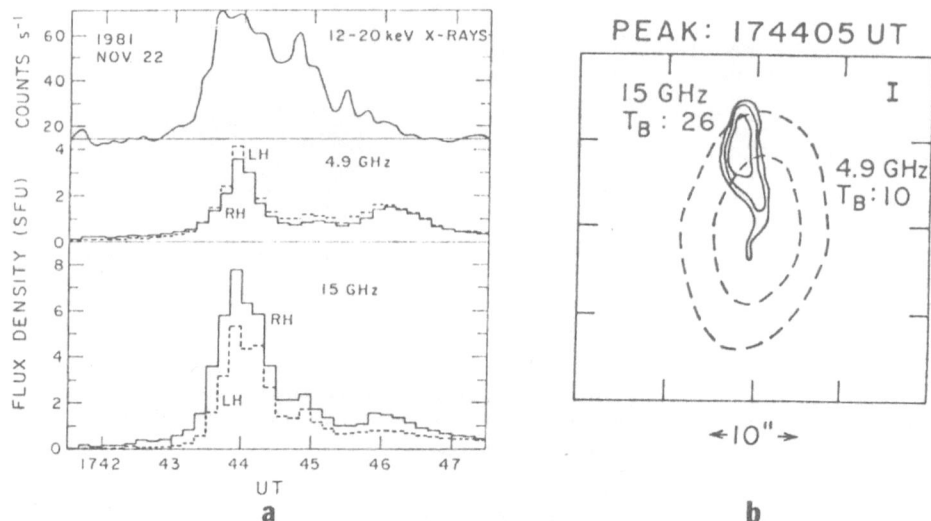

Fig. 3. (a) The time profiles of the hard X-ray, 6 and 2 cm fluxes during the burst on 1981 November 22. (b) The locations of the 6 and 2 cm burst sources.

The brightness temperature peaked at only 10^7 K at 6 cm and 3×10^6 K at 2 cm, whereas the X-ray temperature was about 2×10^7 K. Since the optically thick 6 cm emission is lower in temperature than the X-rays, a simple homogeneous model does not fit these observations. This event can be modelled by a constant-density inhomogeneous source in which the temperature is greatest at the middle of a gaussian-shaped region and decreases towards the edges. The emission mechanism for all three components is thermal bremsstrahlung (gyroresonance is ruled out by the spectrum, and the low brightness temperature argues against non-thermal emission). The 6 cm emission is optically thick throughout, while the 2 cm emission is optically thin. The magnetic field can be estimated from the polarization of the 2 cm source to be around 200 G.

We note that these authors also observed a burst at 2 cm which was absent at 6 cm. They noticed that the radiation had to pass over a strong sunspot on the way to the earth, and concluded that the longer wavelength radiation was absorbed. Without the availability of imaging data showing the location of the sunspot, the interpretation of this event based on the spectrum alone would probably have been erroneous. This illustrates the potential dangers of non-imaging data.

4. Footpoint + Loop-Top Events

Figure 4(a) shows the time evolution of a burst observed at 6 and 2 cm on 1981 November 13 (Shevgaonkar and Kundu, 1985). The 2 cm and hard X-ray profiles are very similar in the rise phase, and the 6 cm profile shows a similar shape but is delayed with respect to the other two profiles by about 20 s. The maps (Figure 4(b) and 4(c)) show two sources interpreted as loop footpoints at 2 cm, and a single large featureless source at 6 cm which is interpreted as the loop top (here by 'loop top' we mean essentially the whole of the volume in which, in the conventional loop picture, the loop turns over, and not just a small volume at the very centre of the loop). The explanation here is that in the higher magnetic fields of the footpoints the emissivity of higher frequency radiation is greater. One deduces magnetic fields in the 6 cm source below 200 G, and can show that the 6 cm-emitting electrons would not be seen at 2 cm,

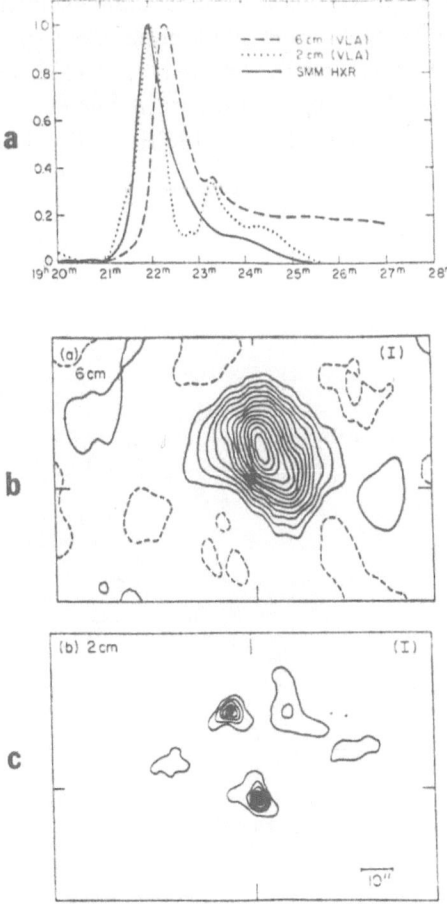

Fig. 4. (a) The time profiles of the hard X-ray, 6 and 2 cm fluxes of the burst on 1981 November 13, all normalized to the same peak value for comparison. (b) The 6 cm map of the burst source. (c) The 2 cm map of the burst source.

provided that the electron energy spectral index exceeds 4. The delay of the 6 cm peak relative to the 2 cm and hard X-ray peaks can be due to a propagation time, or to the electrons being different populations with different production times: thus, depending on the conditions, the 6 cm emission could be due to thermal electrons which require longer to be produced, while the 2 cm electrons are non-thermal electrons produced in the impulsive phase.

5. Simultaneous 6 and 20 cm Observations

The usual spectrum of a homogeneous non-thermal gyrosynchrotron source with a field of several hundred gauss peaks at around 4 cm, and thus the flux at 20 cm from such a source tends to be small. For this reason the sources seen at 20 cm in conjunction with a burst might be expected to show the influence of something other than the non-thermal electrons, as opposed to simultaneous 2 and 6 cm observations, where both wavelengths are close to the peak of the non-thermal burst spectrum. A burst on 1980 May 15 was observed at 6 and 20 cm by Velusamy *et al.* (1987) and showed complexity in both the time profile and in the mapped structure. Figure 5 shows the time profile of the event in two hard X-ray channels from ISEE-3, the GOES soft X-ray flux and

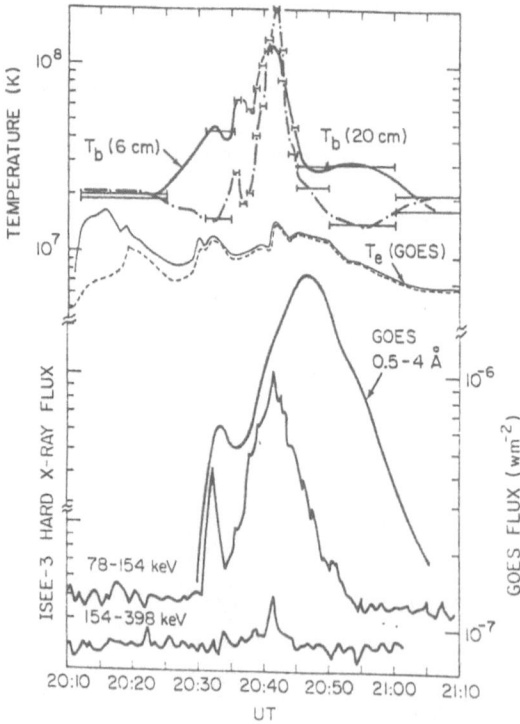

Fig. 5.   The time profiles of the 2 and 6 cm peak brightness temperatures, the X-ray electron temperature deduced from the fluxes in two GOES soft-X-ray bands, the flux in one of the GOES bands, and the hard X-ray fluxes in two of the channels of the ISEE-3 spectrometer.

electron temperature profiles, and the variation of the maximum brightness temperature in maps made at the VLA. The 20 cm profile is seen to peak later than the higher frequencies, and shows less structure. The 20 cm radio burst (Figure 6(a)) shows a source predominantly immediately above the flare region on the disk, coincident with the 6 cm burst source (Figure 6(b), with the corresponding field of view indicated by a

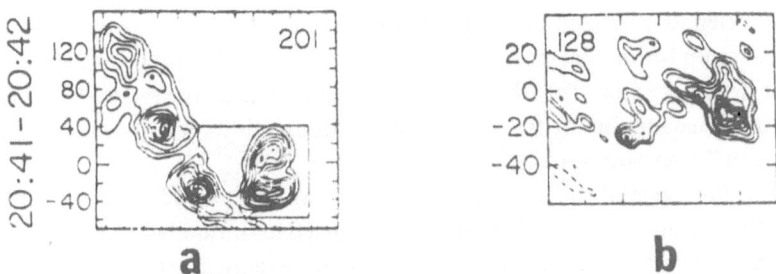

Fig. 6(a, b). The maps of the 20 and 6 cm burst sources, respectively, at one time (20 : 41–20 : 42). Note that the 20 cm structure varied considerably during the burst. The box in (a) indicates the field of view of the 6 cm map in (b).

box on Figure 6(a)), together with an apparently much higher and weaker source well above the limb. The complex time profile of the flare is due to the appearance of multiple sources at both 6 and 20 cm, which are apparent in Figures 6(a) and 6(b). In general, it is difficult to discuss the energetics of this event due to the complexity of the spatial structure. It is interpreted as the interaction of multiple loops in the active region, including some very high loops responsible for the 20 cm emission seen above the limb. Both 6 cm and 20 cm brightness temperatures exceed 10^8 K, and non-thermal emission is implied at both wavelengths.

Melozzi, Kundu, and Shevgaonkar (1985) also observed several bursts at both 6 and 20 cm. They occurred in three groups spread over an hour of observations. The three bursts showed a range of behaviours, although coming from the same active region: all three were seen at 6 cm, but one was completely absent at 20 cm, one was a factor of 4 lower in flux at 20 cm compared to 6 cm, and the third was stronger at 20 cm than at 6 cm. In addition, the latter burst showed more fine time structure at 20 cm than at 6 cm. In some of the bursts the 20 cm sources were larger and simpler in appearance than the 6 cm sources, but generally coincident, so that the 20 cm sources could have been outlining the tops of the loops with the 6 cm sources coming from lower down. The third burst referred to above seemed to be a double source at both wavelengths, but with the two wavelengths not coincident as if the 6 cm outlined one loop and the 20 cm another. The polarization structure of the maps was in agreement with this idea. We note that 'spike' bursts seem to occur more commonly at 20 cm than at 6 cm during the rise phase of flares but cannot be recognized without high-time-resolution equipment (e.g., see the review by Benz, 1986). Such spikes could well explain fine structure seen at 20 cm but not at 6 cm.

6. Discussion

For the average microwave burst the peak wavelength is around 4 cm. For such a burst one does not expect to see the same structures at 6 and 2 cm, since the shorter wavelength is optically thin and the longer wavelength is optically thick. Only when two wavelengths are both optically thin would one expect to see the same source structure, as indeed was the case in the 1980 October 1 event, since one is seeing the same electrons at both wavelengths. When both wavelengths are optically thick, the optically thick layer will be higher up for the longer wavelength. Since one is looking at two different layers, again one does not expect that the observed structure will be the same. The observations confirm this: usually the 20 cm burst sources are much larger than the 6 cm sources, and show different structures.

Thus, on the basis of the small number of examples available, we conclude that the assumption of a simple source structure usually used in the analysis of nonimaging radio data is occasionally adequate, in the sense that essentially the same population of electrons is producing the burst at the two wavelengths, if one is looking at two wavelengths close to the burst peak (in the solar case, 2 and 6 cm). However, if one is looking at a wavelength far from the burst spectral peak, such as 20 cm in most solar bursts, then often other sources are present which may be unrelated to those at higher wavelengths. Conclusions made by comparing data at the two wavelengths in this case are clearly misleading.

For stellar observations of M dwarfs, these conclusions are not particularly relevant since it is well known that radio flares on these stars are due to an as yet unidentified coherent mechanism (e.g., Dulk, 1985). However, radio flares on RS CVn stars are probably due to non-thermal gyrosynchrotron emission, and thus are more similar to solar flares. If the frequency spectra of these flares are particularly simple, then the usual analysis is probably valid. That is, a single population of electrons is probably responsible for the emission at all wavelengths, and one can sensibly use information gained at different wavelengths to deduce information about that electron population. VLBI observations (Mutel et al., 1985) show that the size of the RS CVn radio sources is comparable to the size of the star, at least, and this implies that large loops trap the electrons. We have no information on the structure of these loops, and thus cannot say whether they are truly simple. That this is not the case for solar radio bursts needs to be borne in mind when interpreting stellar observations.

The VLA can now observe at 3.7 cm (8.4 GHz). What would we expect to see in solar observations at this wavelength based on this survey of previous observations? Since 8.4 GHz is close to the typical peak of a non-thermal burst, it will most likely resemble the 6 cm optically thick source structure rather than the 2 cm optically thin source. However, in principle it should be smaller than a 6 cm source, since the latter is at a greater height, and one might hope to see a transition from optically thin to optically thick during the rise phase of a burst.

Acknowledgements

This research was supported by NSF grant ATM-87–17157, NASA grant NAGW-1541, and NASA contract NAG 5–969. The computions were done at the University of Maryland's Computer Science Center.

References

Batchelor, D. A., Crannell, C. J., Wiehl, H. J., and Magun, A.: 1985, *Astrophys. J.* **295**, 258.
Benz, A. O.: *Solar Phys.* **104**, 99.
Dulk, G. A.: 1985, *Ann. Rev. Astron. Astrophys.* **23**, 169.
Dulk, G. A., Bastian, T. S., and Kane, S. R.: 1986, *Astrophys. J.* **300**, 438.
Kundu, M. R., Velusamy, T., and White, S. M.: 1987, *Astrophys. J.* **321**, 593.
Melozzi, M., Kundu, M. R., and Shevgaonkar, R. K.: 1985, *Solar Phys.* **97**, 345.
Mutel, R. L., Lestrade, J.-F., Preston, R. A., and Phillips, R. B.: 1985, *Astrophys. J.* **289**, 262.
Shevgaonkar, R. K. and Kundu, M. R.: 1985, *Astrophys. J.* **292**, 733.
Velusamy, T., Kundu, M. R., Schmahl, E. J., and McCabe, M.: 1987, *Astrophys. J.* **319**, 984.

RADIO EMISSION FROM STELLAR FLARES

JAN KUIJPERS

Sterrekundig Instituut, Utrecht University, P.O. Box 80000, 3508 TA Utrecht, The Netherlands

Abstract. An overview is given of the observations of stellar radio 'flares', defined as radio emission which is both variable in time and created by explosive releases of magnetic energy. The main sources of such flares are late-type Main-Sequence stars, classic close binaries, X-ray binaries, and pre-Main-Sequence stars.

We summarize the interpretations of these observations in terms of the various incoherent and coherent emission mechanisms. The possible importance of a coherent emission process in electrostatic double layers is pointed out.

We briefly indicate the diagnostic importance of radio emission for the flare process in classic and compact stars. In particular we discuss the possible production of radio flares from interactions between an accretion disk and the magnetic field of the central object.

1. Introduction

We define stellar *radio flares* as variable radio emission from stellar mass objects which is possibly produced by explosive releases of magnetic energy. Such flares have been found in (1) late-type Main-Sequence stars including the classic flare stars (dMe and dKe stars), (2) detached, semi-detached, and contact close binaries (RS CVn and, respectively, Algol-type and W UMa-type stars), (3) X-ray binaries, (4) pre-Main-Sequence stars (T Tauri and protostars), (5) magnetic Ap/Bp stars, and (6) late-type giants and supergiants (Hjellming and Gibson, 1980; Gibson, 1985). In this review we shall discuss the first four groups. For the recently discovered variability (time-scale of months) in Ap/Bp stars we refer to (Drake *et al.*, 1987; André *et al.*, 1988). A recent catalogue of the more extended group of (variable and nonvariable) continuum radio stars (but excluding pulsars, compact H II regions and supernovae) has been compiled by Wendker (Wendker, 1987). Recent reviews of radio stars in general including O, B stars (bremsstrahlung, gyrosynchrotron, shocks), M giants, symbiotic stars are given in Dulk (1985, 1988), Kuijpers (1985b), Bookbinder (1988), and Hjellming (1988).

In general the observed radio emission from stellar flares (a few hundreds of MHz to a few tens of GHz) is negligible in comparison to the emission at X-ray, UV, and optical frequencies. However, since part of the radio emission originates at the very site of the primary energy release, it serves as an important *diagnostic tool for the study of the energy release and acceleration mechanisms in the flare*. As in the case of solar flares this works best if the source is spatially resolved at the various frequencies and in both modes of circular polarization. The spatial extent, obtained from direct measurements or, in case of a fluctuating signal, indirectly from light-travel time estimates, and the flux determine the brightness temperature at that frequency and polarization, given by

$$T_b(f) \simeq 4.8 \times 10^{10} \frac{S_f}{100 \,\text{mJy}} \left(\frac{0.48 \,\text{mas}}{\alpha} \frac{5 \,\text{GHz}}{f} \right)^2 \text{K} , \qquad (1)$$

Solar Physics **121**: 163–185, 1989.

where the source has a circular projection of radius α in milliarcsec (mas) and a polarized (of the specific mode) flux density S_f at frequency f (1 mJy = 1 millijansky = 10^{-26} erg cm^{-2} Hz^{-1} s^{-1}; the apparent radius of the Sun at a distance of 10 pc is $\alpha = 0.48$ mas). The *brightness temperature* together with *polarization, temporal and spectral characteristics* then forms an important constraint on the radiation mechanism and the distribution of the radiating particles in momentum space.

The only stellar flare sources other than the Sun in which spatial structure has been observed (with Very Long Baseline Interferometry), are the relatively strong (> 100 mJy) and extended nearby (distance less than 100 parsec) classic close binaries (of the RS CVn and Algol type; angular size 1–3 milliarcsec) and jets of some X-ray binaries and pre-Main-Sequence objects (protostars and T Tauri stars). Recently detailed frequency information of radio flares from stars other than the Sun has become available in the form of multiple-frequency observations and so-called dynamic radio spectra (Bastian and Bookbinder, 1987).

The main observational problem of course is that the active flaring state in most known flaring objects is relatively short; for the classic flare stars the flare frequency is often much less than one per day; therefore, some courage is required to allocate time to observe stellar flares. However, such observations, and in particular when they are done simultaneously in the X-ray, ultraviolet, optical and infrared domain, can be of great value for the fascinating subject of *explosive, magnetically controlled, energy release in astrophysics* (Uchida, 1986).

2. dMe Stars

This is the group of the classic flare stars (dMe type) first studied by Lovell at radio frequencies (Lovell, 1971). The flaring activity of these stars is the most conspicuous of the entire group of active Main-Sequence stars with convective envelopes (spectral type later than F2; but most of which have not yet been detected in the radio (Bookbinder and Walter, 1988; Seaquist and Taylor, 1985; Willson, 1987; Bastian, Dulk, and Chanmugan, 1988), including our Sun with its comparatively weak flaring activity. The properties of the radio flares are summarized in Table I.

2.1. COHERENT EMISSION PROCESSES

Some of the burst emission is probably of coherent origin because of:

Narrow-band emission, in dynamic radio spectra with a relative frequency extent < 0.002 (Bastian and Bookbinder, 1987) and < 0.06 (Jackson, Kundu, and White, 1987); in multiple frequency observations with a relative bandwidth of a few percent or less (Lang and Willson, 1986a, 1988; White, Kundu, and Jackson, 1986; Willson, 1987).

High brightness temperature. Occasionally much larger radio luminosities than in Table I are observed for bursts which are often, but not always (Kundu and Shevgaonkar, 1988), *highly circularly polarized* ($> 75\%$ up to 100%) (Gibson and Fisher, 1981; Gibson, 1984; Lang and Willson, 1988; Kundu *et al.*, 1988; Jackson, Kundu, and White, 1987; Willson, 1988; White, Kundu, and Jackson, 1986; Kundu

TABLE I

Stellar radio flares

	Sun	dMe stars	Close binaries	X-ray binaries					Young stellar objects
				AE Aqr	Cyg X-3	SS433	GX 13+1	GX 17+2	
References	1	2–6	7–17	18	19–22	23–24	25	26–27	28–30
$\log L$ (5 GHz) erg s^{-1} Hz^{-1} [a]	↑13	13.3–15.6	14.3–18.1	17	22–24.3	21.3	20	21	17.4–18.8
L_{rad}/L_{bol}	10^{-12}	10^{-7}	10^{-7}			10^{-9} [b]			
L_{radio}/L_{x}	10^{-6}	10^{-3}	$10^{-3.5}$	10^{-6}	$10^{-3.5}$	10^{-4}	$10^{-7.5}$	$10^{-7.5}$	$10^{-3.5}$
t (5 GHz)	10^{3} s	<1 s–days	30 s–1 day	hr	hr–days	days	hr–days	hr–··	hr–months
log (total flare energy (ergs))	32	32	35	33.5	43		39.5	35.7[c]	≥34
log (characteristic dimension (cm))	10	9.5	11.2	9–10.5	13	15	14		

[a] Values are characteristic for the *unpolarized* flux.

[b] We used the estimate $L_{bol} \sim L_{kin} \sim 10^{40}$ erg s^{-1}.

[c] The energy in one quasi-periodic-oscillation.

References

(1) Kundu and Woodgate, 1986.
(2) Gibson, 1983a.
(3) Gibson, 1985.
(4) Kundu and Shevgaonkar, 1985.
(5) Slee et al., 1984.
(6) Willson, 1988.
(7) Hjellming and Gibson, 1980.
(8) Gibson, 1985.
(9) Mutel and Lestrade, 1985.
(10) Mutel et al., 1987.

(11) Morris and Mutel, 1988.
(12) Drake et al., 1987.
(13) Slee et al., 1987b.
(14) Owen and Gibson, 1978.
(15) Slee et al., 1984.
(16) Feldman et al., 1978.
(17) Willson and Lang, 1987.
(18) Bastian et al., 1988a.
(19) Johnston et al., 1986.
(20) Spencer et al., 1986.

(21) Molnar et al., 1984.
(22) Molnar et al., 1988a.
(23) Vermeulen et al., 1988.
(24) Kawai et al., 1988.
(25) Garcia, 1988.
(26) Penninx et al., 1988.
(27) Lewin et al., 1988.
(28) Montmerle, 1988.
(29) Feigelson, 1986.
(30) Stine et al., 1988.

et al., 1987). Brightness temperatures are $7 \times 10^{12}/\beta^2$ K at 330 MHz (Kundu and Shevgaonkar, 1988); $3 \times 10^{12}/\beta^2$ K at 408 MHz (Davis *et al.*, 1978); $1.8 \times 10^{11}/\beta^2$ K at 1.5 GHz (Kundu *et al.*, 1988); $8.5 \times 10^9/\beta^2$ K at 5 GHz (Slee *et al.*, 1981) (β is defined as the source linear dimension divided by twice the stellar radius). Sometimes *strong temporal variability* (down to 0.005 s (Lang and Willson, 1986b)) can be used to find an even more stringent limit to the brightness temperature: 10^{13} K at 1.4 GHz taking the product of the observed *e*-folding time of 0.2 s and the speed of light as an upper limit to the source dimension (Lang *et al.*, 1983) and 3.6×10^{15} K (rise time 0.02 s) at 430 MHz (Bookbinder *et al.*, 1989). Such brightness temperatures have also been observed in the case of the Sun at metric and decimetric wavelengths (Slottje, 1978; Benz, 1986; Dröge, 1967; Stähli and Magun, 1986).

Because of the strong polarization, the high brightness temperature and the narrow frequency extent, these bursts can only come from a coherent plasma process. (Note, however, that coherent radiation can be completely unpolarized and/or have a relatively large bandwidth.)

2.1.1. *Is the Electron Cyclotron Maser the Universal Explanation of Coherent Emission from Flare Stars?*

For a correct treatment of a linear plasma instability near a *cyclotron resonance* ($\omega - s\omega_{ce}/\gamma - k_{\parallel} v_{\parallel} = 0$) it is essential to take into account the relativistic dependence of the cyclotron frequency on the particle velocity (Blanken, Stix, and Kuckes, 1969; Zheleznyakov and Zlotnik, 1975). Following the early work on a direct instability of electromagnetic waves at (harmonics of) the electron-cyclotron frequency (Twiss, 1958; Ramaty, 1969; Melrose, 1973), an impressive amount of progress has been reached on the electron-cyclotron maser (Stepanov, 1978; Wu and Lee, 1979; Holman, Eichler, and Kundu, 1980; Melrose, Hewitt, and Dulk, 1984; White, Melrose, and Dulk, 1986; Aschwanden, Hewitt, and Dulk, 1989; McKean, Winglee, and Dulk, 1989), and its relativistic version (McCray, 1966; Zheleznyakov, 1967; Louarn, Le Quéau, and Roux, 1987), as an explanation for the above discussed strongly polarized bursts. This direct radiation instability occurs when energetic electrons are released in a magnetic flux tube (magnetic bottle) and develop a (one-sided) *loss-cone distribution* at the magnetic mirrors at the footpoints and if the ratio of electron-cyclotron frequency to plasmafrequency is sufficiently large (Winglee, Dulk, and Pritchett, 1988; larger than 3, or in the related synchrotron maser for MeV electrons larger than 0.3 (Louarn, Le Quéau, and Roux, 1989)). Since the surface coverage with kilogauss magnetic fields in flare stars is much larger than on the Sun (Saar and Linsky, 1985; Saar, 1988), the cyclotron maser probably occurs relatively often in flares on dMe stars. However, it is probably only one of several and not the universal coherent radiation process as is often suggested: In the weaker field regions (ratio of cyclotron to plasma frequency less than three) loss-cone distributions of trapped electrons do not directly produce radiation at one or twice the electron cyclotron frequency but cause the growth of upper hybrid waves which lead to radiation at once (Kuijpers, 1974, 1975; Zheleznyakov and Zlotnik, 1975; Kuijpers, van der Post, and Slottje, 1981) or twice (Stepanov, 1974; Winglee and Dulk, 1986) the upper hybrid frequency (Holman, 1983). Further, in the case of the Sun flares exhibit a rich variety of radio bursts (McLean and Labrum, 1985; Krüger, 1979) with different plasma instabilities. A large number of coherent radiation processes has been proposed for intense radiation observed from members of the solar system (see for a review, for example, Kuijpers, 1985a, b, 1980; Melrose, 1986; Papadopoulos and Freund, 1979; Tsytovich, 1973). Finally, from solar radio astronomy it is well-known that to determine the kind of burst reliably, in general dynamic spectra of high resolution both in frequency (down to 100 kHz) and time (down to millisec) are required. Therefore, although coherent emission processes certainly operate in flare stars, the existing identifications with specific processes, e.g., at the cyclotron frequency (White, Melrose, and Dulk, 1986), are at best tentative; often radiation at the upper hybrid frequency is equally well possible, as is observed in a class of solar bursts (the stationary type IVdm radiation from magnetic traps with continued injection (Kuijpers, 1980)).

Of course these high-frequency instabilities of loss-cone distributions for various ratios of cyclotron to plasma frequency form one physical family. Both the electron-cyclotron maser and the upper hybrid instability are simply opposite extremes of the *magnetically controlled phenomenon of trapped energetic particles*. Observationally, the characteristic emitted frequency is expected to vary from once or twice the cyclotron frequency to the upper hybrid frequency as the ratio electron cyclotron to plasma frequency shifts from values larger than unity to values smaller than unity; therefore, for order of magnitude estimates the observed frequency can be put equal to the *upper hybrid frequency*.

2.1.2. *Coherent Emission from Electrostatic Double Layers*

We now briefly describe a possible coherent radiation mechanism from electrostatic double layers, which has not appeared in the literature before but which may be of potential importance for magnetic flares in various objects.

Electrostatic double layers were first proposed in an astrophysical context by Alfvén (Alfvén, 1958) and since then have been advocated in a variety of cosmic high-energy phenomena (Alfvén, 1981). The reason is that double layers have dynamic properties which are characteristic for explosive phenomena in the Universe: *the acceleration of electrons, of ions and of mass motions*. Electrostatic double layers are localized (dimension of 10–100 Debye lengths) electrostatic potential drops along the magnetic field. The structure is generated by charge separation, which is self-consistently maintained by the motion of particles accelerated in this potential drop (Raadu and Rasmussen, 1988). We restrict ourselves to strong double layers in which the potential drop corresponds to several times the thermal particle energy. They have been observed to occur in current carrying plasmas in the laboratory (Torvén, 1982; Hershkowitz, 1985; Torvén, Lindberg, and Carpenter, 1985). A substantial part of the global magnetic energy of the electric circuit is released locally in double layers and in a highly nonthermal manner. Therefore, they are of interest for astrophysical flares of magnetic origin: the potential drop accelerates particles of both charges and at the same time allows relaxation of the magnetic stresses by the local unwinding of the magnetic field (Jacobsen and Carlqvist,

1964; Alfvén and Carlqvist, 1967; Raadu, 1984). Still many problems in our understanding of double layers remain to be solved, in particular on the question of their formation and on double layers of the explosive type. Under stellar, in contrast with laboratory, conditions it is not likely that the drift velocity of the current particles surpasses the thermal velocity. Rather an alternative may be local reconnection of magnetic fields which changes the path of an electric current system; because of the large dimension the current system has a huge inductance and a large inductive voltage is generated locally, perhaps leading to the (transient) formation of a double layer. In this case one should think of a *spatially distributed large collection of many transient double layers* (Hénoux, 1987).

Coherent radiation is expected to come from double layers at frequencies somewhat above the ambient plasma frequency (Kuijpers *et al.*, 1989). The reason is that electrons are accelerated in the double layer and, therefore emit *'linear accelerator'* radiation. The radiation is expected to be of the masering type somewhat similar to the *free electron laser*. Now, however, the beam is strongly inhomogeneous, not necessarily relativistic and instead of the periodic wiggle of the electron laser the spatial structure of the strong electrostatic potential of the double layer appears. The radiation should be strongest in directions perpendicular to the electric field: in a static structure the resonance condition $(\omega_t - \mathbf{k}_t \cdot \mathbf{v} = \omega - \mathbf{k} \cdot \mathbf{v}$, where ω, $\mathbf{k}$ are the frequency and wave vector of the electromagnetic (subscript t) and of the electrostatic (no subscript) wave) reduces to $\omega_t = -\mathbf{k} \cdot \mathbf{v}$ which for a double layer is of the order of the electron plasma frequency outside. Perhaps this mechanism can explain the short intense bursts of high-frequency radiation which is observed in disrupting double layers in the laboratory (Lindberg, 1987, 1988).

2.2. SPECTRAL CORRELATION

There is a *lack of detailed correlation* in time between emission in various frequency domains:

– Between X-ray (0.02–2 keV) and optical observations (Doyle *et al.*, 1988). This could be caused by a flare occurring in a closed magnetic structure low in the chromosphere, or alternatively, by heating by a proton beam instead of an electron beam, inhibiting the rise of heated material ('evaporation') by the directed momentum of the proton beam and explaining the absence of microwave emission by the absence of energetic electrons (van den Oord, 1988).

– Between emission at different radio frequencies (Kundu and Shevgaonkar, 1988; Kundu and Hjellming, 1985). This may, however, be caused by a complex narrow-band structure as is observed for solar radio flares.

– Between soft X-rays (0.04–2 keV) and radio (6 and 20 cm) observations (Kundu *et al.*, 1988). This may be caused by a number of effects: (i) Observing at the wrong (too soft) X-ray frequency: on the Sun only the hard X-ray flare (> 20 keV) and microwave (1.8 cm) time profiles are very similar (Dennis, 1988). (ii) Gyroresonance absorption, Razin suppression near the plasma frequency or free-free absorption; such effects have been demonstrated to be important for solar flares (Klein, 1987), where in one case the gyrosynchrotron radiation from energetic electrons responsible for a hard X-ray burst observed up to 200 keV was negligible below 30 GHz! (iii) A column depth above the transition region larger than 1.7×10^{20} cm^{-2} or a rapidly converging magnetic field (Lu and Petrosian, 1988). (iv) A variety of particle spectra: in some (impulsive) hard X-ray flares microwaves (17 GHz) are emitted mainly by < 200 keV electrons with a steeply decreasing energy spectrum and which precipitate in lower layer of high (900 G) field strength; in other (extended) flares microwaves are emitted mainly by MeV electrons trapped in a coronal loop of relatively low (< 100 G) field strength, and the correlation is best for very hard (> 360 keV) X-rays (Kosugi, Dennis, and Kai, 1988).

More case studies of flares from these stars with simultaneous coverage at many frequencies are required to show that the X-ray and radio activity are not the combined result of the same flares (Holman, 1985) or that flares from dMe stars would be physically very different from solar flares.

2.3. QUIESCENT EMISSION AND MICROFLARING

Some flare stars show unpolarized or weakly polarized slowly varying (time-scale hour–days) emission or a background of such emission (Gary and Linsky, 1981; Linsky and Gary, 1983; Gary, 1985; Stewart *et al.*, 1988). The radiation cannot be bremsstrahlung (because of the magnitude of the observed X-ray flux) and either gyroresonance emission of a thermal plasma or gyrosynchrotron emission is at play (Gary, 1985). The gyroresonance interpretation requires a large source size ($R = 3R_*$, $T = 3 \times 10^7$ K, $B = 200$ G, $n = 2 \times 10^8$ cm^{-3}, $s = 10$) and sometimes meets some difficulties: one-temperature model not adequate (Kundu and Shevgaonkar, 1985); precipitation losses are two orders of magnitude above the X-ray luminosity (Kundu *et al.*, 1987). The *gyrosynchrotron interpretation* requires a less dense gas of higher particle energy over a smaller volume, either in the form of a tenuous high-temperature component or a non-thermal population ($R = R_*$, $T = 2.7 \times 10^8$ K, $n = 10^{5-6}$ cm^{-3}, $B = 300$ G); it also requires less severe energy losses (precipitation losses $1.8 \times 10^{29} H$ erg s^{-1}, where $H \ll 1$ is a filling factor for the regions of strong diffusion (Kundu *et al.*, 1987)), and explains well the spectral shape of the quiescent emission from dMe stars (magnetic dispoles significantly smaller in size than a stellar radius; typical turnover frequency of the calculated spectra between 4–30 GHz (White, Kundu, and Jackson, 1989)).

To sustain this so-called quiescent emission the continued production of accelerated particles is required for several hours (Kundu *et al.*, 1987; Willson, 1987) or once every hour (Kundu and Shevgaonkar, 1985). Then this leaves open the possibility that the radiation is plasma radiation (Holman, Bookbinder, and Golub, 1985) as in the case of a class of stationary solar metric continua (which can be highly polarized).

Further the distinction between this form of activity and flares becomes rather vague. It may well be that the quiescent emission is caused by a sequence of *nanoflares* (Schadee, de Jager, and Švestka, 1983; Schadee, 1986; Parker, 1988) or *microflares* (Lin *et al.*, 1984; Athay, 1984; Canfield and Metcalf, 1987) as in the solar case. Evidence for stellar *microflares* exist (Butler *et al.*, 1986; Whitehouse, 1985), occurring over tens of seconds to ten minutes with individual energies of 2×10^{30} erg (van den Oord and Barstow, 1988) up to $5 \times 10^{30}-10^{31}$ erg (Kundu *et al.*, 1988).

2.4. Stable Magnetic Structures?

For YZ CMi all observations at various frequencies show the same sense of polarization (Kundu and Shevgaonkar, 1988). Possibly this implies the existence of asymmetric leading-trailing stellar spots and a stable orientation (and a viewing angle well out of the magnetic equator) over the observed number of years (Gibson, 1983b).

2.5. The solar connection

The main observational difference then between solar and dMe flares is at present only the much larger flare luminosities and total energies for these stellar flares. Theoretically this is also expected if the flares are of magnetic origin and if the magnetic energy content of the stellar coronae are much larger than for the Sun (Kuijpers, 1985b). For the latter substantial evidence exists in the form of observed strong surface fields (kG) with large filling factors (Saar, 1988).

3. Close Binaries

About 100 close binaries (1–30 days period) of the RS CVn (detached) and Algol (semi-detached) type have been detected at radio frequencies (see Table I). It now appears that the radio luminosity (8 GHz) is clearly correlated with the rotation rate (Stewart *et al.*, 1988). The spectrum sometimes peaks between 1–15 GHz (Mutel *et al.*, 1987). For individual sources the spectral index between 1.5–5 GHz ($S_f \sim f^{\alpha}$) increases with radio luminosity from -1 to $+1$ (Mutel *et al.*, 1987). Usually the degree of circular polarization is between 0 and 30%, but sometimes higher. For some, non-eclipsing, systems the sense of circular polarization ('helicity') reverses between 1.4 and 4.9 GHz (Mutel *et al.*, 1987). For individual systems the fractional circular polarization decreases with increasing luminosity at 5 GHz (Mutel *et al.*, 1987).

A few contact binaries (W UMa systems) have been detected in the radio: a flare in VW Ceph showed radio emission arriving a few minutes before the soft X-rays, very similar to the solar flare case (Vilhu *et al.*, 1988); FK Comae Berenices has been detected at 2, 6, and 20 cm and the emission requires accelerated electrons (Hughes and McLean, 1987).

3.1. Core-halo gyrosynchrotron source

Dual-frequency VLBI of several of these sources (Lestrade, 1988; Mutel *et al.*, 1985; Lestrade *et al.*, 1988) shows the presence of a variable core-halo structure. For the best measurements the core is smaller than the K-subgiant (assumed to be the active star, see figure in Mutel *et al.*, 1985) and the corresponding brightness temperature at 5 GHz is at least $1-3 \times 10^{10}$ K, requiring electron kinetic energies of 1.3–3.9 MeV in the case of incoherent emission; the size of the halo is comparable to the dimension of the binary (10^{12} cm) and the corresponding brightness temperature of the halo at 5 GHz is typically $(5-10) \times 10^8$ K. The observed brightness temperatures, dimensions, spectral index, and frequency turnover, circular polarization and helicity reversal are best explained as *gyrosynchrotron emission* of electrons with Lorentz factors 2–10 in magnetic fields of strengths 30–200 G (Owen, Jones, and Gibson, 1976; Spangler, Owen, and Hulse, 1977; Spangler, 1977; Feldman, 1983; Borghi and Chiuderi-Drago, 1985; Kuijpers and van der Hulst, 1985; Lestrade *et al.*, 1988). The characteristic frequency dependence of the source size agrees well with models of gyrosynchrotron emission from inhomogeneous distributions in magnetic loops in close binaries (Borghi and Chiuderi-Drago, 1985; Klein and Chiuderi-Drago, 1987; Lestrade *et al.*, 1988). The radiation *from the core is self-absorbed and unpolarized gyrosynchrotron emission in contrast with the optically thin radiation from the halo.* As the luminosity increases the degree of circular polarization decreases, essentially since the luminosity increase is coming from the self-absorbed core (Mutel *et al.*, 1987).

3.2. CONTINUOUS ACCELERATION

When VLBI observations are absent a flare is traditionally modelled by an *expanding, initially optically thick, gyro-synchrotron source* (Mutel *et al.*, 1985). In this model the emission becomes optically thin, and the maximum flux density decreases, at progressively lower frequencies as the blobs expand. In the case of a radio flare from Algol, however, dual-frequency VLBI observation (Lestrade *et al.*, 1988; Lestrade, 1988) showed the flux increase not to be caused by expansion of an energetic cloud or loop, but by an increase in brightness temperature of a source with dimension three times the sub-giant radius (which went up to 3×10^{10} K at 8.4 GHz) and the *continued production of high-energy electrons.* Such an acceleration of electrons occurring over prolonged periods of time of order hours or more was already indicated by the general radio flux behaviour in time of these sources, e.g., Kuijpers and van der Hulst (1985).

3.3. COHERENT EMISSION

(Gyro)synchrotron emission, however, cannot always be the cause of the observed emission. Occasionally a highly polarized (Mutel and Weisberg, 1978) intense outburst or variations on the time-scales of minutes (Brown and Crane, 1979; Newell *et al.*, 1980; Mutel *et al.*, 1987; Lestrade *et al.*, 1988; Willson and Lang, 1987) have been observed. The only explanation for these cases of highly circularly polarized and intense bursts is some form of *coherent* or *'plasma' radiation* as in (Melrose and Dulk, 1982); see Section 2.1.

3.4. SPECTRAL CORRELATION

As is the case for dMe stars there is an absence of detailed correlation between activity at different frequencies: an ultraviolet flare with no radio (1.5 and 4.9 GHz) emission (Lang and Willson, 1988); activity in X-rays and in the radio with non-correlated variations (White, Kundu, and Jackson, 1986; van den Oord *et al.*, 1989). Because this lack of detailed correlation can easily be explained by a variety of effects (see Section 2.2), we conclude that energetically it seems likely that the activity is causally related.

3.5. Global magnetic structures

That large magnetic structures on these stars can exist over a long duration follows from the sense of circular polarization and the helicity reversal between 1.4–5 GHz, which, for two RS CVn systems, stays the same over ten years (Mutel et al., 1987). Because of their small separation (5–10 stellar radii) the binary components are tidally locked and rotate nearly synchronously with the orbital motion, they have high rotational velocities (periods of the order of a few days) and, if they also have a convective envelope, both stars probably have a dynamo magnetic field. Thus it is not surprising if magnetic structures would exist which connect both stars ('magnetospheres'). Such is also indicated by the observed core-halo structure. Further, the observed radio emission shows no clear dependence on phase, indicating that the halo extends over much more than a stellar radius out of the orbital plane. Finally the mean circular polarization in a few binaries increases from 0 to 30% as the angle of the orbital plane with the line of sight varies from 0° to 90° (Mutel et al., 1987); again this points to a large-scale magnetic structure which has some symmetry about the orbital plane.

3.6. Duplicity-enhanced activity

Even if the stellar rotation is synchronous with the orbit, one still expects violent relaxations between the individual evolving stellar magnetic fields (Bahcall, Rosenbluth, and Kulsrud, 1973; DeCampli and Baliunas, 1979; Simon, Linsky, and Schiffer, 1980; Uchida and Sakurai, 1983; Lamb et al., 1983; van den Oord, 1988). The extended halo sources with dimensions comparable to the separation of the binary constitute the observational evidence that such interactions are important. Further since in Algol, which is a semi-detached system with transient accretion gas rings (Kaitchuk and Park, 1988), an X-ray flare and radio outbursts have been observed very similar to those in RS CVn (detached) systems, this perhaps means that accretion also leads to *driven magnetic reconnection*.

3.7. Energy budget

The required rate of energy injection in relativistic particles is of course dependent on the assumed radiation process and on source conditions such as magnetic field strength and background density. For σ^2 CrB the energy in high-energy electrons in the synchrotron interpretation (Lorentz factor 3, magnetic field strength 35 G) is 10^{30} erg, which is a fraction 10^{-6} of the energy in the X-ray gas (Kuijpers and van der Hulst, 1985). Both in the case of σ^2 CrB (Kuijpers and van der Hulst, 1985) and of Algol (van den Oord et al., 1989) the mildly relativistic electrons responsible for the gyro-synchrotron emission could be considered as the high-energy end of an initially co-spatial power-law distribution emitting the X-rays; the fraction of the energy put into the particles with energies above 10 keV would be only 5×10^{-2} for Algol.

On the basis of the observed core-halo structures, their temporal variability and the energetics of the radio and X-ray emission we, therefore, adopt the working hypothesis that the *radio emitting electrons are accelerated at the same time as the X-ray gas is energized*. This can occur in two ways:

(1) In the form of *strong and impulsive flares* as seen in X-rays with dimensions smaller than a stellar radius: total energy in X-rays of 2.4×10^{34}–1.0×10^{35} erg, from loops of length 4.5×10^{10}–1.6×10^{11} cm ($= 0.3$–0.6 stellar radius) and total volume 6.9×10^{29}–1.4×10^{31} cm^3 and density 6×10^{11}–3.1×10^{11} cm^{-3} (van den Oord, Mewe, and Brinkman, 1989; van den Oord and Mewe, 1988). In this case the high-energy particles must be able to escape to an extended region, comparable to the binary separation (Massi *et al.*, 1989), and of low density in view of variations in the radio flux at 1.4 GHz of the order of one day (background density $< 10^7$ cm^{-3} for σ^2 CrB (Kuijpers and van der Hulst, 1985)).

(2) On a more gradual scale in the form of *many local reconnections of magnetic field in between the binary components.*

The large flare energy in these classic close binaries as compared to dMe stellar flares, which again are much larger than solar flares, can be understood if they are all of magnetic origin: in the case of dMe stars and of classic close binaries the filling factor of photospheric magnetic fields is large compared to the solar case; further for similar photospheric field strengths the coronal volume, and, therefore, also the magnetic energy, is much larger for classic close binaries than for dMe stars.

4. X-Ray Binaries

A small number of X-ray binaries with compact stellar objects show strong and variable radio emission: 13 sources at 5 GHz out of 117 X-ray sources (Nelson and Spencer, 1988); see the tables in Hjellming and Johnston (1985) and Chanmugan (1987). Since the spectral index of the observed flux varies between $+1$ and -1 and the brightness temperatures estimated from the observed time variability are moderate (e.g., $10^{6.7}$ K at 5 GHz and $10^{9.1}$ K at 1.5 GHz for GX 13 + 1 (Garcia *et al.*, 1988); 10^9–10^{12} K at 5 GHz for the cataclysmic variable AE Aqr (Bastian, Dulk, and Chanmugan, 1988a)) the radio emission is usually ascribed to the (*gyro*) *synchrotron process* (Dulk, 1986).

The white dwarf binaries (cataclysmic variables) detected in the radio are *magnetic cataclysmic variables* (coherent pulsations in the optical light curve; strong magnetic field guiding the accreting mass near magnetic poles on the white dwarf), *dwarf novae* (optical outbursts in a disk around the white dwarf), *novae*, and *recurrent novae*. We shall not discuss (recurrent) novae since their radio emission is probably not caused by the release of magnetic energy (Chanmugan, 1987; Bookbinder and Lamb, 1987) (bremsstrahlung or (gyro/synchrotron radiation from shock accelerated electrons due to the ejected gas).

4.1. LOCATION OF CENTRAL RADIO EMISSION

A small source can be detected only if its brightness temperature is large enough. Further, radio signals of too low frequencies cannot propagate in dense magnetoplasma near the compact object or are substantially absorbed in intervening gas. As a result radio observations are a diagnostic probe of the outer regions only. Below we shall use field strengths estimated from assumed magnetic dipoles at the location of the compact objects; estimates for gas densities are based on hypothetical isothermal and isotropic mass loss rates, although of course in reality the distribution of the gas around X-ray binaries is very inhomogeneous, with accretion disks, jets and streams. The following physical effects are important:

(1) *Brightness temperature*: an unpolarized source at a distance d and brightness temperature T_b at frequency f produces a flux S at frequency f it its radius exceeds

$$R = 3.1 \times 10^{10} \left(\frac{S}{1\ \mathrm{mJy}}\right)^{1/2} \left(\frac{f}{10^9\ \mathrm{Hz}}\right)^{-1} \left(\frac{T_b}{10^{12}\ \mathrm{K}}\right)^{-1/2} \frac{d}{100\ \mathrm{pc}}\ \mathrm{cm}\ . \qquad (2)$$

Note that the brightness temperature is not limited by the conventional inverse Compton limit of 10^{12} K if the radiation is not synchrotron radiation.

(2) *Electron plasma frequency*: radiation can only escape above the electron plasma frequency. A location at a distance r from the central object with a wind (or jet) $\dot{M}$ with velocity v is visible only above a frequency

$$f > 4.5 \left(\frac{\dot{M}_{-7}}{v_8}\right)^{1/2} r_{11}^{-1}\ \mathrm{GHz}\ . \qquad (3)$$

Here $\dot{M}_{-7}$ is the mass loss rate in units of $10^{-7}\ M_0\ \mathrm{yr}^{-1}$, v_8 the velocity in units of $10^8\ \mathrm{cm\ s}^{-1}$ and r_{11} the radius in units of 10^{11} cm. The observed expansion speed is of order of the escape velocity and varies from a few thousand km s^{-1} for a white dwarf to a hundred thousand km s^{-1} for a neutron star. The mass loss and mass accretion rates vary considerably even within the same group of compact objects. From Equations (2) and (3) it is clear that in general only the outer regions of X-ray binaries ($> 10^{10}$ cm) can be probed at radio frequencies, and at best in the far infrared; the inner regions can be detected in the radio only if both a (coherent) mechanism is at play with brightness temperatures substantially above 10^{12} K and if the central region is not obscured by disk or jet.

(3) *Cyclotron frequency*: the presence of a magnetic field prevents the escape of radiation below the electron-cyclotron frequency, and causes severe absorption of the escaping radiation in the overlying layers due to *cyclotron resonances at the lower harmonics of the cyclotron frequency*, sf_{ce}. This absorption increases with the electron density and the temperature of the overlying gas and the magnetic scale height (van den Oord et al., 1989). Estimating the magnetic field for an assumed dipole at the center of the compact object with polar field strength B_0 at radius R_0 (magnetic moment $0.5B_0R_0^3$), a magnetic neutron star with $B_0 = 10^{12}$ G and $R_0 = 10^6$ cm can only be observed at frequencies above

$$f > 2.8 r_9^{-3} s\ \mathrm{GHz}\ , \qquad (4)$$

whereas a magnetic white dwarf with $B_0 = 10^6$ G and $R_0 = 10^9$ cm, can only be observed above

$$f > 2.8 r_{10}^{-3} s\ \mathrm{GHz}\ . \qquad (5)$$

(4) *Free-free absorption*: an estimate of the optical thickness from radius r to infinity is obtained from $\tau = n_e^2 kL$, where the electron density is, as before, derived from a wind

with constant speed, the length is approximated by the density scale height, $L = r/2$, and the expression for the absorption coefficient is taken from Mezger and Henderson (1967) with $a(f, T) = 1$,

$$\tau_{ff} = 0.35 \left(\frac{\dot{M}_{-7}}{v_8} \right)^2 f_9^{-2.1} T_4^{-1.35} r_{14}^{-3} . \qquad (6)$$

Free-free absorption prohibits the mapping of the inner regions of an X-ray binary in the radio and infrared for values $\dot{M}_{-7}/v_8 \geq 10^{-3}$ unless the circumstellar mass distribution is extremely dependent on angle.

(5) *Scattering*: there are two effects: *scattering by single electrons and by electron inhomogeneities*.

Structural details are smeared out by free electron scattering. The optical thickness is

$$\tau_{es} = 0.85 \, \frac{\dot{M}_{-7}}{v_8} \, r_9^{-1} . \qquad (7)$$

Scattering by inhomogeneities along the line of sight are particularly important for sources in the galactic plane (Rickett, 1977). The change of the refractive index increases with the square of the wavelength. For Cyg X-3 at 10 kpc the scattering size at 23 GHz is estimated to be as large as 0.7 mas (Molnar, Reid, and Grindlay, 1988a), corresponding to a linear scale of 10^{14} cm at that distance.

From these effects we conclude that *free-free absorption of the overlying gas or scattering by electron density inhomogeneities along the line of sight* probably puts the most stringent lower limit on the structural details which can be observed at radio frequencies near the compact companion in a binary. Coherent radiation at far infrared wavelengths from the inner regions may be observable if not obscured by disk or jet.

4.2. FLARE ORIGIN IN CATACLYSMIC VARIABLES

Radio flares have been observed in three cataclysmic variables: AM Her (Bastian, Dulk, and Chanmugan, 1985), AE Aqr (Bookbinder and Lamb, 1987; Bastian, Dulk, and Chanmugan, 1988), and AO Psc (Wendker, 1987). Probably the radio flares are caused by accretion since their intensity is large compared to the radio flares of single late-type Main-Sequence dwarfs or compared to the weak radio emission from the cataclysmic variable progenitor V471 Tau (Bastian, Dulk, and Chanmugan, 1988). Further, only gravitational and not nuclear energy liberation plays a role, as any nuclear burning on the surface of the white dwarf in AM Her proceeds in a stationary non-impulsive manner. As the observed radio active cataclysmic variables are magnetic, probably also *magnetic fields* are important. AE Aqr has a relatively small field (10^5–10^6 G) allowing the formation of an accretion disk and the flares possibly result from interactions between the white dwarf magnetic field and the disk (see Section 4.8 and Bookbinder and Lamb (1987)). AM Her, on the other hand, also has a magnetic field but of sufficient strength (few $\times$ 10^7 G at the white dwarf surface) that no accretion disk forms (Lamb and Melia, 1986). Its radio emission sometimes shows appreciable circular polarization (25%) and is explained as gyrosynchrotron emission with a source size of radius 10^{11} cm (electrons with average Lorentz factor 2 in a magnetic field of 40 G; brightness temperature at 4.9 GHz of 3×10^9 K; harmonic number 40) (Chanmugan and Dulk, 1982; Bastian, Dulk and Chanmugan, 1985). The presence of polarization implies the existence of a magnetic field on a scale comparable to the size of the orbit and the flaring is perhaps caused by distortion of the white dwarf magnetic field in the outer region by the accreting material and subsequent magnetic reconnection with the field of the Main-Sequence companion which initially permeates the accreting material. Why no radio emission has been detected from all other magnetic cataclysmic variables is a problem, but it may be related to obscuration (Fürst *et al.*, 1986; see Section 4.1).

4.3. COHERENT EMISSION

In the cataclysmic variable AM Her a strong ($T_b > 10^{10}$ K at 4.9 GHz) and probably completely circularly polarized radio flare must have been produced by a coherent process (Dulk, Bastian, and Chanmugan, 1983).

In some dwarf novae radio emission has been discovered, with one to two days delay after the optical maximum (Benz and Güdel, 1989). The authors propose that the radiation is coherent cyclotron emission from electrons accelerated near the disk and radiating above the poles of the white dwarf (1000–2000 G).

Infrared (2.0–2.4 μm) flare events of few minutes duration in Cyg X-3 without clear counterparts in X-rays (1.5–10.0 keV) (Mason, Cordova, and White, 1986), are interpreted by these authors as free-free emission from hot clumps ejected from the system in jets and forming the precursors of radio flares (typical parameters of a blob: 10^{11} cm, $n = 10^{14}$ cm^{-3} and $T = 10^6$ K, as compared to small radio flares with a dimension of 6×10^{12} cm (Molnar, Reid, and Grindlay, 1984)).

We would like to suggest an alternative of coherent emission in the infrared produced by energetic particles at early stages of the flares.

4.4. EXPANDING SYNCHROTRON BLOBS OR SUSTAINED PARTICLE ACCELERATION

In some of the X-ray binaries the observed temporal and spectral behaviour can be well described as *gyrosynchrotron emission from relativistic electrons in expanding plasmoids mixed with nonrelativistic gas* (Shklovski, 1965; van der Laan, 1966; Kellermann and Pauliny-Toth, 1968) (see also Section 3.2). In particular the model works well for the giant flares (reaching 20 Jy) (Gregory *et al.*, 1972; Seaquist and Gregory, 1977) and the low-state radio flares (Molnar, Reid, and Grindlay, 1984) in Cyg X-3. However, this model is not valid for all sources. Even in the case of the giant outbursts from Cyg X-3 the flux density can vary by factors of three over a time-scale of hours to days (Johnston *et al.*, 1986) indicating that several repeated injections of energetic particles contribute to the large expanding plasmoid. In the case of GX 13 + 1 (Garcia *et al.*, 1988) and of Sco-X1 the radio outbursts show no systematic time delay at 6 and 20 cm wavelengths nor does the flux density reach higher values at higher frequencies, as expected in the expanding plasmoid model. In these sources *continued production of relativistic particles* is required. For a particular striking flare radio event in AE Aqr the initial size of the cloud in this model is at least 5×10^{10} cm or, alternatively, the fast particles are injected over a longer period of time (Bastian, Dulk, and Chanmugan, 1988).

4.5. JETS

Some of the supposed neutron star binaries show radio jets (SS433, Cyg X-3, Sco X-1). In the case of SS433 the jets consist of individual knots travelling at constant speed of $0.26c$ along a precessing axis (twin corkscrews) (Hjellming and Johnston, 1985; Romney et al., 1987; Vermeulen et al., 1988). The central source is unresolved ($< 10^{15}$ cm) with brightenings which are not always associated with outgoing blobs. These individual blobs attain their maximum radio brightness at a distance of a few 10^{15} cm from the nucleus, and subsequently fade away on a time-scale of days as the knots travel outwards; the radio spectral index is always between -0.6 and -0.8 and indicates optically thin synchrotron emission. The observed brightenings in time and the discrete nature of the radio structures require *repetitive flare-like production of relativistic particles*; these properties are not explained by the 'cylindrical supernova' model of Hjellming and Johnston (1988). Cyg X-3 shows some similarities to SS 433: LBI observations of giant (10 Jy at 5 GHz; decay time of days) radio outbursts can be interpreted as expanding double sources (projected bulk velocities of $0.35c$ at a distance of 10 kpc) (Spencer et al., 1986); VLBI observations of two moderate (1 Jy at 23 GHz; decay time of hours) flares can be interpreted as an expanding and decaying double source (projected bulk velocities of $0.16c$–$0.31c$ at a distance of 10 kpc) (Molnar, Reid, and Grindlay, 1988a). Again in this source repeated injection of accelerated particles (see Section 4.4) is required.

4.6. CORRELATED X-RAY AND RADIO FLUX

In some of the radio emitting X-ray binaries, Cir X-1 (Haynes, Lerche, and Murdin, 1980), LSI $+ 61°303$ (Taylor and Gregory, 1984), Cyg X-3 (Molnar, Reid, and Grindlay, 1984, 1985) the radio flares are correlated with the X-ray emission, which is modulated on an orbital time-scale either because of variable mass transfer (Band and Grindlay, 1984) or a periodically varying viewing geometry.

In other sources, GX 13 + 1 (Garcia et al., 1988) and Sco X-1, no evidence for a correlated X-ray and radio flux is found. It is noteworthy, however, that both objects show a bimodal hardness ratio diagram in X-rays, and that, at least for Sco X-1, radio flares occur only when the source is on the less luminous branch (Priedhorsky et al., 1986). Recently also GX 17 + 2 has been found to emit strong radio emission when its X-ray state is on the hard non-X-ray-flaring branch (Penninx et al., 1988). These systems all have quasi-periodic oscillations in X-rays and are believed to be low mass X-ray binaries radiating at the Eddington limit. The observed absence of correlation can be understood in principle if the central object is a fast spinning weakly magnetized neutron star which acts as a flywheel, temporally storing accretion energy (Priedhorsky, 1986). At high mass accretion rates the luminosity is dominated by X-rays from accretion; at low-mass accretion rates the luminosity is supplied electrodynamically from *interaction between the central object and the disk*. Only in the latter case are important radio flares to be expected (Aly and Kuijpers, 1989).

4.7. ENERGY BUDGET

To explain the radio emission of GX 13 + 1 in the synchrotron interpretation the required energy injection rate in relativistic electrons is of order 10^{36} erg s^{-1} (or 10^{-2} times the X-ray luminosity) for the optically thin, minimum energy case (Garcia *et al.*, 1988); the equipartition field strength is 1 G.

For the cataclysmic variable AE Aqr the averaged required energy injection into relativistic electrons is of order one to a few times 10^{30} erg s^{-1} (or 10^{-3} of the bolometric luminosity of the system) at a field strength of 25 G (Bastian, Dulk, and Chanmugan, 1988).

Characteristically less than *one percent of the accretion energy is required for particle acceleration*.

4.8. A PERIODIC FLARE MODEL

A powerful flare mechanism for disk accretion is the formation of *magnetic links between the compact object* (or in case of a central black hole: between a region near the horizon) *and the disk* (Aly and Kuijpers, 1989). If the magnetic field in the disk is highly non-uniform as is observed for the solar photosphere and for the galactic disk, the interaction between central field and disk field is expected to take place in the disk corona by stochastic reconnections leading to the formation of magnetic links. Since, in general, the footpoints of such a link rotate with different angular speeds kinetic energy is transformed electrodynamically into the field structure of the link. If the Alfvén travel time along the coronal link is short in comparison with the shearing time the link evolves through a series of force-free equilibria, expands away from the disk and is expected to cause a flare, similar to the solar case (Aly, 1985; Low, 1986). When the outer footpoint is outside the corotation radius the energy is extracted from the rotation of the central object; if on the other hand the link is within the corotation radius (which occurs for small accretion rates) the energy is delivered by the accreting material. In the former case accretion is hindered, so that the magnetic flare energy is important, while in the latter case the link promotes accretion, so that the luminosity from the flare is relatively less important. The total luminosity from these magnetic flares in radiation, particle acceleration and explosive motion is on average (Aly and Kuijpers, 1989)

$$L_m = 1.3 \times 10^{35} B_{10}^2 R_{*6}^6 R_7^{-4} l_6 \frac{\Omega_B}{200 \text{ Hz}} \text{ erg s}^{-1},\tag{8}$$

where B_{10} is the polar field strength at the star in units of 10^{10} G, R_{*6} is the stellar radius in units of 10^6 cm, R_7 the distance of the outer footpoint in 10^7 cm, l_6 the thickness of the link in 10^6 cm and Ω_B is the angular beat frequency between both footpoints. Equation (8) shows that in principle this process can release the required acceleration energies for a magnetized neutron star and, adapting Equation (8) for a white dwarf, also for a magnetic cataclysmic variable.

Finally we like to point out that this linking process leads to a *quasi-periodic production of flares* on a period determined by the beat period.

5. Young Stellar Objects

Young stellar (or pre-Main Sequence) objects of low-mass show strong flaring activity. Depending on their infrared emission in the band 1–100 μm they form a sequence characterized by the progressive disappearance of circumstellar dust and the optical appearance of the central object (T Tauri, FU Ori, and post-T Tauri stars) (Lada, 1988; Adams, Lada, and Shu, 1987); this sequence probably corresponds to an evolutionary succession, whereby the gas and dust envelope of the protostar, and ultimately also the gaseous disk, disappears. T Tauri and FU Ori stars have flares and strong variability (10^{34} ergs in 10 minutes) in the optical (Gahm, 1986, 1988). The variability of T Tauri stars is caused by a combination of *variable obscuration by circumstellar dust, gas infall and the presence of a disk and boundary layer* (Lynden-Bell and Pringle, 1974; Brown, 1988; Bertout, Basri, and Bouvier, 1988). The strong flares in FU Ori objects are probably the result of enhanced mass accretion in a disk (Kenyon, Hartmann, and Hewett, 1988; Hartmann and Kenyon, 1988). The cause of the enhanced mass accretion is, however, not clear and may well be related to *magnetic interactions between disk and star* as in X-ray binaries (see Section 4.8).

5.1. RADIO EMISSION

Dramatic eruptions in X-rays ($> 10^{34}$ erg, 4×10^{31} erg s^{-1}) in the Rho Oph dark cloud focused attention on young stellar objects as flaring stars (Montmerle, 1988; Feigelson, 1986). Subsequent radio observations of this and other dark clouds revealed the existence of stellar flaring objects in the radio (André, Montmerle, and Feigelson, 1987; André, 1986; Brown, 1987; Montmerle and André, 1988; Stine *et al.*, 1988). The time-scales of the variable strong radio emission are so short (hours) that the radiation cannot be explained as free-free radiation from circumstellar gas (with corresponding time-scales of months to years). The radio flare fluxes (up to 1.5×10^{18} erg s^{-1} Hz^{-1}) are comparable to those of the RS CVn systems; it is possible that their quiescent emission is usually not observed as they are at a larger distance (> 130 pc) than the observed RS CVn stars (30–50 pc). Most ($> 60\%$) have substantial X-ray emission but the reverse is not true (André, Montmerle, and Feigelson, 1987). In a radio survey of T Tauri stars a low detection rate was found and a strong correlation with sources with outflow (Cohen, 1987; Bieging, Cohen, and Schwartz, 1984; Cohen and Bieging, 1986). This suggests that the strong radio emitters are older than deeply embedded infrared sources, but younger than classical T Tauri stars.

A particularly interesting example of such a radio flaring object is the K0 pre-Main-Sequence X-ray star DoAr 21; its variable radio emission can be modelled with *gyro-synchrotron radiation* from a magnetic loop extending from one to ten stellar radii, a field strength of 1.5×10^2 G at the base and 8×10^{-2} G at the top, a power-law electron energy spectrum with index 2, a high-energy cut-off at 4.5 MeV and a column density of energetic particles of 3.4×10^{17} cm^{-2} (Feigelson and Montmerle, 1985; André, 1986).

5.2. JETS

Young stellar objects are often associated with jets: Many infrared sources have bipolar outflows (Lada, 1988); Some T Tauri stars have highly collimated visual jets (Bührke, Brugel, and Mundt, 1986; Mundt, 1988). The few T Tauri stars that are radio emitters are relatively strong and often associated with jet-like structures: These T Tauri stars, which are the least luminous young stellar objects, produce the largest fraction of radio to total luminosity (Hamann, Simon, and Ridgway, 1988). One object, DG Tau, shows a time-variable 5–15 GHz structure with a bipolar morphology pointing towards a nearby Herbig–Haro knot (200 km s^{-1}, 10^2–10^4 cm^{-3}) (Cohen, 1987) in the optical jet (Cohen and Bieging, 1986). Another object, L1551, shows two unresolved radio components perpendicular to an extended elongated radio source (Bieging and Cohen, 1985) and possibly corresponding to an ionized inner part of a confining torus with projected radius of 25 AU (Rodríguez *et al.*, 1986). A similar radio structure of two point-like sources and two extended radio lobes is centered on the deeply embedded infrared source IRS 7 (Brown, 1988).

Since T Tauri stars sometimes have magnetic fields with a large surface filling factor, it is quite well possible that the flaring behaviour of these and younger stars is caused by magnetic interactions between star and disk.

6. Conclusion

Increasing evidence exists that at least some of the violent radio outbursts in dMe stars, classic close binaries, T Tauri stars and their predecessors, magnetic cataclysmic variables and luminous X-ray binaries are the result of magnetic flaring activity. The connection with solar flares is that the outbursts probably originate in force-free magnetic structures which are bounded, energized and finally driven to instability by a differentially rotating dense high-beta atmosphere. The differences are related partly to different drivers (single star, star-star, star-disk, star-cloud) and partly to different magnetic surface fluxes and dimensions (with consequently larger stored and released amounts of energy). Much of the (micro)physics leading to the observed energy partitioning (motion/heating/acceleration/coherent radiation) remains to be sorted out and probably varies greatly going from non-compact single stars to compact accreting stars (role of radiation field, relativistic effects). Finally although radio emission can be obscured by accretion and excretion flows, it appears that the relatively small number of radio active X-ray binaries mirrors intrinsic differences in state of activity, related to the relative mass accretion rate and the stellar magnetic moment.

References

Adams, F. C., Lada, C. J., and Shu, F. H.: 1987, *Astrophys. J.* **312**, 788.
Alfvén, H.: 1958, *Tellus* **10**, 104.

Alfvén, H.: 1981, *Cosmic Plasma*, D. Reidel Publ. Co., Dordrecht, Holland.

Alfvén, H. and Carlqvist, P.: 1967, *Solar Phys.* **1**, 220.

Aly, J.-J.: 1985, *Astron. Astrophys.* **143**, 19.

Aly, J.-J. and Kuijpers, J.: 1989, *Astron. Astrophys.* (submitted).

André, Ph.: 1986, in T. Montmerle and C. Bertout (eds.), *Protostars and Molecular Clouds*, CEA, Saclay, France, p. 143.

André, Ph., Montmerle, T., and Feigelson, E. D.: 1987, *Astron. J.* **93**, 1182.

André, Ph., Montmerle, T., Feigelson, E. D., Stine, P. C., and Klein, K.-L.: 1988, *Astrophys. J.* **335**, 940.

Aschwanden, M. J., Hewitt, R. G., and Dulk, G. A.: 1989, in B. M. Haisch and M. Rodonò (eds.), 'Solar and Stellar Flares', presented at *IAU Colloq.* **104** (unpublished).

Athay, R. G.: 1984, *Solar Phys.* **93**, 123.

Bahcall, J. N., Rosenbluth, M. N., and Kulsrud, R. M.: 1973, *Nature Phys. Sci.* **243**, 27.

Band, D. L. and Grindlay, J. E.: 1984, *Astrophys. J.* **285**, 702.

Bastian, T. S. and Bookbinder, J. A.: 1987, *Nature* **326**, 678.

Bastian, T. S., Dulk, G. A., and Chanmugan, G.: 1985, in R. M. Hjellming and D. M. Gibson (eds.), *Radio Stars*, D. Reidel Publ. Co., Dordrecht, Holland, p. 225.

Bastian, T. S., Dulk, G. A., and Chanmugan, G.: 1988, *Atrophys. J.* **324**, 431.

Bastian, T. S., Dulk, G. A., and Slee, O. B.: 1988, *Astron. J.* **95**, 794.

Benz, A. O.: 1986, *Solar Phys.* **104**, 99.

Benz, A. O. and Güdel, M.: 1989, in G. Belvedere (ed.), *Accretion Disks and Magnetic Fields in Astrophysics*, Noto, Sicilia (in press).

Bertout, C., Basri, G., and Bouvier, J.: 1988, *Astrophys. J.* **330**, 350.

Bieging, J. H. and Cohen, M.: 1985, *Astrophys. J.* **289**, L5.

Bieging, J. H., Cohen, M., and Schwartz, P. R.: 1984, *Astrophys. J.* **282**, 699.

Blanken, R. A., Stix, T. F., and Kuckes, A. F.: 1969, *Plasma Phys.* **11**, 945.

Bookbinder, J. A.: 1988, in O. Havnes, B. R. Pettersen, J. H. M. M. Schmitt and J. E. Solheim (eds.), *Activity in Cool Star Envelopes*, Kluwer, Dordrecht, Holland, p. 257.

Bookbinder, J. A. and Lamb, D. Q.: 1987, *Astrophys. J.* **323**, L131.

Bookbinder, J. A. and Walter, F. M.: 1988, in J. L. Linsky and R. E. Stencel (eds.), *Cool Stars, Stellar Systems and the Sun*, Fifth Cambridge Workshop, Springer-Verlag, New York, p. 260.

Bookbinder, J. A., Bastian, T. S., Dulk, G. A., and Davis, M. M.: 1989, presented at *IAU Colloq.* **104** (unpublished).

Borghi, S. and Chiuderi-Drago, F.: 1985, *Astron. Astrophys.* **143**, 226.

Brown, A.: 1987, *Astrophys. J.* **322**, L31.

Brown, A.: 1988, in J. L. Linsky and R. E. Stencel (eds.), *Cool Stars, Stellar Systems and the Sun*, Fifth Cambridge Workshop, Springer-Verlag, New York, p. 466.

Brown, R. L. and Crane, P. C.: 1979, *Astron. J.* **83**, 1504.

Bührke, T., Brugel, E. W., and Mundt, R.: 1986, *Astron. Astrophys.* **163**, 83.

Butler, C. J., Rodonò, M., Foing, B. H., and Haisch, B. M.: 1986, *Nature* **321**, 679.

Canfield, R. C. and Metcalf, T. R.: 1987, *Astrophys. J.* **321**, 586.

Chanmugan, G.: 1987, *Astrophys. Space Sci.* **130**, 53.

Chanmugan, G. and Dulk, G. A.: 1982, *Astrophys. J.* **255**, L107.

Cohen, M.: 1987, in W. Kundt (ed.), *Astrophysical Jets and Their Engines*, D. Reidel Publ. Co., Dordrecht, Holland, p. 91.

Cohen, M. and Bieging, J. H.: 1986, *Astron. J.* **92**, 1396.

Davis, R. J., Lovell, B., Palmer, H. P. and Spencer, R. E.: 1978, *Nature* **273**, 644.

DeCampli, W. M. and Baliunas, S. L.: 1979, *Astrophys. J.* **230**, 815.

Dennis, B. R.: 1988, *Solar Phys.* **118**, 49.

Doyle, J. G., Butler, C. J., Byrne, P. B., and van den Oord, G. H. J.: 1988, *Astron. Astrophys.* **193**, 229.

Drake, S. A., Simon, T., and Linsky, J. L.: 1987, *Astron. J.* **91**, 1229.

Drake, S. A., Abbott, D. C., Bastian, T. S., Bieging, J. H., Churchwell, E., Dulk, G. A., and Linsky, J. L.: 1987, *Astrophys. J.* **322**, 902.

Dröge, F.: 1967, *Z. Astrophys.* **66**, 176.

Dulk, G. A.: 1985, *Ann. Rev. Astron. Astrophys.* **23**, 169.

Dulk, G. A.: 1986, *Adv. Space Res.* **6**, 95.

Dulk, G. A.: 1988, in J. L. Linsky and R. E. Stencel (eds.), *Cool Stars, Stellar Systems and the Sun*, Fifth Cambridge Workshop, Springer-Verlag, New York, p. 72.

Dulk, G. A., Bastian, T. S., and Chanmugan, G.: 1983, *Astrophys. J.* **273**, 249.

Feigelson, E. D.: 1986, in T. Montmerle and C. Bertout (eds.), *Protostars and Molecular Clouds*, CEA, Saclay, France, p. 123.

Feigelson, E. D. and Montmerle, T.: 1985, *Astrophys. J.* **289**, L19.

Feldman, P. A.: 1983, in P. B. Byrne and M. Rodonò (eds.), 'Activity in Red-Dwarf Stars', *IAU Colloq.* **71**, 429.

Feldman, P. A., Taylor, A. R., Gregory, P. C., Seaquist, E. R., Balonek, T. J., and Cohen, N. L.: 1978, *Astron. J.* **83**, 1471.

Fürst, E., Benz, A., Hirth, W., Kiplinger, A., and Geffert, M.: 1986, *Astron. Astrophys.* **154**, 377.

Gahm, G. F.: 1986, in P. M. Gondhalekar (ed.), *Flares: Solar and Stellar*, Rutherford Appleton Lab., U.K.

Gahm, G. F.: 1988, in A. K. Dupree and M. T. V. T. Lago (eds.), *Formation and Evolution of Low-Mass Stars*, Kluwer, Dordrecht, Holland, p. 295.

Garcia, M. R., Grindlay, J. E., Molnar, L. A., Stella, L., White, N. E., and Seaquist, E. R.: 1988, *Astrophys. J.* **328**, 552.

Gary, D. E.: 1985, in R. M. Hjellming and D. M. Gibson (eds.), *Radio Stars*, D. Reidel Publ. Co., Dordrecht, Holland, p. 185.

Gary, D. E. and Linsky, J. L.: 1981, *Astrophys. J.* **250**, 284.

Gibson, D. M.: 1983a, in P. B. Byrne and M. Rodono (eds.), 'Activity in Red-Dwarf Stars', *IAU Colloq.* **71**, 273.

Gibson, D. M.: 1983b, in S. L. Baliunas and L. Hartmann (eds.), *Cool Stars, Stellar Systems and the Sun*, Springer-Verlag, Berlin, p. 197.

Gibson, D. M.: 1984, in P. F. Gott and P. S. Riherd (eds.), *Proc. Southwest Regional Conf. for Astron. Astrophys. No. 9*, p. 35.

Gibson, D. M.: 1985, in R. M. Hjellming and D. M. Gibson (eds.), *Radio Stars*, D. Reidel Publ. Co., Dordrecht, Holland, p. 213.

Gibson, D. M. and Fisher, P. L.: 1981, in P. F. Gott and P. S. Riherd (eds.), *Proc. Southwest Regional Conf. for Astron. Astrophys. No. 6*, p. 33.

Gregory, P. C., Kronberg, P. P., Seaquist, E. R., Hughes, V. A., Woodsworth, A., Viner, M. R., Retallack, D., Hjellming R. M., and Balick, B.: 1972, *Nature (Phys. Sci.)* **239**, 114.

Hamann, F., Simon, M., and Ridgway, S. T.: 1988, *Astrophys. J.* **326**, 859.

Hartmann, L. and Kenyon, S.: 1988, in A. K. Dupree and M. T. V. T. Lago (eds.), *Formation and Evolution of Low-Mass Stars*, Kluwer, Dordrecht, Holland, p. 163.

Haynes, R. F., Lerche, I., and Murdin, P.: 1980, *Astron. Astrophys.* **87**, 299.

Hénoux, J.-C.: 1987, in V. E. Stepanov and V. N. Obridko (eds.), *Solar Maximum Analysis*, VNU Science Press, Utrecht, p. 109.

Hershkowitz, N.: 1985, *Space Sci. Rev.* **41**, 351.

Hjellming, R. M.: 1988, in G. L. Verschuur and K. I. Kellermann (eds.), *Galactic and Extragalactic Radio Astronomy*, Springer-Verlag, Berlin, p. 381.

Hjellming, R. M. and Gibson, D. M.: 1980, in M. R. Kundu and T. E. Gergely (eds.), *Radio Physics of the Sun*, *IAU Colloq.* **86**, D. Reidel Publ. Co., Dordrecht, Holland, p. 209.

Hjellming, R. M. and Johnston, K. J.: 1985, in R. M. Hjellming and D. M. Gibson (eds.), *Radio Stars*, D. Reidel Publ. Co., Dordrecht, Holland, p. 309.

Hjellming, R. M. and Johnston, K. J.: 1988, *Astrophys. J.* **328**, 600.

Holman, G. D.: 1983, *Adv. Space Res.* **2**, No. 11, 181.

Holman, G. D.: 1985, *Astrophys. J.* **293**, 584.

Holman, G. D., Eichler, D., and Kundu, M. R.: 1980, in M. R. Kundu and T. E. Gergely (eds.), 'Radio Physics of the Sun', *IAU Colloq.* **86**, 457.

Holman, G. D., Bookbinder, J., and Golub, L.: 1985, in R. M. Hjellming and D. M. Gibson (eds.), *Radio Stars*, D. Reidel Publ. Co., Dordrecht, Holland, p. 35.

Hughes, V. A. and McLean, B. J.: 1987, *Astrophys. J.* **313**, 263.

Jackson, P. D., Kundu, M. R., and White, S. M.: 1987, *Astrophys. J.* **316**, L85.

Jacobsen, C. and Carlqvist, P.: 1964, *Icarus* **3**, 270.

Johnston, K. J., Spencer, J. H., Simon, R. S. *et al.*: 1986, *Astrophys. J.* **309**, 707.

Kaitchuk, R. H. and Park, E. A.: 1988, *Astrophys. J.* **325**, 225.

Kawai, N., Matsuoka, M., Pan, H. C., and Stewart, G. C.: 1989, *Publ. Astron. Soc. Japan* (in press).
Kellermann, K. I. and Pauliny-Toth, I. I. K.: 1968, *Ann. Rev. Astron. Astrophys.* **6**, 417.
Kenyon, S. J., Hartmann, L., and Hewett, R.: 1988, *Astrophys. J.* **325**, 231.
Klein, K.-L.: 1987, *Astron. Astrophys.* **183**, 341.
Klein, K.-L. and Chiuderi-Drago, F.: 1987, *Astron. Astrophys.* **175**, 179.
Kosugi, T., Dennis, B. R., and Kai, K.: 1988, *Astrophys. J.* **324**, 1118.
Krüger, A.: 1979, *Introduction to Solar Radio Astronomy and Radio Physics*, D. Reidel, Publ. Co., Dordrecht, Holland.
Kuijpers, J.: 1974, *Solar Phys.* **36**, 157.
Kuijpers, J.: 1975, *Astron. Astrophys.* **40**, 405.
Kuijpers, J.: 1980, in M. R. Kundu and T. E. Gergely (eds.), 'Radio Physics of the Sun', *IAU Colloq.* **86**, 341.
Kuijpers, J.: 1985a, in J. Fanta and G. Pantoflíček (eds.), *Trends in Physics*, Proc. of the Sixth General Conf. of the Europ. Phys. Soc., Prague, p. 170.
Kuijpers, J.: 1985b, in R. M. Hjellming and D. M. Gibson (eds.), *Radio Stars*, D. Reidel Publ. Co., Dordrecht, Holland, p. 3.
Kuijpers, J. and van der Hulst, J. M.: 1985, *Astron. Astrophys.* **149**, 343.
Kuijpers, J., van der Post, P., and Slottje, C.: 1981, *Astron. Astrophys.* **103**, 331.
Kuijpers, J., Lindberg, L., Raadu, M. A., Stenflo, L., and Torvén, S.: 1989, in preparation.
Kundu, M. R. and Hjellming, R. M.: 1985, in R. M. Hjellming, and D. M. Gibson (eds.), *Radio Stars*, D. Reidel Publ. Co., Dordrecht, Holland, p. 397.
Kundu, M. R. and Shevgaonkar, R. K.: 1985, *Astrophys. J.* **297**, 644.
Kundu, M. R. and Shevgaonkar, R. K.: 1988, *Astrophys. J.* **334**, 1001.
Kundu, M. R. and Woodgate, B. (eds.): 1986, *Energetic Phenomena on the Sun*, NASA Conf. Publ. 2439, GSFC, Maryland, Ch. 2, Ch. 5.
Kundu, M. R., Jackson, P. D., White, S. M., and Melozzi, M.: 1987, *Astrophys. J.* **312**, 822.
Kundu, M. R., Pallavicini, R., White, S. M., and Jackson, P. D.: 1988, *Astron. Astrophys.* **195**, 159.
Lada, C. J.: 1988, in A. K. Dupree and M. T. V. T. Lago (eds.), *Formation and Evolution of Low-Mass Stars*, Kluwer, Dordrecht, Holland, p. 93.
Lamb, D. Q. and Melia, F.: 1986, in K. O. Mason, M. G. Watson, and N. E. White (eds.), *The Physics of Accretion onto Compact Objects*, Springer-Verlag, Berlin, p. 113.
Lamb, F. K., Aly, J., Cook, M., and Lamb, D. Q.: 1983, *Astrophys. J.* **274**, L71.
Lang, K. R. and Willson, R. F.: 1986a, *Astrophys. J.* **302**, L17.
Lang, K. R. and Willson, R. F.: 1986b, *Astrophys. J.* **305**, 363.
Lang, K. R. and Willson, R. F.: 1988, *Astrophys. J.* **326**, 300.
Lang, K. R., Bookbinder, J., Golub, L., and Davis, M.: 1983, *Astrophys. J.* **272**, L15.
Lestrade, J.-F.: 1988, in M. J. Reid and J. M. Morgan (eds.), 'The Impact of VLBI on Astrophysics and Geophysics', *IAU Symp.* **129**, 265.
Lestrade, J.-F., Mutel, R. L., Preston, R. A., and Phillips, R. B.: 1988, *Astrophys. J.* **328**, 232.
Lewin, W. H. G., van Paradijs, J., and van der Klis, M.: 1988, *Space Sci. Rev.* **46**, 273.
Lin, R. P., Schwartz, R. A., Kane, S. R., Pelling, R. M., and Hurley, K. C.: 1984, *Astrophys. J.* **283**, 421.
Lindberg, L.: 1987, *XVIII Int. Conf. on Phenomena in Ionized Gases*, Swansea, U.K.
Lindberg, L.: 1988, *Astrophys. Space Sci.* **144**, 3.
Linsky, J. L. and Gary, D. E.: 1983, *Astrophys. J.* **274**, 776.
Louarn, P., Le Quéau, D., and Roux, A.: 1987, *Solar Phys.* **111**, 201.
Louarn, P., Le Quéau, D., and Roux, A.: 1989, in B. M. Haisch and M. Rodonò (eds.), *IAU Colloq.* **104**, 'Solar and Stellar Flares', Poster Papers, Publ. Catania Astrophys. Obs., Special Volume, p. 329.
Lovell, B.: 1971, *Quart. J. Roy. Astr. Soc.* **12**, 98.
Low, B. C.: 1986, *Astrophys. J.* **307**, 205.
Lu, E. T. and Petrosian, V.: 1988, *Astrophys. J.* **327**, 405.
Lynden-Bell, D. and Pringle, J. E.: 1974, *Monthly Notices Roy. Astron. Soc.* **168**, 603.
Mason, K. O., Cordova, F. A., and White, N. E.: 1986, *Astrophys. J.* **309**, 700.
Massi, M., Felli, M., Pallavicini, R., Tofani, G., Palagi, F., and Catarzi, M.: 1989, *Astron. Astrophys.* (in press).
McCray, R.: 1966, *Science* **154**, 1320.
McKean, M. E., Winglee, R. M., and Dulk, G. A.: 1989, in B. M. Haisch and M. Rodonò (eds.), *IAU Colloq.* **104**, 'Solar and Stellar Flares', Poster Papers, Publ. Catania Astrophys. Obs., Special Volume, p. 333.

McLean, D. J. and Labrum, N. R. (eds.): 1985, *Solar Radiophysics*, Cambridge Univ. Press, Cambridge.
Melrose, D. B.: 1973, *Australian J. Phys.* **26**, 229.
Melrose, D. B.: 1986, *Instabilities in Space and Laboratory Plasmas*, Cambridge Univ. Press, Cambridge.
Melrose, D. B. and Dulk, G. A.: 1982, *Astrophys. J.* **259**, 844.
Melrose, D. B., Hewitt, R. G., and Dulk, G. A.: 1984, *J. Geophys. Res.* **89**, 897.
Mezger, P. G. and Henderson, A. P.: 1967, *Astrophys. J.* **147**, 471.
Molnar, L. A., Reid, M. J., and Grindlay, J. E.: 1984, *Nature* **310**, 662.
Molnar, L. A., Reid, M. J., and Grindlay, J. E.: 1985, in R. M. Hjellming and D. M. Gibson (eds.), *Radio Stars*, D. Reidel Publ. Co., Dordrecht, Holland, p. 329.
Molnar, L. A., Reid, M. J., and Grindlay, J. E.: 1988a, *Astrophys. J.* **331**, 494.
Molnar, L. A., Reid, M. J., and Grindlay, J. E.: 1988b, in M. J. Reid and J. E. Moran (eds.), 'The Impact of VLBI on Astrophysics and Geophysics', *IAU Symp.* **129**, 279.
Montmerle, T.: 1988, in E. H. Schröter and M. Schüssler (eds.), *Solar and Stellar Physics*, Springer-Verlag, Berlin, p. 117.
Montmerle, T. and André, Ph.: 1988, in A. K. Dupree and M. T. V. T. Lago (eds.), *Formation and Evolution of Low-Mass Stars*, Kluwer, Dordrecht, Holland, p. 225.
Morris, D. H. and Mutel, R. L.: 1988, *Astron. J.* **95**, 204.
Mundt, R.: 1988, in A. K. Dupree and M. T. V. T. Lago (eds.), *Formation and Evolution of Low-Mass Stars*, Kluwer, Dordrecht, Holland, p. 257.
Mutel, R. L. and Lestrade, J.-F.: 1985, *Astron. J.* **90**, 493.
Mutel, R. L. and Weisberg, J. M.: 1978, *Astron. J.* **83**, 1499.
Mutel, R. L., Lestrade, J.-F., Preston, R. A., and Phillips, R. B.: 1985, *Astrophys. J.* **289**, 262.
Mutel, R. L., Morris, D. H., Doiron, D. J. and Lestrade, J.-F.: 1987, *Astron. J.* **93**, 1220.
Nelson, R. F. and Spencer, R. E.: 1988, *Monthly Notices Roy. Astron. Soc.* **234**, 1105.
Newell, R. T., Gibson, D. M., Becker, R. M., and Holt, S. S.: 1980, in P. F. Gott and P. S. Riherd (eds.), *Proc. Southwest Regional Conf. for Astron. Astrophys.* **5**, 13.
Owen, F. N. and Gibson, D. M.: 1978, *Astron. J.* **83**, 1488.
Owen, F. N., Jones, T. W., and Gibson, D. M.: 1976, *Astrophys. J.* **210**, L27.
Papadopoulos, K. and Freund, H. P.: 1979, *Space Sci. Rev.* **24**, 511.
Parker, E. N.: 1988, *Astrophys. J.* **330**, 474.
Penninx, W., Lewin, W. H. G., Zijlstra, A. A., Mitsuda, K., van Paradijs, J., and van der Klis, M.: 1988, *Nature* **336**, 146.
Priedhorsky, W.: 1986, *Astrophys. J.* **306**, L97.
Priedhorsky, W., Hasinger, G., Lewin, W. H. G., Middleditch, J., Parmar, A., Stella, L., and White, N. E.: 1986, *Astrophys. J.* **306**, L91.
Raadu, M. A.: 1984, in R. Schrittwieser and G. Eder (eds.), *Proc. Second Symposium on Plasma Double Layers and Topics*, Innsbruck, p. 3.
Raadu, M. A. and Rasmussen, J. J.: 1988, *Astrophys. Space Sci.* **144**, 43.
Ramaty, R.: 1969, *Astrophys. J.* **158**, 753.
Rickett, B. J.: 1977, *Ann. Rev. Astron. Astrophys.* **15**, 479.
Rodríguez, L. F., Cantó, J., Torrelles, J. M., and Ho, P. T. P.: 1986, *Astrophys. J.* **301**, L25.
Romney, J. D., Schilizzi, R. T., Fejes, I., and Spencer, R. E.: 1987, *Astrophys. J.* **321**, 822.
Saar, S. H.: 1988, in J. L. Linsky and R. E. Stencel (eds.), *Cool Stars, Stellar Systems and the Sun*, Fifth Cambridge Workshop, Springer-Verlag, New York, p. 10.
Saar, S. H. and Linsky, J. L.: 1985, *Astrophys. J.* **299**, L47.
Schadee, A.: 1986, *Adv. Space Res.* **6**, No. 6, 41.
Schadee, A., de Jager, C., and Švestka, Z.: 1983, *Solar Phys.* **89**, 287.
Seaquist, E. R. and Gregory, P. C.: 1977, *Astrophys. Letters* **18**, 65.
Seaquist, E. R. and Taylor, A. R.: 1985, *Astron. J.* **90**, 2049.
Shklovski, I. S.: 1965, *Soviet Astron-AJ* **9**, 22.
Simon, T., Linsky, J. L., and Schiffer III, F. H.: 1980, *Astrophys. J.* **239**, 911.
Slee, O. B., Haynes, R. F., and Wright, A. E.: 1984, *Monthly Notices Roy. Astron. Soc.* **208**, 865.
Slee, O. B., Touky, I. R., Nelson, G. J., and Renie, C. J.: 1981, *Nature* **292**, 220.
Slee, O. B., Nelson, G. J., Stewart, R. T., Wright, A. E., Innis, J. L., Ryan, S. G., and Vaughan, A. E.: 1987, *Monthly Notices Roy. Astron. Soc.* **229**, 659.
Slottje, C.: 1978, *Nature* **275**, 520.

Spangler, S. R.: 1977, *Astron. J.* **82**, 169.

Spangler, S. R., Owen, F. N., and Hulse, R. A.: 1977, *Astron. J.* **82**, 989.

Spencer, R. E., Swinney, R. W., Johnston, K. J., and Hjellming, R. M.: 1986, *Astrophys. J.* **309**, 694.

Stähli, M. and Magun, A.: 1986, *Solar Phys.* **104**, 117.

Stepanov, A. V.: 1974, *Soviet Astron.* **17**, 781.

Stepanov, A. V.: 1978, *Soviet Astron. Letters* **4**, 103.

Stewart, R. T., Innis, J. L., Slee, O. B., Nelson, G. J., and Wright, A. E.: 1988, *Astron. J.* **96**, 371.

Stine, P. C., Feigelson, E. D., André, P., and Montmerle, T.: 1988, *Astron. J.* **96**, 1394.

Taylor, A. R. and Gregory, P. C.: 1984, *Astrophys. J.* **283**, 273.

Torvén, S.: 1982, *J. Phys. D: Appl. Phys.* **15**, 1943.

Torvén, S., Lindberg, L., and Carpenter, R. T.: 1985, *Plasma Phys. Contr. Fusion* **27**, 143.

Tsytovich, V. N.: 1973, *Ann. Rev. Astron. Astrophys.* **11**, 363.

Twiss, R. Q.: 1958, *Australian J. Phys.* **11**, 564.

Uchida, Y.: 1986, *Adv. Space Res.* **6**, No. 8, 29.

Uchida, Y. and Sakurai, T.: 1983, in P. B. Byrne and M. Rodonò (eds.), 'Activity in Red-Dwarf Stars', *IAU Colloq.* **71**, 629.

van den Oord, G. H. J.: 1988, *Astron. Astrophys.* **207**, 101.

van den Oord, G. H. J. and Barstow, M. A.: 1988, *Astron. Astrophys.* **207**, 89.

van den Oord, G. H. J. and Mewe, R.: 1989, *Astron. Astrophys.* (submitted).

van den Oord, G. H. J., Mewe, R., and Brinkman, A. C.: 1989, *Astron. Astrophys.* (in press).

van den Oord, G. H. J., Kuijpers, J., White, N. E., van der Hulst, J. M., and Culhane, J. L.: 1989, *Astron. Astrophys.* **209**, 296.

van der Laan, H.: 1966, *Nature* **211**, 1131.

Vermeulen, R. C., Schilizzi, R. T., Icke, V., Fejes, I., and Spencer, R. E.: 1988, in M. J. Reid and J. M. Moran (eds.), 'The Impact of VLBI on Astrophysics and Geophysics', *IAU Symp.* **129**, 275.

Vilhu, O., Caillault, J.-P., Neff, J., and Heise, J.: 1988, in O. Havnes, B. R. Pettersen, J. H. M. M. Schmitt, and J. E. Solheim (eds.), *Activity in Cool Star Envelopes*, Kluwer, Dordrecht, Holland, p. 179.

Wendker, H. J.: 1987, *Astron. Astrophys. Suppl. Ser.* **69**, 87.

White, N. E., Culhane, J. L., Parmar, A. N., Kellett, B. J., Kahn, S., van den Oord, G. H. J., and Kuijpers, J.: 1986, *Astrophys. J.* **301**, 262.

White, S. M., Kundu, M. R., and Jackson, P. D.: 1986, *Astrophys. J.* **311**, 814.

White, S. M., Melrose, D. B., and Dulk, G. A.: 1986, *Astrophys. J.* **308**, 424.

White, S. M., Kundu, M. R., and Jackson, P. D.: 1989, *Astrophys. J.* (submitted).

Whitehouse, D. R.: 1985, *Astron. Astrophys.* **145**, 449.

Willson, R. F.: 1987, *Proc. of Sac Peak Workshop on Coronal Structure and Dynamics* (in press).

Willson, R. F. and Lang, K. R.: 1987, *Astrophys. J.* **312**, 278.

Willson, R. F., Lang, K. R., and Foster, P.: 1988, *Astron. Astrophys.* **199**, 255.

Winglee, R. M. and Dulk, G. A.: 1986, *Astrophys. J.* **307**, 808.

Winglee, R. M., Dulk, G. A., and Pritchett, P. L.: 1988, *Astrophys. J.* **328**, 809.

Wu, C. S. and Lee, L. C.: 1979, *Astrophys. J.* **230**, 621.

Zheleznyakov, V. V.: 1967, *Soviet Phys. JETP* **24**, 381.

Zheleznyakov, V. V. and Zlotnik, E. Y.: 1975, *Solar Phys.* **43**, 431.

SOLAR AND STELLAR MAGNETIC FIELDS AND STRUCTURES: OBSERVATIONS

JEFFREY L. LINSKY

Joint Institute for Laboratory Astrophysics, National Institute of Standards and Technology, and the University of Colorado, Boulder, CO 80309–0440, U.S.A.

"If the Sun did not have a magnetic field, it would be as uninteresting a star as most astronomers believe it to be."

attributed to ROBERT B. LEIGHTON

"Magnetic fields are to astrophysics what sex is to psychoanalysis."

HENK VAN DE HULST (1988)

Abstract. This review of stellar magnetic field measurements is both a critique of recent spectral diagnostic techniques and a summary of important trends now appearing in the data. I will discuss both the Zeeman broadening techniques that have evolved from Robinson's original approach and techniques based on circular and linear polarization data. I conclude with an ambitious agenda for developing self-consistent models of the magnetic atmospheres of active stars.

1. Perspective

Six years ago the topic of solar and stellar magnetic fields was the centerpiece of two IAU meetings – Colloquium No. 71 'Activity in Red Dwarf Stars' (Catania) and Symposium No. 102 'Solar and Stellar Magnetic Fields: Origins and Coronal Effects' (Zürich). As I reread the review papers by Marcy (1983), Golub (1983), and Linsky (1983a, b), I was struck by the enormous progress made in this field in the few years since then. Six years ago few direct measurements were available and the spectral diagnostic techniques were rudimentary. Now we have available a much richer data set from which sophisticated diagnostic methods are extracting more reliable magnetic field parameters for many late-type stars.

In view of this rapid progress, it is important to review and critique the spectral diagnostic techniques now employed and to understand the propagation of systematic and random errors into the derived magnetic parameters. Hartmann's (1987) thoughtful discussion of several of these problems is a useful introduction. I will then summarize what I believe are the important trends emerging from the observations, but I encourage the reader to consult earlier reviews by Linsky (1985), Saar (1987b), and Gray (1988). Finally, as a challenge to both theoreticians and observers, I will lay out an ambitious agenda for the next 6 years to develop a self-consistent model for the magnetic atmospheres of active stars that should explain the magnetic field, X-ray, ultraviolet, and radio data in a manner consistent with the dynamics, energetics, and geometry of these atmospheres. This may appear to be an unreachable goal, but it is no more ambitious

Solar Physics **121**: 187–196, 1989.

than what has been accomplished during the past 6 years. One motive for this review is to stimulate the development of such comprehensive models.

2. Spectral Diagnostics of Stellar Magnetic Fields

Solar physicists have long known that magnetic fields play critical roles in heating the chromosphere and corona, determining the geometry of structures in these regions, and otherwise influencing the dynamics and energetics of the diverse phenomena that are called 'solar activity'. The spatial correlation between magnetic fields and phenomena on specific regions on the Sun is well established, because the proximity of the Sun permits spatially-resolved observations. X-ray, ultraviolet, and radio observations of dwarf stars of spectral type F–M and certain subgiant and giant stars, such as components of RS CVn-type binary systems, indicate active phenomena on these stars as well, but often orders of magnitude more energetic, indicating that these stars probably also have strong pervasive magnetic fields.

Magnetic fields in the solar photosphere are typically measured from the difference in absorption line shapes obtained in opposite circular polarizations for magnetically-sensitive transitions (i.e., large Landé g factors). This procedure works because the fields very likely have the same direction in the small regions on the Sun defined by the instrumental aperture and seeing. However, the measured quantity is the magnetic flux rather than the field strength because the magnetic elements may only partially fill the aperture. Application of analogous methods for measuring magnetic field properties in solar-type dwarf stars have yielded null results (e.g., Babcock, 1958; Vogt, 1980; Borra, Edwards, and Mayor 1984). The classical method has failed for these stars because for unresolved stellar observations the contributions of oppositely-directed field elements cancel to high precision, just as they do in integrated sunlight. Thus to measure magnetic fields on late-type stars, one must first devise a better diagnostic procedure.

2.1. ZEEMAN BROADENING TECHNIQUES USING UNPOLARIZED LIGHT

Robinson (1980) proposed that the average magnitude of the stellar photospheric magnetic field could be derived from a careful study in unpolarized light of the enhanced Zeeman broadening of a magnetically-sensitive line (high Landé g factor) compared with another spectral line very similar in shape and formation, but with smaller magnetic sensitivity. Extreme care must be taken in applying this diagnostic technique, because the splitting of a simple Zeeman triplet from line center is only 42 mÅ or 2.1 km s^{-1} for a 6000 Å line with $g = 2.5$ in 1000 G field. The splitting is small compared with the typical width of stellar line profiles; the magnetic field slightly broadens the profile in the inner wings. Since the Zeeman broadening increases as the square of the wavelength, infrared lines should have more pronounced broadening. Indeed, Saar and Linsky (1985) have resolved the Zeeman triplet pattern in Ti I lines located near 2.2 microns in the spectrum of the dM3.5e flare star AD Leo, which they interpret as due to a field of 3800 $\pm$ 260 G covering 0.73 $\pm$ 0.06 of its surface. Stellar observations of the 12 micron Mg I lines detected in the solar spectrum (Brault and Noyes, 1983) should reveal completely split Zeeman patterns.

Robinson's technique requires observation of a pair of spectral lines carefully selected to have similar equivalent widths, central intensities, heights of formation, and temperature sensitivities, but with very different magnetic sensitivities. He proposed to derive the magnetic field strength from the excess width determined from a Fourier analysis of the magnetic/nonmagnetic line pair. This technique was initially used by Robinson, Worden, and Harvey (1980) to derive magnetic field strengths (B) and fractional disk filling factors (f) for two stars: ξ Boo A (G8V) and 70 Oph A (K0V). Later Gray (1984) used a modified Fourier analysis technique to derive magnetic field parameters in 7 of 18 dwarf stars studied, and Marcy (1984) used a profile fitting variation of this technique to determine field parameters in 19 of 29 dwarf stars observed.

Because the observed quantity is a subtle increase in line width, the inferred magnetic field properties depend crucially upon the accuracy of the diagnostic technique, and systematic errors can be critical. Several authors, in particular Hartmann (1987), have raised the following questions:

(1) Weak line blends, especially those located in the inner wings of the magnetically-sensitive line, can mimic spurious magnetic fields especially in the coolest stars where line blending is nearly ubiquitous. Saar (1988) has evaluated this effect quantitatively. His solution for the problem (cf., Saar, Linsky, and Beckers, 1986) is a line difference technique in which one subtracts the profile of the same magnetically-sensitive line in a less active star from that of a more active star of the same spectral type, after adjusting the profiles of the two stars for differences in their nonmagnetic broadening parameters. The difference profile is then analyzed for the excess Zeeman broadening, but some limitations to this technique are discussed below.

(2) The spectral lines commonly analyzed for Zeeman broadening are not optically thin. Typically their equivalent widths place them near or on the flat part of the curve of growth, so their shapes depend upon line optical depths and on the line/continuum opacity ratio. The Robinson (1980) technique implicitly assumes that both the magnetically-sensitive and insensitive lines are optically thin, but Hartmann (1987) showed that the difference between two lines with different degrees of saturation can be appreciable in the inner line wings and thus produce a spurious magnetic signature. This problem may be ameliorated by comparing observed line profiles with computed profiles that include line saturation effects. Saar (1988) has employed an analytical solution to the radiative transfer equation in which the LTE line source function depends linearly upon optical depth, and the line/continuum opacity ratio is independent of depth. Basri and Marcy (1988) have instead solved for the Stokes vector in an LTE model atmosphere in which all parameters were allowed to vary with depth. Since their technique yields magnetic field parameters fairly similar to those found by Saar for stars in common, Saar's simpler analytical technique appears to be approximately valid. Nevertheless, the complete model atmosphere approach is preferred when line saturation is a concern.

(3) The magnetic field parameters inferred from a comparison of observed and computed line profiles depend sensitively on the assumed stellar rotational velocity, and microturbulent and macroturbulent broadening (Hartmann 1987; Saar 1988). Thus even the model atmosphere technique has its limitations to the extent that the nonmagnetic broadening parameters are uncertain.

(4) The comparison of profiles of the same magnetically-sensitive line in two different stars does cancel the effects of line blends to first order, but the two stars may have somewhat different thermal structures and abundances and thus different line saturation and widths: these effects must be compensated for in order for the diagnostic procedure to be reliable.

(5) The measurement of small differences in line width due to Zeeman broadening requires high quality observations. Saar (1988) showed that for lines in the optical spectrum, the minimum requirements are signal/noise > 80, $v \sin i < 8$ km s^{-1}, and a spectral resolution $> 70\,000$.

(6) The thermal structure of the magnetic regions of a stellar atmosphere may differ considerably from that of the nonmagnetic regions, whereas all approaches to date have assumed that the two atmospheric regions have the same thermal structures. If the magnetic regions are hotter with a brighter continuum adjacent to the magnetically-sensitive line (analogous to solar faculae as observed near the limb), then the magnetic regions will contribute disproportionately to the disk-integrated line profile, and the true magnetic filling factors will be smaller than computed for a homogeneous atmosphere. This effect may partly explain the large filling factors (as large as $f = 0.9$) that Saar, Linsky, and Giampapa (1987) have deduced from infrared spectra of M dwarf stars. Mathys and Solanki (1988) provide evidence that the magnetic regions for ε Eri are indeed hotter. On the other hand, if the magnetic regions are cooler than the surrounding photosphere (e.g., pores on the Sun), then the filling factors have been underestimated.

(7) The distribution of magnetic flux across a stellar surface is assumed homogeneous in all widely used modeling techniques. If this assumption is invalid, then the derived magnetic flux will be overestimated when the flux is concentrated near disk center and underestimated when the flux is concentrated near the limb.

(8) What is the proper physical interpretation of the derived magnetic field strengths? Because starspots are very dark in the visible, the disk-integrated line profiles include very little contribution from spots even when they cover a large portion of the stellar surface as is the case for dMe stars. The inferred fields must, therefore, refer to penumbrae of spots or to localized bright magnetic regions perhaps analogous to solar plages or faculae. The empirical correlation of observed field strengths with the values computed by balancing magnetic with gas pressure forces in the photosphere (Saar and Linsky, 1986; Saar, Linsky, and Giampapa, 1987) suggests that convective motions enhance the photospheric fields until pressure equilibrium is achieved. In the solar photosphere typical field strengths of 1200–1500 G in small structures (Tarbell and Title, 1977) are consistent with this explanation. Thus the inferred field strength represents an average of the 'pressure-equilibrium' field over the whole stellar disk. The computed line profiles that are compared with observed profiles to determine the field parameters should thus be proper averages over the stellar disk rather than profiles computed at an average viewing angle (e.g., Bopp et al., 1989).

Mathys and Solanki (1988) discuss a technique that analyzes statistically the shift in the center of gravity of a large number of spectral lines observed in unpolarized light. This technique, pioneered with solar data by Stenflo and Lindegren (1977), requires high-resolution spectra of many lines to infer field parameters from correlations of absorption line depths, widths, and equivalent widths (below specified line depths) with atomic parameters. Mathys and Solanki (1988) have applied this technique to four dwarf stars, obtaining values of fB and B significantly larger than those obtained with the line broadening technique (Saar, 1988) for the two stars in common (ε Eri and 40 Eri A). The explanation for the differences is not known. For ε Eri, Mathys and Solanki (1988) deduced that the magnetic regions are hotter than the nonmagnetic regions, so that the filling factor is less than inferred with the usual assumption that the atmospheric models for the magnetic and nonmagnetic regions are the same.

2.2. Circular and Linear Polarization Techniques

While most recent empirical studies of magnetic fields in late-type stars are based on measurements of the excess Zeeman broadening of line profiles in unpolarized light, polarization techniques have been pushed to their limits to obtain information on the fields complementary to the broadening analyses. For example, Kemp et al. (1987) detected variable broad-band circular polarization of amplitude 0.002–0.004% in observations of the single-lined RS CVn system λ And (G8III–IV + ?). They interpreted the net polarization signal as due to ordered magnetic fields, perhaps in large spots near the rotational pole and thus always near the limb, that are not fully cancelled in the disk-averaged flux because of their concentration near the limb.

Broad-band circular polarization measurements include many spectral lines and the continuum. The interpretation is necessarily complex, but Murset, Solanki, and Stenflo (1988) have provided some insight by simulating broad-band measurements (roughly 100 A bandpass) from high spectral resolution solar Stokes V profiles. They explain the wavelength dependance and center-to-limb variation of their simulated broad-band circular polarization measurements as due to line rather than to continuum polarization. The net line polarization could be due to the presence of a large bipolar region on the disk with each component at a different projection angle μ. The two polarities do not cancel either due to different angles of the magnetic flux lines to the line-of-sight or to a change in the asymmetry in the Stokes V profile with μ. Murset, Solanki, and Stenflo (1988) also say that differences in temperature or other properties between the magnetic elements of the two polarities could give rise to a net circular polarization signal. They interpret the Kemp et al. (1987) circular polarization observations of λ And as due to the rotation across the disk of either a single spot at intermediate latitude or to the rotation of a plage and a spot of the same polarity but different longitudes.

Linear polarization from magnetic regions distributed across the disk does not cancel in integrated starlight, and has been recorded in broad-band measurements for a few stars (e.g., Huovelin, Saar, and Tuominen, 1988). While the interpretation of these data in terms of magnetic parameters is difficult and not unique (cf. Landi Degl'Innocenti, 1982), the polarization amplitude is a measure of the net tangential component of the magnetic field and will be largest when sufficient magnetic flux is concentrated asymmetrically near the stellar limb. In contrast, the magnetic filling factor derived from unpolarized line broadening measurements is weighted towards disk-center regions because of projection and limb-darkening effects. Thus the combination of simultaneous broad-band linear polarization data and line broadening data for the same star provides positional information from which a 'magnetic image' might be assembled. Saar *et al.* (1988) have constructed a magnetic image of ξ Boo A (G8V) on the basis of a coordinated observing campaign in June 1986.

3. Important Trends Emerging from the Stellar Magnetic Field Measurements

Altogether, magnetic field parameters have been obtained using some variant of the Robinson (1980) technique for about 50 late-type stars by Marcy (1984), Marcy and Bruning (1984), Gray (1984, 1985), Saar (1987a), Saar and Linsky (1986), Saar, Linsky, and Beckers (1986), Saar, Linsky, and Giampapa (1987), Bruning, Chenoweth, and Marcy (1987), and Basri and Marcy (1988). Linsky (1985) and Saar (1978b) have summarized the important trends emerging from these recent measurements. Here I will bring the summaries up to date.

(1) There is a trend of increasing magnetic field strength with decreasing effective temperature for late-type dwarfs. Since the gravity and thus the photospheric pressure at continuum optical depth unity at 5000 Å both increase systematically towards the lower Main Sequence, it is not clear *a priori* whether the effective temperatures or gravity is responsible for the observed trend. Saar and Linsky (1986) find that $B \sim P_{gas}^{1/2}$ fits the data well and is consistent with the physically plausible situation in which the magnetic pressure in the magnetic flux tubes is the dominant factor balancing the photospheric gas pressure in the surrounding nonmagnetic region. The explanation is simplistic to the extent that the magnetic regions have some internal gas pressure and the height at which the magnetic field is measured lies above that at which the photospheric gas pressure is computed. Nevertheless, since the crude scaling law makes physical sense and fits the data, we adopt it as a working hypothesis. One consequence of the scaling law is that magnetic field strengths should be small (and thus more difficult to measure) in stars located above the Main Sequence, because of their lower gravities and thus their lower photospheric pressures. Indeed, Marcy and Bruning (1984) failed to detect fields in 8 active giants, and the marginal detection of a field of about 800 G in λ And (G8III–IV + ?) by Gondoin, Giampapa, and Bookbinder (1985) supports the scaling law. However, Bopp *et al.* (1989) have reported a field of $B = 2000 \pm 300$ G in VY Ari (K0IV–Ve + ?): possibly the first detection of a field in a pre-Main-Sequence star, and perhaps a counterexample to the scaling law.

(2) The derived magnetic filling factors for the nonspot fields are not correlated with B; thus Gray's (1985) suggestion that fB is a constant has not been supported by subsequent data (e.g., Saar and Linsky, 1986). Instead, f increases with angular rotation rate such that for $\omega \gtrsim 0.25$ days^{-1}, among the dMe stars, f approaches 0.80 and the stellar surface becomes saturated with magnetic regions. On the other hand, the inactive, slowly-rotating dM stars show no evidence of magnetic flux and likely have $f < 0.2$ (Saar, Linsky, and Giampapa, 1987). The dependence of f on rotation rate and its saturation are reasonably matched by the dynamo theory of Skumanich and MacGregor (1986).

(3) The spatial correlations of bright ultraviolet emission lines and X-ray flux with magnetic flux on the Sun implies that similar correlations should also exist for stars. Schrijver *et al.* (1989) and Saar and Schrijver (1987) show that the correlations of stellar X-ray flux and Ca II flux with magnetic flux (fB) are sensibly tight and are consistent with the solar correlations derived from spatially-resolved data. This implies that monitoring of active stars which have asymmetric distributions of active regions across their disks should show the rise and fall of ultraviolet emission, X-ray flux and magnetic flux in phase as the major active regions rotate into and out of view. This prediction has been confirmed at a low level for ε Eri (K2V) by Saar, Linsky, and Duncan (1986) and at a much higher level for ξ Boo A (G8V) by Saar *et al.* (1988). Additional rotational modulation studies are needed to derive definitive correlations for a range of stars.

4. An Ambitious Agenda for the Next 6 Years of Stellar Magnetic Field Research

I conclude with a rather ambitious agenda for the next 6 years of stellar magnetic field research. Now that we have developed the basic diagnostic techniques and have acquired some confidence in their application to measuring stellar magnetic fields, it is useful to ask where we should go from here. I believe that we should aim towards developing self-consistent multicomponent models for the magnetic atmospheres of active stars that incorporate both the essential physics and the basic phenomenology that we are observing in these stars. Such models should include the following aspects:

(1) High-resolution solar observations indicate that magnetic atmospheres are highly structured and far from spherical symmetry. An approximate geometry would be one in which the hot magnetically-heated plasma is confined by closed magnetic loops with the surrounding material cooler and not confined by closed field lines. Since the gas pressure likely decreases with height (outside the loops) more rapidly than the magnetic pressure, the field lines should diverge with height above the photosphere, so that somewhere in the chromosphere the field lines become nearly horizontal for inactive stars like the Sun with small values of f in their photospheres, or they essentially fill the available volume for active stars like dMe stars with large values of f in their photospheres. In either case at least two atmospheric temperature structure models are required in a complex geometry.

(2) Transition zone lines formed in solar magnetic regions are often redshifted (e.g., Brueckner, 1981), implying that systematic downflows occur in these regions, presumably guided by the magnetic geometry. Since the magnetic fields are presumably closed and the flows cannot cut across the field lines, the observed flows must be transient or perhaps represent circulation patterns in which the downflow component is brighter than the upflowing component. Mariska (1988) has proposed an elegant explanation in which the concentration of heating close to one footpoint of a closed loop induces a syphon flow along the loop for which the temperature gradient in the upflow (from the footpoint closest to where the heating is concentrated) is much steeper than in the downflow. Thus the emission measure of the downflowing plasma far exceeds that of the upflowing plasma at temperatures below 200 000 K and the integrated spectrum

shows a net downflow velocity of about 10 km s^{-1} at 100 000 K for all orientations of the loop. Transition zone emission lines in active dMe and RS CVn stars (Ayres, 1984; Ayres, Jensen, and Engvold, 1988; Elgaroy *et al.*, 1988) show global redshifts with comparable velocities. Thus realistic models of magnetic atmospheres should include flows along the loops.

(3) Thermal models for each component should be self-consistent with mechanical heating balanced by radiative cooling, thermal conduction, and the enthalpy of the flowing plasma. At present we have little information concerning the detailed heating processes; thus the energy equation can only be used to infer the rate of mechanical heating and its location in the atmosphere.

(4) Solar spectra in the 4.6 micron fundamental vibration-rotation bands of the CO molecule indicate that the lower chromosphere has two basic thermal structures – one with a hot temperature minimum (about 5000 K) and steep temperature rise in which CO is not present, and a second component with a cool temperature minimum (below 3700 K) and very little if any temperature rise in which CO is an important species (Ayres and Testerman, 1981; Ayres, Testerman, and Brault, 1986). Ayres (1981), Kneer (1983), Muchmore and Ulmschneider (1985), and others have interpreted this thermal bifurcation of the solar atmosphere as due to a thermal instability driven by the rapid increase in CO formation with decreasing temperature and the efficiency with which the CO infrared bands can cool the atmosphere. Thus the nonmagnetic regions are kept cool by the CO and the regions of strong magnetic heating remain hot as there is no CO to cool them. Stellar model atmospheres should include the thermal bistability due to CO, and to other molecules like SiO for stars much cooler than the Sun.

(5) Atmospheric models should be consistent with the measured photospheric magnetic flux (fB), f, and B. For the Sun one can extrapolate the measured fields upwards into the chromosphere and corona using, for example, the current-free approximation (e.g., Poletto *et al.*, 1975). This is not feasible for stars, but one may instead use the measured filling factor in the photosphere and a plausible estimate of its divergence with height to estimate the field strength in the magnetic component with height. Measurement of fB using lines formed at different heights will help in estimating these effects.

(6) Another set of constraints on multicomponent atmospheric models is that they must be consistent with such observables as the radio, X-ray, Ca II, Mg II, and other ultraviolet emission line fluxes. If either the X-ray or Ca II flux are not available, they may be estimated from the magnetic flux using empirical scaling laws (Schrijver *et al.*, 1989). Aside from the radio emission, the other observables may be computed from the emission measure distribution. The radio emission is typically gyrosynchrotron emission from mildly relativistic electrons, and is thus not simply related to the thermal distribution of electrons but is a function of the coronal magnetic fields. At present we must treat the distribution of nonthermal electrons and their volume as free parameters.

(7) Finally, the models should include the sizes and locations of starspots obtained from photometry (e.g., Rodono *et al.*, 1986) and Doppler imaging studies (e.g., Vogt and Penrod, 1983). The sizes and locations of active regions in the chromospheres and transition zones derived from emission-line Doppler imaging studies (e.g., Walter *et al.*,

1987; Neff *et al.*, 1989) are particularly useful in establishing the gross geometry of a stellar atmosphere.

Acknowledgements

This work is supported in part by NASA grants NGL 06–003–057, W-15103, and NAG5–82. I am indebted to Steve Saar and Tom Ayres for their comments on the manuscript. I would like to thank the Organizing Committee of IAU Colloquium No. 104 for the opportunity to review and rethink this interesting topic in a stimulating environment.

References

Ayres, T. R.: 1981, *Astrophys. J.* **244**, 1064.
Ayres, T. R.: 1984, *Astrophys. J.* **284**, 784.
Ayres, T. R. and Testerman, L.: 1981, *Astrophys. J.* **245**, 1124.
Ayres, T. R., Testerman, L., and Brault, J. W.: 1986, *Astrophys. J.* **304**, 542.
Ayres, T. R., Jensen, E., and Engvold, O.: 1988, *Astrophys. J. Suppl.* **66**, 51.
Babcock, H. W.: 1958, *Astrophys. J. Suppl.* **3**, 141.
Basri, G. and Marcy, G. W.: 1988, *Astrophys. J.* **330**, 274.
Bopp, B. W., Saar, S., Feldman, P., Dempsey, R., Allen, M., Ambruster, C., and Barden, S. P.: 1989, *Astrophys. J.* (in press).
Borra, E. F., Edwards, G., and Mayor, M.: 1984, *Astrophys. J.* **284**, 211.
Brault, J. and Noyes, R.: 1983, *Astrophys. J.* **269**, L61.
Brueckner, G. E.: 1981, in F. G. Orrall (ed.), *Solar Active Regions*, Colorado Assoc. Univ. Press, Boulder.
Bruning, D. H., Chenoweth, R. E., Jr., and Marcy, G. W.: 1987, in J. L. Linsky and R. E. Stencel (eds.), *Cool Stars, Stellar Systems, and the Sun*, Springer-Verlag, Berlin, p. 36.
Elgaroy, O., Joras, P., Engvold, O., Jensen, E., Pettersen, B. R., Ayres, T. R., Ambruster, C., Linsky, J. L., Clark, M., Kunkel, W., and Marang, F.: 1988, *Astron. Astrophys.* **193**, 211.
Gray, D. F.: 1984, *Astrophys. J.* **277**, 640.
Gray, D. F.: 1985, *Publ. Astron. Soc. Pacific* **97**, 719.
Gray, D. F.: 1988, *Lectures on Spectral-line Analysis: F, G, and K Stars*, The Publisher, Arva, Ontario.
Golub, L.: 1983, in J. O. Stenflo (ed.), 'Solar and Stellar Magnetic Fields: Origins and Coronal Effects', *IAU Symp.* **102**, 345.
Gondoin, P., Giampapa, M. S., and Bookbinder, J. A.: 1985, *Astrophys. J.* **297**, 710.
Hartmann, L.: 1987, in J. L. Linsky and R. E. Stencel (eds.), *Cool Stars, Stellar Systems, and the Sun*, Springer-Verlag, Berlin, p. 1.
Huovelin, J., Saar, S. H., and Tuominen, I.: 1988, *Astrophys. J.* **329**, 882.
Kemp, J. C., Henson, G. D., Kraus, D. J., Dunaway, M. H., Hall, D. S., Boyd, L. J., Genet, R. M., Guinan, E. F., Wacker, S. W., and MCook, G. P.: 1987, *Astrophys. J.* **317**, L29.
Kneer, F.: 1983, *Astron. Astrophys.* **128**, 311.
Landi Degl'Innocenti, E.: 1982, *Astron. Astrophys.* **110**, 25.
Linsky, J. L.: 1983a, in J. O. Stenflo (ed.), 'Solar and Stellar Magnetic Fields: Origins and Coronal Effects', *IAU Symp.* **102**, 313.
Linsky, J. L.: 1983b, in P. B. Byrne and M. Rodono (eds.), 'Activity in Red Dwarf Stars', *IAU Colloq.* **71**, 39.
Linsky, J. L.: 1985, *Solar Phys.* **100**, 333.
Marcy, G. W.: 1983, in J. O. Stenflo (ed.), 'Solar and Stellar Magnetic Fields: Origins and Coronal Effects', *IAU Symp.* **102**, 3.
Marcy, G. W.: 1984, *Astrophys. J.* **276**, 286.
Marcy, G. W. and Bruning, D. H.: 1984, *Astrophys. J.* **281**, 286.

JEFFREY L. LINSKY

Mariska, J. T.: 1988, *Astrophys. J.* **334**, 489.

Mathys, G. and Solanki, S. K.: 1988, *Astron. Astrophys.* (in press).

Muchmore, D. and Ulmschneider, P.: 1985, *Astron. Astrophys.* **142**, 393.

Murset, U., Solanki, S. K., and Stenflo, J. O.: 1988, *Astron. Astrophys.* **204**, 279.

Neff, J. E., Walter, F. M., Rodono, M., and Linsky, J. L.: 1989, *Astron. Astrophys.* (in press).

Poletto, G., Vaiana, G. S., Zombeck, M. V., Krieger, A. S., and Timothy, A. F.: 1975, *Solar Phys.* **44**, 83.

Robinson, R. D.: 1980, *Astrophys. J.* **239**, 961.

Robinson, R. D., Worden, S. P., and Harvey, J. W.: 1980, *Astrophys. J.* **236**, L155.

Rodono, M. *et al.*: 1986, *Astron. Astrophys.* **165**, 135.

Saar, S. H.: 1987a, Ph.D. Thesis, University of Colorado.

Saar, S. H.: 1987b, in J. L. Linsky and R. E. Stencel (eds.), *Cool Stars, Stellar Systems, and the Sun*, Springer-Verlag, Berlin, p. 10.

Saar, S. H.: 1988, *Astrophys. J.* **324**, 441.

Saar, S. H. and Linsky, J. L.: *Astrophys. J.* **299**, L47.

Saar, S. H. and Linsky, J. L.: 1986, *Adv. Space Phys.* **6**, No. 8, 235.

Saar, S. H. and Schrijver, C. J.: 1987, in J. L. Linsky and R. E. Stencel (eds.), *Cool Stars, Stellar Systems, and the Sun*, Springer-Verlag, Berlin, p. 38.

Saar, S. H., Linsky, J. L., and Beckers, J. M.: 1986, *Astrophys. J.* **302**, 777.

Saar, S. H., Linsky, J. L., and Duncan, D. K.: 1986, in M. Zeilik and D. M. Gibson (eds.), *Cool Stars, Stellar Systems, and the Sun*, Springer-Verlag, New York, p. 275.

Saar, S. H., Linsky, J. L., and Giampapa, M. S.: 1987, in L. Delbouille and A. Monfils (eds.), *Observational Astrophysics with High Precision Data*, Univ. de Liège, Liège, p. 103.

Saar, S. H., Huovelin, J., Giampapa, M. S., Linsky, J. L., and Jordan, C.: 1988, in O. Havnes *et al.* (eds.), *Midnight Sun Workshop on Activity in Cool Star Envelopes*, D. Reidel Publ. Co., Dordrecht, Holland, p. 45.

Schrijver, C. J., Coté, J., Zwaan, C., and Saar, S. H.: 1989, *Astrophys. J.* **337**, 964.

Skumanich, A. and MacGregor, K.: 1986, *Adv. Space Phys.* **6**, No. 8, 151.

Stenflo, J. O. and Lindegren, L.: 1977, *Astron. Astrophys.* **59**, 367.

Tarbell, T. D. and Title, A. M.: 1977, *Solar Phys.* **52**, 31.

van de Hulst, H.: 1988, in *Modeling the Stellar Environment How and Why*, Fourth IAP Astrophysics Meeting (to appear).

Vogt, S. S.: 1980, *Astrophys. J.* **240**, 567.

Vogt, S. S. and Penrod, G. D.: 1983, *Publ. Astron. Soc. Pacific* **95**, 566.

Walter, F. M., Neff, J. E., Gibson, D. M., Linsky, J. L., Rodono, M., Gary, D. E., and Butler, C. J.: 1987, *Astron. Astrophys.* **186**, 241.

THE ASSOCIATION OF FLARES TO CANCELLING MAGNETIC
FEATURES ON THE SUN

SILVIA H. B. LIVI

Universidade Federal do Rio Grande do Sul, 90 049 Porto Alegre RS, Brazil

SARA MARTIN and HAIMIN WANG

Big Bear Solar Observatory, California Institute of Technology, Pasadena, CA 91125, U.S.A.

and

GUOXIANG AI

Beijing Astronomical Observatory, Beijing, China

Abstract. Previous work relating flares to evolutionary changes of photospheric solar magnetic fields are reviewed and reinterpreted in the light of recent observations of cancelling magnetic fields. In line-of-sight magnetograms and H-alpha filtergrams from Big Bear Solar Observatory, we confirm the following 3 associations: (a) the occurrence of many flares in the vicinity of emerging magnetic flux regions (Rust, 1974), but only at locations where cancellation has been observed or inferred; (b) the occurrence of flares at sites where the magnetic flux is increasing on one side of a polarity inversion line and concurrently decreasing on the other (Martres *et al.*, 1968; Ribes, 1969); and (c) the occurrence of flares at sites where cancellation is the only observed change in the magnetograms for at least several hours before a flare (Martin, Livi, and Wang, 1985). Because cancellation (or the localized decrease in the line-of-sight component of magnetic flux) is the only common factor in all of these circumstances, suggest that cancellation is the more general association that includes the other associations as special cases. We propose the hypothesis that cancellation is a necessary, evolutionary precondition for flares. We also confirm the observation of Martin, Livi, and Wang (1985) that the initial parts of flares occur in close proximity to cancellation sites but that during later phases, the flare emission can spread to other parts of the magnetic field that are weak, strong, or not cancelling.

1. Review of Previous Work

The association of flares with observed photospheric magnetic fields has been made previously in terms of configurations and evolutionary changes. The earliest studies of flare positions relative to photospheric magnetic fields by Severny (1958, 1960) showed that flares occurred near polarity inversion lines (previously also called neutral lines or $H = 0$ lines). The centering of flares around polarity inversion lines was confirmed by Martres *et al.* (1968a) and Smith and Ramsey (1967). Michard (1971) also noted that when the initial H-alpha brightenings have more than one knot, they are located on two different polarities, on both sides of the inversion line, rather than directly on it.

Specific flare-related magnetic field changes were reported by Martres *et al.* (1968a, b). They studied an active region in which all the flares occurred where magnetic flux was increasing on one side of the polarity division line while it was decreasing on the other side.

The frequent association of flares with emerging magnetic flux regions was first noted by Rust (1972, 1974), verified by Vorpahl (1973), and subsequently elaborated on by these and many other authors (Martin *et al.*, 1983; Priest *et al.*, 1986; and references therein). When opposite polarity fields come into close contact because of the emergence of new magnetic flux in pre-existing active regions, a steep magnetic field gradient builds on the magnetic inversion line between the new flux and the pre-existing flux. The association of flares with high magnetic field was also made by Severny (1960) and confirmed by Martres, Michard, and Soru-Iscovici (1966).

Although many flares happen in association with new flux, it is also known that flares occur in the absence of emerging flux (Martin *et al.*, 1984). Martin, Livi, and Wang (1985) studied a decaying region and found that all of the observed flares began at sites where magnetic flux was cancelling. Cancellation is the gradual and mutual decrease of magnetic flux at the boundary between closely-spaced opposite polarity magnetic fields as seen in line-of-sight photospheric magnetograms (Martin, 1984; Livi, Martin, and Wang, 1985). Magnetic flux is observed to gradually decrease in both polarities as the magnetic fields migrate together and a high magnetic field gradient is observed as long as the fields are cancelling. In many cases the fragment with less magnetic flux completely disappears. To date cancellation has only been observed in magnetograms of the line-of-sight component which leaves the physical interpretation of cancellation open to several possible interpretations (Zwaan, 1987).

2. The Data

We illustrate examples of flares and cancelling magnetic fields from observations taken from 8–11 July, 1988 in an active region that crossed the central meridian during this interval. The data obtained on this active region are especially well-suited for the study of magnetic field changes and flares because: (1) the magnetograms were of high quality due to good seeing; (2) collaborative observations of the magnetic fields were taken at the Huairou Solar Observing Station of the Beijing Astronomical Observatory and at the Big Bear Solar Observatory; (3) the active region was located near the Sun's central meridian which is favorable for the acquisition and interpretation of line-of-sight magnetograms; and (4) the active region produced many small flares and a few large ones during the observing hours at Big Bear Solar Observatory.

The magnetograms used in the illustrations are mostly from the Big Bear Solar Observatory because the study is centered around flares observed at the Big Bear Solar Observatory. Unfortunately, Hα filtergrams are not yet taken at the Huairou Observatory. However, the videomagnetograms taken at Huairou are important in the evaluation of long-term changes before and after major flares; magnetograms from both sites have been matched in scale and sensitivity during the processing of the data. On all of the magnetograms negative magnetic polarity is presented in black and positive in white. Observing hours at Huairou Observatory are from approximately 01:00 until 12:00 UT and observing hours at Big Bear Observatory are from approximately 15:00 until 02:00 UT during the early summer.

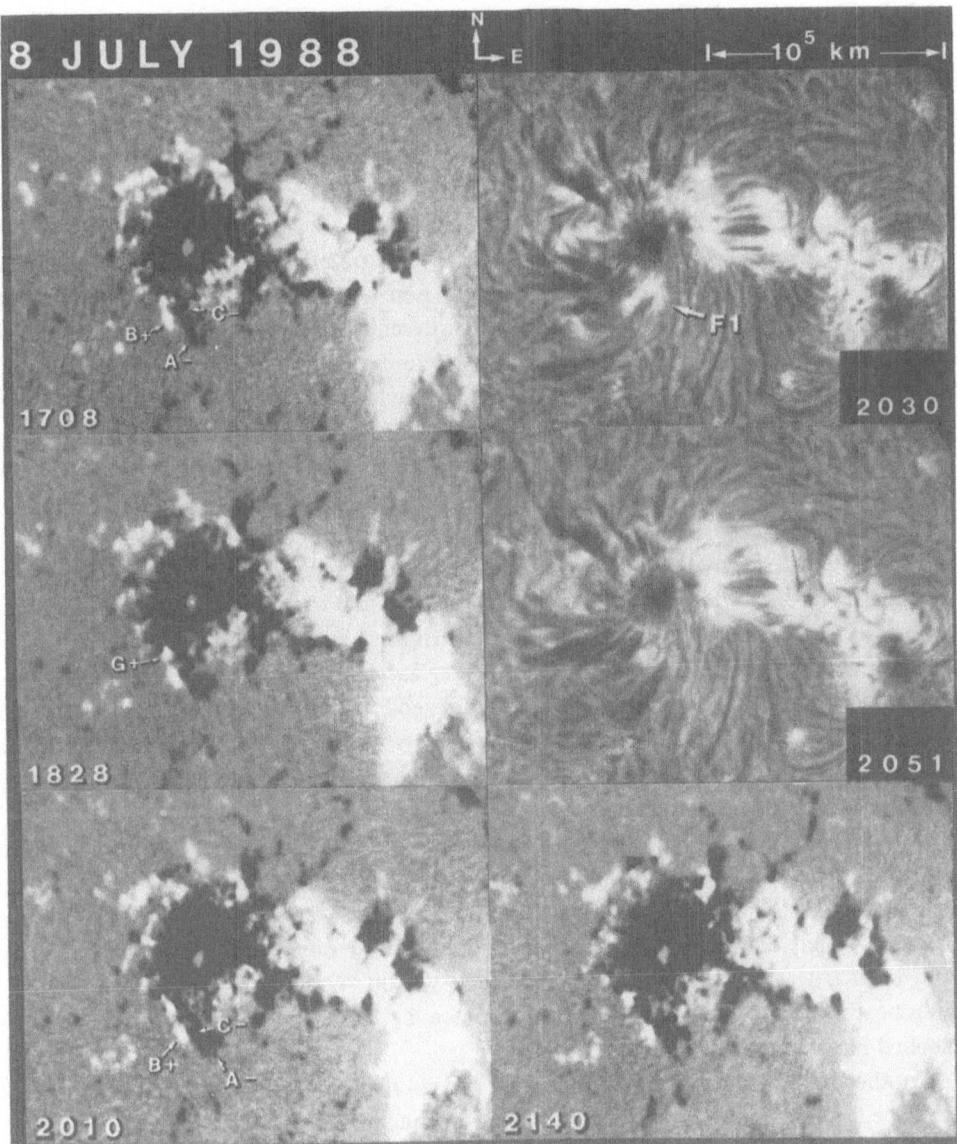

Fig. 1. The flare, F1, in the upper right frame, corresponds in position to the site of converging patches of magnetic field labelled $A-$, $B+$, and $C-$. All of these patches of flux are identified as part of a system of Moving Magnetic Features (MMFs) which originate near the penumbral boundary and flow approximately radially away from the associated sunspot. As they move away from the sunspot, $A-$, $B+$, and $C-$ converge. $B+$ and $C-$ begin to cancel each other when they come into contact but the slow reduction in magnetic flux becomes apparent only in the continuation of this series in Figure 2 where it is seen that the area of $B+$ is decreasing. The expected equivalent loss of flux in $C-$ is not seen because it merges with $A-$ and other negative magnetic flux closer to the sunspot.

3. New Examples of Flares and Their Relationship to Cancelling Magnetic Fields

In Figure 1, Hα filtergrams in the upper right and middle right reveal a flare, labelled F1, at 20:30 UT and the aftermath of the flare at 21:40. The flare occurs just below the sunspot near the middle of the images. The videomagnetograms in the left side show the magnetic field configuration for over 3 hours prior to the flare. The final magnetogram in the lower right corresponds to approximately one hour after the flare. We have put arrows and labels on the illustrations to identify the patches of photospheric magnetic flux that correspond to the site of the flare. Then we trace the same magnetic fields backward and forward in the time to see their evolution before and after the flare. The flare, F1, occurs above the photospheric magnetic flux patches marked $A-$, $B+$, and $C-$. Our system of labelling in this paper is to use letters to designate specific patches of magnetic field and to follow the letter with $+$ or $-$ to eliminate any ambiguity about whether the arrow points to a positive polarity patch (white) or a negative polarity patch (black). Tracing $A-$, $B+$, and $C-$ back in time we see that a convergence of these elements occurs between 17:08 and 18:28. This convergence brings $B+$ and $C-$ into contact. The convergence of magnetic flux of opposite polarity usually leads to cancellation at about the time that patches of magnetic flux appear to come into contact (Martin, 1984; Martin, Livi, and Wang, 1985). In this circumstance in Figure 1, it is not clear that cancellation has begun until after the flare. Although the visible effect of the cancellation is marginal in Figure 1, a definite reduction in the area of $B+$ is evident in the continuation of this time series of magnetograms in Figure 2. The cancellation is not yet conspicuous in the negative polarity because of the convergence of additional negative flux from the sunspot moat.

In Figure 2, a second flare, F2, is visible at 23:04. One part of this second flare F2a coincides with flare F1 in Figure 1. In Figure 2 at 22:21, negative patch $D-$ has moved into juxtaposition with $B+$ to form a new cancellation site. At 23:41, another new patch, $E+$, has coalesced from smaller patches of new positive flux. $D-$ is cancelling with both $B+$ and $E+$. By the end of series at 00:42, $D-$ has almost completely disappeared. These changes are taking place at the site of the part of the flare labelled F2a.

Another part of the flare, F2b, is to the left and closer to the sunspot. This part of the flare lies above and adjacent to a cluster of magnetic fields that have also emerged in the sunspot moat. To the upper left of $E+$, lie two other patches, $F-$ and $G+$. Tracing the previous evolution of $G+$ back through the magnetograms in Figure 1, we see that it was previously adjacent to $B+$ at 18:28 but has moved nearly tangential to the spot and has initially gained flux. This behavior differs from most of the small positive and negative magnetic knots around the spot that are called Moving Magnetic Features or MMFs (Harvey and Harvey, 1973). Most MMFs form near the outer penumbra of the sunspot and flow radially away from the spot. We re-examined the time-lapse videomagnetogram film and found that $G+$ and $C-$ comprised a new bipole (an ephemeral region) whose $+$ and $-$ components move away from each other as their fluxes increase. Thus $G+$, while growing, was also moving perpendicular to the outward

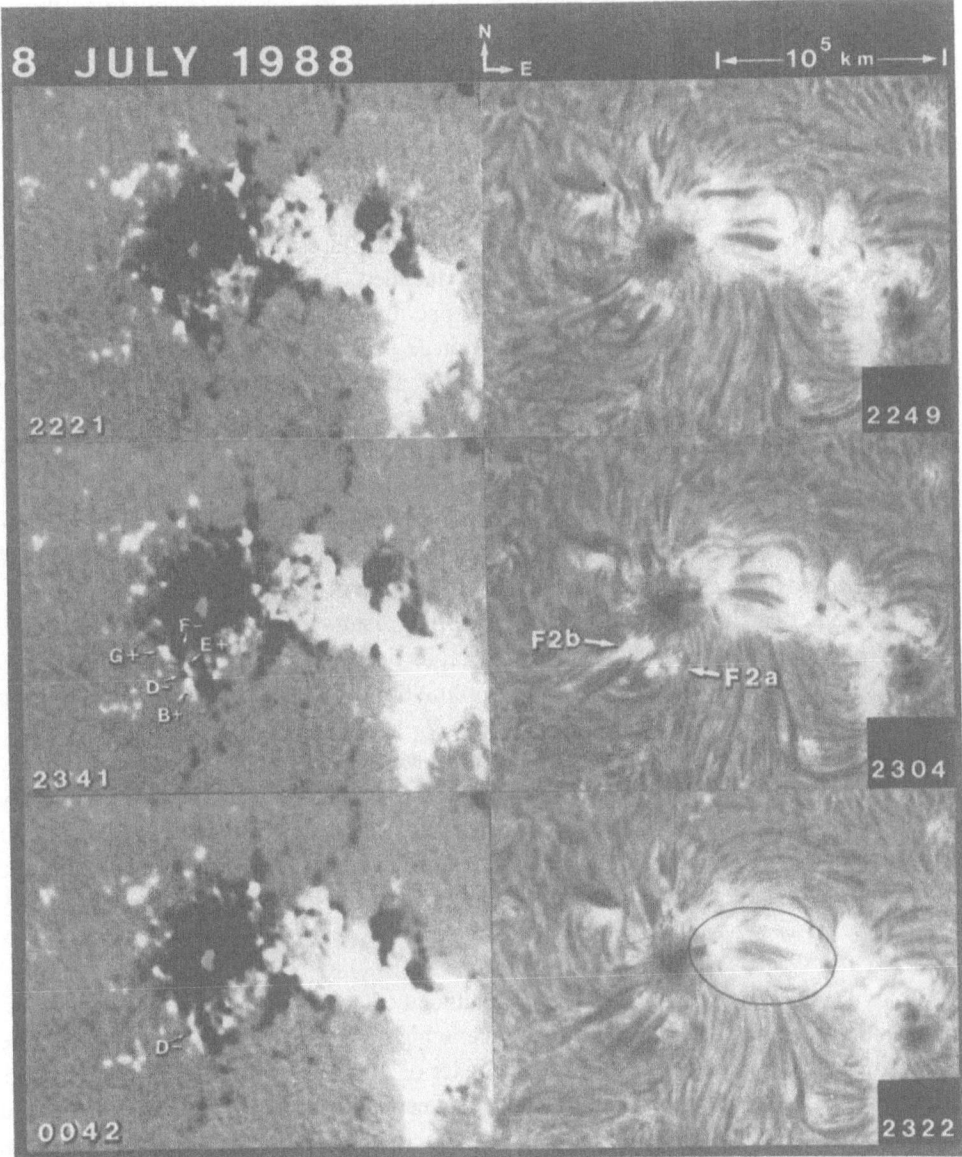

Fig. 2. Another flare, F2 at 2304, is seen at and adjacent to the site of the Flare F1 in Figure 1. The part labelled F2a is at the same site of cancelling magnetic features as in Figure 1. However, a new patch, $D-$, has moved into position to cancel with $B+$. $D-$ is also cancelling with another new patch, $E+$, and has almost disappeared by the end of the sequence. The other part of the flare, F2b, corresponds to magnetic patches $G+$ and $F-$. $F-$ is negative polarity Moving Magnetic Feature (MMF) that is flowing away from the sunspot. However, $G+$ is component of a small bipole, an ephemeral region, whose other initial negative pole is the patch labelled $C-$ in Figure 1. As the ephemeral region grows, its poles, $G+$ and $C-$, migrate in opposite directions approximately tangential to nearly circular moat of MMFs that radially emanate from the sunspot. $G+$, therefore, is moving approximately orthogonally to the MMFs that originate near the penumbral boundary. It encounters and cancels with negative polarity MMF labelled $F-$ before, during and after the flare.

flow of other patches within the moat. Since most of the nearby encountered flux is negative, we infer that $G+$ was probably growing and cancelling at the same time, as we have observed for ephemeral regions on the quiet Sun (Livi, Martin, and Wang, 1985; Martin, 1988). We conclude that this part of the flare, F2b, corresponds to a special dynamic circumstance within the sunspot moat including flux growth, flux cancellation and unusual motions. The other part of the flare, F2a, lies above photospheric magnetic fields that are simply converging and cancelling.

In Figure 3, we illustrate the sites of two other small flares, F3 and F4, in the H-alpha filtergrams. F3 coincides with flux patches, labelled $J+$, $I-$, and $H-$ in the first magnetogram before the flare. In the time-lapse film, we found that $J+$ and $I-$ are MMFs that are flowing away from the sunspot. $H-$ is a small part of the larger of two, magnetically complex emerging flux systems that are encompassed by ovals in the last H-alpha image. (The emerging flux system can also be identified in Figures 1 and 2 from the east–west aligned system of arch filaments. In the last H-alpha image in Figure 2 an oval is drawn around the emerging flux system.) $H-$ is growing and moving towards the sunspot to its left. As it does so, it encounters and merges with $I-$ moving to the right. $J+$, also moving to the right away from the sunspot, and $H-$, moving to the left toward the sunspot, also encounter each other. This forceful encounter of opposite polarity magnetic fields moving towards each other results in cancellation. This example of cancellation is very noticeable from the rapid decrease and disappearance of $J+$, respectively, during and after the time of the small flare. Any loss of flux in $H-$ during the encounter with $J+$ cannot be seen because $H-$ is a large and growing clump of flux. Hence, this example is like the cases studied by Martres *et al.* (1968a, b) in which they were able to associate flares with flux that was increasing on one side of a polarity inversion line and decreasing on the other. In observations from Big Bear Solar Observatory, we find the pattern observed by Martres to be the general case in situations where emerging flux develops in existing active regions. Thus, the association made by Martres *et al.* is synonymous to the association of many flares with emerging flux subsequently discussed and illustrated by Rust (1974).

We make the new point that in circumstances of emerging flux as just illustrated, cancellation also occurs and this occurrence is highly predictable. For example, in the last frame in Figure 3, we note that a new cluster of MMFs is approaching $H-$. Hence, a new cancellation site can be anticipated between the cluster of positive MMFs and the new flux $H-$. Figure 4 shows the development of the new cancellation site and a new corresponding patch of bright plage.

Flare F4, seen in the H-alpha filtergram at 18:39, corresponds to the tiny fragments $K+$ and $L-$ and $M-$. All are MMFs moving away from the positive-polarity trailing sunspot seen just above the time insert in the lower right of the H-alpha images. It appears that $L-$ and $M-$ simply overtake $K+$ and cancel with it. By approximately two hours after the flare, $K+$ no longer exists and only residual flux of $L-$ and $M-$ can be seen at the site of the little flare.

Flares F5 and F6 are shown in $H\alpha$ filtergrams in the right side of Figure 4. Flare F5 has two components that lie near, but not on, a small cancelling field, $N-/O+$. The

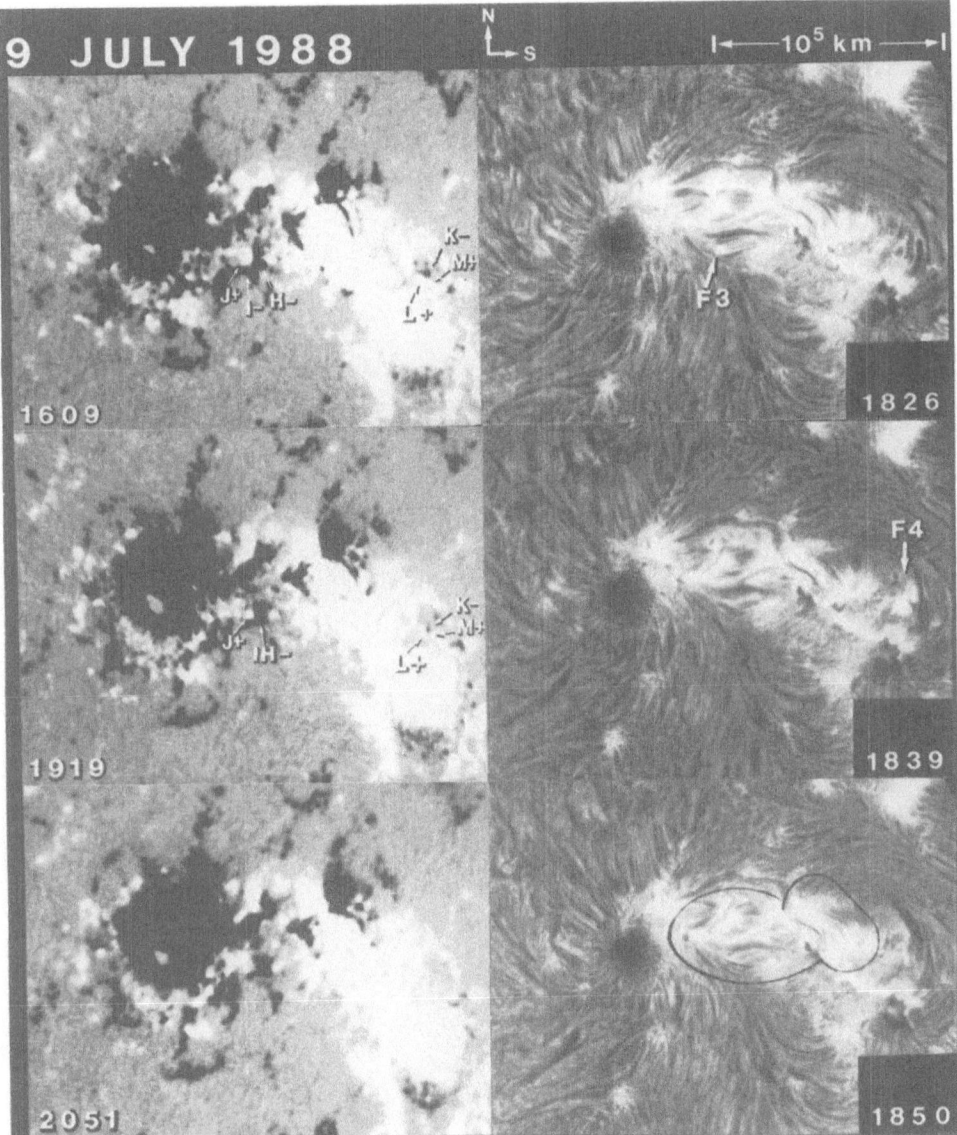

Fig. 3. Subflares F3 and F4 in the upper right and right middle frames are typical of other sites in this active region where small flares can occur. F3 corresponds to the magnetic field patches labelled $J+$, $I-$, and $H-$ in the first magnetogram. $J+$ and $I-$ are both MMF patches while $H-$ is part of the larger of two complex emerging flux systems that are enclosed within the ovals in the lower right. The larger emerging flux system was already present on the previous day and is the area within the oval in the lower right of Figure 2. The encounter of $J+$ with the merged fields of $I-$ and $H-$ results in flux cancellation which is seen by the reduction and disappearance of $J+$ by the time of the last magnetogram at 2051. Subflare F4 corresponds to the cancelling MMFs, $L+$ and $M+$ with $K-$. $K-$ also disappears completely by the end of the sequence.

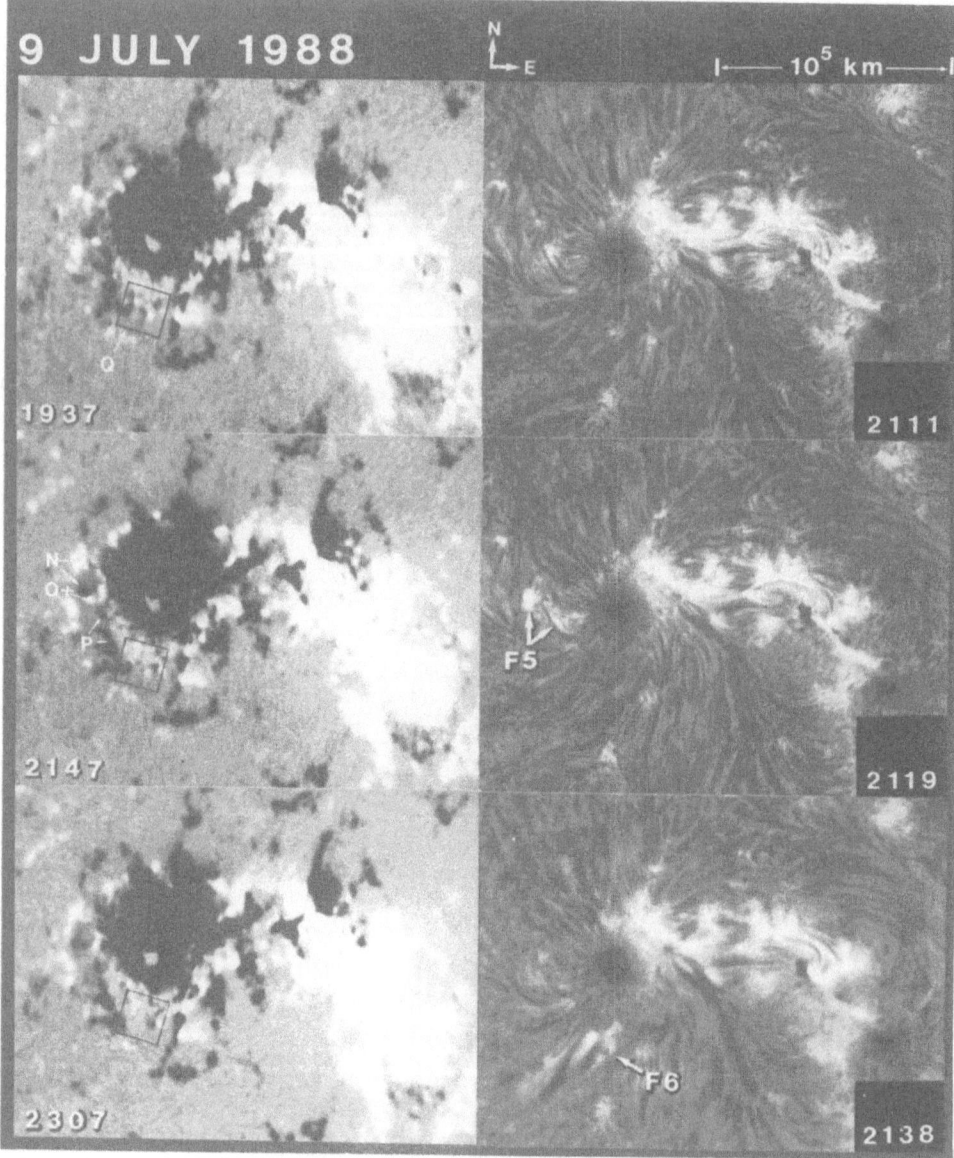

Fig. 4.   The subflare F5 in the middle right is another example of a small flare which is related to changes in MMFs to the left of the negative polarity sunpot. The larger flare segment lies just to the left of cancelling feature, $N - /O +$, and the smaller segment corresponds to $P -$. Thus these two chromospheric components of the flare occur on opposite sides of a cancelling feature rather than coinciding with it. The other flare, F6, seen below the sunspot in the last frame, corresponds to a cluster of very small MMFs with the square labelled Q. Both of these examples, F5 and F6, show that flares are not necessarily coincident with strong magnetic fields.

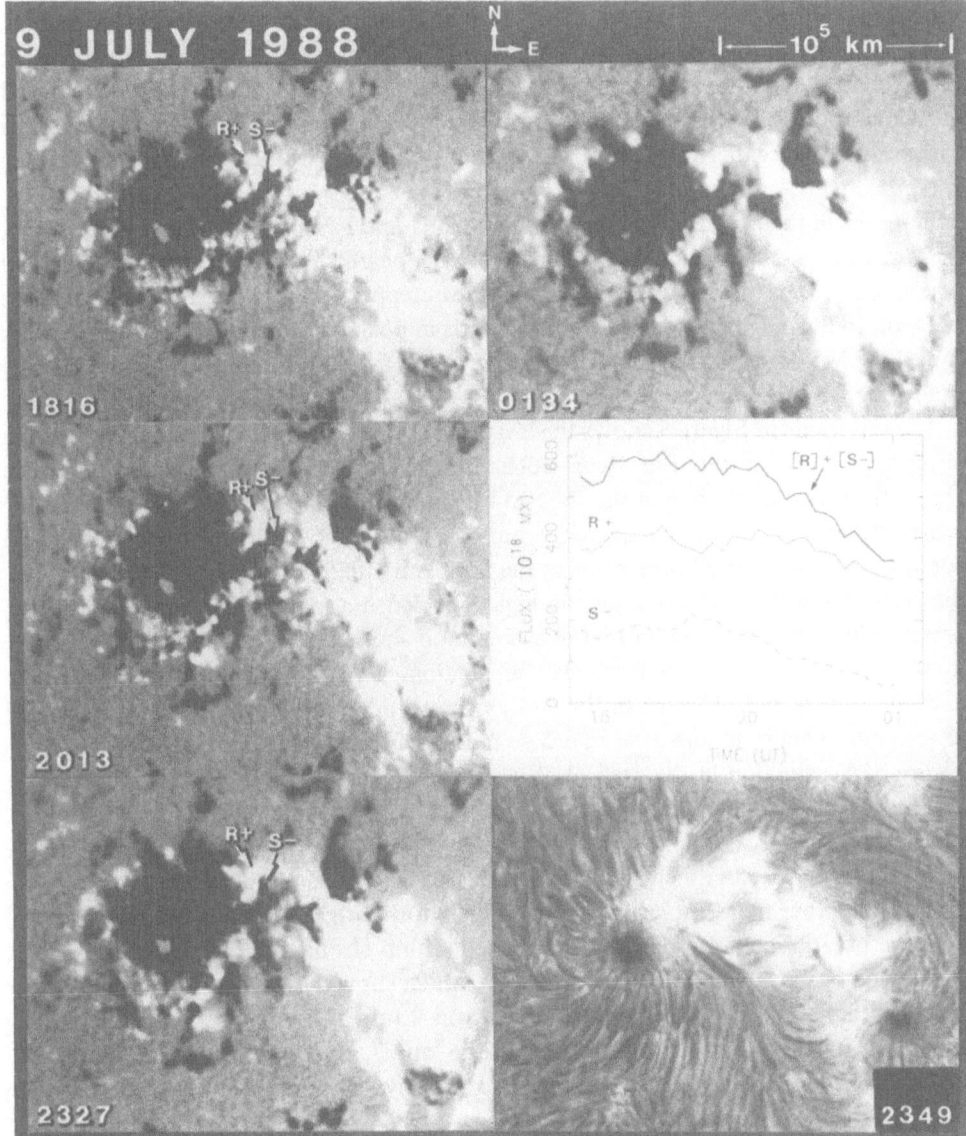

Fig. 5.   The large flare in the lower right has spread to all of the areas of enhanced plage seen in the previous figures. It corresponds to all of the complex magnetic configuration within both emerging flux regions shown within the ovals in the lower right of Figure 2. In this illustration we label just two patches of magnetic flux, $R+$ and $S-$ that are cancelling during this flare. The loss of flux can be seen in the reduced area of $S-$ and the lower part of $R+$. Measurements of the rate of cancellation of $R+$ and $S-$ are shown in the graph. From 20:00 until 01:00 UT, the mean rate of cancellation in just this small area is 3×10^{19} Mx hr^{-1}.

left component of F5 lies just left of $N-$ where there appears to be no magnetic field. The time-lapse film, however, shows that this is a site of the convergence of weak positive flux. The other flare component corresponds to the negative fragment, $P-$. If the two chromospheric parts of this flare are connected by loops in the corona, then this flare straddles the cancelling feature, rather than coinciding with either or both cancelling components, $N-$ and $O+$.

Flare F6, in Figure 4, is near the sites of flares F1 and F2 in Figures 1 and 2. However, the previous cancelling fields have completely disappeared and new ones have developed among the MMFs that have emanated from the outer penumbral border of the associated sunspot. In the magnetograms, F6 corresponds in position to the cluster of small, weak cancelling MMFs within the square labelled Q. By the end of this series, only one tiny cancelling feature remains.

The last frame in Figure 5 shows the largest flare observed in this active region during observing hours at the Big Bear Solar Observatory. The image at 23:40 is at $H\alpha - 0.6$ Å during the rise of the flare to maximum and the image at 23:49 is at line center at flare maximum. The magnetograms in Figure 5 show the preflare magnetic field configuration for over 5 hours prior to the flare and a final image about one hour after the start of the flare. Because of the high flux density and the large amount of flux present, this interval shown in Figure 5 is still too short to illustrate most of the growth from emerging flux and simultaneous cancellation of flux at the boundaries where the new flux meets the pre-existing flux of opposite polarity. However, a conspicuous cancellation occurs at site $R+$ and $S-$ marked in the upper parts of the frames in Figure 5. A reduction in the area of $S-$ and the lower part of $R+$ can be seen. $R+$ and $S-$ are also sufficiently separated from adjacent flux that they could be measured. A gradual loss of flux at a mean rate of 3×10^{18} Mx hr^{-1} is shown in the graph on the right side of Figure 5. A much longer time series is needed to see the effects of cancellation around the other major polarity inversion lines to the right where two larger areas of negative polarity field are partially embedded within strong positive fields. Therefore, we include two illustrations, Figures 6 and 7, which show the long-term evolution of the magnetic flux for nearly 48 hours from early on 9 July until the end of 10 July. Using Figures 1 and Figure 8 along with Figures 6 and 7, one can trace the continuous evolution of the active region fields for approximately 80 hours except during a time gap during the first 15 hours of 11 July.

Figure 6 begins with images taken at the Huairou Solar Observatory and ends with images from Big Bear Solar Observatory nearly 24 hours later. Examples of the disappearance of magnetic flux are seen in negative polarity patches $T-$, $U-$, and $V-$. All three patches disappear before 16:09. Concurrently, several negative polarity patches grow in the middle of the region and converge to form the patch $W-$. Another new negative patch, $X-$, associated with the second new flux system, is first apparent at 20:13. $X-$ grows and merges with the large patch of negative flux to its upper left, while also cancelling with the intermediate area of positive flux. Note also in Figure 7, that $W-$ (from Figure 6) grows until about 15:27. Thereafter $W-$ begins to fragment and cancel with the adjacent flux. It is reduced in area by about 50% by the end of

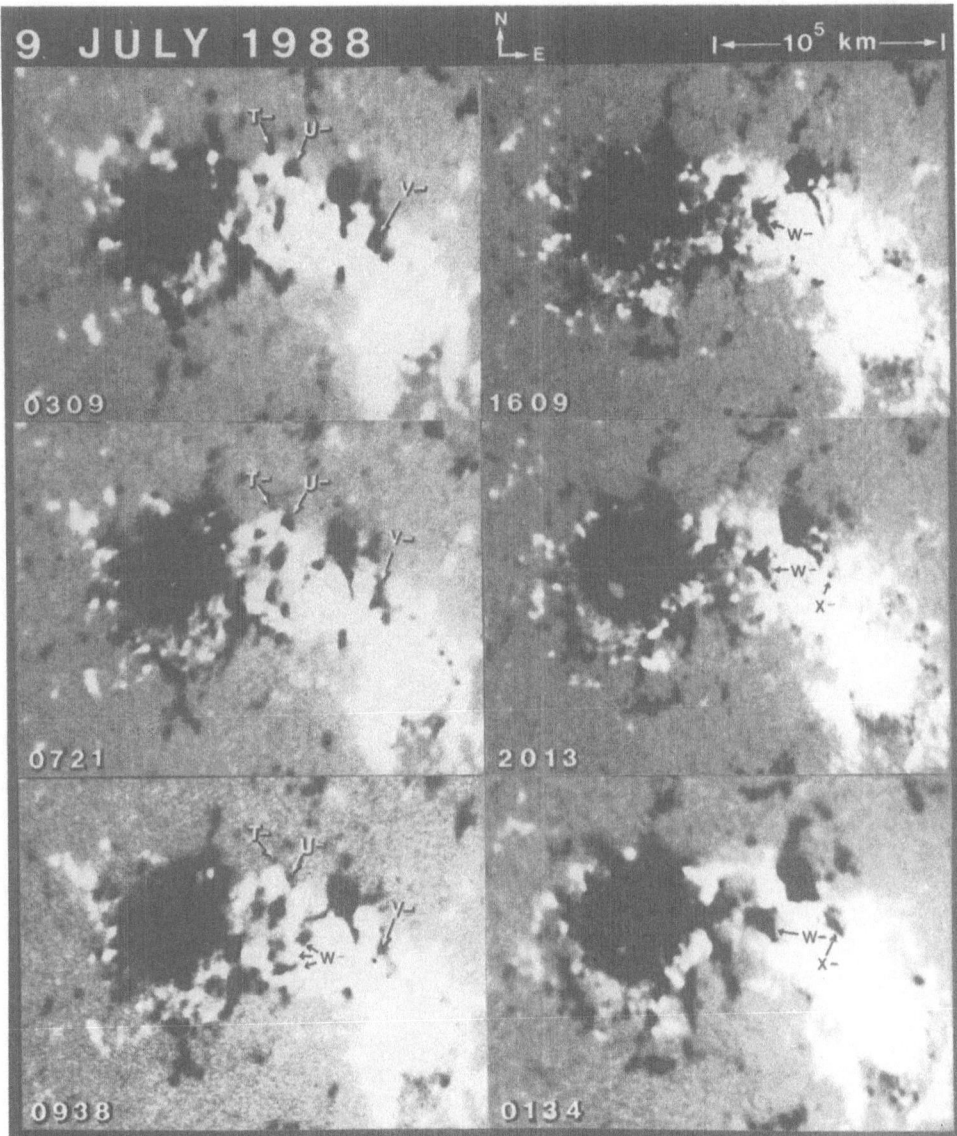

Fig. 6.   This illustration encompasses a longer time interval around the major flare shown in Figure 5 in order to show the type of changes in flux that both precede and occur during that major flare. In this series from early on 9 July until early on 10 July, it is possible to identify many sites where the magnetic flux is either cancelling or growing because of the emergence of the new flux regions shown within the ovals in the lower right of Figure 3. Sites $T-$, $U-$, and $V-$ are negative field patches where the magnetic flux is decreasing and $W-$ and $X-$ are other negative patches where the magnetic flux is seen to increase. The first 3 images on the left are from the Huairou Observatory and the last 3 images on the right are from the Big Bear Solar Observatory.

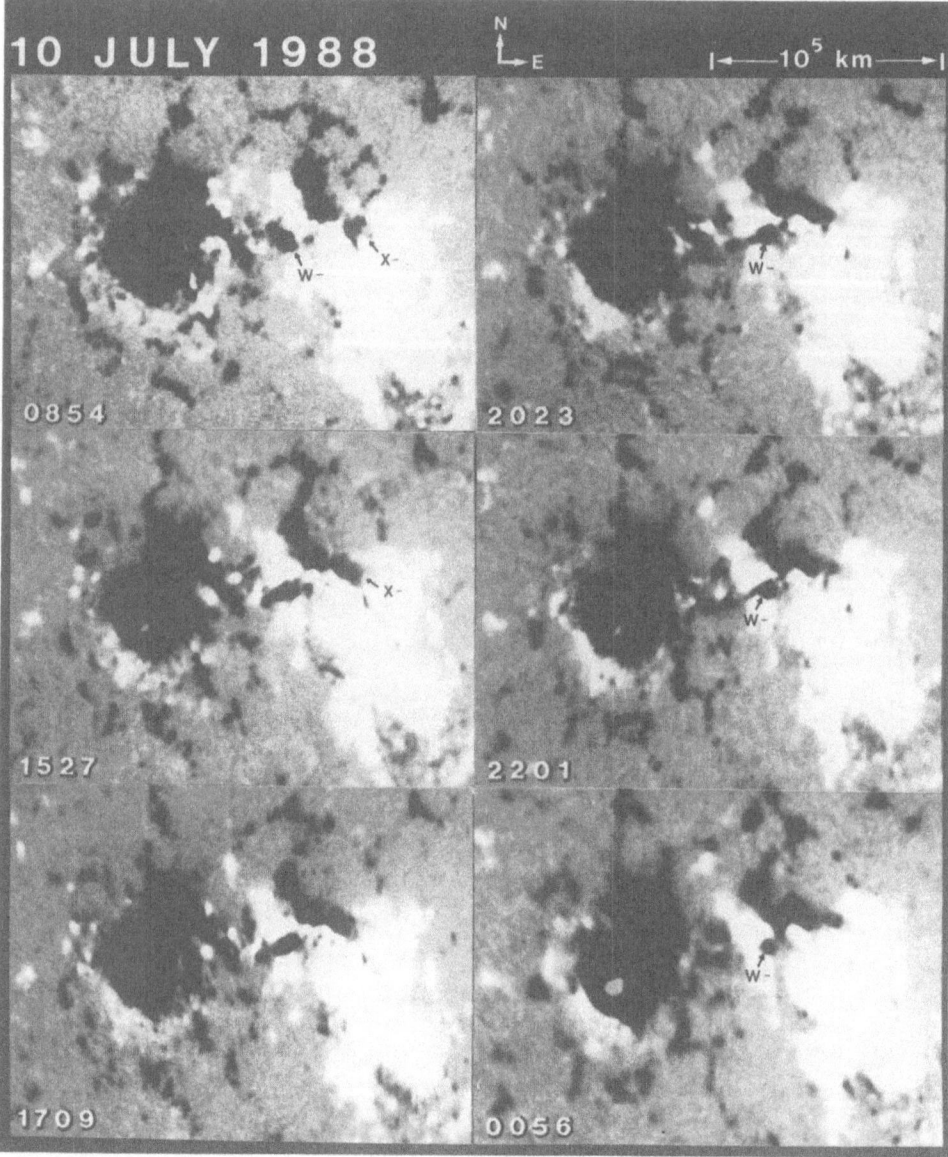

Fig. 7.   This series of magnetograms continues from Figure 6. The area $W-$ continues to grow until about 15:27. Even during the growth stage, cancellation can be seen at the left border of $W-$, with small patches of neighboring positive polarity flux. Magnetic flux disappears to the left of $W-$ from about 20:13 in the preceding series in Figure 6 until the end of Figure 7. $W-$ ceases its overall growth at about 15:27 and thereafter it slowly shrinks as it is cancelled by the surrounding positive polarity flux. The comparable reduction in the positive flux around $X-$ is not conspicuous until the next day seen in Figure 8. Then it is seen in Figure 8 that most of the positive and negative flux in between the sunspots on the left and right has cancelled prior to the occurrence of the next observed major flare.

10 July. By the next day, 11 July, seen in Figure 8, it has completely disappeared along with the majority of the other negative polarity flux in the trailing polarity of this active region.

In Figures 6 and 7, we have labelled only the largest and most conspicuous sites of flux growth and disappearance. Many smaller sites of growth and disappearance can be found. The purpose was to show that much flux emergence and concurrent disappearance is taking place in the active region in the general area of the major flare shown in Figure 5. These changes are taking place before, during and after the major flare. In such complex situations as illustrated in Figures 6 and 7, it is not readily apparent that the magnetic flux disappears by cancellation of negative and positive flux. This process is much more clearly seen either in small active regions (Livi, Wang, and Martin, 1985) or during the decay phase of active regions (Martin *et al.*, 1985), illustrated in Figure 8 for example.

Figure 8 exemplifies the cancellation of flux associated with a major flare in a magnetically simple situation. Because the active region has decayed and the overall flux density is lower in the middle of the active region, we chose to make the final illustration from the original magnetograms. The contours are generated by reversing the color (black to white and *vice versa*) each time that the 8-bit memory is filled in the image processor. Hence, the polarity is determined by the color (black or white) outside of the lowest contour. Where the flux density is not too high, the original contoured magnetograms show the changes in magnetic flux just as well as the reduced magnetograms for which the contours have been removed.

In Figure 8, positive patch $AA +$ and negative patch $BB -$ slowly and simultaneously diminish during the 7-hour interval shown. As they diminish, tiny fragments such as $a1 +$ and $a2 +$ break away from the patch $AA +$. Other tiny fragments from neighboring patches of opposite polarity flux similarly separate from larger patches such as $CC -$ and $DD -$. Examples are $c1 -$, $c2 -$, and $d1 -$.

All of these fragments except $c2 -$ have cancelled with neighboring flux by the end of the day: $c2 -$ migrated toward $BB +$ as $c1 -$ was cancelling; $c2 -$ then replaced $c1 -$ and began cancelling with $BB +$ between 20:17 and 22:28. The study of Martin *et al.* (1985) demonstrated that this process of fragmentation and cancellation can take place continuously along a primary polarity inversion zone within a decaying active region. Where the flux density and magnetic field gradients are high, such as between $AA +$ and $BB -$, the fragmenting elements are usually not resolved. However, the rate of cancellation is measurable. The cancellation is then seen as simply a steady, slow decrease in the area and the magnetic flux. $AA +$ and $BB -$ clearly dminish throughout the day while maintaining approximately the same magnetic field gradient across their common boundary.

The flare in Figure 8 began around the primary cancellation site in the middle of the active region. Then the flare spread to the areas of single polarity both east and west of the cancellation site.

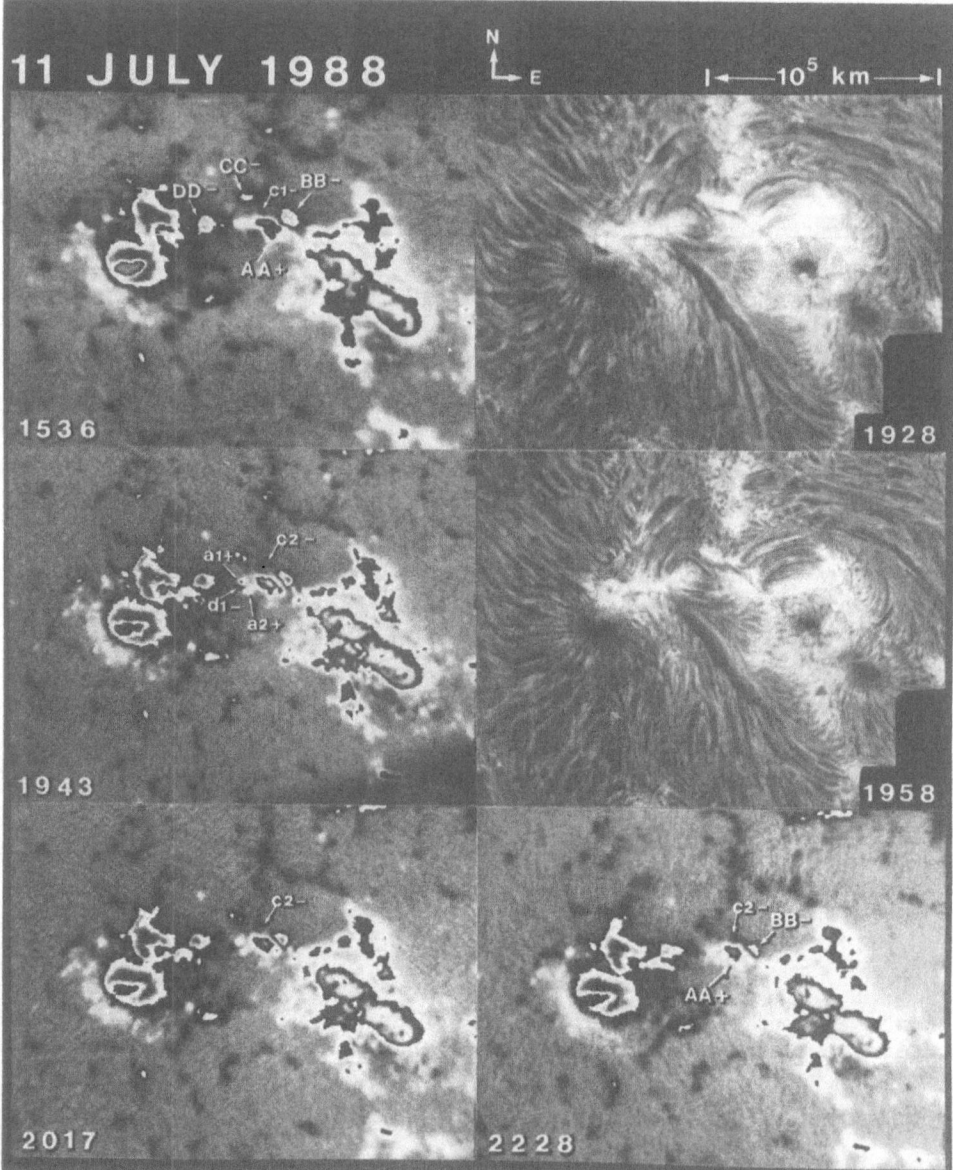

Fig. 8.   This series continues from Figure 7 but the magnetograms are shown in their original format as taken at the Big Bear Solar Observatory. The strongest fields of each polarity lie within the contours which are created by reversing the color from black to white at successive levels of saturation of the magnetograph signal. Hence, the polarity is determined by whether the area around the perimeter of each contour is white (positive) or black (negative). The major flare in the upper right is shown at maximum at 19:28 and in its decaying phase at 19:58. The one remaining major site of cancellation is between $AA+$ and $BB-$. The cancellation is seen from the shrinking of the contours within $AA+$ and $BB-$. In addition, minor cancellation sites develop around the periphery of $AA+$ and $BB-$ as very small magnetic field fragments split off of $AA+$ and $BB-$ and from the neighboring areas such as the ones labelled $CC-$ and $DD-$. The cancellation of $d1-$ with $a1+$ and $a2+$ is seen to the left of $AA+$. The successive motion of small fragments, $c1-$ and $c2-$, toward $AA+$ are seen prior to their cancellation with $AA+$.

4. Discussion

The different circumstances that have been previously considered relevant in flare production might appear contradictory or irrelevant: adjacent decreasing and increasing evolving magnetic features (Ribes, 1969), emerging magnetic flux (Rust, 1974), cancelling magnetic flux (Martin, Livi, and Wang, 1985), magnetically complex regions (Smith and Howard, 1968), high magnetic field gradients (Severny, 1960), bright regions (Dizer, 1969), delta spots (Zirin and Liggett, 1987), sunspot motions (Zirin and Lazareff, 1975), to cite just a few early representative papers. However, there are common factors in all of these situations: they either indicate cancellation sites or show the collision of opposite polarity features which will lead to cancellation. Tanaka (1975), referring to flares in August 1972 and July 1974, writes: 'all sunspot motions indicate a collision between the two polarities'. Evidence now suggests that sunspot motion or emerging flux alone will not lead to flares. Their role is the forcing together of opposite polarity fields which in turn induces cancellation. Higher cancellation rates are expected with faster motions and higher concentrations of magnetic flux.

As illustrated above, a spatial relationship between cancellation and flares is now becoming more clear: flares begin at or near opposite polarity features that are cancelling. Flares often occur when magnetic flux is emerging, but now we think that they only occur if emerging fields also collide with opposite polarity leading to cancellation. Previous observations of emerging flux regions were unable to reveal the loss of flux that occurs on the same side of a polarity inversion line where growth or increase is also seen; they could only show the loss in flux on the opposite side of the polarity inversion line from the emerging side (Martres *et al.*, 1968a, b; Ribes, 1969). It was, therefore, observed that flares occurred around the polarity inversion lines where flux increased on one side and decreased on the other. The many previous associations of flares with emerging flux and the more specific association of flares at sites of both increasing and decreasing flux are entirely valid and still apply to many of our present-day observations, as in some of the examples above, which have saturation effects, inadequate resolution or inadequate sensitivity to detect the decrease in flux on both sides of polarity inversion lines. The new association of flares with cancelling magnetic fields thus does not invalidate previous results. It is only a more general association that encompasses the previous associations.

Since the paper of Martin, Livi, and Wang (1985), which first discussed the association of cancelling fields to flares, we have not yet found any flares at sites where cancellation was not observed or inferred, provided that we have acquired observations with sufficiently high resolution and sensitivity (on order of 10 G). Therefore, we propose that cancellation is a necessary, evolutionary condition for the occurrence of flares. This does not imply that all cancellation necessarily leads to flares nor that it is the only necessary condition.

The significance of the association of flares to cancelling magnetic fields is most clearly understood by studying flares that happen at cancellation sites in the absence of emerging flux and comparing these to circumstances when emerging flux is present.

From these comparisons, we have come to understand that emerging flux is not the primary reason for flares but that in many circumstances it plays an important, but secondary, role in forcing opposite polarity flux together. It is not yet known when a flare should happen during the cancellation process or if flares happen in all circumstances where cancellation takes place.

We speculate that gradual releases of energy might happen in all cancellation sites because they tend to be brighter than similar isolated magnetic features of single polarity. It might be that cancellation leads to flares only in special circumstances. There might be other necessary conditions such as a sheared magnetic configuration which has not been discussed in this paper. This is even expected because of the observation that filaments also form at cancellation sites (Martin, 1986) and filaments are generally recognized to represent sheared magnetic field configurations. In addition, there are many studies that have shown associations between filament orientation, sheared configurations, filament eruptions, and flares. These topics are outside the scope of this paper but are still fertile areas of research in understanding flare build-up (Gaizauskas and Švestka, 1987).

A new theory on the formation and eruption of prominences by van Ballegooijen and Martens (first presented at this Colloquium; unpublished) was stimulated by previous observations of cancellation and the formation of filaments at cancellation sites. By interpreting cancellation as magnetic reconnection at the photosphere, van Ballegooijen and Martens developed a model whereby part or all of the disappearing photospheric line-of-sight component is reconfigured into an increased transverse magnetic field component and, hence, disappears.

The site of increased transverse component becomes the filament. With continued cancellation, the magnetic field in the filament expands outward, eventually becomes unstable, and erupts. The instability that triggers the eruptions can be the same instability that results in solar flares. In this scenario, cancellation is a key part of the preflare build-up. At present, this theory and our observations of the relationship of cancelling fields to solar flares are remarkably consistent.

5. Summary

The examples cited above and many others found during our search through the time-lapse movies from the Big Bear Solar Observatory, show that cancellation happens with magnetic fields spanning a wide range of magnetic field strengths. It is shown that flares of all magnitudes begin adjacent to cancellation sites, whether the associated active region as a whole is developing or decaying. Both small and big flares are initiated near cancelling sites, from the microflares associated with ephemeral regions to the kernels of the great flares.

By reinterpreting previous results on emerging or increasing magnetic flux regions in terms of their possibility to induce cancellation, the apparent conflict between the various circumstances of flares with changing magnetic fields is resolved. Cancelling magnetic flux is observed or deduced to be the common denominator among all observed associations of flares to changing magnetic fields. In particular, flares have been observed when cancellation has been observed or inferred to occur concurrently with emerging magnetic flux, as well as in circumstances of verified absence of increasing magnetic fields during the decay of active regions. Additionally, flares cease occurring in decayed active regions when cancellation sites disappear. Therefore, we propose cancelling magnetic fields to be a necessary evolutionary condition for the initiation of solar flares. However, cancellation is still considered an indirect precondition to flares because the time-scale of cancellation is slower than the time-scale of flares. More studies will be necessary to know if all flares are preceded by cancellation and if observed cancellation corresponds to physical processes that result in stored energy which can be later released in flares.

Acknowledgements

S.H.B.L. expresses appreciation to Dr Harold Zirin for providing the opportunity to work in his solar astronomy group. Support for the Big Bear Solar Observatory staff who took the data for this paper came from NASA grant NGL 05 002 034 and NSF grant ATM-8513577 and Contract AFOSR-87–0023. The travel of S.H.B.L. to the IAU Colloquium 104 was supported by a grant from CNPq (Brazil). The contributions of S.F.M. and H.W. were made under Contract AFOSR-87–0023 from the Air Force Office of Scientific Research.

References

Dizer, M.: 1969, *Solar Phys.* **10**, 416.
Gaizauskas, V. and Švestka, Z.: 1987, *Solar Phys.* **114**, 389.
Harvey, J. H. and Harvey, K. L.: 1973, *Solar Phys.* **28**, 61.
Livi, S. H. B., Wang, J., and Martin, S. F.: 1985, *Australian J. Phys.* **38**, 855.
Marsh, K.: 1978, *Solar Phys.* **59**, 105.
Martin, S. F.: 1980, *Solar Phys.* **68**, 217.
Martin, S. F.: 1984, S. Keil (ed.), *Small-Scale Dynamic Processes in Quiet Stellar Atmospheres*, p. 30.
Martin, S. F.: 1986, *Coronal and Prominence Plasmas*, NASA Conf. Publ. 2442, p. 73.
Martin, S. F.: 1988, *Solar Phys.* **117**, 243.
Martin, S. F., Livi, S. H. B., and Wang, J.: 1985, *Australian J. Phys.* **38**, 929.
Martin, S. F., Dezsö, L., Antalová, A. Kučera, A., and Harvey, K. L.: 1983, *Adv. Space Res.* **2**, 39.
Martin, S. F., Bentley, R. D., Schadee, A., Antalová, A., Kučera, A., Dezsö, L., Gesztelyi, L., Harvey, K. L., Jones, H., Livi, S. H. B., and Wang, J.: 1984, *Adv. Space Res.* **4**, 61.
Martin, S. F., Livi, S. H. B., Wang, J. and Shi, Z.: 1985, *Measurements of Vector Magnetic Fields*, NASA Conf. Publ. 2374, May 1984, Marshall Space Flight Center, Alabama, NASA, Washington, DC, p. 403.
Martres, M. J., Michard, R., and Soru-Iscovici, I.: 1966, *Ann. Astrophys.* **29**, 249.
Martres, M. J., Michard, R., Soru-Iscovici, I., and Tsap, T.: 1968, in K. O. Kiepenheuer (ed.), 'Structure and Development of Solar Active Regions', *IAU Symp.* **35**, 318.
Martres, M. J., Michard, R., Soru-Iscovici, I., and Tsap, T.: 1968b, *Solar Phys.* **5**, 187.
Michard, R.: 1971, in R. F. Howard (ed.), 'Solar Magnetic Fields', *IAU Symp.* **43**, 359.
Priest, E. R., Gaizauskas, V., Hagyard, M. J., Schmahl, E. J., and Webb, D. F.: 1986, *Energetic Phenomena on the Sun*, NASA Conf. Publ. 2439, pp. 1-1 to 1-79.
Ribes, E.: 1969, *Astron. Astrophys.* **2**, 316.

Rust, D. M.: 1972, *Solar Phys.* **25**, 141.

Rust, D. M.: 1974, *Flare-Related Magnetic Field Dynamics*, HAO Conference held in Boulder, Colorado, p. 243.

Severny, A. B.: 1958, *Izv. Krymsk. Astrofiz. Obs.* **20**, 22.

Severny, A. B.: 1960, *Izv. Krymsk. Astrofiz. Obs.* **22**, 12.

Smith, S. F. and Howard, R. F.: 1968, in K. O. Kiepenheuer (ed.), 'Structure and Development of Solar Active Regions', *IAU Symp.* **35**, 33.

Smith, S. F. and Ramsey, H.: 1967, *Solar Phys.* **2**, 158.

Tanaka, K.: 1975, *Solar Phys.* **47**, 247.

Vorpahl, J.: 1973, *Solar Phys.* **28**, 115.

Zirin, H.: 1988, *Astrophysics of the Sun*, Cambridge University Press, Cambridge, p. 343.

Zirin, H. and Lazareff, B.: 1975, *Solar Phys.* **41**, 425.

Zirin, H. and Liggett, M. A.: 1987, *Solar Phys.* **113**, 267.

Zwaan, C.: 1987, *Ann. Rev. Astron. Astrophys.* **25**, 83.

MASS MOTIONS ASSOCIATED WITH SOLAR FLARES

SARA F. MARTIN

Big Bear Solar Observatory, California Institute of Technology, Pasadena, CA 91125, U.S.A.

Abstract. Mass motions are a principal means by which components of solar flares can be distinguished. Typical patterns of mass motions in Hα are described for chromospheric flare ribbons, remote chromospheric flare patches, flare loops, flaring arches, surges, erupting filaments and some expanding coronal features. Interrelationships between these phenomena are discussed and illustrations of each are presented.

1. Introduction

Throughout this review, solar flares are discussed as consisting of a number of components. As listed in Table I, these include loops, ribbons, arches, remote patches, surges, erupting filaments, and other expanding coronal features. Mass motions play an integral role in the existence of these components. It is the distinctly different velocity and spatial patterns of the mass motions of these flare components that has lead investigators to put them into separate categories and to assign them different names.

In Table I, the most closely related flare components in the chromosphere and corona are shown on the same line but in separate columns. The term 'coronal' in this paper refers to the volume of solar atmosphere above the chromosphere rather than to the typical temperature regime of that atmosphere.

TABLE I

Coronal flare components	Chromospheric flare components
Flare loops	Ribbons
Flaring arches	Remote patches
Surges	
Erupting filaments	
Expanding coronal features	

The majority of flares have at least the two components listed in the first row in Table I, flare loops and chromospheric ribbons at the footpoints of the flare loops. Both are found in a wide variety of sizes and degrees of compactness. In Hα the ribbons are seen many minutes before the appearance of flare loops if and when conditions render the loops visible in Hα. However, it is now well known that the loops are always present in X-rays whenever the ribbons are seen in Hα. The other phenomena apparently require specific additional conditions in order to occur in association with the loops and ribbons.

The literature on the mass motions involved in these many flare components is too extensive for an in-depth discussion on each one in the space allotted for this review. This presentation is limited to two specific objectives. The first goal is to summarize the general character of the mass motions of each of the above flare components by referring

Solar Physics **121**: 215–238, 1989.
© 1989 *Kluwer Academic Publishers.*

only to a few key papers and to previously unpublished, representative examples of each. The selection of key papers on a topic usually includes the first known papers, comprehensive papers, review papers and very recent papers, especially those which include references to earlier work. The second goal is to discuss and illustrate some of the relationships of these phenomena to each other as understood to date.

The illustrations in this paper are primarily Hα filtergrams. In depicting mass motions, Hα observations offer three advantages: (1) superior sharpness of detail and contrast (compared to current observations at most other wavelengths) afforded by the width and depth of the Hα line as observed with narrow band birefringent filters of $\frac{1}{4}$ or $\frac{1}{2}$ Å passband, (2) a wide selection of events from an abundance of time-lapse observations taken in Hα over the last 3 solar cycles, and (3) a means of identifying the coronal flare components seen in projection against the chromospheric background.

The flare components that lie in the corona have parts or phases that include both very high temperatures (invisible in Hα) and relatively low temperatures (visible in Hα). The requirements for observing Hα flare components in the corona but against the chromospheric background is the existence of a plasma with a density of at least 10^{10} particles cm^{-3} and a temperature in the range of about $6–15 \times 10^3$ K. A great many of the flare-associated mass events pass through this physical regime sometime during their evolution. Hence, very few flare components have been discovered in the UV or in X-rays whose existence was not already at least partially known or deduced from Hα observations. For these reasons, Hα is a good starting point for discussing mass motions. This review does not attempt to comprehensively discuss mass motions outside of the visible part of the spectrum. The author hopes this approach of using Hα observations will lay the groundwork for a series of more extensive reviews on each of the particular types of mass motions and their signatures as detected throughout the whole spectrum from X-rays to radio wavelengths.

2. Mass Motions within Chromospheric Flare Components

2.1. CHROMOSPHERIC RIBBONS

The ribbons are distinguishable from the other forms of flare emission because they form simultaneously in the chromosphere on either side of a polarity inversion line. A polarity inversion line can be identified directly in a magnetogram displaying only the line-of-sight component of the chromospheric or photospheric magnetic field. Alternatively, many polarity inversion lines can be identified indirectly by the presence of a filament in filtergrams or spectroheliograms taken in chromospheric lines. The chromospheric ribbons are not usually symmetrically positioned on either side of a polarity inversion line. Also the ribbon patterns can be complex when several polarity inversion lines are involved in a single flare. This means that there can be multiple associated loop systems.

During most of their lifetimes, the flaring elements within the chromospheric flare ribbons have very small Doppler shifts on the order of 0–0.2 Å, corresponding to motions up to about 10 km s^{-1}. However, the Doppler shifts of very bright elements, within the flare ribbons, known as 'flare kernals', can be as high as 1–2 Å for very intense flares. During their development, chromospheric flare elements are almost always Doppler-shifted or asymmetric towards the red (Ellison, 1952; Švestka, Kopecký, and Blaha, 1962; Tang, 1983; Ichimoto and Kurokawa, 1984).

In this paper, I follow some of the above authors in making a distinction between red-asymmetric profiles discovered by Waldmeier (1941) and red-shifted flare profiles later recognized by Teske (1962). According to Ellison (1952), who first used the term 'red asymmetry', the peak intensity is at line center and the red wing is brighter and more extensive than the blue wing. In contrast, a red-shifted profile is one in which the peak intensity is in the red wing. A red-shifted profile can be unambiguously interpreted as a Doppler shift representing motion away from the observer and hence is the more relevant parameter for this discussion. Teske (1962) was the first to find a clear example of a red-shifted profile in Hα spectra taken at the McMath Hulbert Observatory. A typical red-shifted profile at the growing boundary of a two-ribbon flare is illustrated in Švestka, Martin, and Kopp (1980). An unusually clear example of a flare profile with a large red-shift is seen in Figure 20 of Zirin and Tanaka (1973). More typical examples are shown in Figure 1 of Ichimoto and Kurokawa (1984).

In their comprehensive study of flare spectra, Ichimoto and Kurokawa (1984), unlike the previous authors, used the terms red-shift and red asymmetry interchangeably but they primarily show examples of red-shift. This usage of terms seems justifiable if the red asymmetry comes from a combination of red-shifted and non-shifted profiles which are not spatially resolved. In specific studies, such as those of Teske (1962) and Švestka, Martin, and Kopp (1980), it is clear that flare elements of medium and faint intensity do not exhibit red-asymmetry but do exhibit red-shift. At this writing, it is still not evident, from previous studies of bright flares and flare kernals, whether the red-asymmetry results from a combination of unshifted and shifted profiles or from the Stark-effect or other phenomenon independent of red-shift (Švestka, 1976). However, Ichimoto and Kurokawa (1984) interpret all red-asymmetry and red-shift as downward motion.

Studies from filtergrams alone do not allow one to distinguish between red-asymmetry and red-shift, but they do give a more complete picture of the evolution of flare elements than most flare spectra with the exception of time-lapse spectra. Studies of time-lapse spectra (Ichimoto and Kurokawa, 1984), and studies from time-lapse filtergrams of mostly chromospheric flare elements (Tang, 1983), show the red-shifts (or red-asymmetries) of individual flare elements to be a time and energy dependent characteristic of the flare fine structure rather than of the flare as a whole. When fine structure can be resolved in flares, it is seen that chromospheric flares consist of a succession of many elementary localized brightenings. Even what we call 'flare kernals' are usually a whole envelope of these successively forming and decaying flare elements. This is exemplified in the flares of 5 November, 1980 illustrated in Wu et al. (1986). Only minutes after a flare element forms, its red-shift disappears and is supplanted by a nearly symmetric flare profile. The red-shift occurs only while the flare or any specific part of a flare is increasing in optical intensity (Tang, 1983; Ichimoto and Kurokawa, 1984).

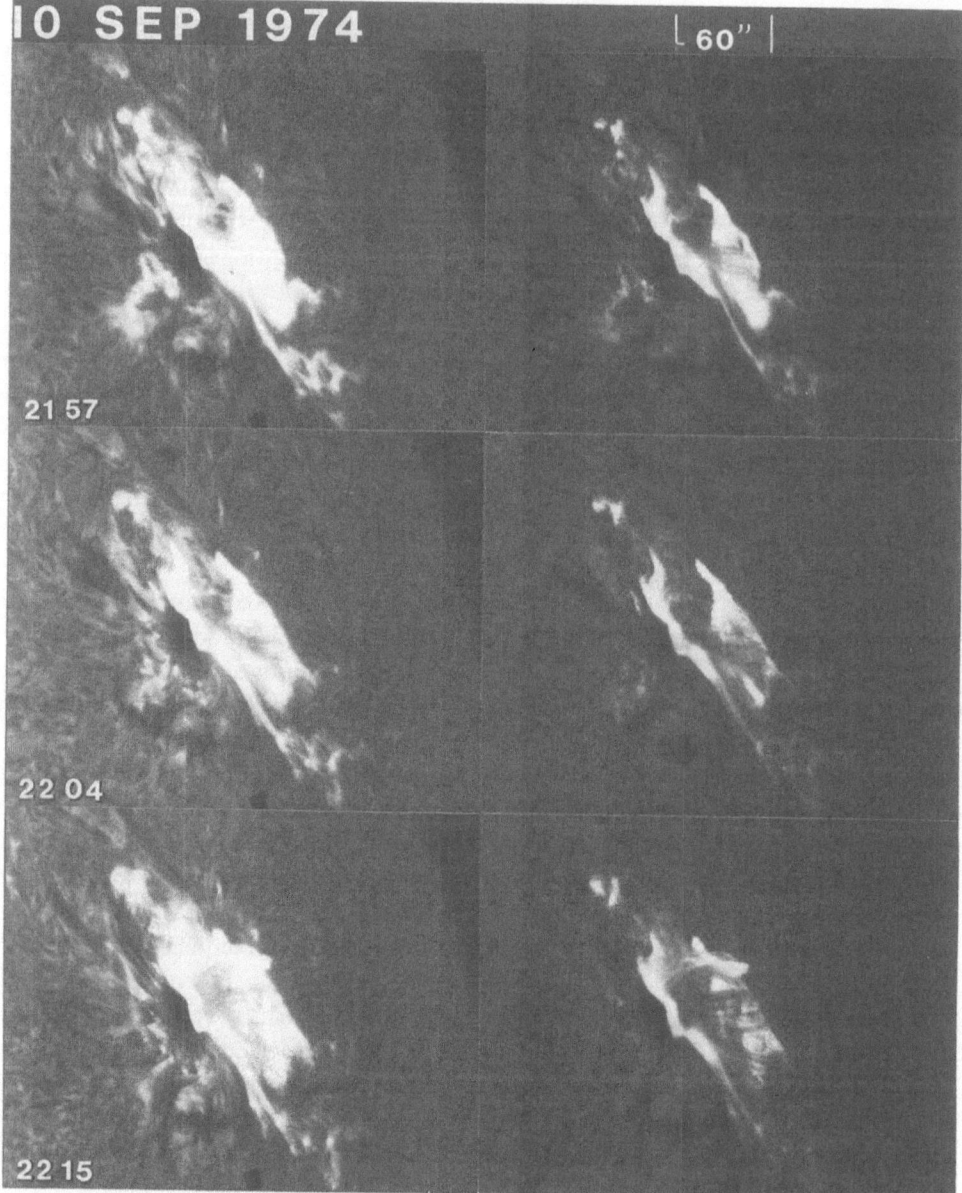

Fig. 1a.   The early development of flare loops are shown in normally exposed prints on the left and in heavily exposed (× 4) prints at the same times on the right. In the heavily exposed prints, the near legs of the loops appear as approximately parallel, dark striations superposed against the chromospheric flare ribbons.

The red-shift is most clearly seen in spectra of flares that cover large areas of the chromosphere (Švestka, Martin, and Kopp, 1980). In such flares, it is also clear that the red-shifted flare elements occur only in newly formed flare elements at the outer perimeter of the flare and occur simultaneously on both sides of the polarity inversion

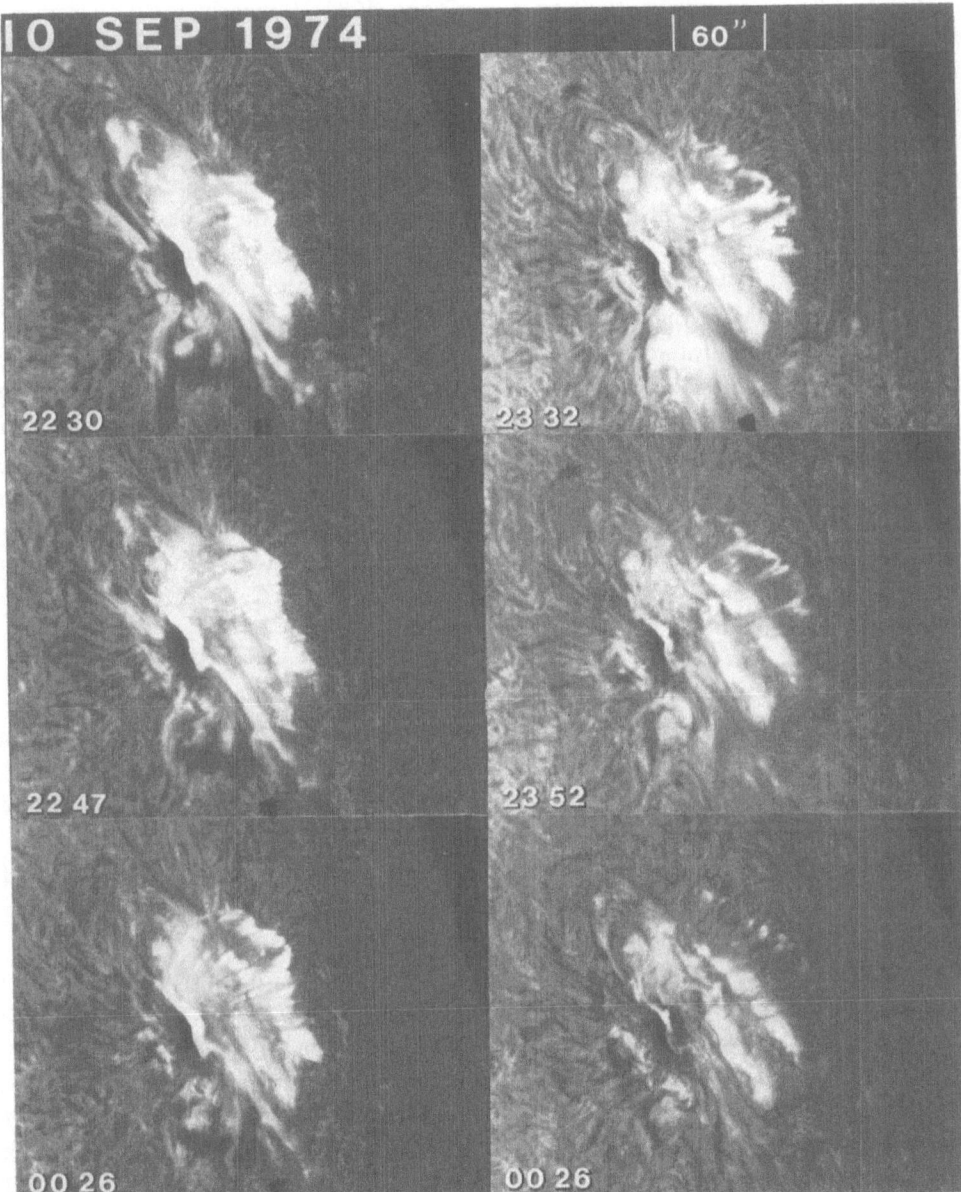

Fig. 1b.   As the chromospheric flare ribbons separate, the rising system of loops, that connect the ribbons of opposite polarity, become more conspicuous at normal exposures for the chromosphere. The tops of the loops have concentrations of emission while the legs are less bright and appear in absorption against the flare ribbons.

line that separates the flare ribbons (Rust and Bar, 1973). Because a whole flare does not form simultaneously, it is difficult (and often inaccurate) to describe flares as having 'phases'. Even during the so-called 'decay phase' of extensive flares, newly forming red-shifted flare elements continue to occur (Tang, 1983). Similarly, the flare elements that form early in a flare can already be decaying before the flare as a whole has reached maximum intensity or area.

The accuracy of description, and hence the physical interpretation of flares and their mass motions, are highly dependent upon the resolution of the fine structure of flares. It is likely that all flare ribbons are composed of sub-arc sec structures. However, very often photographic techniques limit its visibility in addition to the limited spatial resolution of our telescopes and image degradation from our atmosphere. As illustrated in Figure 1(a), the fine structure in flares can sometimes be made more visible by heavily exposing prints to bring out detail captured on film which has been exposed to optimize chromospheric detail rather than detail in flares.

As will be seen in the later sections of this paper, the mass motions of the chromospheric flare ribbons differ substantially from the mass motions of the coronal components. Chromospheric flare elements have a very small range of redshifts while the associated coronal components such as erupting filaments and surges typically reveal both blue shifts and red shifts and have a much higher velocity range. These differences are exemplified in the analyses of the energy content in the chromospheric and coronal components of the flare of 5 September, 1973, respectively, by Canfield *et al.* (1980) and Webb *et al.* (1980).

2.2. REMOTE CHROMOSPHERIC FLARE COMPONENTS

Some flares also have peripheral chromospheric flare components in addition to the ribbons that are centered around a polarity inversion zone. These peripheral flare patches have previously been designated as 'remote brightenings' (Tang and Moore, 1982). They can be tens to a few hundred thousand kilometers from the primary flare site. To my knowledge there are yet no published papers that give Doppler shift measurements of these peripheral parts of flares. It is expected, however, that these remote brightenings would not have Doppler shifts in excess of the Doppler shifts of the flare kernals. In many circumstances it has been shown that the remote flare components derive their energy from the source flare via the propagation of a disturbance along coronal arches connecting to the source flare. Such disturbances have been detected as: (1) flaring arches-mass flowing in coronal arches and seen in emission in Hα in projection against the disk (Mouradian, Martres, and Soru-Escaut, 1983) following the flow of X-ray emission (Martin and Švestka, 1988) or (2) soft X-ray emission with no visible emission in Hα (Rust and Webb, 1977; Tang and Moore, 1982), or (3) radio emission (Tang and Moore, 1982).

3. Coronal Flare Components

3.1. FLARE LOOPS

Flare loops in Hα are interpreted as just the late phase of flare loops that form at the ouset of a flare (Moore *et al.*, 1980). The X-ray experiments on board Skylab firmly established that flares in X-rays, almost invariably have the form of coronal loops that link the positive and negative chromospheric flare kernals and ribbons (summary by Sturrock, 1980). The spatial resolution of most X-ray images of flares, acquired to date, does not allow us to resolve the many tightly-packed, individual loops that must exist as a flare rises to maximum intensity. However, we know that this finely-structured system of loops exists from observations in Hα during the maximum and post-maximum phases of flares. An example of flare loops is shown in Figure 1(a); to better identify the loops, the development of the flare is shown at 2 print densities. In the left column of images, the exposures are made to show the chromospheric structure and the general outline of the flare. In the right column of images, longer exposures, by a factor of 4, reveal the early appearance and fine structure of the legs of the loop system superposed against the rest of the flare. As the brightness of the flare declines, as shown in Figure 1(b), the loops become apparent at print densities appropriate for the chromosphere. Because the site of this flare is quite near the limb, it can be seen that new loops form at successively higher altitudes. Less apparent, in these still images, than in the time-lapse movie of this event, is the continuous downflow of mass from the tops of the loops to the flare ribbons below. At this stage, the loops actually connect the inner decaying part of the chromospheric flare ribbons (Rust and Bar, 1973). Above these loops in Hα are loops visible in X-rays, EUV, and UV wavelengths. The X-ray loops are the top of the hierarchy (Moore et al., 1980). They connect the newly-forming outer parts of the flare ribbons while the EUV and UV wavelengths connect to the intermediate parts of the ribbons.

It is common for the chromospheric flare ribbons to appear to move away from each other with time. This is an apparent rather than a real motion that is due the new flare elements forming on the outer boundaries of the flare ribbons concurrently with the formation of the coronal flare loops in X-rays (Rust and Bar, 1973; Moore et al., 1980).

In Hα the loops typically reveal downflows in both legs of the loops in the range of 50–100 km s^{-1}. Downflows of this magnitude have little or no ability to depress or brighten the chromosphere. Flare spectra confirm the absence of red-shift during the late phase (the Hα phase). The red-shifts, discussed in Section 2.1, occur concurrently with the formation of the loops in X-rays. This implies that either high speed particles and/or possible shock waves are responsible for the red shifts that occur during the early development of the chromospheric flare elements. The loops in Hα are only an aftermath of a much more energetic process. From the presence of the flare loops in Hα, we can also deduce that another invisible phase of mass motion must take place between the initial phase which causes the red-shifted chromospheric elements and the late phase when Hα loops are seen. The Hα loops are too dense for their mass to be condensed out of the corona. Therefore, the filling of the loops with dense mass must be initiated at their footpoints; such upflows have recently been observed in Hα Schmieder et al. (1987). Substantial evidence of the upflow of mass into flare loops has also been observed at shorter wavelengths as discussed in Doschek et al. (1986). Thus flare loops have at least three stages of mass motions: (1) an inferred rapid downflow of high speed, possibly relativistic particles, (2) an upflow, now often called 'chromospheric evaporation' in the range of 100–500 km s^{-1}, and (3) a slow visible downflow commonly observed in Hα in the range of 50–100 km s^{-1} (and other chromospheric lines) as illustrated in Figures 1(a) and 1(b).

For flares that occur very close to the limb, the loops are readily observed against the background of the sky through appropriate filters.

However, loops in Hα are only seen against the disk if the densities of the loops reaches approximately 10^{11} cm^{-3}. Because of this density requirement, only a minority of flares actually exhibit flare loops in Hα in projection against the disk. Usually these are large and bright flares in Hα. In the less dense loops or in the legs of the loops, the loops appear in absorption rather than emission. In a few flares, the tops of the loops remain bright throughout their whole evolution. However, it is more common to see the

Fig. 2.   Flare loops both in absorption and in emission extend between the chromospheric flare ribbons. In this case we see only the tops of the loops. The legs and footpoints of the loops cannot be seen because their Doppler shifts exclude them from the passband of the filter used to take these images. The filter passband was 0.25 Å centered at approximately Hα − 0.5 Å.

loops gradually change in time from emission to absorption or to see them only in absorption (Martin, 1979) in projection against the disk.

A seen through narrow band Hα filters (0.5 Å or less) in line center, one rarely sees the entire loop because the Doppler shifts are usually in excess of the bandwidth. However, with a rapidly tunable filter, it is possible to observe the full range of loop velocities. If viewed near disk center, only the tops of loops are typically observed such as in Figure 2. In this example, some of the loops are in emission and another set are in absorption and many others are invisible. Because the density of the successively formed loops decreases rapidly with time, they become invisible, in projection against the disk, usually before the chromospheric ribbons have decayed. In observations at the limb, however, it is well known that the loops often are observed many hours after the chromospheric elements have disappeared in Hα.

3.2. FLARING ARCHES

The term 'flaring arches' was introduced by Mouradian, Martres, and Soru-Escaut (1983) to refer to parts of flares that have not commonly been discussed in the literature. Martin and Švestka (1988) and Mouradian and Soru-Escaut (1989, this Colloquium) have more specifically defined the term flaring arches to describe a component of some flares in which emitting mass is observed first to flow upward into the corona, traverse an arch-like path, and descend to another point in the chromosphere. An example, seen in projection against the disk, is illustrated in Figures 3(a) and 3(b). Figure 4 shows a flaring arch at the limb. The energy for the upflow is somehow derived from processes associated with the formation of flare loops and ribbons.

Flaring arches can vary over a wide range of dimensions and intensities at the observed wavelengths. Martin and Švestka (1988) reported that flaring arches have an X-ray phase, usually preceding the bulk of Hα mass that travels through the arch. They also found the early excitation of the secondary footpoint seen both before the arrival of the Hα mass, and surprisingly, even before the arrival of the first detectable hard X-rays. But it should be noted that additional brightenings of the secondary footpoint were also seen at the time the X-ray emission reached the secondary footpoint.

The complex example in Figure 3(a) shows a series of arches and a series of secondary footpoints rather than a single arch and a single secondary footpoint as in the events described by Martin and Švestka (1988) and Švestka et al. (1989). The flare is in progress in the first frame of Figure 3(a) and an ejection of mass can be seen protruding from the flare. In successive frames, it is seen that the mass ejecta follows a curved trajectory and intersects the chromosphere first at P1 at a distance of about 1 arc min from the primary footpoint, and subsequently, at P2 and P3, about 2 arc min from the source. Note that P1, P2, and P3 all brighten before the successive appearances of visible mass in the various arches. Following the initial bright ejecta confined to the arch system connecting to P1, less bright mass continues to pour from the same initial site towards P2 and P3. This mass is seen partly in emission and partly in absorption in frames at 19:37 and 19:45. Because the closest part of the limb is toward the top of the frames, the longer absorption arches apparently also have a higher trajectory than the initial arch in emission. Although there were no spatially resolved X-ray images acquired, we suspect from the similarity of this event to the ones already studied, that the visible ejection is undoubtedly preceded by X-rays that traverse the arches at a higher

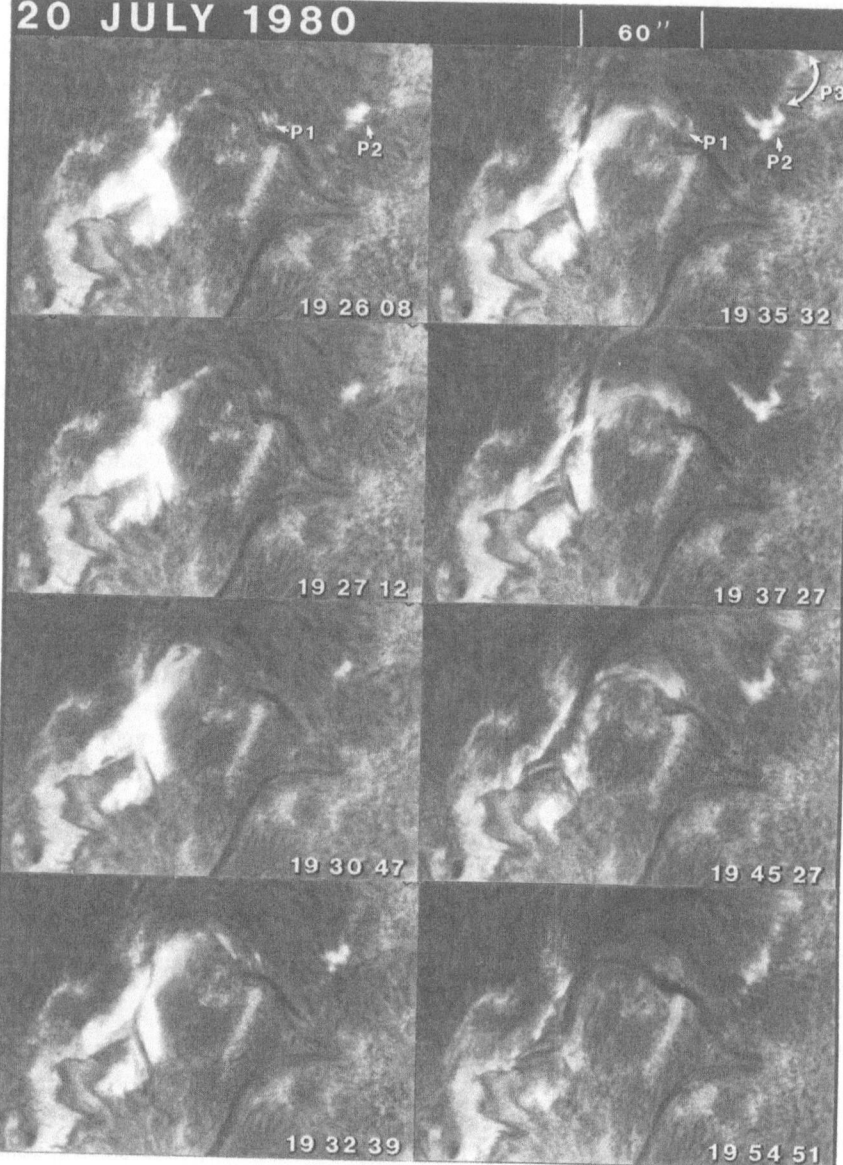

Fig. 3a.   The beginning of the flaring arch is the emission protruding upward from the flare in the first frame. In successive frames in the first column, this emission follows an arch-shaped trajectory through the corona to a chromospheric footpoint, P1, approximately 1 arc min to the upper right the primary flare site. The emission phase of this 'flaring arch' is followed by the flow of additional mass, mostly seen in absorption, from the flare site to more distant chromospheric footpoints, P2, and adjoining row of fainter footpoints, P3, in the upper right of each frame.

speed than these Hα components; the X-ray arches are expected about the same time as the brightening of the secondary footpoints.

Figure 3(b) illustrates the Doppler shifts in the same flaring arches on 20 July in the wavelength range from − 1 to + 1 Å. In the first row are images at the center of the Hα line and in the second and third rows are images showing Doppler-shifted components at − and + 0.5 Å and − and + 1.0 Å, respectively. Blue wing images showing the upflow are on the left and red wing images showing the downflow towards the secondary footpoint are on the right.

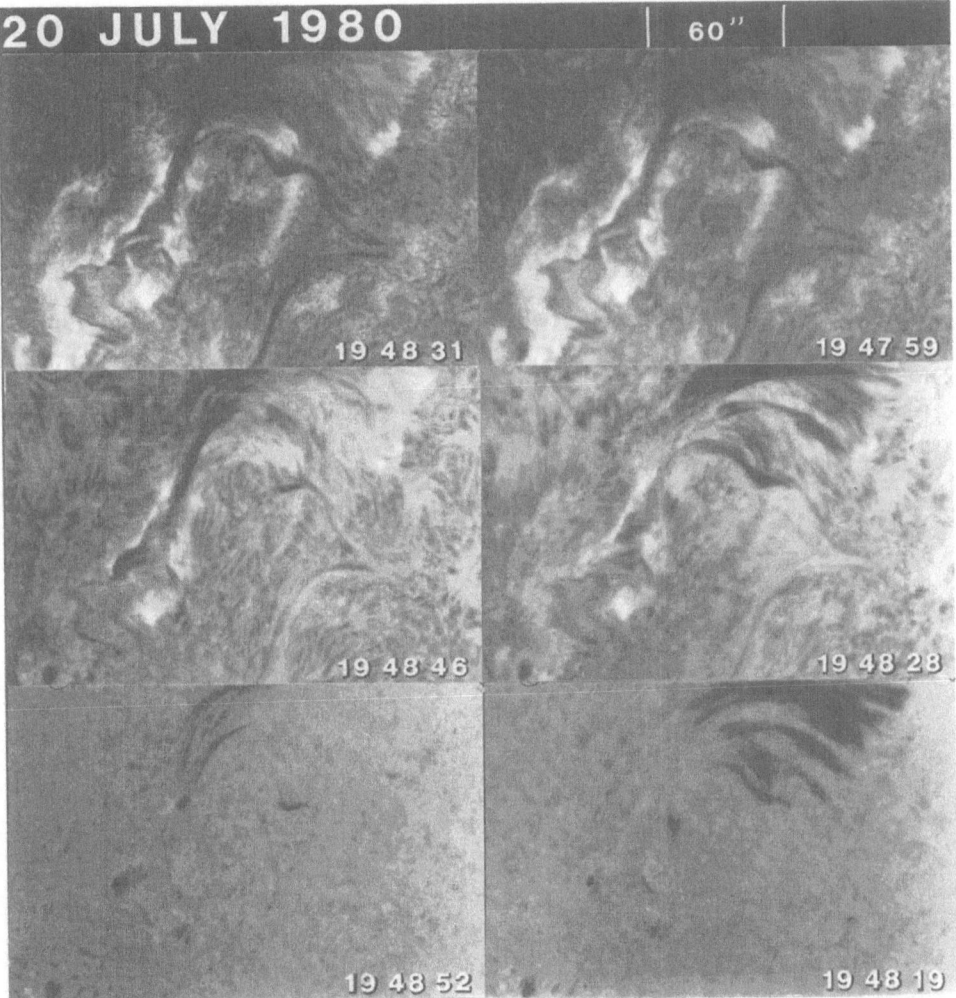

Fig. 3b.   The Doppler-shifted mass within the same flaring arches as in Figure 3(a) is illustrated in the first column in images at Hα centerline, Hα − 0.5 Å and Hα − 1.0 Å and in the second column at Hα centerline, Hα + 0.5 Å and at Hα + 1.0 Å. This series of images, taken in a two-minute interval, shows the pattern of Doppler shifts at about the same time during the event. They confirm the visual impression seen in the time-lapse film of upward motion from the main flare site along an arch-shaped path and then downward to the distant secondary footpoints in the upper right of each frame.

An example of a flaring arch at the limb is shown in Figure 4 in images recorded at the former observing site of the Lockheed Solar Group. The limb observations were made through a 10 Å interference filter. The event has a duration of about one hour although the full extent of the arch is seen about 20 minutes after the start of the event. The source flare is not seen but that is not surprising because it is probably small and

15 JULY 1966

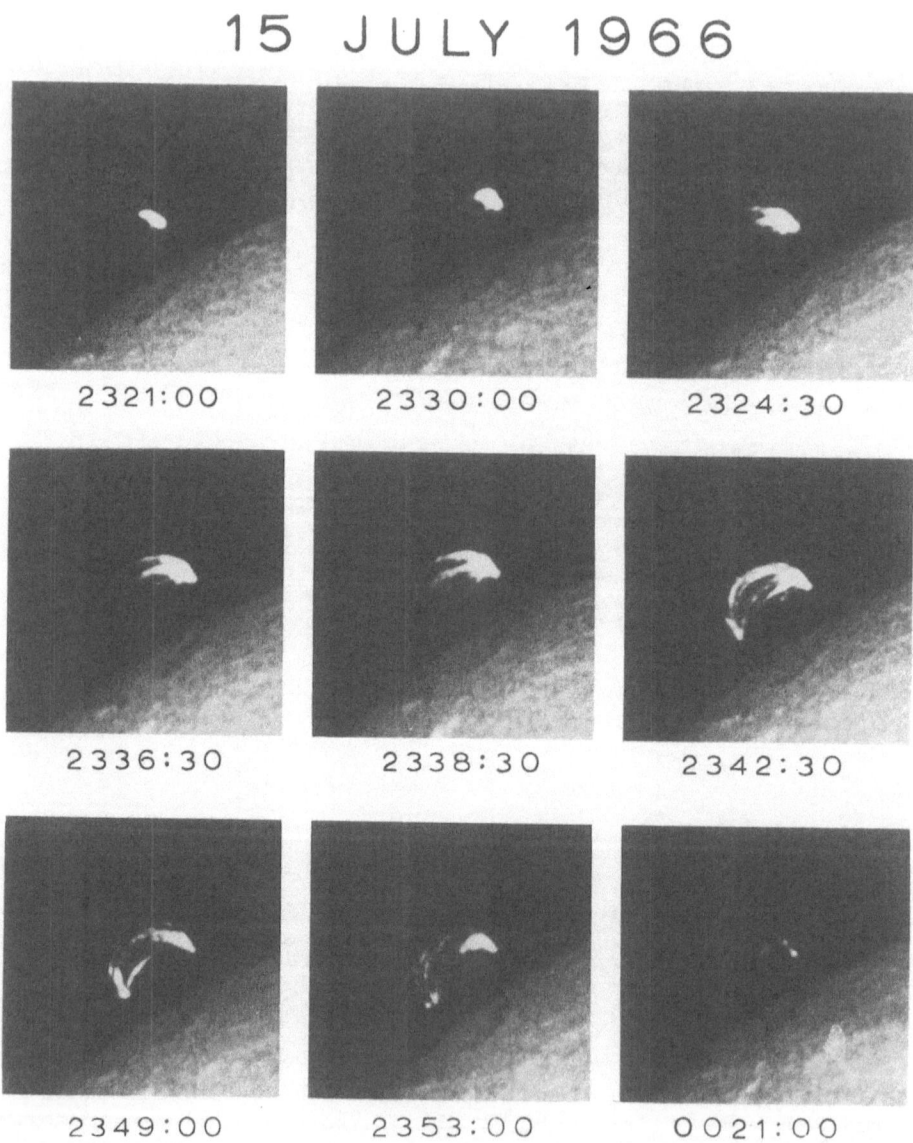

Fig. 4. This example of a flaring arch at the limb was photographed through a 10 Å filter at the former site of the Lockheed Solar Group at Saugus, California, U.S.A. The disk of the Sun, however was photographed through a filter of 0.5 Å passband. The mass in the flaring arch travels from right to left.

obscured by the occulting disk. This event is remarkably similar to the 6 November, 1980 event published by Martin and Švestka (1988) which was associated with a very small source flare.

Flaring arches and flare loops have been shown here to be very different in their pattern of mass motions and in their spatial relationship to the polarity inversion line separating the flare ribbons. Surges on the other hand are more similar to flaring arches than they are different. Both flaring arches and surges follow pre-existing lines of force – unlike flare loops systems (Martin, 1979) – and both can occur in homologous series on the order of tens of successive events.

3.3. SURGES

Surges are the most common of all flare-associated mass motions. Surges are traditionally defined as spike-shaped mass ejecta that move from the chromosphere into the corona, slow with time, stop, and flow back to the chromosphere in the reverse direction from their outflow. This definition is somewhat too restrictive to fit all surges and should be understood as a very general definition rather than an exact definition. Exceptions are observed in which not all of the surge mass falls back to the originating site. Additionally, the paths of outflow and subsequent inflow are generally parallel but not necessarily coincident. Occasionally, some of the mass of a surge, like a flaring arch, is seen to traverse the entire distance of a coronal arch from the initiating site to a secondary site. Like flaring arches, surges can be entirely in emission, on part in emission and part in absorption, although most surges are seen entirely in absorption. Like flaring arches, surges also tend to follow existing lines of force and they most frequently occur in homologous series.

Figure 5 displays an example of a surge that is in emission when it begins (first 3 frames, 22:30–22:35, arrow points to surge). Thereafter it gradually becomes an absorbing feature as seen projected against the chromosphere. The outflow phase lasts until $\sim 22:45$. The inflow phase becomes conspicuous at 23:02. This surge is terminated by the next homologous surge originating from nearly the same site at $\sim 23:12$.

Although some studies have found an absence of X-rays specifically at the sites of surges (Schadee and Martin, 1986), other studies have found emission in the UV and X-rays to be coincident with surges (Rust, 1980; Rust et al., 1977; Schmieder et al., 1988). From the differences in these studies, it seems that surges in absorption are probably at the opposite end of the energy spectrum from surges in emission. Surges in absorption are unlikely to have related X-ray emission while surges in emission are likely to have X-ray emission. There has been no study of just bright surges to find if they are generally associated with X-ray or UV radiation. The study of Rust et al. (1980) indicates that many surges might have a UV component.

As noted above, there are minimal differences between flaring arches and surges in emission; the major difference for flaring arches is that the dense ejected mass successfully reaches the top of a pre-existing coronal loop and hence some or all of the mass falls (or is driven) down the other end of the loop instead of returning to its initial site. I, therefore, suggest that flaring arches are the equivalent of 'super' surges. A principal question that needs to be addressed about surges (and flaring arches) is how their energy is derived.

Although it is sometimes stated that surges are not always associated with flares such as in the paper by Schmieder *et al.* (1988), the question of flare association has much

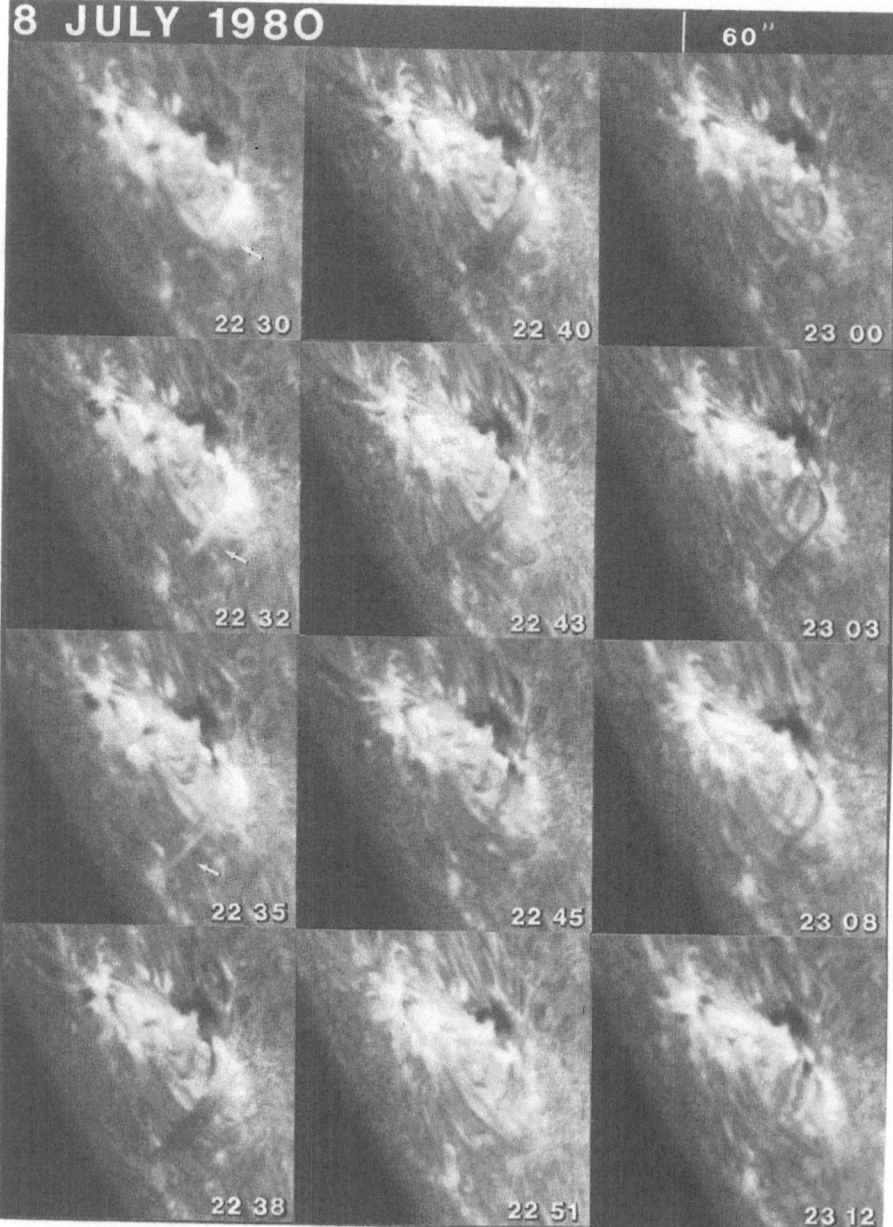

Fig. 5.   The arrows in the first 3 frames point to a surge in emission. As the mass extends into the corona, the surge gradually changes from emission to absorption. Surges seen in emission occasionally have been recognized in UV and X-ray images.

to do with how one defines both a flare and a surge. In some cases, the flares at the base of a surge may be very small in comparison with the associated surge or even obscured by the surge or mistaken as part of the surge.

3.4. ERUPTING FILAMENTS AND FLARE-ACTIVATED FILAMENTS

Three classes of activated filaments initiated by or with flares are discussed and illustrated:

(1) the preflare activation and subsequent eruption,

(2) the surge-like activation,

(3) activation via impact from ejected mass or a shock wave produced by a neighboring flare.

The mass motions associated with each of the above types of activations follow different patterns.

The first category is the only one that should be considered a 'primary' activation because it precedes and accompanies the emission components of a flare in a pattern that suggests that it is an intimate part of the process of energy release of the flare. Hence, this category of activation is sometimes considered (as in this paper) to be an integral part of the flare process while the other categories are secondary processes that occur after or because of the primary energy release.

The 'preflare activation and subsequent eruption' is a continuous process that is often seen to begin during the hour preceding the initiation of the flare loops and chromospheric ribbons. The preflare activation is an outward expansion of either the middle or one end of a filament usually accompanied by increased absorption. The preflare activation irreversibly leads to the complete eruption of the filament, and usually to the expulsion of some of its mass into the interplanetary medium.

The preflare activation and eruption of filaments with major flares is described in detail by Smith and Ramsey (1964) and statistics are presented in Martin and Ramsey (1972). Specific examples are cited in a review by Martin (1980). Relationships of the preflare activation to other preflare conditions and changes at other wavelengths are discussed in Schmahl et al. (1986) and by Gaizauskas (1989, this issue).

Three widely differing examples of the preflare activation and eruption of filaments are shown in Figures 6 and 7. The first 4 frames of Figure 6 (1st column) show the slow preflare ascent of a filament observed on 7-8 June, 1985. The preflare component of motion in the plane of the sky can be seen in Figure 6 only in the upper (southern) end of the filament from 00:05 until the start of the flare at 00:15. As the outward motion of the filament accelerates during the flare, it is clear that the southern end of the filament is rising more rapidly than the northern end. This leads to a 'whip-like' eruption as opposed to the also common 'symmetric' eruption in which the middle of a filament rises while both ends temporarily retain a connection to the chromosphere. The chromospheric connection is maintained as long as there is mass flow down one or both legs of the filament. During the whip-like eruption in Figure 6, the more rapidly ascending end of the filament disconnects with the chromosphere while downflow of some of the mass of the filament simultaneously occurs in the other end of the filament. The combination of bodily outward motion and internal downflow keeps this lower leg of the erupting filament (EF) visible until 00:32 in Figure 9. The disappearance occurs as successive parts of filament exceed the Doppler shifts that are visible within the filter

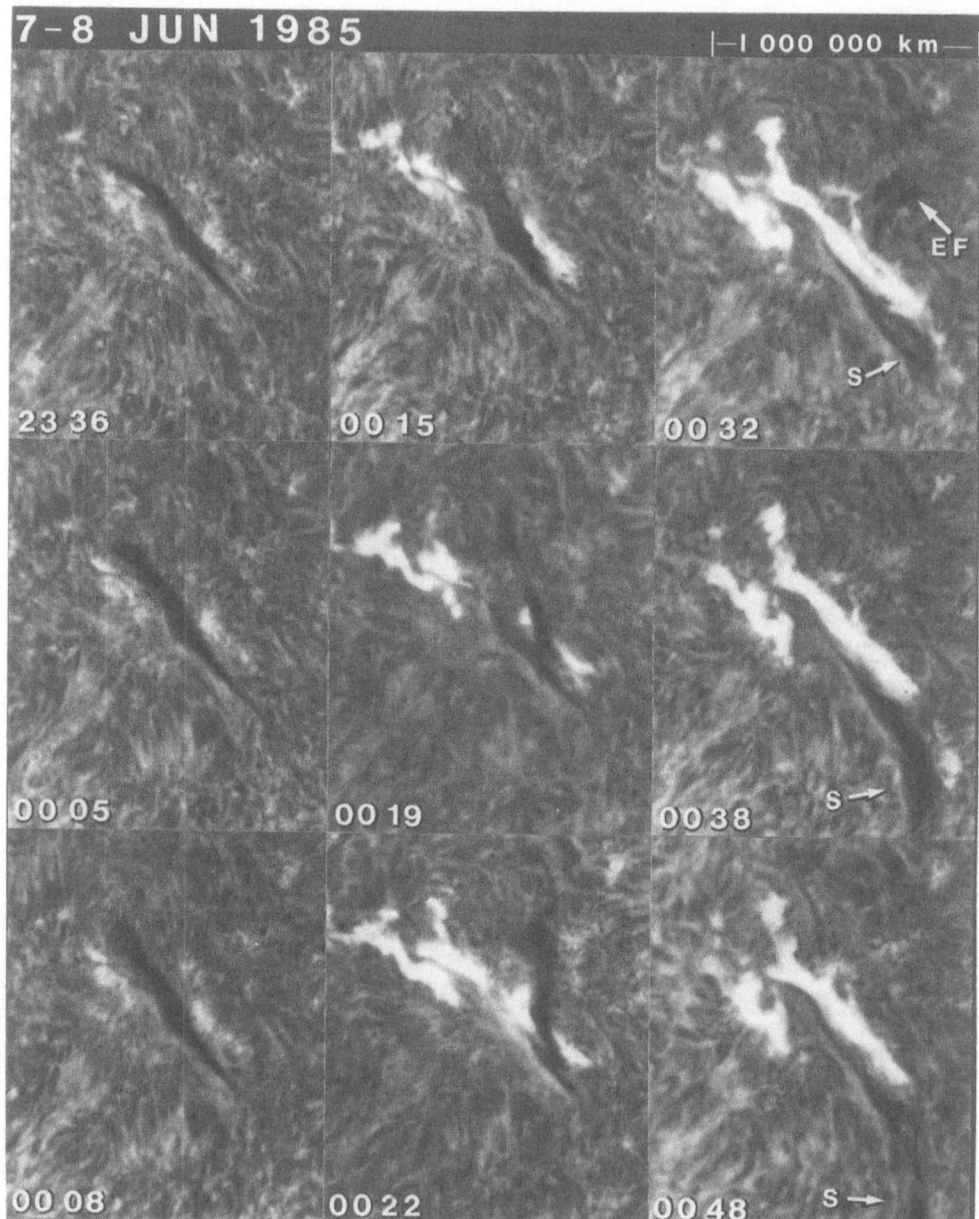

Fig. 6.   This example of an erupting filament with a flare has the form of a whip as the upper end ascends more rapidly into the corona than the lower end. A non-erupting low segment of the filament is ejected as a surge. This mass subsequently flows back into the polarity inversion zone (the channel or site previously occupied by the erupted filament), where it contributes to the formation of a new filament.

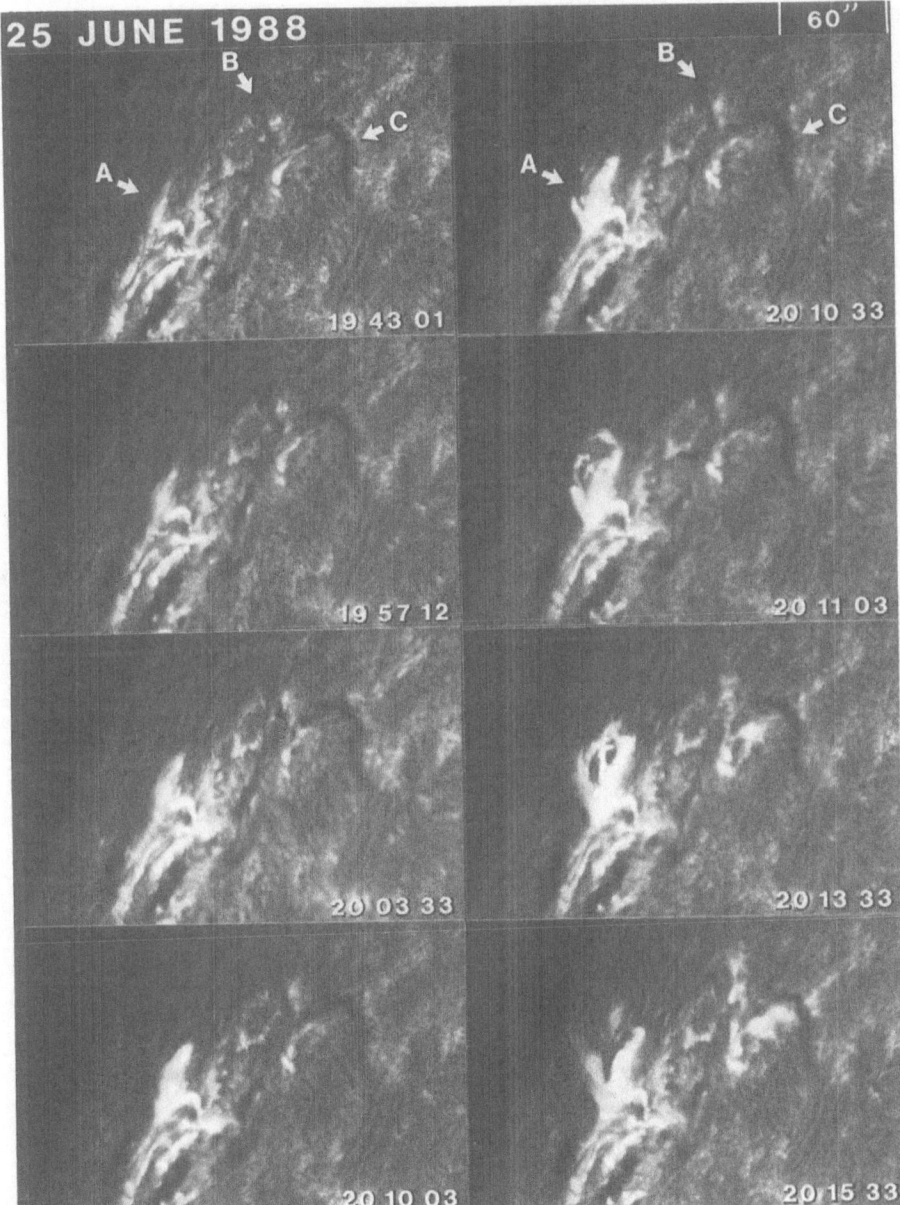

Fig. 7a.   The two major flares in this series are each accompanied by the eruption of the filaments labelled
A and *B* in the top row. Although filament *A* appears to be of flare intensity, the corresponding images at
Hα − 0.6 Å in Figure 7(c) show the filament to be a separate structure from the smaller underlying
chromospheric flare. Filament *B* also changes from an absorbing feature to an emission feature in Hα as
it erupts. A mass ejecta from the flare associated with erupting filament *B* impacts filament *C* at about 20 : 15,
the last frame.

passband. In the case in Figure 6, the passband of the filter is 0.25 Å (halfwidth at half
maximum) centered at approximately 0.25 Å in the blue wing of the Hα line. The mass

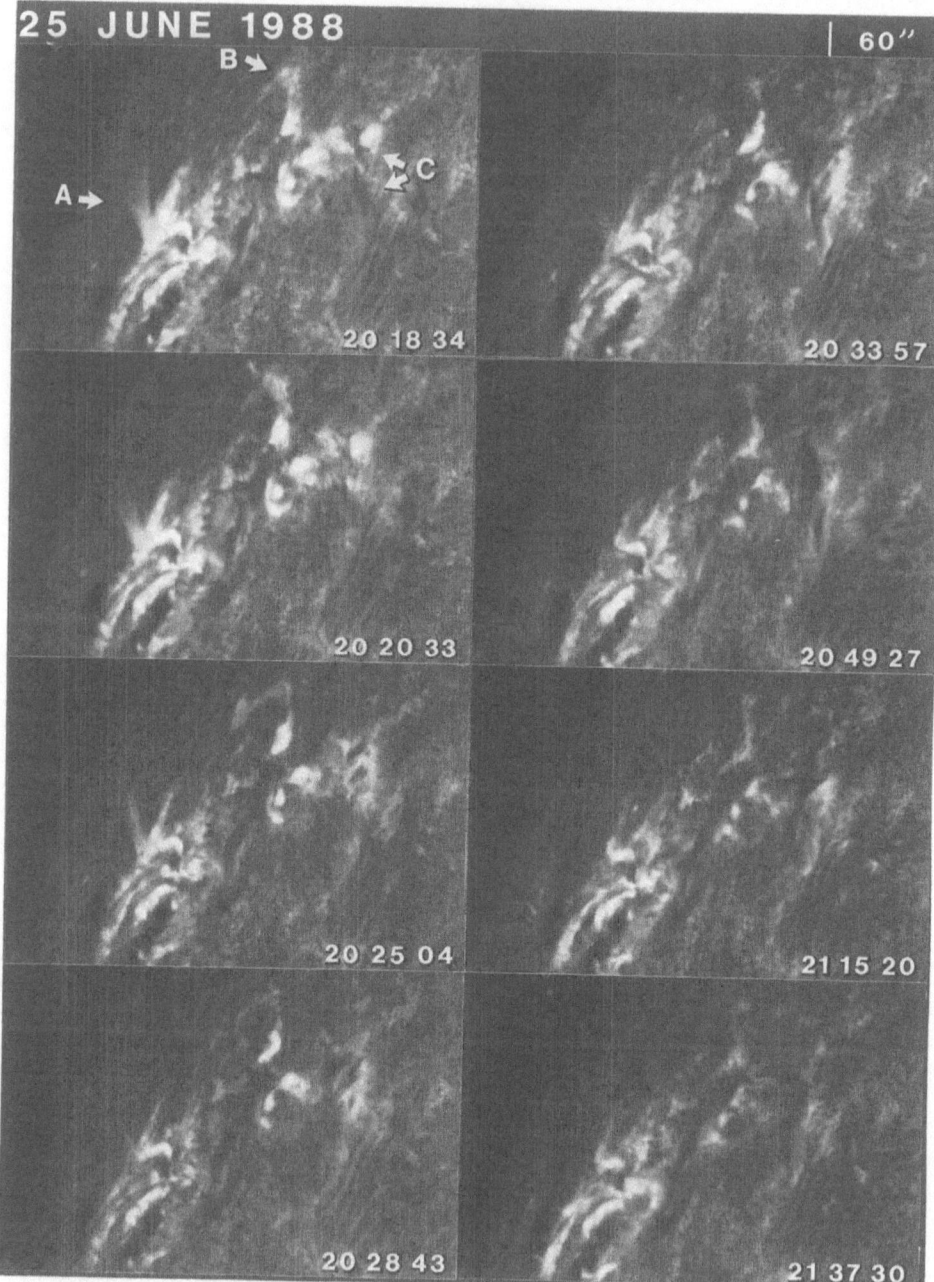

Fig. 7b.   The erratic activation of filament *C* is seen to begin abruptly after the impact of a mass ejection
from the flare associated with erupting filament *B*. After about one hour of turbulent motion, filament *C*
assumes its approximate former configuration.

of the filament is visible in the approximate wavelength range of twice the half-width
of the filter, or from 0 to -0.5 Å, corresponding to a maximum line of sight velocity
of 45 km s^{-1}.

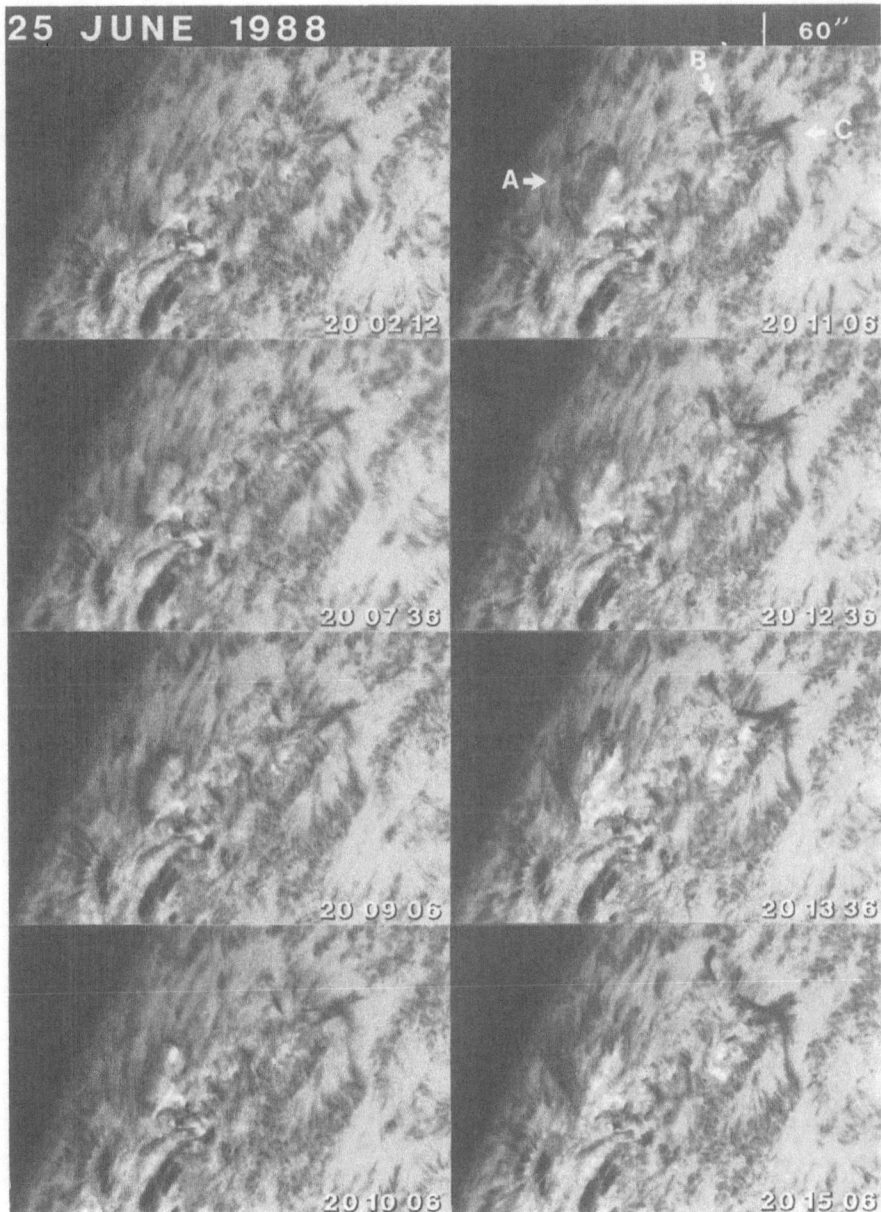

Fig. 7c. These images at Hα − 0.6 Å allow one to more clearly distinguish the activated filaments *A* and *B* from other chromospheric features than in the Hα centerline images in Figures 7(a) and 7(b); the active filaments have high contrast relative to the chromosphere and most of the flare emission cannot be seen this far into the wing of the line.

The eruption shown in Figure 6 is also of the type described by Tang (1986) in which only the upper part of a filament actually erupts into the corona. Upper and lower segments are seen at 00:15 and 00:22. The upper part of the erupting filament is last

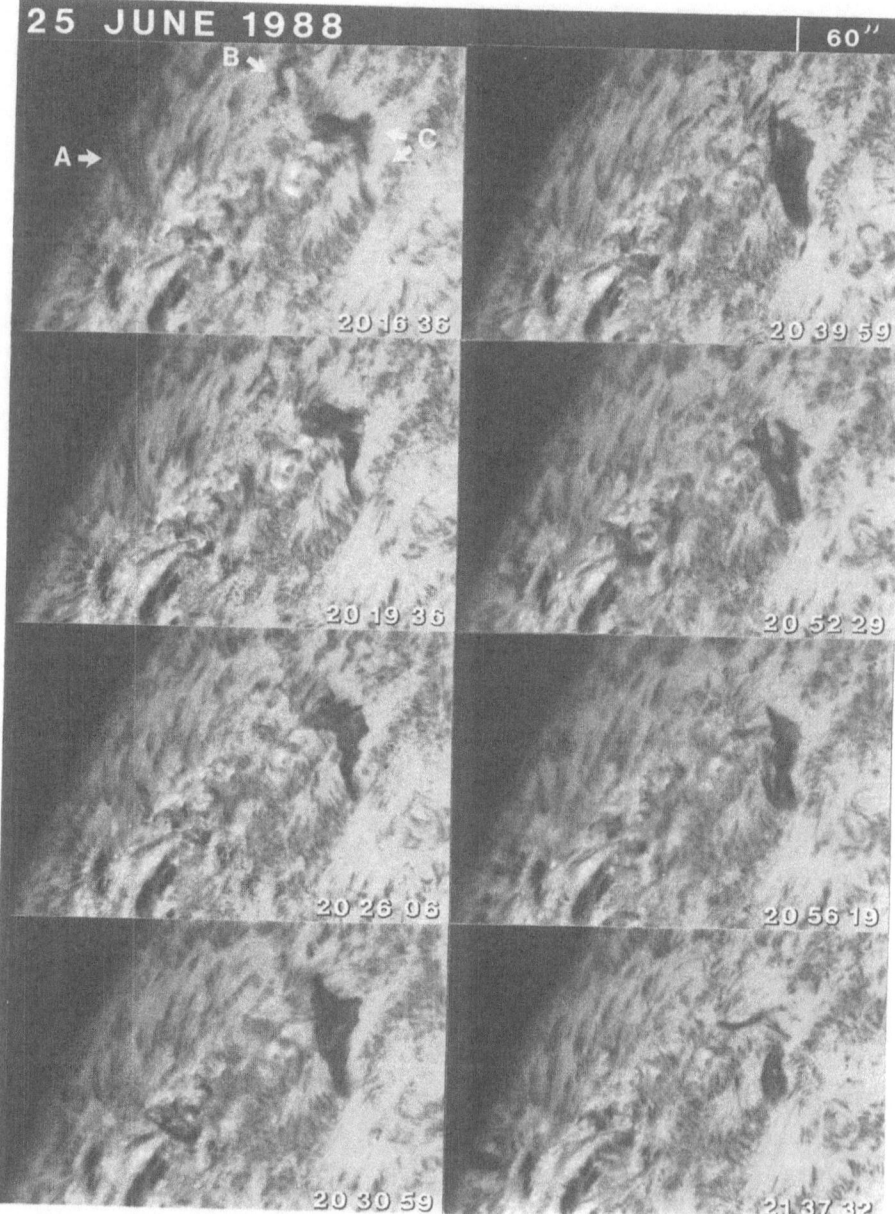

Fig. 7d.   The activation of filament C and the ending of the eruptions of filaments A and B are shown in these images at Hα − 0.6 Å in continuation of the time series in Figure 7(c).

seen above and to the right of the flare in the frame at 00:32. The lower part is activated and becomes the surge (S) seen at 00:32, 00:38 and 00:48 in the lower part of each frame. Mass flows from the filament channel away from the flare. Around 00:48, the flow reverses and this mass then contributes to the formation of a new filament at the site of the eruption and activation.

Surge-like activations within filaments occur frequently even in filaments which are not ready to erupt. Surge-like activations are most commonly triggered by small flares that occur near one end of a filament. A mass ejection is then observed to propagate along the filament. Such surge-like activations, triggered by small flares, were observed in the filament illustrated in Figure 6 several hours before the major flare that accompanied the eruption of the filament. In such a case, the mass flow typically runs along the filament first away from the flare site and then in the reverse direction. In addition, there can be highly turbulent flow patterns as were observed in this same filament at the Ottawa River Solar Observatory (V. Gaizauskas, personal communication).

Figures 7(a–d) are an unusual series of images which depict, in a single field of view on 25 June, 1988, two erupting filaments with flares and the activation of a third filament from a mass ejecta associated with one of the flares. The filaments are identified as A, B, and C in the upper frames of Figures 7(a–d). Different aspects of preflare activation and eruption are seen in filaments A (19:43–23:10) and B (19:43–23:11). The activation due to impact of a mass ejecta takes place in filament C (20:16–21:37) during the flare with the eruption of filament B. The culmination of these events is shown in images at the center of the Hα line in Figure 7(b). Figures 7(c) and 7(d) show the same sequence of events at Hα – 0.6 Å.

Noteworthy features in this series of events are itemized below according to the nature of the activation.

3.4.1. *Preflare Activation and Flare-Related Eruption*

(a) The flare (20:10–20:15 Figure 7(a)) with filament eruption A is an example in which some of the chromospheric and coronal parts of the flare are difficult to distinguish from each other. The starting time of the chromospheric ribbons below the ascending filament is uncertain (between 19:57 and 20:10) but the slow ascent of filament A before its complete disruption is clearly revealed from 19:43 until 23:10. Its position at various times is confirmed by the blue wing observations in Figure 7(c). In the original time-lapse film from which these images were selected, mass from the southern (upper) end of the filament appears to be ejected over the limb while down flow is seen in the northern leg of the filament until approximately 20:33. The rapid ejection of mass over the limb has in the past been called a 'spray'. Tandberg-Hanssen, Martin, and Hansen (1980) conclude that most 'sprays' have their origin in erupting prominences.

(b) The slow preflare ascent of filament B is not readily seen in Figure 7(a) because of the orientation of the filament to our line-of-sight. However, the filament does darken and exhibit minor structural changes characteristic of the preflare ascending stage. The outward motion is confirmed in the corresponding blue wing (– 0.6 Å) images in Figure 7(c) which show increased absorption from 20:07 until 20:16 and then the sudden disappearance of the filament between 20:16 and 20:19. It is common that the slow preflare ascent is not readily detectable in Hα filtergram images at line center except for events close to the limb. In many cases, however, the filaments observed at line center become noticeably darker during the hour prior to the flare. This is indirect evidence of either Doppler shift or spectral broadening of the filament, or both, during its preflare phase.

(c) Filament *B* changes from absorption feature to an emission feature in the images at Hα line center only during its expulsion phase (Figures 7(a) and 7(b) at 20:15, 20:18 and 20:20). The transition to emission is possible evidence of heating.

(d) The activation of filament *C* begins abruptly as an ejection of mass that hits the filament broadside at 20:15 (Figure 7(A)). The activation of this filament results in highly turbulent motion (Figures 7(b) and 7(d)). In contrast, some activations via impact result in the organized bodily upward and downward motion of the whole filament (Ramsey and Smith, 1966; Harvey, Martin, and Riddle, 1974). This raises the question of whether activations have a different character depending upon whether the activation is initiated by mass ejecta, apparent in this case in Figure 7, or by a flare-related shock wave.

(e) The activation of filament *C* is accompanied by faint emission (Figure 7(b)); such emission is not common for this type of activation.

3.5. EXPANDING CORONAL FEATURES

Observations of the coronal continuum above the limb from the coronagraph/ polarimeter experiments on both the Skylab and Solar Maximum Mission satellites show that most, if not all, erupting filaments are accompanied by a phenomenon now known as a 'Coronal Mass Ejection' (CME). A coronal mass ejection is an outwardly expanding volume of the corona that occupies a large amount of space above and around an erupting filament. CMEs are not discussed here in detail because recent reviews and general discussions of this phenomenon are given in Wagner (1984), Jackson (1985), Zirker (1985), and Hildner *et al.* (1986).

An outwardly expanding phenomenon, possibly related to CMEs, that occurs during the rise of a few impulsive flares, is the propagation of a broad front of diffuse emission away from the core of the flare. The diffuse emission may take either the form of an arc propagating away from a flare or of a brief emission halo around the core of a developing flare. Although these phenomena were first presented and discussed by Smith and Harvey (1971) along with evidence of flare-associated shock waves inferred from the oscillatory motions of filaments and chromospheric fibrils, these authors already gave reasons to doubt that the diffuse emission was an effect of the flare shock wave. They instead suggested that it represented another form of flare-associated mass motion. In 1980, at the Skylab Flare Workshop, I first proposed the hypothesis that the diffuse emission seen in filtergrams in Ca II and discussed in van Hoven *et al.* (1980) and Webb *et al.* (1980) could be a manifestation of a 'coronal transient', the phenomenon now known as a 'Coronal Mass Ejection' (CME), seen in projection against the disk. This idea has never been proven or disproven and no observations have yet been obtained (to my knowledge) in which this phenomenon has been observed concurrently with a CME. A primary problem in obtaining concurrent observations is that the diffuse emission is best seen with flares observed in the middle of the disk while the CME observations favor events near the limb.

Although the diffuse emission is often referred to as a 'Moreton wave', 'blast wave', or just 'flare wave', it is more likely that the real wave is invisible and that the emission is part of a mass ejection that possibly drives a shock wave through the corona. The strongest evidence of the existence of such shock waves in Hα images are the relatively rare observations of successive upward and downward motions of the chromospheric structures in an arc-shaped pattern that rapidly propagates away from a flare during its explosive rise to maximum (Moreton 1960, 1964, Moreton and Ramsey, 1960). This phenomenon, which we will call the 'chromospheric flare wave effect' has only been observed with a small percentage of very impulsive bright flares (Smith and Harvey, 1971). Downward and upward oscillations of distant filaments during flares is also good evidence of the existence of true shock waves from bright flares (Dodson, 1949; Ramsey and Smith, 1966; Harvey, Martin, and Riddle, 1974). It is less certain whether the flare shock wave can trigger point-like brightenings of the chromosphere outside of the flare (Smith and Harvey, 1971) because such brightenings could also be initiated by particles travelling along field lines directly from the associated flare to distant parts of the chromsophere as mentioned in Martin and Harvey (1972) and discussed in Sections 2.2 and 3.2.

4. Epilogue

Solar flares represent the sudden release of energy in a localized parts of the solar atmosphere where oppositely directed magnetic fields are present. This release of energy results in various forms of mass motion whose geometry and velocity depend on the presence and varying configurations of the magnetic field. Some mass motions in the corona, such as those of erupting filaments, coronal mass ejections, and other expanding features, give evidence that the magnetic field is changing in the volume of space within and around the flare while others such as surges and flaring arches suggest that mass in these components is guided along a relatively static magnetic field configuration. Still other mass motions, in the chromospheric flare ribbons and adjacent chromosphere, are most likely the result of the impact of high speed particles and/or shock waves. Studies of these many forms of mass motion provide a means of deducing information about the changes in the magnetic field in the corona which cannot be directly detected in our present day magnetograms of the photosphere and chromosphere. The observed mass motions and related magnetic fields also provide constraints on the possible processes of energy release that take place during solar flares.

For a more comprehensive understanding of solar flares, it is hoped that this limited review of flare-related mass motions in the optical part of the spectrum will be followed by more comprehensive reviews of the different types of mass motions that are observed in flare components throughout the electromagnetic spectrum.

Acknowledgements

This review was accomplished under the sponsorship of the Air Force Office of Scientific Research under contract AFOSR-87–0023. The acquisition of new observational material presented herein was possible through support of the Big Bear Solar Observatory by NASA grant NGL05002034 and NSF grant ATM-8513577.

References

Canfield, R. C., Cheng, C. C., Dere, K. P., Dulk, G. A., and Schoolman, S. A.: 1980, in P. A. Sturrock (ed.), *Solar Flares, A Monograph from Skylab Solar Workshop II*, Appendix A, pp. 451–469.
Dodson, H. W.: 1949, *Astrophys. J.* **110**, 382.

Doschek, G. A. and 13 co-authors: 1986, in M. Kundu and B. Woodgate (eds.), *Energetic Phenomena on the Sun, The Solar Maximum Mission Flare Workshop*, NASA Conference Publ. 2439, pp. 4-1 to 4-42.

Ellison, M. A.: 1952, *Publ. Roy. Obs. Edinburgh* **1**, 75.

Gaizauskas, V.: 1989, *Solar Phys.* **121**, 135 (this issue).

Harvey, K. L., Martin, S. F., and Riddle, A. C.: 1974, *Solar Phys.* **36**, 151.

Hildner, E. and 20 co-authors: 1986, *Energetic Phenomena on the Sun, The Solar Maximum Mission Flare Workshop Proceedings*, NASA Conference Publ. 2439, Chapter 6.

Ichimoto, K. and Kurokawa, H.: 1984, *Solar Phys.* **93**, 105.

Jackson, B. V.: 1985, *Solar Phys.* **100**, 563.

Martin, S. F.: 1979, *Solar Phys.* **64**, 165.

Martin, S. F.: 1980, *Solar Phys.* **68**, 217.

Martin, S. F. and Harvey, K. L.: 1973, in A. J. Hundausen and Gordon Newkirk, Jr. (eds.), *Flare-Produced Shock Waves in the Corona and in Interplanetary Space*, Report on Conference held in Boulder, Colorado, 11–14 September 1972, High Altitude Observatory, National Center for Atmospheric Research.

Martin, S. F. and Ramsey, H. E.: 1972, in P. McIntosh and M. Dryer (eds.), *Solar Activity, Observations and Predictions*, MIT, Cambridge, p. 371.

Martin, S. F. and Švestka, Z.: 1988, *Solar Phys.* **116**, 91.

Moore, R. L. and 16 co-authors: 1980, in P. A. Sturrock (ed.), *Solar Flares, A Monograph from Skylab Solar Workshop II*, pp. 341–409.

Moreton, G. E.: 1960, *Astrophys. J.* **65**, 494.

Moreton, G. E.: 1964, *Astron. J.* **69**, 145.

Moreton, G. E. and Ramsey, H. E.: 1960, *Publ. Astron. Soc. Pacific* **72**, 357.

Mouradian, Z. and Soru-Escaut, I.: 1989, in B. M. Haisch and M. Rodonò (eds.), *IAU Colloq.* **104**, *Solar and Stellar Flares*, Poster Papers, Publ. Catania Astrophys. Obs., Special Volume, p. 227.

Mouradian, Z., Martres, M. J., and Soru-Escaut, L.: 1983, *Solar Phys.* **87**, 309.

Ramsey, H. E. and Smith, S. F.: 1966, *Astron. J.* **71**, 197.

Rust, D. M. and Bar, V.: 1973, *Solar Phys.* **33**, 445.

Rust, D. M. and Webb, D. F.: 1977, *Solar Phys.* **54**, 403.

Schadee, A. and Martin, S. F.: 1986, in D. Neidig (ed.), *The Lower Atmosphere of Solar Flares*, National Solar Observatory, Sacramento Peak Observatory, Sunspot, New Mexico, U.S.A., p. 360.

Schmahl, E. J., Webb, D. F., Woodgate, B., Waggett, P., Bentley, R., Hurford, G., Schadee, A., Schrijver, J., Harrison, R., and Martens, P.: 1986, *Energetic Phenomena on the Sun*, NASA Conference Publ. 2439, p. 1-48 to p. 1-72.

Schmieder, B., Simnett, G. M., Tandberg-Hanssen, E., and Mein, P.: 1988, *Astron. Astrophys.* (in press).

Schmieder, B., Forbes, T. G., Malherbe, J. M., and Machado, M. E.: 1987, *Astrophys. J.* **317**, 956.

Schoolman, S. A. and Ganz, E. D.: 1981, *Solar Phys.* **70**, 363.

Smith, S. F. and Harvey, K. L.: 1971, in Macris (ed.), *Physics of the Solar Corona*, pp. 156–167.

Smith, S. F. and Ramsey, H. E.: 1964, *Z. Astrophys.* **60**, 1.

Sturrock, P. A.: 1980, in P. A. Sturrock (ed.), *Solar Flares*, p. 1-14.

Švestka, Z.: 1976, *Solar Flares*, D. Reidel Publ. Co., Dordrecht, Holland, pp. 76–80.

Švestka, Z., Kopecký, M., and Blaha, M.: 1962, *Bull. Astron. Inst. Czech.* **13**, 37.

Švestka, Z., Martin, S. F., and Kopp, R. A.: 1980, in M. Dryer and E. Tandberg-Hanssen, *Solar and Interplanetary Dynamics*, p. 217.

Švestka, Z., Fárník, F., Fontenla, J., and Martin, S. F.: 1989, *Solar Phys.* (in press).

Tandberg-Hanssen, E., Martin, S. F., and Hansen, R. T.: 1980, *Solar Phys.* **65**, 357.

Tang, F.: 1983, *Solar Phys.* **83**, 15.

Tang, F.: 1986, *Solar Phys.* **105**, 399.

Tang, F. and Moore, R. L.: 1982, *Solar Phys.* **77**, 263.

Teske, R.: 1962, *Astrophys. J.* **136**, 534.

van Hoven, G. and 18 co-authors: 1980, in P. A. Sturrock (ed.), *Solar Flares*, A Monograph from Skylab Workshop II, pp. 46–48.

Wagner, W. J.: 1984, *Ann. Rev. Astron. Astrophys.* **22**, 267.

Waldmeier, M.: 1941, *Ergebnisse und Probleme der Sonnenforschung*, p. 197.

Webb, D. F., Cheng, C. C., Dulk, G. A., Edberg, S. J., Martin, S. F., McKenna-Lawlor, S., McLean, D. J.: 1980, in P. A. Sturrock, *Solar Flares, A Monograph from Skylab Workshop II*, Appendix B, pp. 471–499.

Wu, S. T. and 19 co-authors: 1986, *Energetic Phenomena on the Sun*, NASA Conference Publ. 2439, pp. 5-57 to 5-60.

Zirin, H. and Tanaka, K.: 1973, *Solar Phys.* **32**, 202.

Zirker, J. B.: *Solar Phys.* **100**, 281.

SOLAR AND STELLAR FLARES: QUESTIONS AND PROBLEMS

D. J. MULLAN

Bartol Research Institute, University of Delaware Newark, DE 19716, U.S.A.

Abstract. Although progress has been made in understanding certain aspects of the physics of solar and stellar flares, there are a number of topics which, in the author's opinion, still pose a problem. We summarize these topics here.

1. Introduction

The purpose of this article is to go beyond a review of solar and stellar flares which appeared in this journal more than a decade ago (Mullan, 1977), and to touch upon certain aspects of solar and stellar flares where, in the author's opinion, problems still persist in the physical interpretation of the phenomena. The approach will be to focus on matters which can be discussed quantitatively for *both* solar and stellar flares. As a result, certain topics which are of great current interest in solar flares (e.g., chemical anomalies in solar energetic particles, high-time resolution observations of X-rays, directivity of gamma-ray emission, neutron production, and decay, etc.) will not be discussed here because current stellar data cannot contribute to a resolution of the problems.

2. Physical Conditions in Stellar Flares

In order to discuss meaningfully the nature of physical problems in flares, the first question we need to ask is: what are the physical conditions in flares? For the sake of definiteness, we shall refer to the parameters derived by Haisch (1983) for a sample of eight flares which were observed by the Einstein IPC detector. For each flare, Haisch analyzed the time behavior of the soft X-ray emission during the time period following maximum intensity. During this decay phase, he extracted three quantities from the IPC data: the emission measure (EM), the temperature (T), and the decay time (τ). Three assumptions were made in order to interpret the data: (a) each flare was assumed to occur in a loop of length L and aspect ratio 0.1; (b) the magnetic field B in the loop was assumed to be strong enough to contain the flare plasma (hence, $B^2 > 8\pi p_g$, where $p_g = 2N_e kT$ is the gas pressure in the flare plasma); and (c) the radiative and conductive cooling times of the flare plasma were assumed to be equal. From these, Haisch derived values of L and N_e, and a lower limit on B.

As an example, for flare number 2 on Proxima Centauri in Haisch's table, we find $EM = 10^{51}$ cm^{-3}, volume $V = 10^{27}$ cm^3, $N_e = 9 \times 10^{11}$ cm^{-3}, $L = 5 \times 10^9$ cm, $T = 4 \times 10^7$ K, and $B > 500$ G. For future reference, we note that the Alfvén speed in this case has a value $v_A > 1200$ km s^{-1}, and because of assumption (b), the lower limit on v_A depends only on T.

Solar Physics **121** (1989) 239–259.
© 1989 *by Kluwer Academic Publishers.*

We may ask: how reliable are these parameters? To answer this question, we make the following three points. (i) Schmitt *et al.* (1987) have examined data from Einstein IPC during time intervals when the IPC happened to be scanning the Earth. Time profiles of the X-ray emission during such intervals have revealed certain events which look like solar flare profiles. Schmitt *et al.* interpret these events indeed as solar flare X-rays scattered off the Earth's atmosphere. They then subject the events to exactly the analysis used by Haisch (1983) and find values of N_e and L which are quite consistent with the values derived directly for compact loop flares in the Sun. (ii) Are radiative and conductive cooling times really equal in flares? Strong *et al.* (1986) have used SMM data to show that in fact these two time-scales may be quite different. In six flares which they analyzed, they extracted the two time-scales from inferred values of N_e, T, and L, and found the ratio of the time-scales to vary between 0.2 and 20. However, the estimates of the time-scales are uncertain because of unknown filling factors and unknown extent of inhibition of conduction. (iii) Reale *et al.* (1988) have constructed a hydrodynamical model of a flare in a loop, and applied it to flare No. 2 on Proxima Centauri. These authors make no assumption about equality between radiative and conductive cooling times. They fit the decay of the X-rays with a model which has the same T and EM as Haisch (1983), but with N_e lower by a factor of about 5, and L larger by a factor of 2–3. Since T is unchanged, $v_A > 1200$ km s^{-1} as before.

In view of these results, it seems acceptable to use the physical parameters derived by Haisch (1983) in order to make order-of-magnitude estimates of various effects in stellar flares.

3. Radio Flares

The first problem we turn to concerns radio emission. Radio flares in M dwarfs, as well as microwave 'spike bursts' in solar flares, are observed to have brightness temperatures well in excess of 10^{10} K (e.g., Mullan, 1985), indicating the presence of a coherent emission process. The most widespread model for explaining the coherent emission is the electron cyclotron maser (ECM) (Melrose and Dulk, 1982). Electrons are supposed to be accelerated somehow in the initial flare release, and they stream towards the footpoints of the magnetic loop. Some of the electrons reach the chromosphere, and are lost there. The remaining electrons mirror, and create a loss-cone distribution in the loop. If physical conditions are right, this distribution is unstable to the growth of electromagnetic (*em*) waves at the electron cyclotron frequency Ω_e (and possibly harmonics thereof): the waves can tap the free energy available in the loss-cone distribution. The mode of the *em* waves in a cold plasma is x-mode if the ratio $R = \omega_p/\Omega_e$ is < 0.3, where ω_p is the electron plasma frequency. The x-waves can escape and be observed directly: this is the most direct mechanism for producing coherent radio emission. In a warm plasma, the upper limit on R increases with increasing temperature (Winglee, 1985), but this is not a significant factor for the values of T mentioned above. A beneficial side product of x-mode emission (as well as explaining directly the coherent radio emission) is that coronal heating also occurs by means of these x-modes: they can escape the source region and be absorbed as a second harmonic elsewhere.

For $0.3 < R < 1.3$, *em*-waves no longer dominate in the instability: rather, electrostatic modes do. Since these are not *em*-waves, they cannot explain directly the observed radio waves: some mechanism must first convert them to *em*-radiation. For R values of a few (say, 3–4), maser emission becomes more difficult in the context of ECM, but not impossible (e.g., see Louarn *et al.*, 1987). For R of order 10 or more other sources of maser action must be considered (e.g., Kuijpers, 1985).

The upper limit on R for direct *em*-maser emission is equivalent to a lower limit on the Alfvén speed. Thus, $R < R_c = 0.3$ corresponds to $v_A > v_{Ac} = 23\,000$ km s^{-1}. Hence, direct *em*-maser radiation can occur only in regions of very high Alfvén speed.

Do such regions exist in solar and/or stellar flares? In the Sun, VLA data allow one to measure both B and N_e. In a sunspot loop, Lang (1983) reported $B = 600$ G and $N_e = 10^9$ cm^{-3}, corresponding to $v_A = 42\,000$ km s^{-1}. In such a loop, direct ECM emission is possible. However, a sunspot loop is a region of exceptionally large B and unusually low N_e, and is, therefore, the most favorable site for finding large v_A: other loops in solar active region will undoubtedly have lower values of v_A. For example, results by Lang *et al.* (1987) for an active region loop indicate v_A of less than 6000 km s^{-1}. Such a loop would not be a candidate for direct ECM emission.

In stellar loops, the results of Haisch (1983) indicate lower limits on the values of v_A which are in all cases of order 1000 km s^{-1}. Unless the lower limits are very far from the true values, the conclusion is that direct ECM emission cannot be operating in flare star loops. If we wish to save the hypothesis of direct ECM emission for the Haisch flares, we require either an increase in B by a factor of 23, or a decrease in N_e by a factor of more than 500. Thus, in Proxima Centauri, we require $B = 11.5$ kG (if the density remains unchanged), or N_e must be reduced to less than 2×10^9 cm^{-3}. Now, fields of order 10 kG may very well exist on flare star surfaces (Mullan, 1984a): in fact, fields of 5–6 kG have already been detected in some such stars (Saar *et al.*, 1987), and even higher fields are almost certainly be present (but currently undetectable) in cool spots. However, increasing B to more than 1700 G means that emission at Ω_e would emerge at frequencies too high to be detected by the VLA at 6 cm. And as regards the densities, Katsova *et al.* (1987) have shown from X-ray data that the mean densities at the base of coronae in cool dwarfs are in the range 10^9–10^{10} cm^{-3}: therefore, the loop on which the ECM should occur would have a density no greater than the average coronal density. This seems unlikely in a loop which is the site of a flare: such loops in the Sun are found to have densities which are considerably enhanced relative to the average. For example, Canfield (unpublished paper at this conference) has reported that preflare densities in solar flare loops are 100–1000 times the average density. Moreover, in order to create any ECM, a loss-cone distribution must be set up: this requires that electrons in the loss-cone be already removed by the chromosphere. The latter will be heated by the loss-cone electrons, and so, before the ECM occurs, the loop will already be filling up with material evaporated off the chromosphere, there by further enhancing the density in the loop.

We conclude that, using Haisch's stellar loop parameters for post-maximum conditions in stellar flares, it is unlikely that direct *em*-maser emission at Ω_e (as occurs in the simplest version of ECM) is at work. Undoubtedly *some* maser process is at work, perhaps involving nonlinear conversion of plasma waves into *em*-radiation (see Kuijpers, 1985), and perhaps involving different processes in different flares (Bastian and Bookbinder, 1987), but direct *em*-maser emission should not be assumed as a universal explanation of coherent radio emission from dMe flares.

On the other hand, radio bursts from RS CVn stars (which are sometimes discussed as an analog of radio bursts from dMe stars) do not usually need a maser mechanism to account for them. Therefore, the present discussion is not applicable to RS CVn bursts. It should not be assumed uncritically that bursts in RS CVn stars necessarily involve the same physical processes as those on dMe stars (Mullan, 1985).

4. Penetration of Particle Beams and Photons

A problem related to the topic of electron beams arises in connection with the electron densities discussed above. The excitation of an optical flare in a solar/stellar chromosphere is believed to depend on the propagation of a disturbance downward from the site of initial energy release in the corona: when this disturbance penetrates into the chromosphere, the optical flare can begin. Candidates for the disturbance are beams of charged particles or photons. In this section, we are mainly concerned with the following question: can an electron beam penetrate to the chromosphere of a flare star? At the end of this section, we mention briefly the effects of photons, and we return in Section 5 to the question of proton beams.

The answer depends on the column density ξ through which the beam must pass between its source and the chromosphere. An electron of energy E keV can penetrate to $\xi_p(E) = 6 \times 10^{19}(E/20)^2$ cm^{-2} if the only stopping mechanism is Coulomb collisions (Brown, 1971): this figure is reduced somewhat if allowance is made for excitation of plasma oscillations by the passing beam (Hamilton and Petrosian, 1987).

In the Sun, values of ξ on different loops can be estimated from typical solar loop parameters: with $L = (0.3-3) \times 10^9$ cm and $N_e = (1-100) \times 10^9$ cm^{-3}, solar loops have ξ ranging from 3×10^{17} to 3×10^{20} cm^{-2}. Thus, for some loops, ξ is less than $\xi_p(E = 20)$, and in such loops, an electron beam of energy 20 keV (which is the energy at which non-thermal electron beams in the Sun seem to contain most of their energy) can indeed reach the chromosphere. If the electron beam has a large enough flux, the chromosphere is explosively evaporated, and the Hα line develops strong Stark wings (Canfield *et al.*, 1984). We may refer to this case as 'an electron beam flare' in the chromosphere. On the other hand, some loops contain so much material that ξ is in excess of $\xi_p(E = 20)$: in such loops, 20 keV electrons are stopped before they reach the chromosphere. They deposit their energy in the corona, and then a thermal conduction front propagates down into the chromosphere. In this case, the Hα line is narrow, with no Stark wings, and no central reversal (Canfield *et al.*, 1984). We may refer to this as a 'thermal conduction flare' in the chromosphere.

In the case of flare stars, using the ranges of loop parameters listed by Haisch (1983), namely $L = (0.2-6) \times 10^{10}$ cm   and $N_e = (1-30) \times 10^{12}$ cm^{-3},   we   find   that

$\xi = 2 \times 10^{21}$–2×10^{24} cm^{-2}. In all cases, such loops have column densities greatly in excess of $\xi_p(E = 20)$. This suggests that in stellar flares, it may be difficult to find an example of a pure 'electron beam flare' in the chromosphere: rather, most chromospheric flares in M dwarfs may be thermal conduction flares. This is consistent with a suggestion proposed some years ago that thermal conduction is the primary physical agent which couples coronal plasma to the chromospheric emitting material in stellar flares (Mullan, 1976). Of course, if stellar flares produce most of their beam energy as electrons of significantly higher energy (say 1 MeV, although there is no evidence to support this) or protons with energies of tens of MeV, then such beams may penetrate to the chromosphere in all but the densest of the loops we have considered here.

In the present discussion, we have used Haisch's flare parameters which pertain to the post-maximum phase. We may, therefore, be overestimating somewhat the loop column densities in the early stages of the flares. But our overestimates would have to be as large as 30–30 000 in order to alter the conclusion that thermal conduction flares dominate optical flares in M dwarfs. In this regard, in a sample of flares observed with high spectral resolution by Schneeberger *et al.* (1979), the width of Hα did not increase, although the overall intensity did. In these flares, there was also no strong indication of any central reversals in Hα. Such behavior in Hα is consistent with the thermal conduction class of flare discussed by Canfield *et al.* (1984).

Let us turn briefly to photons as a possible agent in the process whereby primary energy release in the corona is communicated to the chromosphere/photosphere in order to cause the 'optical flare'. The importance of X-ray emission in this regard in solar flares can be seen most readily by examining data from one subset of solar flares, namely, the white-light flares (WLF; see Neidig, this volume). In these flares, continuum emission originates in the lower chromosphere and/or the upper photosphere. If the emission occurs in the photosphere, Neidig rules out electron beams, proton beams, thermal conduction, and soft X-rays as agents to power a white light flare. But the light curve of such flares tracks the hard X-ray emission (at energies of 50–100 keV) during both the impulsive and the gradual phases of the flare: this suggests that hard X-rays may play an important role in initiating the optical flare. A non-LTE radiative transfer model which explains the correlation between hard X-rays and white light solar flares has been proposed by Aboudarham and Henoux (this volume).

In stellar flares, the possibility that X-ray photons can contribute significantly to the optical light curve was demonstrated some years ago (Tarter and Mullan, 1977).

A hybrid mechanism can be imagined in which both a particle beam and the photons which the beam creates as it enters the dense lower atmosphere may be responsible for communicating the original energy release down to the optical flare region. This is obviously a highly nonlinear process. See Aboudarham and Henoux (this volume) for a solar flare model involving an electron beam plus its hard X-rays, and see Grinin and Sobolev (1989) for a stellar flare model involving a proton beam plus its photons. In both cases, the particle beam first penetrates a certain distance, and then the photons take over and penetrate considerably deeper into the atmosphere. In particular, in a case considered by Grinin and Sobolev, although the beam itself (composed of 10 MeV protons) produces maximum heating rates at a column density of about 10^{22} cm^{-2}, the heating due to the photons produced by the proton beam reach their maximum rate at deeper than 10^{24} cm^{-2}. Moreover, the amplitude of the photon heating exceeds the direct particle heating by about one order of magnitude.

5. Momentum Balance in Stellar Flares?

An interesting question has arisen in solar physics recently concerning momentum balance in flare plasma. Once the initial energy is released in a flare, it flows downward towards the chromosphere, either as a conduction front or as an electron beam. The chromosphere is heated and as a result, material is evaporated upwards into the corona. To balance the upward momentum, chromospheric material also moves downward. Is there evidence that the upward and downward momenta are balanced?

Canfield *et al.* (1987) have evaluated both upward and downward momenta in a solar flare. The downward momentum can be evaluated from red-shifted Hα data:

$$p_d = \mu m_{\mathrm{H}} N_c S_c v_d^2 \tau,$$

where μ is the mean molecular weight, m_{H} is the mass of a hydrogen atom, N_c is the chromospheric density prior to the flare, S_c is the area of the chromospheric region where Hα is red-shifted, v_d is the downflow velocity of the Hα material, and τ is the duration of the red-shifted phase in Hα. The upflow momentum can be evaluated from blue-shifted data from the soft X-ray lines emitted by the evaporated plasma:

$$p_u = \mu m_{\mathrm{H}} v_u (\mathrm{EM} \times V)^{0.5},$$

where v_u is the blue shift of the soft X-ray plasma, EM is its emission measure, and V is the volume of the soft X-ray plasma. In the solar flare studied by Canfield *et al.*, it appears that indeed p_d is equal to p_u to better than one order of magnitude.

Let us now apply the argument to stellar flares. Using Haisch's data, we have EM $= 10^{51}$ and $V = 10^{27}$ c.g.s. No direct observations of blue-shifted X-ray lines have been reported for stellar flares. But it is expected that upward expansion velocities will be no more than a few times the local sound speed. A value of $v_u = 600$ km s^{-1} has been suggested by Reale *et al.* (1988). We adopt 100–1000 km s^{-1} here. Then the upward momentum per unit mass is 10^{46}–10^{47} c.g.s.

For the downward momentum, red-shifted Hα has been observed to persist for up to $\tau = 120$ s in large stellar flares (e.g., Bopp and Moffett, 1973). The amount of redshift is difficult to quantify. But the red wing of Hα is observed to extend to velocities of about 1100 km s^{-1} (Bopp and Moffett, 1973). This suggests that the peak of the red-shifted component in Hα may lie at velocities of several hundred km s^{-1}. We adopt $v_d = 300$–500 km s^{-1} here. For the area of the chromosphere which is participating in the downflows, we note that Cram and Woods (1982) require from their Hα modelling that as much as 10–20% of the visible disk area must be emitting in Hα. We take 10% here, and therefore find that for a star such as Proxima Centauri (with radius 10^{10} cm^2),

$S_c = 3 \times 10^{19}$ cm^2. For flare stars of larger mass, a representative flare area may be taken to be 10^{20} cm^2. (This is certainly larger than solar flare counterparts: further arguments in favor of larger area in stellar flares can be found in Neidig (this conference).) Finally, for the pre-flare chromospheric density, we note that in general, the chromosphere/corona of flare stars are denser than their solar analogs by factors which may be as large as 10–100 (e.g., Mullan, 1977). In the solar flares analyzed by Canfield (unpublished paper at this conference), the preflare densities were found to be 2–3×10^{13} cm^{-3}: scaling these upwards by factors of a few, we suggest $N_c = 10^{14}$ cm^{-3} in the pre-flare loops on flare stars. Chromospheric densities outside flares have been evaluated by Pettersen (1989) for a sample of 8 flare stars of comparatively early spectral types (K5–M1): he finds densities of 5×10^{11}–5×10^{12} cm^{-3}. Canfield has stressed that chromospheric densities in a pre-flare loop are higher than quiescent values by factors of 100–1000. Hence, our choice of $N_c = 10^{14}$ cm^{-3} for the pre-flare chromosphere seems conservative.

Are our choices of flare parameters consistent with other information on stellar flares? To answer this, we note that with the above choices of N_c and v_d, the flare energy flux which drives chromospheric evaporation must be about $F_{ce} = 4 \times 10^{13}$ ergs cm^{-2} s^{-1} (see Fisher, 1987). Therefore, over a flare area of S_c, a total energy of $E_{ce} \approx 10^{34}$ ergs is available for chromospheric evaporation in a flare lasting ≈ 10 s. The total output of flare energy cannot be smaller than this (cf. Fisher, 1987). Are such energies reasonable for stellar flares? Energies of order 10^{33} ergs are observed in the B band alone in large flares of solar neighborhood flare stars: even larger energies are observed in Orion and Pleiades flare stars (Shakhovskaya, 1979). The total optical energies of stellar flares in larger than B-band energy by a factor of 4.2. Also, bolometric energies in large flares must also include X-ray, EUV, and mass motions. The latter numbers are very uncertain, but may be as much as 10–100 times the optical energy (Gershberg and Shakhovskaya, 1983). Hence, total energy releases of a few times 10^{33} ergs are expected to be available in flares with B-band energies of 10^{31}–10^{32} ergs. Flares of such energies are not rare events on solar neighborhood flare stars: they are observed with frequencies of once every 1–10 hours (Shakhovskaya, 1979). Hence, our choice of parameters are not excessive from the point of view of energetics.

With the above choices, we find that the downward momentum per unit mass is $10^{50.5}$–10^{51} c.g.s. These figures suggest that there is a discrepancy between upward and downward momenta in stellar flares in the sense that the downward momentum exceeds upward momentum by a factor which may be as large as 5 orders of magnitude.

Significant revisions must be made in one or more of the above physical parameters if this discrepancy is to be removed. Perhaps a beam of *protons* rather than electrons is created in stellar flares: this would provide downward momentum while slowing down the process of chromospheric evaporation (Van den Oord, 1987). Independent arguments for the possible presence of proton beams in stellar flares have been presented by Simnett (1989).

6. Sub-Surface Source of Flare Power: Solar-Stellar Differences

We now turn to the problem of energizing the activity in a flare star atmosphere. (By 'activity' we refer to both flare activity and 'quiescent' coronal heating.) The ultimate source of activity is mechanical energy which creates stresses in the atmospheric magnetic fields. The questions we ask here are: what flux of mechanical energy is required in flare star atmospheres? and is that flux available from the convection?

To answer the first question, we need to evaluate the non-radiative energy budget of the atmosphere both in its quiescent condition and also during flares. It is convenient to express the energy requirements of the various parts of the atmosphere as a ratio R_f of the total power output of the star.

Vilhu and Walter (1987) have examined a sample of F–M dwarfs and have determined radiative losses in UV lines from the chromosphere, in transition region (TR) lines, and in coronal X-rays. Expressed in terms of R_f, these quantities are found to span ranges of several orders of magnitude for stars of a given spectral class. However, there are apparently 'saturated' levels above which no star in the sample was found to lie. For the UV chromospheric lines, the TR lines, and the X-rays, the saturated values of R_f were found to be about 10^{-3}, 10^{-4}, and $10^{-2.5}$, respectively. In dMe stars, Balmer line emission may also contribute, making R_f as large as 10^{-3}–10^{-2}. There may also be significant mechanical energy deposited into the upper photosphere of M dwarfs. This could re-emerge as enhanced radiation in the H^- continuum or in the large number of weak spectral lines which are formed near the temperature minimum (Rutten et al., 1989): of the latter contributions to the mechanical energy budget we have no current knowledge. Hence, in quiescent conditions, the mechanical energy flux which must be supplied to the atmosphere of a dMe star almost certainly saturates at a value which is at least of order 1% of the total power of the star.

Estimates of the time-averaged power in flares are subject to large uncertainties. For example, the amount of emission in Lyman lines and continuum is a complete unknown, as is the amount of mechanical energy associated with mass ejections (Gershberg and Shakhovskaya, 1983). In view of the work of Kahler et al. (1988), the lack of information on mechanical energy is particularly serious. The maximum energy released in a stellar flare will remain uncertain until the mechanical energy can be evaluated. In optical light, the upper limits on flare energy in the Sun are about 10^{32} ergs, and in the stars, 10^{34}–10^{35} ergs. Mechanical energy requirements cause the solar value to be increased by a factor of about 10 (Webb et al., 1980). If the stellar flares have a comparable correction, then the maximum energy in a stellar flare may be of order 10^{36} ergs. Combining these results with the rate at which flares are observed to occur, we find that the mean power in stellar flares may be a few percent of the total stellar power. This result depends on the bolometric correction for flare light, and the averaged flare power may be as large as 10% of the total stellar power output (Mullan, 1977).

We, therefore, ask: is there enough mechanical energy in an M dwarf to amount to 1–10% of the total power output from the star and to provide up to 10^{36} ergs in the largest flares? We can summarize two different approaches to this problem, one based on a general discussion of convection, the other based on specific models of the stellar envelope.

First, according to standard models of convection, based on the Boussinesq approximation (in which the convecting gas is assumed incompressible except for the buoyancy effects), the flux of kinetic energy associated with convection is about 1% of the thermal flux, and is directed upwards (Mullan, 1984b). In fact, the mixing-length theory of convection assumes that the flux of kinetic energy in the convection zone is negligible: the only important flux is supposed to be the upward flux of heat. Hence, a 1% level for the kinetic energy flux is acceptable from a consistency point of view. But it barely suffices to supply the mechanical energy needs of the flare stars.

Second, McClymont and Fisher (1988) have evaluated the amount of mechanical energy available to drive solar flares in three different scenarios: (a) photospheric dynamo, (b) coronal storage, and (c) energy available in an erupting flux tube. To evaluate (a) and (b), they propose that mechanical energy can be supplied from the convection zone in the form of Alfvén waves (the flux rope acting as a conduit): the difference between (a) and (b) is that in (a), Alfvén waves emerge only over the course of the flare itself (assumed to last one hour), whereas in (b), the waves emerge over the course of one day prior to the flare and are stored in the corona. Therefore, they integrate the available mechanical energy flux in convection down to the level inside the convection zone from which Alfvén waves could have propagated to the surface in a time of (a) one hour, (b) one day. For the convection zone, they adopt a standard mixing-length model. They find that in (a), the available power suffices to power only the subflares, whereas in (b), flares with energies up to $(1-3) \times 10^{31}$ ergs can be powered, i.e., enough for a major flare. As for (c), the energy available in an active region of area 10^{19} cm^2 and field strength 1000 G is 10^{33} ergs, enough for the largest flares (including mechanical energy requirements). We have repeated their calculations for the case of two flare stars, one with mass 0.6 M(Sun) (see model in Schwarzschild, 1958, where again, a standard mixing-length model is assumed), and the second with a mass so small that the star is completely convective (for which we assume a polytrope with index 1.5). Then for (a), we find maximum powers of up to 10^{28} ergs s^{-1}, sufficient to power a very small stellar flare. For (b), the energy which can be tapped in one day is found to be 10^{34} ergs for the 0.6 M(Sun) star, and about 10^{36} ergs for the completely convective star (given a surface area of the flux tube of 10^{20} cm^2: Cram and Woods, 1982). The reason that these energies are larger than solar is a combination of the higher density in the deep convection zone and the larger surface area of the flux tube. As for (c), using the available magnetic flux (Saar et al., 1987), we find that total energies of 10^{36} ergs are available. Thus, it appears that, using one-dimensional mixing-length models of convection zones, the mechanical energy in the convection zone of lower main sequence stars is sufficient (but only just!) to power the largest stellar flares.

However, this conclusion depends on adopting a mixing length theory of convection, i.e., a one-dimensional model. But the convection zone is in fact composed of compressible gas, and this has been taken into account only recently in 3-D modelling of convection (Chan and Sofia, 1984). The 3-D calculations show that the upflows and downflows are very different in character. Because of the density stratification, downflowing gas becomes concentrated into strong plumes whereas the upflows spread out over large areas. The downdrafts are sites of high-speed flows: hence, when one takes a horizontal average over a plane in the convection zone, one finds that the flux of kinetic energy is no longer the negligible amount which had been assumed in mixing-length theory: the KE flux now rises to the startling value of about 50% of the upward heat flux. Even more surprising, the KE flux is directed *downwards*, and the heat flux must increase locally to 50% larger values to compensate for the downdrafts. Hence, in contrast to mixing-length theory, the flux of KE in stellar convection is by no means negligible compared with the heat flux.

This remarkable new view of the convection zone leads to new questions about the heating of flare star atmospheres. How does the KE flux which is mainly downward couple to stresses in the surface magnetic fields which give rise to magnetic activity? Can electrodynamic coupling still be used to estimate the coupling effects (Mullan, 1984b)? How do the strong downdrafts interfere with the buoyancy of magnetic flux ropes? Is it acceptable to use the mixing length model of the convection in calculating thermal shadowing effects of a flux tube? In a completely convective star, do the downdrafts extend all the way to the center of the star? (As to where exactly on the main sequence stars become completely convective, there are still uncertainties: the critical mass may be lower than previously suspected, cf. Cox *et al.*, 1981.) In a star with a radiative core, how do overshoots of the strong downdrafts affect the physical conditions at the core-envelope interface? Do they interfere with dynamo activity at the interface? Do they give rise to significant *g*-mode pulsations in the core? And finally, how will the estimates of available flare power (based on McClymont and Fisher (1988) arguments) be altered? (The total energy in erupting flux will not be altered.)

7. Kinetic Energy Flux: Mass Loss?

Although the kinetic energy flux *inside* the convection zone is of interest as the origin of mechanical energy (see Section 6), a more directly observable manifestation of KE flux *outside* the star is associated with mass loss. Here we ask the questions: do flare stars lose mass at a rate which is significant for the interstellar medium? and do flare stars necessarily lose mass via the same mechanism as the Sun?

In the Sun, the KE flux $F(KE)$ associated with mass loss (10^{12} g s^{-1} at speeds of 300 km s^{-1}) is about 10^{27} ergs s^{-1}. Compared with the flux of mechanical energy required to heat the solar corona ($F(Cor) = $ a few times 10^{28} ergs s^{-1}; cf. Holzer, 1980), the Sun diverts less than 10% of its coronal energy supply into kinetic form. To understand why this percentage is so small, we note that the mass loss from the Sun occurs mainly as a result of the gradient in thermal pressure: thus, mechanical energy emerging from the convection zone is first converted into disorganized form (i.e., heat), and then a steady organized flow is driven by the gradients in thermal pressure. This is an inherently inefficient method of organizing flow, and it is therefore not surprising that the ratio of $F(KE)$ to $F(Cor)$ is rather small in the Sun.

However, in the solar wind, there are also transients which may carry up to 10% of the mass flux of the wind: this figure is based on estimates from coronagraph data (Howard *et al.*, 1985), and refers only to transients with masses in excess of about 10^{14} g. No knowledge is currently available as to the contributions which small transients (masses less than 10^{14} g) make to the overall mass flux, although it is possible that the fraction may be large in coronal holes (cf. Holt and Mullan, 1987). The large transients detected by the coronagraphs are mostly due to filament eruptions: in these events, a magnetic configuration is somehow driven (presumably by convective pushing of the footpoints) to a condition where magnetostatic equilibrium can no longer exist. With the breakdown of equilibrium, the unbalanced magnetic forces drive outflows with high efficiency. Certain coronal transients were initially classified as being caused by flares rather than being associated with eruptive filaments. However, recent studies suggest that certain flares may actually be a response to a filament eruption (i.e., coronal transient), rather than the reverse (cf. Kahler *et al.*, 1988). In fact, the bulk of the energy release in a solar flare may not be in the visible flare at all, but rather in the coronal mass ejection (Webb *et al.*, 1980). It appears that most (if not all) coronal transients originate in a failure to find magnetic equilibrium. If this is so, then coronal transients rely directly on magnetic forces to drive mass loss.

The conclusions of Kahler *et al.* (1988) have an important implication for our study of flare stars: they suggest that for every stellar flare, there may be an associated mass ejection which actually contains most of the energy release. Now, in the Sun, there is a well-known anti-correlation between the sites of magnetic activity and the sites of mass loss: active regions contain mainly closed magnetic loops, and the associated averaged mass loss rate in transient activity is small, whereas solar mass loss occurs mainly via thermal expansion in magnetically quiet regions (coronal holes). This anti-correlation may have biassed our expectations of the mass loss process in other stars so much that we may have overlooked an important point: the Sun, from the viewpoint of magnetic activity, is a very poor specimen compared with many lower Main-Sequence stars.

As an indication that the magnetic activity on the Sun is at a low level compared with the levels on dMe stars, we may cite the magnetic fluxes on M stars: they are larger than solar by several orders of magnitude, with average fields stronger (up to 5–6 kG), and areal coverage factors much larger (60–90%) (Saar *et al.*, 1987). Hence, the fact that magnetically driven mass loss constitutes only a small fraction of the total mass loss from the Sun does *not* mean that the same will apply to flare stars. Let us explore the possibility that in fact, *mass loss from M dwarfs may be qualitatively distinct from the solar case*: in the M dwarfs, we speculate that the mass loss rate due to magnetic forces (i.e., coronal transients) may be much greater than the mass loss rate due to thermal driving (i.e., steady coronal expansion). Can we find support for this proposal?

Since the M dwarfs are as a whole much more magnetically active than the Sun, it seems likely that magnetically driven mass loss in flare stars will be of considerably greater significance than in the Sun. Thus, if equipartition of sorts exists in the coronae of flare stars, we may have $F(KE)$ of the same order as $F(Cor)$. And if reconnection is responsible for driving mass (see Waldron and Mullan, 1987), then $F(KE)$ may actually exceed $F(Cor)$: at a reconnection site, it is important to recognize that magnetic energy is converted directly into *kinetic* form, and the appearance of thermal energy is only secondary. (For a discussion of how kinetic energy may be converted ultimately to thermal energy in the context of a flare, see Bornmann, 1987.)

Suppose flare stars have $F(\text{KE}) = F(\text{Cor})$. Then since flare stars have $F(\text{Cor})$ up to 1% of the bolometric power, a star of spectral class, say dM4–5e, with $M_V = 11\text{–}12$ and $M_{\text{bol}} = 9\text{–}10$ will have $F(\text{KE})$ of order 4×10^{29} ergs s^{-1}, i.e., some 400 times the solar value. The wind speed is expected to be comparable to the escape speed v_e from the stellar surface: on the lower main sequence, where stellar radii scale almost exactly with stellar mass, v_e remains essentially unchanged. Hence, the mass loss from the above M dwarf is expected to be some 400 times solar, i.e., about 6×10^{-12} solar masses yr^{-1}.

The significance of mass loss from M dwarfs was pointed out by Coleman and Worden (1976). Since M dwarfs are the most numerous population in the galaxy (there are 10^{11} of them), they supply a significant amount of material to the interstellar medium (ISM) if each M dwarf loses on average 10^{-12} solar masses yr^{-1}. (O stars, Wolf–Rayet stars, and planetary nebulae all contribute about 0.1 solar masses yr^{-1} to ISM.) With the above estimates of mass loss from a dM4–5e star, it is apparent that in fact M dwarfs might be a significant (or dominant) contributor to ISM.

Is there any evidence for mass loss from cool dwarfs? As far as I know, there are no reports of detection of mass loss from individual M dwarfs. However, a K2 dwarf in the binary V471 Tauri has recently been found to be losing mass (Mullan *et al.*, 1988; Mullan *et al.*, 1989). The K2 dwarf does *not* overflow its Roche lobe: the Roche surface is larger than the stellar radius by a factor of about 40%. We are, therefore, talking of 'ordinary' mass loss from the corona of the K2 dwarf. Discrete absorption features have been detected in lines of Mg I, Mg II, Fe I, and Fe II in high-resolution IUE spectra. The discrete features are variable in strength on time-scales ranging from days to months. Analysis of the Mg II absorption allows us to derive a lower limit on the mass loss rate in the discrete features: it is 10^{-11} solar masses yr^{-1}. Moreover, the discrete absorption features are observed in ions which are formed at remarkably low temperatures (no more than a few times 10^4 K). In the solar wind, low temperatures are a characteristic signature of coronal transients (presumably because they are magnetically isolated from the effects of thermal conduction, and because of adiabatic cooling). The fact that the discrete features are observed to be time variable in V471 Tauri is also consistent with transient behavior. Arguments can be made that there is no significant mass loss rate in higher temperature gas (Mullan *et al.*, 1989): thus, the wind from the K2 dwarf in V471 Tauri seems to be dominated by coronal transient material, rather than by matter which has expanded from a thermal corona. If this conclusion can be substantiated, the wind from this star is quite different from the solar wind.

Now, the K2 dwarf in V471 Tauri is rotating rapidly (70–80 km s^{-1}), and might be expected to have strong magnetic fields. In fact, on the basis of observed period changes, the field at the base of the convection zone can be estimated (Applegate and Patterson, 1987): it is indeed large, almost 10^6 G. There are detectable flares (Young *et al.*, 1983), which is unusual in a star of such early spectral type: activity must be at a very high level in this star. The high level of magnetic activity indicates that magnetic flux loops on the surface of the K2 dwarf are frequently out of equilibrium: hence, the optical magnetic activity may be only the tail of a distribution of magnetic energy releases, with most of the energy being released as transient mass loss. The ratio of magnetically driven mass flux to coronal expansion in this star appears to be much higher than in the Sun.

Rapidly rotating K dwarfs have been reported also in the Pleiades: the rotational velocities are 100 km s^{-1} or more (Van Leeuwen and Alphenaar, 1983). The existence of such rapid rotation in young stars is a consequence of contraction along the evolutionary track towards the Main Sequence. Since the 'rotation-activity' connection applies equally to single stars and binary members (Basri, 1987), we expect that the rapid K rotators in the Pleiades will also lose mass at a rate some orders of magnitude greater than solar. In fact, the fastest rotating Pleiades K dwarf has Hα emission which suggests a mass loss rate of up to 10^{-9} solar masses yr^{-1} (Marcy et al., 1985).

As regards M dwarfs, the only evidence so far for mass loss occurs in cataclysmic variables. In these systems, the measured quantity is the mass capture rate by the white dwarf: presumably the mass loss rate by the red dwarf is larger than the transfer rate by a few orders of magnitude. In these systems, the transfer rates can be as large as 10^{-7} solar masses yr^{-1} (Patterson, 1984): but in those cases, Roche-lobe overflow is probably responsible for the extremely large mass loss rates. These results, therefore, tell us very little about mass loss from the corona of an individual M dwarf. But the argument presented above suggests that attention should be paid to the possibility that M dwarfs with efficient coronal heating may be important suppliers of mass to the ISM.

To summarize this section, we have proposed that the mass loss process in active stars may be dominated by magnetically driven transients, rather than by thermal expansion of a hot corona. Our proposal is at odds with what is known about the solar wind, but this is not necessarily critical: after all, the Sun is, from a magnetic standpoint, a comparatively inactive and uninteresting star. If, as we propose, mass loss from active dwarfs is dominated by magnetic loops which have lost equilibrium, then it is not necessary for all of the material on an erupting loop to be carried out in the wind: some of it may fall back to the surface of the star. This would give rise to red-shifted material in the spectrum: such material is indeed observed in the spectra of V471 Tauri (Mullan et al., 1989). There is also a report of both blue-shifted and red-shifted material in the surroundings of T Tauri stars (Mundt, 1984): these stars are also candidates for preferential mass loss by magnetic driving, since their coronal emission is frequently very weak, suggesting that thermally-driven expansion in these systems may contribute little to mass outflow.

8. Flare Energy Release: Reconnection Modelling

It is very likely that the release of energy in a solar/stellar flare is related to magnetic reconnection in some way. The question we ask here is: how is the process of magnetic reconnection to be modelled in a flare?

Starting with the initial work by Sweet and Parker in the 1950's, and until very recently, the modelling has been entirely in two dimensions. In 2-D, steady reconnection can occur in the vicinity of an X-type neutral point: magnetic islands form and saturate when they reach equilibrium, after which they evolve resistively. An extensive discussion of the various scenarios which have been proposed for steady 2-D reconnection has been provided by Forbes and Priest (1986).

Addition of fluid turbulence to 2-D reconnection has been modelled by Matthaeus and Lamkin (1985): in this case, local potential wells form in the flow and enable efficient electric field acceleration on the neutral line. In the context of solar flares, acceleration to energies of order 1 GeV is possible.

In reality, however, magnetic reconnection will occur as a 3-D process. In this case, magnetic islands may overlap, and if they do, the field lines will wander stochastically, so that no equilibrium is possible. If the overlapping islands happen to have opposite helicities, tearing mode turbulence (TMT) will occur. Spicer (1976) was the first to provide a discussion of 3-D effects and how they affect the energy release in solar flares. More recently, Strauss (1988) has used his 'reduced MHD equations' (which are written for the approximation of long thin flux tubes and strong axial fields) to obtain approximate numerical estimates of certain aspects of TMT in the context of solar flares and solar coronal heating.

The effects of TMT are to lead to relaxation of the current *gradients*, in contrast to the effects of resistivity, which lead to a relaxation of the currents themselves. In the presence of TMT, Ohm's law for the mean magnetic field includes not only the usual terms for induction electric field and Joule heating, but also a term representing diffusive decay of the current gradient: the 'diffusion coefficient' in this term Strauss refers to as 'hyperresistivity'. The energy of the mean field decreases with time not only because of Joule heating but also (and more especially, in solar coronal conditions) because energy is converted into magnetic and kinetic energy of TMT by the hyperresistivity. When the turbulence level becomes high enough, the effects of hyperresistivity dominate the resistive term and determine the growth rates of TMT. In this limit, Strauss finds an approximate expression for the hyperresistivity which is consistent with laboratory and simulation data. With this estimate, he finds that decay of magnetic energy due to hyperresistivity is more rapid than that due to turbulent resistivity by a factor of 10^9 in the solar corona. The heating which is produced by TMT is significantly larger than previous estimates of coronal heating rates in reconnection sites. Moreover, the onset of TMT leads to rapid expansion of the current sheet to a thickness of order ML (where M is the inflow Mach number, typically 0.03–0.3, and L is the length of the sheet). Hence, the volume of the coronal material which can participate in reconnection is no longer confined to a singular line, but is finite and large.

Further work remains to be done in the context of solar and stellar flares in order to determine how important the effects of hyperresistivity in fact are in helping us to understand the rapid release of magnetic energy in 3-D reconnection.

9. 3-D Geometry of Current Sheets

What is the geometrical structure of the flare site in a stellar atmosphere?

In the simplest model of 2-D reconnection, the initial release of flare energy occurs in a current sheet close to an X-type neutral line, and is, therefore, in principle confined to an infinitesimal volume. From that small volume, the release of energy must propagate elsewhere to ensure that a sufficiently large volume of the atmosphere contributes to the energy release which is observed. There may be a problem, however, with the propagation phase: not only must a certain (large) amount of energy be released in the flare, but it also must be released on a time-scale which is short (less than 1 s in stellar flares). The question is: are the transport properties of the solar/stellar atmosphere adequate to handle the rapid transfer of triggering information from the neutral line to the entire volume of flaring plasma which must be 'processed' if the total energy release of the flare is to be accounted for? The answer is that, in at least some cases, the plasma properties must be pushed to extremes to do this (Low and Wolfson, 1988).

It seems preferable, therefore, to imagine the reconnection site in terms of a 3-D structure from the beginning. (This is especially true now that detailed modelling of reconnection processes are being done in 3-D, cf. Strauss, 1988.) Rather than starting with the concept of an X-type neutral *line* as the site of initial energy release, the relevant entity in modelling flare physics should be the separatrix *surface*: this is the surface which partitions the magnetic field in the stellar atmosphere into flux cells. Each cell is distinguished by a unique field line connectivity, and when one crosses the separatrix surface, there is an abrupt reorientation of magnetic field vectors. The shape of the separatrix surface is determined by magnetic tension and pressure forces which arise when the photospheric motion causes the foot points of magnetic field lines to move in various directions: as long as the foot points remain rooted in a given flux cell (i.e., as long as global connectivity is preserved), a current sheet forms over the entire separatrix surface. The current sheet is formed even if there were no neutral points present in the initial field configuration (Low and Wolfson, 1988).

Lines along which two separatrices intersect are called separator lines: these are potential sites for initiation of reconnection and double layers (Baum and Bratenahl, 1980). However, the existence of current sheets over the entire surface of the intersecting separatrices is an important aspect of ensuring that a finite volume can be processed quickly.

Thus, from the point of view of modelling flares, it would be helpful to know the 3-D structure of separatrix surfaces and their intersections. Using a small personal computer, Baum and Bratenahl (1980) have provided an instructive example. Since that time, I would have anticipated that the availability of graphics packages and CAD/CAM routines for personal computers should have opened up a much more extensive vista on the shapes and geometry of separatrix surfaces. As far as I know, however, this has not happened in the astrophysical literature. In my opinion, it would be a worthwhile exercise to compile an atlas of representative surfaces and intersections for comparison with imaged flare data from (say) SMM.

10. Coronal Heating and Flares

Is there a continuum between coronal heating and flares? In terms of the separatrix surfaces discussed above, Low and Wolfson (1988) propose an answer to this question. They propose that if the magnetic free energy in a current sheet is small, then the currents can dissipate resistively, and Joule heating can occur in a non-explosive manner, thereby heating the 'quiet' corona. If, on the other hand, the magnetic free energy in the sheet is large, the sheet may go unstable by tapping into the free energy to drive tearing modes, leading to an explosive dynamical phase in which the bulk of the stored free energy is released rapidly. The latter case would appear as a flare.

Recent results by Machado *et al.* (1988) have a bearing on this suggestion. Machado *et al.* have surveyed a number of flares for which SMM provided images and for which magnetogram data were available. They find that in all cases, multiple loops were involved in the flares, with interactions between neighboring loops. The loops participating in the flares were classified in two categories, active and passive. The most pronounced energy releases occurred in loops which, according to the magnetograms, had been sheared prior to the flare, thereby building up a store of magnetic free energy: these were referred to as active loops. Other loops seemed to serve merely as repositories of energy injected from outside: these loops were found to have little free energy stored (according to the magnetograms), and were referred to as passive. Initially, energy release occurs in a single loop or at the interaction site between two loops. In the impulsive phase of the flare, the initiating loop and the impacted loop show strong brightenings simultaneously. Most of the total energy in the impulsive phase is released inside the initiating loop and/or inside one or more adjacent loops, rather than at the interaction site. Thus, interaction of loops is important for triggering a flare, but most of the energy released comes not from the triggering stie: rather, it comes from a reservoir throughout the loops.

These results seem to be very consistent with the suggestions of Low and Wolfson as far as flares in the Sun are concerned. However, we note that the theoretical question as to how magnetic energy is actually stored in coronal currents is not yet settled (cf. Chiueh and Zweibel, 1989). In the coronal heating problem, it may be equally important to consider not only the current sheets, but also concentrated vorticity structures: 3-D simulations of turbulent magnetofluids suggest that large amounts of kinetic energy are also dissipated in vortices (Dahlburg *et al.*, 1988).

Can flare stars help us in addressing the question of the connection between flaring and coronal heating? Several authors have suggested that coronal heating in flare stars is due to microflaring (e.g., Doyle and Butler, 1985; Skumanich, 1985; Katsova *et al.*, 1987) although these suggestions have been questioned by Ambruster *et al.* (1987).

In this regard, a further question arises on the basis of recent infrared data: namely, it appears that even in 'quiescent' coronal conditions, relativistic electrons may be present in the corona of dMe stars. To explain this claim, we note that a recent study of flare stars using the IRAS data (Mullan, Stencel, and Backman, 1988) has found that, of 74 flare stars observed by IRAS, 15 have been detected at a wavelength of

100 microns with fluxes which exceed the photospheric values by factors of 10^2–10^3. Moreover, the fluxes at 100μ exceed those at 60μ by a factor of at least 2–3 on average. Referring to work by Ohki and Hudson (1975) on possible infrared signatures of solar flares, the IRAS results suggest that a possible candidate for the infrared emission is synchrotron radiation. No other emission mechanism can cause a significant increase between 60 and 100μ. Now, the IRAS scans were made at random during the one-year lifetime of the satellite: each scan lasted no more than a few seconds, and a total of 10–20 were made on each star. Thus, it seems unlikely that IRAS would have 'caught' 20% of the stars in a flaring state. Instead, the IRAS data probably refer to quiescent conditions. If this is true, then there must be significant populations of relativistic electrons in the quiet coronae of flare stars. Hence, coronal heating in these stars must involve efficient acceleration of relativistic electrons.

The possibility that superthermal electrons are present in the quiescent coronae of flare stars has previously been discussed on the basis of microwave emission. How are such electrons accelerated? A current sheet model for accelerating a nonthermal population of electrons in quiescent flare star coronal has been formulated by Holman (1986): if the X-ray emitting coronal plasma is heated by current sheet dissipation, an electric field of about 3% of the runaway field accelerates sufficient nonthermal electrons to account for the observed microwave emission. However, it is not yet clear that the same population of electrons extends to the relativistic energies with sufficiently large fluxes to explain the observed infrared fluxes.

If, in fact, relativistic electrons are present in quantity in the corona of a flare star in its quiescent state, the hypothesis that broadband optical polarization may be due to synchrotron emission (Mullan, 1975) should be re-examined.

The presence of mildly relativistic electrons in the Sun's atmosphere outside flares has been discussed by Chiuderi Drago et al. (1987). Quantitative estimates were made of the number density of such electrons (10^{-4} times the ambient number density), and of the magnitude of the electric field required to do so. The electric field is found to be about 5% of the runaway value: this is comparable to the value estimated in flare stars by Holman (1986).

11. Flare Magnetic Fields: Origins?

The final question we ask here is: where do the magnetic fields which give rise to solar/stellar flares originate? Are they produced by a dynamo in the outer convection zone? As far as energetics are concerned, we have already seen (Section 6 above) that there is just enough mechanical power in convection to power the largest solar and stellar flares: the margin is uncomfortably small. Since flare energy release cannot be 100% efficient, it would have been preferable to have, say, one order of magnitude excess of mechanical energy: but our estimates suggest that the convection zone does not seem capable of supplying such an excess. This is not the only difficulty with a convection zone dynamo. (For a summary of difficulties encountered in the Sun, see the article entitled 'The Dynamo Dilemma' by Parker (1987).) In the case of the Sun, the following points can be made.

(a) Helioseismological data suggest that the angular velocity gradient in the convection zone has the wrong sign to make the simplest $\alpha\omega$-dynamo consistent with the observed equatorward drift of sunspots (Duvall and Harvey, 1984). (b) Large flares appear in favored longitudes which persist in solid-body rotation for 20–30 years (Bai, 1988): this suggests that the fields responsible for the large flares are not affected by latitudinal differential rotation (LDR). How deeply into the Sun does the observed LDR penetrate? Duvall *et al.* (1986) find that the surface LDR persists essentially unchanged throughout the entire convection zone. Bai's results, therefore, suggest that the flare fields may originate from below the convection zone. (c) Flare periodicity at about 152 days has now been confirmed in a variety of data sets, showing phase coherence through at least two solar cycles. Bai (1987) argues that the underlying cause of the periodicity involves in some way the entire Sun, although rotational beating of g-modes is not an acceptable explanation. (d) Dicke (1982) has argued that the temporal distribution of solar cycles does not show the random distribution of phases which one would expect if each cycle were due to the appearance of new flux erupting randomly through the surface. Instead, the timing of the solar cycles appears to be controlled by a high-Q oscillator deep inside the Sun. (e) R. Davis and collaborators have reported a possible anti-correlation between solar neutrino flux and sunspot number (e.g., Bahcall *et al.*, 1988): although the statistical significance is rather small, the existence of such a correlation would indicate that surface fields are somehow coupled to the innermost core of the Sun.

In the case of stellar dynamos, it should be pointed out that observations of the Ca K fluxes are now available for about 100 stars ranging in spectral type from F to M. (The sample contains mostly G and K stars: there is only one M dwarf.) Main-Sequence stars in this range show a very large range in the properties of the convection zone, with Rossby numbers spanning a range of about 30. However, when the cycle periods are plotted as a function of Rossby number, they exhibit no systematic behavior whatever (Baliunas, 1986). It seems natural to expect that if the convection zone is the seat of the stellar dynamo, there ought to be a clearly discernible trend in the period as the convection zone properties vary so dramatically along the Main Sequence. Yet such a trend is not apparent in the cycle periods which are currently available.

On the basis of these points, we are led to ask: could the flare fields in the Sun and stars originate in a region other than the convection zone? For example, the core of the Sun may have a large magnetic field (up to 10^8 G) without violating any observational limits on surface oblateness or neutrino fluxes. Dicke (1979) has shown that stars of solar type (i.e., having radiative cores) can have stable magnetic cores provided that the rotation is fast enough to stabilize gravitational perturbations. (The fields must be contained in the nuclear generation regions so that gradients of molecular weight are present.) Perhaps the solar cycle may be modelled as an oscillation of some kind in this core field.

Some years ago (1949–1979), various suggestions were made ascribing the solar cycle to the solar core, but so far, no quantitative modelling of any such model has been produced (cf. Parker, 1987). The recent emergence of arguments against the simplest

$\alpha\omega$-dynamo models in the Sun and flare stars may perhaps spur research activity in such alternative directions. For example, the persistent flare longitudes (Bai, 1988) may reflect preferred axes of the core magnetic field. A displacement of the core field slightly from the equatorial plane could explain why north-south asymmetries are observed in solar activity: thus, of 850 flares observed in 1975, northern hemisphere flares predominate in all categories of flare grouping at greater than 99% confidence level (Wilson, 1987), and solar rotation also shows a long-lasting north-south asymmetry (Bieber, 1988). More than one symmetry axis in the solar core is required to account for distortions in the solar surface (Dicke, 1982), perhaps indicating the presence of a quadrupole field in the core. The presence of a strong field in the core would help to explain why the properties of the Sun along the rotation axis are discernibly different from those in the equatorial plane (Duvall et al., 1986). And if a strong field in the nuclear-energy generating core of the Sun undergoes an oscillation of some sort, the change in local thermodynamic quantities may help to explain the solar-cycle dependence of neutrino emission (Bahcall et al., 1988).

From this viewpoint, we would expect that, as long as a radiative core exists in a lower Main-Sequence star, the core field would survive and drive activity, independent of the properties of the convection zone (as observed). Only when the core disappears altogether would we need to switch to a dynamo action rooted in the convection zone itself. However, we note that the observational signatures, if any exist, of an alteration in emission characteristics at the onset of complete convection are, to say the least, confusing at the present time. For example, the distribution of X-ray luminosity appears to change markedly at $R - I = 1.3$, corresponding to spectral types M2–M3 (Bookbinder et al., 1986), whereas X-ray variability amplitude does not seem to undergo a transition until the spectral class is as late as, or later than, M5 (Ambruster et al., 1987). On the other hand, the chromospheric properties of M dwarfs, as seen in Hα, do not appear to undergo any abrupt changes in nature at the transition to complete convection (Giampapa and Liebert, 1986).

Finally, since a critical rotation must be exceeded to stabilize the magnetic core (Dicke, 1979), we propose that the distinction between dMe stars and dM stars (which rotate faster and slower than 5 km s^{-1}, respectively, cf. Bopp et al., 1981) may give rise to a novel 'rotation-activity connection': we suggest that the dMe stars have been successful in retaining their strong magnetic cores because their rotational velocities are large, whereas the dM stars have been unable to retain such cores.

Acknowledgements

I am indebted to Drs B. Haisch and M. Rodono for making it possible for me to attend this Colloquium. This work has been supported partially by NASA Grant NAGW-1295.

References

Ambruster, C. et al.: 1987, Astrophys. J. Suppl. **65**, 273.
Applegate, J. and Patterson, J.: 1987, Astrophys. J. **322**, L99.

Bahcall, J. *et al.*: 1988, *Nature* **334**, 487.

Bai, T.: 1987, *Astrophys. J.* **318**, L85.

Bai, T.: 1988, *Astrophys. J.* **328**, 860.

Baliunas, S.: 1986, in M. Zeilik and D. Gibson (eds.), *Cool Stars, Stellar Systems, and the Sun*, Springer-Verlag, Berlin, p. 3.

Basri, G.: 1987, *Astrophys. J.* **316**, 377.

Bastian, T. and Bookbinder, J.: 1987, *Nature* **326**, 678.

Baum, P. and Bratenahl, A.: 1980, *Solar Phys.* **67**, 245.

Bieber, J.: 1988, *J. Geophys. Res.* **93**, 5903.

Bookbinder, J. *et al.*: 1986, in M. Zellik and D. Gibson (eds.), *Cool Stars, Stellar Systems, and the Sun*, Springer-Verlag, Berlin, p. 97.

Bopp, B. and Moffett, T.: 1973, *Astrophys. J.* **185**, 239.

Bopp, B. *et al.*: 1981, *Astrophys. J.* **249**, 210.

Bornmann, P.: 1987, *Astrophys. J.* **313**, 449.

Brown, J. C.: 1971, *Solar Phys.* **18**, 489.

Canfield, R. C. *et al.*: 1984, *Astrophys. J.* **282**, 296.

Canfield, R. C. *et al.*: 1987, *Nature* **326**, 165.

Chan, K. and Sofia, S.: 1986, *Astrophys. J.* **307**, 222.

Chiuderi Drago, F. *et al.*: 1987, *Solar Phys.* **112**, 89.

Chiueh, T. and Zweibel, E.: 1989, *Astrophys. J.* (in press).

Coleman, G. and Worden, S. P.: 1976, *Astrophys. J.* **205**, 475.

Cox, A. *et al.*: 1981, *Astrophys. J.* **245**, L37.

Cram, L. and Woods, T.: 1982, *Astrophys. J.* **257**, 269.

Dahlburg, R. *et al.*: 1988, *Astron. Astrophys.* **198**, 300.

Dicke, R.: 1979, *Astrophys. J.* **228**, 899.

Dicke, R.: 1982, *Solar Phys.* **78**, 3.

Doyle, J. and Butler, J.: 1985, *Nature* **313**, 378.

Duvall, T. and Harvey, J.: 1984, *Nature* **310**, 19.

Duvall, T. *et al.*: 1986, *Nature* **321**, 500.

Fisher, G.: 1987, *Solar Phys.* **113**, 307.

Forbes, T. and Priest, E.: 1986, *J. Geophys. Res.* **91**, 1289.

Gershberg, R. and Shakhovskaya, N.: 1983, *Astrophys. Space Sci.* **95**, 235.

Giampapa, M. and Liebert, J.: 1986, *Astrophys. J.* **305**, 784.

Grinin, V. P. and Sobolev, V. V.: 1989, in B. M. Haisch and M. Rodonò (eds.), IAU Colloq. 104, *Solar and Stellar Flares*, Poster Papers, Publ. Catania Astrophys. Obs., Special Volume, p. 297.

Haisch, B. M.: 1983, in P. B. Byrne and M. Rodono (eds.), *Activity in Red Dwarf Stars*, D. Reidel Publ. Co., Dordrecht, Holland, p. 255.

Hamilton, R. J. and Petrosian, V.: 1987, *Astrophys. J.* **321**, 721.

Holman, G.: 1986, in M. Zeilik and D. Gibson (eds.), *Cool Stars, Stellar Systems, and the Sun*, Springer-Verlag, Berlin, p. 271.

Holt, R. D. and Mullan, D. J.: 1987, *Solar Phys.* **107**, 63.

Holzer, T.: 1980, in A. Dupree (ed.), *Smithsonian Astrophys. Obs. Spec. Rep.*, No. 389, p. 165.

Howard, R. *et al.*: 1985, *J. Geophys. Res.* **90**, 2314.

Kahler, S. *et al.*: 1988, *Astrophys. J.* **328**, 824.

Katsova, M. *et al.*: 1987, *Astron. Zh.* **64**, 1243.

Kuijpers, J.: 1985, in R. M. Hjellming and D. M. Gibson (eds.), *Radio Stars*, D. Reidel Publ. Co., Dordrecht, Holland, p. 23.

Lang, K. R.: 1983, in P. B. Byrne and M. Rodono (eds.), *Activity in Red Dwarf Stars*, D. Reidel Publ. Co., Dordrecht, Holland, p. 331.

Lang, K. R. *et al.*: 1987, *Astrophys. J.* **322**, 1044.

Louarn, P. *et al.*: 1987, *Solar Phys.* **111**, 20.

Low, B. C. and Wolfson, R.: 1988, *Astrophys. J.* **324**, 574.

Machado, M. *et al.*: 1988, *Astrophys. J.* **326**, 425.

Marcy, G. *et al.*: 1985, *Astrophys. J.* **288**, 259.

Matthaeus, W. and Lamkin, S.: 1985, *Phys. Fluids* **28**, 303.

McClymont, A. and Fisher, G.: 1988, in *Outstanding Problems in Solar System Plasma Physics: Theory and Instrumentation*, Proc. of Yosemite 1988 Conference.

Melrose, D. B. and Dulk, G. A.: 1982, *Astrophys. J.* **259**, 844.

Mullan, D. J.: 1975, *Astrophys. J.* **201**, 630.

Mullan, D. J.: 1976, *Astrophys. J.* **207**, 289.

Mullan, D. J.: 1977, *Solar Phys.* **54**, 183.

Mullan, D. J.: 1984a, *Astrophys. J.* **279**, 746.

Mullan, D. J.: 1984b, *Astrophys. J.* **282**, 603.

Mullan, D. J.: 1985, in R. M. Hjellming and D. M. Gibson (eds.), *Radio Stars*, D. Reidel Publ. Co., Dordrecht, Holland, p. 173.

Mullan, D. J., Stencel, R. E., and Backman, D.: 1989, *Astrophys. J.* (in press, August 1989 issue).

Mullan, D. J. *et al.*: 1988, in E. J. Rolfe (ed.), *A Decade of UV Astronomy with the IUE Satellite*, ESA SP-281, Paris, Vol. 1, p. 423.

Mullan, D. J., Sion, E. M., Bruhweiler, F. C., and Carpenter, K. G.: 1989, *Astrophys. J.* **339**, L33.

Mundt, R.: 1984, *Astrophys. J.* **280**, 749.

Ohki, K. and Hudson, H.: 1975, *Solar Phys.* **53**, 405.

Parker, E. N.: 1987, *Solar Phys.* **110**, 11.

Parker, E. N.: 1988, *Astrophys. J.* **330**, 474.

Patterson, J.: 1984, *Astrophys. J. Suppl.* **54**, 443.

Pettersen, B.: 1989, *Astron. Astrophys.* **209**, 279.

Reale, F. *et al.*: 1988, *Astrophys. J.* **328**, 256.

Rutten, R. G. M. *et al.*: 1989, *Astron. Astrophys.* (in press).

Saar, S. *et al.*: 1987, in *27th Liège Int. Astrophys. Colloquium*.

Schmitt, J. *et al.*: 1987, *Astrophys. J.* **322**, 1033.

Schneeberger, T. *et al.*: 1979, *Astrophys. J.* **231**, 148.

Schwarzschild, M.: 1958, *Structure and Evolution of the Stars*, Princeton Univ. Press, Princeton, p. 257.

Shakhovskaya, N. I.: 1979, *Izv. Krymsk. Astrofiz. Obs.* **60**, 24.

Simnett, G. M.: 1989, in B. M. Haisch and M. Rodonò (eds.), IAU Colloq. 104, *Solar and Stellar Flares*, Poster Papers, Publ. Catania Astrophys. Obs., Special Volume.

Skumanich, A.: 1985, *Australian J. Phys.* **38**, 971.

Spicer, D. S.: 1976, *NRL Report* 8036.

Strauss, H.: 1988, *Astrophys. J.* **326**, 412.

Strong, K. T. *et al.*: 1986, in M. Kundu and B. Woodgate (eds.), *Energetic Phenomena on the Sun*, NASA Conf. Proc. CP-2349, pp. 5–38.

Tarter, C. and Mullan, D.: 1977, *Astrophys. J.* **212**, 179.

Van den Oord, G.: 1987, 'Stellar Flares', doctoral dissertation, Utrecht, p. 108.

Van Leeuwen, F. and Alphenaar, F.: 1983, in P. B. Byrne and M. Rodono (eds.), *Activity in Red Dwarf Stars*, D. Reidel Publ. Co., Dordrecht, Holland, p. 189.

Vilhu, O. and Walter, F.: 1987, *Astrophys. J.* **321**, 958.

Waldron, W. and Mullan, D. J.: 1987, *Astrophys. J.* **319**, 971.

Webb, D. *et al.*: 1980, in P. Sturrock (ed.), *Solar Flares*, Colorado Assoc. Univ. Press, Boulder, p. 471.

Wilson, R.: 1987, *NASA Tech. Paper* 2714.

Winglee, R.: 1985, in R. Hjellming and D. Gibson (eds.), *Radio Stars*, D. Reidel Publ. Co., Dordrecht, Holland, p. 49.

Young, A. *et al.*: 1983, *Astrophys. J.* **267**, 655.

THE IMPORTANCE OF SOLAR WHITE-LIGHT FLARES

DONALD F. NEIDIG

Air Force Geophysics Laboratory, National Solar Observatory/Sacramento Peak, National Optical Astronomy Observatories*, Sunspot, NM 88349, U.S.A.

Abstract. The basic results of white-light flare (WLF) photometric and spectrographic observations are reviewed. WLFs represent the most extreme density conditions in solar optical flares and are similar to stellar flares in many respects. It is shown that WLFs originate in the low chromosphere and upper photosphere, and that their huge radiative losses remain difficult to explain within the context of known mechanisms of energy transport.

1. Introduction

Solar white-light flares (WLFs) are defined as the components of flares that are visible in optical continuum or integrated light. WLF emission appears as patches, waves, or ribbons, often containing smaller (< 3 arc sec) bright kernels (see descriptions and additional references in Neidig and Cliver, 1983a; and Canfield *et al.*, 1986). When observed with small aperture patrol telescopes at $\lambda < 4000$ Å, WLFs occur at a rate ≈ 15 per year near solar maximum (Neidig and Cliver, 1983b). Basic properties of WLFs, including their associated emissions, morphology, apparent lack of polarization, and the solar active regions in which they occur, are summarized in a catalog of events observed since 1859 (Neidig and Cliver, 1983a). Except for the appearance of continuum, WLFs bear no particular morphological or spectral distinction, nor are their associated emissions at X-ray or radio wavelengths unusual in any respect. WLFs are, however, among the more energetic solar flares, and it has been shown that optical continuum becomes visible whenever the flare EUV or soft X-ray luminosity exceeds a relatively large threshold (McIntosh and Donnelly, 1972; Neidig and Cliver, 1983b). Thus optical continuum is probably present in all flares, but attains a detectable level of brightness in relatively few cases. This conclusion carries with it the corollary that WLFs are not fundamentally different from ordinary flares. Nevertheless, WLFs are of great importance in flare research because they are similar in many respects to stellar flares (cf. Worden, 1983) and because they represent the most extreme conditions attained in solar optical flares, thereby presenting a major challenge to atmospheric models and energy transport mechanisms. The following sections review the spectral characteristics, energetics, and physical conditions in WLFs; these topics then lead to

* Operated by the Association of Universities for Research in Astronomy, Inc., under contract with the National Science Foundation. Partial support for the National Solar Observatory is provided by the USAF under a Memorandum of Understanding with the NSF.

a final discussion on the consequences of WLFs to mechanisms of energy transport in highly energetic flares.

2. Gross Spectral Characteristics and Energetics

WLF photometric data are usually obtained from photographic images in broad (10–100 Å) spectral bands at several different wavelengths. Ordinarily the spectral bands are chosen to avoid strong chromospheric flare lines. However, numerous narrow absorption lines formed in the photosphere and low chromosphere are not excluded from these broad bandpasses, and, although relatively few of these lines show emission in WLFs, the photometric data cannot in any case be considered to be pure continuum. Figure 1 shows the WLF contrast, $(I_f - I_0)/I_0$, where I_0 is the intensity of the solar background, for the brightest kernels in four events. The spectra are relatively flat for $\lambda > 4000$ Å, but show a marked increase at $\lambda < 4000$ Å due, in part, to the reduced brightness of the solar background at short wavelengths. Nevertheless, in terms of absolute intensity $(I_f - I_0)$ the WLFs in Figure 1 still average approximately three times brighter at 3600 Å than at 5000 Å. Thus solar WLFs are 'blue', as are stellar flares. The increase in brightness at short wavelengths is attributed to: (1) the presence of Balmer continuum in some cases, (2) the higher temperature of the flare relative to the quiet Sun, (3) the merging of Balmer lines near the Balmer limit (Donati-Falchi, Falciani, and

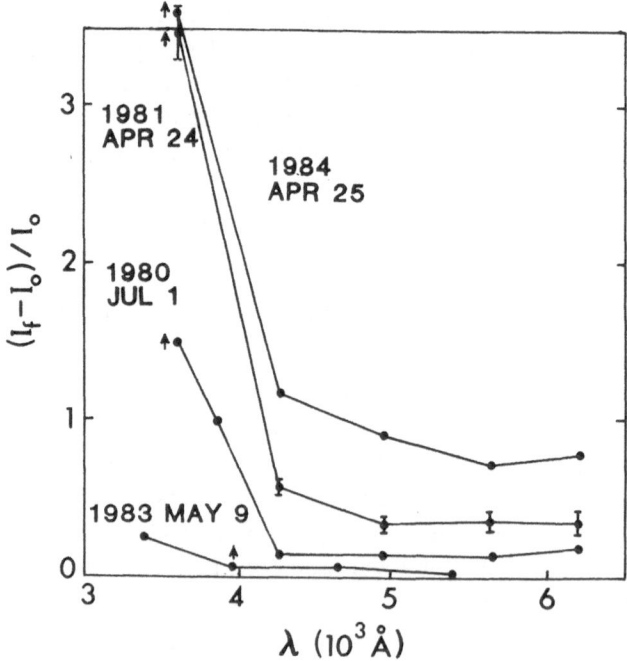

Fig. 1.   Peak contrasts in the brightest kernels of four WLFs, measured in broad (20–50 Å) bands. Arrows identify data points that are lower limits. The 1981 April 24 flare has the most reliable spectral shape. Except for the 1984 April 25 event, data are from Canfield *et al.* (1986).

Smaldone, 1986), and (4) emission in photospheric and low chromospheric lines that become progressively more numerous at short wavelengths.

Peak surface fluxes, luminosities, and total energy radiated in the continuum of WLFs are summarized in Table I. These data assume that the WLF spectrum spans a range $2500-10^4$ Å (Kane *et al.*, 1985; Canfield *et al.*, 1986). It should be noted that even the largest WLFs (Table I) have luminosities only $\approx 10^{-3}$ of the largest flares on dMe stars (cf. Gurzadyan, 1980). With regard to understanding the heating mechanisms of WLFs the most important datum is the peak flux, which, in the 1984 April 25 event, attained a value equivalent to a 100% enhancement over the quiet Sun. As shown in Section 4, the large fluxes observed in WLFs are not easily explained within the context of known mechanisms of flare energy transport.

TABLE I

Summary of WLF energetics

	Flux in bright kernel ($\mathrm{erg\ s^{-1}\ cm^{-2}}$)	Luminosity ($\mathrm{erg\ s^{-1}}$)	Total energy (erg)
Typical WLF[a]	$1-2 \times 10^{10}$	$10^{27}-10^{28}$	10^{30}
Largest WLF[b] (1984 April 25)	6×10^{10}	2×10^{29}	$> 3 \times 10^{31}$

[a] Data from Neidig and Cliver (1983a).
[b] Sacramento Peak data, not published elsewhere.

The ratio of the WLF continuum flux, F_{cont}, to the flux in emission lines at flare maximum may be similar to that found in stellar flares. For example, in the 1981 April 24 event (Neidig, 1983) $F_{cont}/F_{H\alpha} = 60$, which, after allowing a factor of 6 for the flux in emission lines other than $H\alpha$, leads to the conclusion that approximately 90% of the total flux was in continuum – a value that compares well with stellar flares (Gurzadyan, 1980). However, due to the larger area and longer duration of the $H\alpha$ emission the total energy radiated in $H\alpha$ in the WLF can be comparable to that radiated in continuum (Slonim and Korobova, 1975).

3. Physical Conditions Derived from Spectral Analysis

Spectral analysis of WLFs is made difficult by (1) insufficient spatial and temporal resolution, (2) terrestrial atmospheric spectral dispersion, (3) improper placement of the spectrograph slit (the brightest kernels of WLFs have not yet been observed spectrographically!), and (4) inaccurate subtraction of the background solar spectrum (which is spatially irregular and often nearly as bright as the flare itself). In short, the relative nearness of the Sun has not afforded all possible advantages to WLF observations. Nevertheless, several important results have been obtained.

One significant discovery is the presence of Balmer continuum (Hiei, 1982; Neidig, 1983; Donati-Falchi, Falciani, and Smaldone, 1984; Neidig and Wiborg, 1984). The electron density, N_e, in the Balmer-emitting layers, as measured from Stark broadening of the high Balmer lines, was found to be $3\text{--}5 \times 10^{13}\,\text{cm}^{-3}$ for several bright events. Measurements of the optical thickness in the Balmer continuum and wings of low Balmer lines in the 1981 April 24 WLF (Neidig, 1983) yielded column densities of second-level hydrogen atoms $N_2 z = 1\text{--}3 \times 10^{16}\,\text{cm}^{-2}$. Usually, the line spectra of WLFs that show Balmer continuum are quite similar to spectra of bright non-WLFs; this suggests that the canonical temperature ($\approx 10^4\,\text{K}$) of the optical flare applies also to WLFs, with the only essential difference being the somewhat larger values of N_e and $N_2 z$. Allowing for uncertainties in flare temperature and departures from LTE, the geometric thickness, z, of the April 24 WLF is probably 25–250 km. The April 24 event also shows a Paschen jump (Neidig and Wiborg, 1984; absence of a Paschen jump in other WLFs has not been confirmed), indicating that the continuum at visible wavelengths in this flare is dominated by recombination (H_{fb}) emission (thus retracting the original interpretation of H^- emission by Neidig, 1983). This conclusion is corroborated by the observed absence of emission in the cores of metallic lines formed below 600 km in the atmosphere (Neidig, 1986a).

Two WLFs are known in which the Balmer continuum is apparently absent (Machado and Rust, 1974; Boyer et al., 1985); instead, the continuum in these events shows a monotonic increase toward short wavelengths. Machado and Rust (1974) attributed the continuum to H_{fb} emission, although the absence of the Balmer jump presents a difficulty for this interpretation. In the event studied by Boyer et al. (1985) H_{fb} emission was ruled out on the basis of the extremely weak Balmer lines; instead, the authors argued for H^- emission originating in a slightly warmed upper photosphere (evidence for strong heating near the quiet Sun $\tau_{5000} = 1$ level has not been found in any WLF). Thermal bremsstrahlung from a hot ($T > 10^5\,\text{K}$) plasma seems unlikely in such events, as it is incompatible with observed fluxes in soft X-rays and EUV. Even temperatures in the range $2 \times 10^4\text{--}10^5\,\text{K}$ may not contribute significantly to the continuum, as helium lines in WLFs generally indicate small emission measures in this temperature range (Lites, Neidig, and Bueno, 1986).

Boyer et al. (1985) noted the possibility that the two, relatively faint WLFs without Balmer jumps – the spectrograms of which were obtained in the post-impulsive, or gradual, phase of the flare – might represent a phenomenon different from the bright continuum events associated with the impulsive phase (see also Machado and Rust, 1974; Rust, 1986). Machado et al. (1986) formally proposed the existence of two extreme types of WLF: Type I, showing strong Balmer lines and H_{fb} continuum originating in the chromosphere, and Type II, showing weak Balmer emission and H^- continuum originating in the upper photosphere. Spectrograms of WLFs studied by Hiei (1982) and Hiei et al. (1982) would indicate a mixture of both types, as might be true of all WLFs. In this case, when the relative size of the Balmer jump and the temperature are known, it becomes possible to separate the contributions of H_{fb} and H^- emission at wavelengths longward of the Balmer jump if some assumption can be made on the flare optical thickness. In general, however, the relative size of the Balmer jump remains a function of the flare optical thickness when the condition $\tau_{cont} \ll 1$ does not hold (Grinin and Sobolev, 1977). This is an especially important point because large optical thicknesses in flares have been shown to be disguised by filling factors (Švestka, 1972), and the same is apparently true for WLFs as well (Neidig, 1986b).

Precise interpretation of the line and continuum spectra of WLFs requires self-consistent models that can properly simulate the effects of unresolved structure as well as the distributions of mass, temperature, and velocity in the flare atmosphere. Unfortunately, the majority of available models (see references above) are very approximate, often assuming homogeneous, isothermal slabs. Even non-LTE, semi-empirical models (e.g., Avrett, Machado, and Kurucz, 1986; Gan and Fang, 1988), which are able to reproduce line and some continuum features of WLFs, presently suffer from the assumption of hydrostatic equilibrium, which effectively guarantees insufficient loading on the flare chromosphere to produce the emission measures necessary to explain bright continuum. High-pressure equilibrium, with $NT \geq 10^{18}$ cm^{-3} K everywhere within the flare loop above the photosphere, might be a more appropriate initial assumption for models.

4. Energy Transport Mechanisms

A conventional (albeit unproven) assumption of flare physics is that all of the flare optical emission is powered by energy released in the corona and subsequently transported to the high density ($N > 10^{12}$ cm^{-3}) regions. WLFs are important in this context because their large radiative fluxes place severe constraints on known mechanisms of energy transport. Several transport mechanisms are summarized below.

4.1. HEAT CONDUCTION

Assuming Spitzer conductivity and applying the analysis of Shmeleva and Syrovatskii (1973) under conditions of either constant density or constant pressure (the latter seems more likely to apply), it is found that, in order to sustain fluxes exceeding 10^{10} erg s^{-1} cm^{-2}, the temperature gradient must be so steep that the emission measure at $T < 10^5$ K will be too small to radiate the observed WLF flux. The predicted radiative flux in the constant pressure case is a factor $\approx 10^3$ too small. Heat conduction seems unlikely as a power source in stellar flares as well, unless the ratio of flare area to flare luminosity is much larger than in the solar case.

4.2. ELECTRON BEAMS

If solar flare hard X-ray bursts are interpreted according to a 'thick target' model (see Hudson, 1972; Švestka, 1976, and references therein), then the total power residing in nonthermal electrons with energies $E \geq 50$ keV has been shown in several cases to be sufficient to power WLFs in the chromosphere (Kane et al., 1985; Canfield et al., 1986). Electrons with $E \geq 50$ keV are thermalized in the solar atmosphere at densities exceeding $10^{13.5}$ cm^{-3}, which is appropriate for chromospheric WLFs. Moreover, the time profiles of the hard X-ray and white light emissions are approximately similar (Rust and Hegwer, 1975; Canfield et al., 1986), and this appears to be true regardless of whether the emissions are impulsive or gradual in nature (Kane et al., 1985). The 1980 July 1

WLF (Zirin and Neidig, 1981) is a possible exception to electron beam heating, as Canfield *et al.* (1986) have demonstrated that the column density over which the electrons deposited their energy was insufficient to produce the optical continuum. The latter calculation, however, did not make allowance for the full range of column densities appropriate to the range of electron energies that could power the WLF. Therefore, within the limits of various uncertainties, it may be premature to rule out electron beam heating in this WLF. On the other hand, the effects of reverse currents in reducing the effectiveness of beam heating at large densities have not been included in any of the calculations above. With regard to heating the upper photosphere during WLFs, electron beams are hopelessly inadequate, as electron energies ≥ 900 keV are necessary for penetration to these depths, and the total power carried by these electrons is a factor $\geq 10^2$ too small.

A variation of the electron beam model (Livshits *et al.*, 1981; MacNeice *et al.*, 1984) proposes energy transport in two steps: explosive heating of the upper chromosphere by an electron beam (or any other source that produces explosive heating) generates a downward-propagating compression wave that heats the deeper atmosphere and produces the WLF. The expected time delay for the hydrodynamic response is on the order of 10 s, which is not resolved in the WLF/hard X-ray profiles presented in Canfield *et al.* (1986).

4.3. HIGH-ENERGY (>4 MeV) PROTON BEAMS

Quantitative analysis of gamma-ray and white light observations has been published for only one WLF (1980 July 1 (Ryan *et al.*, 1983)), with the conclusion that thick target heating by high-energy protons was energetically insufficient at the time of WLF maximum. Ryan *et al.* (1983) concluded also that during the impulsive phase (which occurred well before the white light maximum was attained) the July 1 flare could have been powered by protons, but the same is true for electrons as well (Canfield *et al.*, 1986). The July 1 event, although not spectrally or morphologically unusual, is the only known WLF whose time profile differs significantly from that of hard X-rays – a fact that might be noted when using this single case to rule out high-energy proton heating in general.

4.4. LOW-ENERGY PROTONS

Simnett (1986) proposed that low energy protons, accelerated by a series of small shocks, are the major energy carriers in the impulsive phase. These protons are stopped at relatively low density in the upper chromosphere, resulting in explosive heating that ultimately produces the WLF by a downward-propagating compression wave. A consequence of this model is that the hard X-rays are produced thermally. Starr *et al.* (1988) argue that thermally-produced hard X-ray bursts might be observationally difficult to distinguish from those produced by nonthermal electrons. Thus, aside from the disadvantage that low-energy protons produce no unique signature of their own, the low-energy proton/hydrodynamic model may suffer no observational contradictions; its evaluation as a transport mechanism in WLFs awaits quantitative predictions relating hard X-ray and white light emissions.

4.5. IRRADIATION BY 1–8 Å X-RAYS

At peak WLF luminosity the power in 1–8 Å X-rays directed toward the solar surface is typically only 10^{-1} of the power radiated by the WLF, and in no case has the 1–8 Å power equaled or exceeded the peak WLF luminosity. This includes even the 1980 July 1 WLF which peaked nearly simultaneously with the soft X-ray event. Ordinarily, the 1–8 Å X-rays peak 1–2 minutes after the WLF maximum.

4.6. IRRADIATION BY 10–1030 Å (EUV) EMISSION

WLFs are associated with large EUV fluxes (McIntosh and Donnelly, 1972), corresponding to EUV luminosities in the range $6–23 \times 10^{27}$ erg s^{-1}. Furthermore, it follows from the known temporal correlations between the WLF and hard X-rays and between hard X-rays and EUV, that the timing of the WLF and EUV emission must be similar as well. It can be speculated, therefore, that strong EUV emission (generated by an electron beam or other energy source in the upper chromosphere) might irradiate the deeper layers of the atmosphere and produce the WLF. A fundamental question, however, is whether EUV photons can attain sufficient range in an ambient atmosphere that is ordinarily opaque to most EUV radiation. This difficulty has been expressed recently by Poland, Milkey, and Thompson (1988).

4.7. ALFVÉN WAVES

The dissipation of Alfvén waves in regions of high resistivity was proposed by Emslie and Sturrock (1982) as a means of explaining the warming of the temperature minimum region in ordinary flares (cf. Machado, Emslie, and Brown, 1978). The calculations of Emslie and Sturrock (1982) were aimed at energy deposition rates ≈ 10 erg s^{-1} cm^{-3}, and made no attempt to account for rates $\approx 10^3$ erg s^{-1} cm^{-3} as would apply in WLFs. Delays of several seconds or more between the hard X-ray emission and the WLF are expected as a result of the finite velocity of the waves.

5. Conclusions

A wide variety of observational data and modeling indicates that (1) WLFs are not fundamentally different from ordinary flares, (2) WLFs are located in the lower chromosphere and upper photosphere, with no evidence for strong heating near the quiet Sun $\tau = 1$ level, (3) the WLF light curve approximately follows the hard (≈ 50 keV) X-ray emission regardless of whether the hard X-rays are associated with the impulsive or gradual phase of the flare, and (4) heat conduction, irradiation by soft X-rays, and heating by high-energy protons are not sufficient to power the WLF at the time of its peak luminosity. Heating by an electron beam, which might be important in chromospheric WLFs, is totally inadequate for the upper photospheric component of WLFs.

WLFs show considerable similarity to stellar flares, with one notable exception being the energetics. If solar and stellar flares can be described by models that are qualitatively similar, then the greater luminosity and total energy of stellar flares must ultimately derive from larger magnetic field energies and faster rates of conversion of the field energy into other forms. Observationally, these questions are probably less accessible than the simpler problem of understanding how the stellar flare atmosphere can sustain optical radiative losses as much as 10^3 times greater than the largest solar WLFs. In answer to the latter it is speculated that, since both solar and stellar flares have temperatures that are nearly optimum ($\approx 10^4$ K) for optical radiative loss, the greater luminosity of stellar flares can be understood only in terms of larger flare area and greater optical thickness than in the solar case.

WLFs are astrophysically important because they challenge our understanding of flare energy transport, and because they illustrate that assumptions of hydrostatic equilibrium are probably inappropriate for flare models. They also remind us that, even in a star as near as the Sun, some flare structures remain unresolved, with the consequence that physical conditions derived from spectral analysis warrant careful interpretation. Finally, observed similarities between WLFs and stellar flares point to the universal nature of the flare phenomenon, and suggest that much can be gained from the study of each.

References

Avrett, E. H., Machado, M. E., and Kurucz, R. L.: 1986, in D. Neidig (ed.), *The Lower Atmosphere of Solar Flares*, National Solar Observatory, p. 216.

Boyer, R., Machado, M. E., Rust, D. M., and Sotirovski, P.: 1985, *Solar Phys.* **98**, 255.

Canfield, R. C., Bely-Dubau, F., Brown, J. C., Dulk, G. A., Emslie, A. G., Enome, S., Gabriel, A. H., Kundu, M. R., Melrose, D., Neidig, D. F., Ohki, K., Petrosian, V., Poland, A., Rieger, E., Tanaka, K., and Zirin, H.: 1986, in M. Kundu and B. Woodgate (eds.), *Energetic Phenomena on the Sun*, NASA CP 2439, Chapter 3.

Donati-Falchi, A., Falciani, R., and Smaldone, L. A.: 1984, *Astron. Astrophys.* **131**, 256.

Donati-Falchi, A., Falciani, R., and Smaldone, L. A.: 1986, in D. Neidig (ed.), *The Lower Atmosphere of Solar Flares*, National Solar Observatory, p. 136.

Emslie, A. G. and Sturrock, P. A.: 1982, *Solar Phys.* **80**, 99.

Gan, W.-Q. and Fang, C.: 1988, *Chin. Astron. Astrophys.* **12**, 146.

Grinin, V. P. and Sobolev, V. V.: 1977, *Astrofizika* **13**, 587.

Gurzadyan, G. A.: 1980, *Flare Stars*, Pergamon Press, New York.

Hiei, E.: 1982, *Solar Phys.* **80**, 113.

Hiei, E., Tanaka, K., Watanabe, T., and Akita, K.: 1982, *Hinotori Symposium on Solar Flares*, Institute of Space and Astronautical Science, Tokyo, p. 208.

Hudson, H. S.: 1972, *Solar Phys.* **24**, 414.

Kane, S. R., Love, J. J., Neidig, D. F., and Cliver, E. W.: 1985, *Astrophys. J.* **290**, L45.

Lites, B. W., Neidig, D. F., and Trujillo Bueno, J.: 1986, in D. Neidig (ed.), *The Lower Atmosphere of Solar Flares*, National Solar Observatory, p. 101.

Livshits, M. A., Badalyan, O. G., Kosovichev, A. G., and Katsova, M. M.: 1981, *Solar Phys.* **73**, 269.

Machado, M. E. and Rust, D. M.: 1974, *Solar Phys.* **38**, 499.

Machado, M. E., Emslie, A. G., and Brown, J. C.: 1978, *Solar Phys.* **58**, 363.

Machado, M. E., Avrett, E. H., Falciani, R., Fang, C., Gesztelyi, L., Hénoux, J.-C., Hiei, E., Neidig, D. F., Rust, D. M., Sotirovski, P., Švestka, Z., and Zirin, H.: 1986, in D. Neidig (ed.), *The Lower Atmosphere of Solar Flares*, National Solar Observatory, p. 483.

MacNeice, P., McWhirter, R. W. P., Spicer, D. S., and Burgess, A.: 1984, *Solar Phys.* **90**, 357.

McIntosh, P. S. and Donnelly, R. F.: 1972, *Solar Phys.* **23**, 444.

Neidig, D. F.: 1983, *Solar Phys.* **85**, 285.

Neidig, D. F.: 1986a, in D. Neidig (ed.), *The Lower Atmosphere of Solar Flares*, National Solar Observatory, p. 142.

Neidig, D. F.: 1986b, in D. Neidig (ed.), *The Lower Atmosphere of Solar Flares*, National Solar Observatory, p. 152.

Neidig, D. F. and Cliver, E. W.: 1983a, Air Force Geophysics Laboratory Technical Report AFGL-TR-83-0257, Hanscom AFB, Mass.

Neidig, D. F. and Cliver, E. W.: 1983b, *Solar Phys.* **88**, 275.

Neidig, D. F. and Wiborg, P. H.: 1984, *Solar Phys.* **92**, 217.

Poland, A. I., Milkey, R. W., and Thompson, W. T.: 1988, *Solar Phys.* **115**, 277.

Rust, D. M.: 1986, in D. Neidig (ed.), *The Lower Atmosphere of Solar Flares*, National Solar Observatory, p. 282.

Rust, D. M. and Hegwer, F.: 1975, *Solar Phys.* **40**, 141.

Ryan, J. M., Chupp, E. L., Forrest, D. J., Matz, S. M., Rieger, E., Reppin, E., Kanbach, G., and Share, G. H.: 1983, *Astrophys. J.* **272**, L61.

Shmeleva, O. P. and Syrovatskii, S. I.: 1973, *Solar Phys.* **33**, 341.

Simnett, G. M.: 1986, *Solar Phys.* **106**, 165.

Slonim, Y. M. and Korobova, Z. B.: 1975, *Solar Phys.* **40**, 397.

Starr, R., Heindl, W. A., Crannell, C. J., Thomas, R. J., Batchelor, D. A., and Magun, A.: 1988, *Astrophys. J.* **329**, 967.

Švestka, Z.: 1972, *Solar Phys.* **24**, 154.

Švestka, Z.: 1976, *Solar Flares*, D. Reidel Publ. Co., Dordrecht, Holland.

Worden, S. P.: 1983, in P. B. Byrne and M. Rodono (eds.), *Activity in Red Dwarf Stars*, D. Reidel Publ. Co., Dordrecht, Holland, p. 207.

Zirin, H. and Neidig, D. F.: 1981, *Astrophys. J.* **248**, L45.

SOLAR AND STELLAR MAGNETIC FIELDS AND
ATMOSPHERIC STRUCTURES: THEORY*

E. N. PARKER

Enrico Fermi Institute, Depts. Physics and Astronomy, University of Chicago, Chicago, IL 60637, U.S.A.

Abstract. This presentation reviews selected ideas on the origin of the magnetic field of the Sun, the dynamical behavior of the azimuthal field in the convective zone, the fibril state of the field at the photosphere, the formation of sunspots, prominences, the spontaneous formation of current sheets in the bipolar field above the surface of the Sun, coronal heating, and flares.

1. Introduction

The subject indicated by the title of this review is the basis for all of stellar activity. A balanced assessment of each of the many substantial theoretical scenarios currently available is not possible in so short a space, so this review concentrates on what seems at the moment to be the most likely point of view on each of a limited number of phenomena. It should be emphasized that the shifting nuances of observation have many times in the past sunk a substantial theoretical ship, and the most likely explanation of today may be found washed up on the beach tomorrow. So the theoretical opinions offered here need extensive observational testing before they can be considered hard scientific explanations of stellar activity.

The discussion centers exclusively on the Sun, because the questions posed by the stellar magnetic field, and the attendant atmospheric structures, are too complex to be answered by serendipity alone. The precise observational details are essential to the construction of a scientific theory. The Sun is the only star that can be seen (resolved) so the Sun is necessarily the rosetta stone for stellar fields and stellar activity. We can only remark that most other stars probably create their fields in the same general way as the Sun, and their flares are evidently of the same general nature as the solar flare. The reader is referred to the other papers presented in this IAU Colloquium No. 104 for a survey of present knowledge of the behavior of the fields and flares of the distant stars.

As a matter of fact, the Sun is none too close for scrutiny, because much of the action on the Sun takes place on scales of the order of 10^2 km, well below the limit of resolution of present ground-based telescopes. Fortunately the march of technical development holds promise for resolving 10^2 km within the next decade. So our somewhat blurred rosetta stone may yet be brought into sharp focus before the end of the century.

Speaking in the most general terms it appears that solar activity is primarily the result

* This work was supported in part by the National Aeronautics and Space Administration under NASA Grant NGL-14-001-001.

of the displacement of magnetic flux tubes by convective fluid motions beneath the surface of the Sun and consequent displacement of the tenuous fluid by the discontinuities and instabilities of the magnetic field above the surface of the Sun. The process must be viewed as a whole if we are to get the physics properly, and this review is structured on that point of view.

2. Origin of Solar Magnetic Fields

The magnetic field of the Sun evidently has its origins in the convective fluid motions driven by the unstable stratification of the outer 2×10^{10} cm of the solar radius (7×10^{10} cm). The variation of the angular velocity Ω across the convective zone is the essential effect for generating the azimuthal field of the Sun. Helioseismology indicates that there is little or no vertical gradient in the angular velocity Ω (Duvall, Harvey, and Pomerantz, 1986) through the convective zone. The principal variation of Ω is with latitude ψ, in the form

$$\Omega(\psi) = \Omega_0(1 - 0.2 \sin^2 \psi)$$

(Newton and Nunn, 1955; Ward, 1965; Howard and Harvey, 1970; Howard, Gilman, and Gilman 1984; Gilman and Howard, 1984a, b, 1985; Schröter, 1985), where Ω_0 is the equatorial angular velocity ($\Omega_0 \cong 2.93 \times 10^{-6}$ rad s^{-1}). The surface rotation varies somewhat with the phase of the magnetic cycle (Howard and Harvey, 1970; Yoshimura, 1972, 1981; Howard and LaBonte, 1980, 1981; Howard, 1984; Gilman and Howard, 1985) but that is a fine point that need not be considered here.

There is some vague evidence for a poleward meridional flow of the order of 20 m s^{-1} (Durney, 1975; Howard, 1979; Duvall, 1979; Beckers, 1979; Perez Garde et al., 1981; Anderson, 1988) as well as giant convective cells of one form or another (Yoshimura, 1971). The granules and supergranules are the surface manifestation of the general turbulent convective overturning. The mixing length representation of convective heat transport (Schwarzschild, 1958) is used to estimate the convective velocities through the convective zone (Spruit, 1974). Such estimates yield r.m.s. turbulent velocities v_t of the order of 40 m s^{-1} across the middle of the convective zone (at a depth of 10^{10} cm), declining only slowly with depth until near the bottom (at 2×10^{10} cm) where the calculated v_t falls abruptly to zero. The calculated equipartition magnetic field $B_e \equiv (4\pi\rho)^{1/2} v_t$ has a broad maximum of about 3×10^3 G across the lower two-thirds of the convective zone, falling to zero abruptly at the bottom.

Yoshimura (1975) points out that the convection involves converging flows in the lowest levels of the convective zone, so that the Coriolis force causes the rising fluid to rotate more rapidly than the surroundings. Higher up in the convective zone the rising convective plumes expand with the declining density, so that the fluid above some intermediate level rotates more slowly than the surroundings (Steenbeck and Krause, 1969). It follows that the helicity $(\mathbf{v} \cdot \nabla \times \mathbf{v})$ of a rising convective column in the northern hemisphere is positive in the lowest levels of the convective zone and negative above, with the opposite signs prevailing in the southern hemisphere. We presume, then, that the associated downflow is confined to the broad regions between the rising columns, so that the updrafts dominate the helicity, with the sense noted above. Schmitt (1987) considers the helicity in geostrophic waves.

There are a number of reasons for believing that the magnetic field is generated in the lowest levels of the convective zone (Parker, 1975, 1987a, b) or in the overshoot region below (Golub *et al.*, 1981; Galloway and Weiss, 1981; Spruit and van Ballegooijen, 1982; van Ballegooijen, 1982; Schmitt and Rosner, 1983; Durney, 1981; Gilman, Morrow, and DeLuca, 1989). The idea that the dynamo action occurs at the bottom and perhaps in the overshoot region originally followed from the sign of the helicity of the convection at those levels, providing the observed equatorward migration of the azimuthal field in combination with the downward decrease of Ω indicated by some of the numerical simulations of the large-scale circulation (Gilman, 1983; Glatzmaier, 1985; DeLuca and Gilman, 1986; Parker, 1987a). Golub *et al.* (1981) suggested that the normal active regions appearing at the surface of the Sun are the consequence of the dynamo in the overshoot region, while the small-scale active regions – X-ray bright points and ephemeral active regions – represent the magnetic 'debris' escaping from the dynamo process.

We may wonder if the current results of helioseismology, that Ω varies but little with radius, somehow miss perhaps a thin layer of vertical shear in the deep convective zone. The more definitive results eventually to be hoped for from the GONG observations might possibly change the picture. But for the moment we have no basis for assuming anything beyond the present evidence that Ω seems not to vary significantly with depth, the principal effect being then a decrease of Ω with increasing latitude, as already noted. The variation of Ω with latitude, in combination with the helicity of the convection, provides dynamo waves that migrate vertically instead of horizontally.

Consider, then, the dynamo that results from this combination of fluid motions. The azimuthal field is produced by the latitudinal shear of the latitudinal, or θ, component of the poloidal field. The poloidal field is a consequence of the interaction of the cyclonic convection with the azimuthal field. Combining these effects somewhere in the lower convective zone produces dynamo waves with a natural tendency to migrate vertically downward (Parker, 1955). The migration is blocked by the bottom of the convective zone, of course, so the waves pile up against the bottom. They are deflected north or south, away from their source, depending on the latitudinal distribution of the cyclonic convection and nonuniform rotation. Since both the cyclonic convection and the gradient of Ω are larger at higher latitudes, the dynamo waves apparently issue from higher latitudes and migrate toward the equator. Formal examples illustrating this effect are available in the literature (cf. Parker, 1971, 1979c; Lerche and Parker, 1972).

This provides a self-contained magnetic dynamo system in the lower part of the convective region. The fields created by the dynamo press downward against the bottom so they are relatively unaffected by the opposite dynamo action at higher levels. The convection is strong, with the equipartition field of the order of 3×10^3 G. The dynamo extends into whatever overshoot region there may be below the conventional bottom of the convective zone (at a depth of about 2×10^{10} cm). It remains to be shown what role is played by meridional circulation *vis-à-vis* the dynamo period.

The azimuthal field is held against the bottom of the convective zone by the combined effect of the dynamo migration and the weight of the cool shadow that forms on the upper side of each band of azimuthal field (Parker, 1987a). To elaborate this picture, one infers that the azimuthal field in the lower convective zone has a strength of 3×10^3 G or more (Parker, 1987a) based on the amount of magnetic flux that appears at the surface (Gaizauskas et al., 1983). Hence, the field is strong enough to suppress the convective heat transport to some significant degree. That is to say, each band of azimuthal field represents a major obstacle to the upward convective heat transport. Hence, each band has a cool shadow above and an accumulation of heat underneath. The heat accumulation lies near the bottom of the convective zone where a substantial fraction of the transport is radiative and where the convective transport requires a much smaller difference between the actual and the adiabatic temperature gradients than in the cool shadow region on top of the azimuthal field. Hence, the cool shadow is the larger effect, suppressing the buoyancy of the magnetic field (Parker, 1987a–d). The net result is that the azimuthal field is held down by both the dynamo migration and the cool shadow. However, the under surface of the azimuthal field is subject to a Rayleigh–Taylor instability as a consequence of the local accumulation of heat (Parker, 1987a–e). The instability initiates thermal plumes that penetrate upward through the field and to the surface of the Sun (Parker, 1988a). The plumes carry some of the field with them to the surface, producing the normal active regions. A plume recurs at a given location at intervals of a week or two based on the time required to accumulate the heat to initiate the Rayleigh–Taylor instability across the lower boundary of the magnetic field (Parker, 1987e). Such recurring eruptions from any one location are responsible for the long-lived activity complex (active longitude) with the continuing intermittent emergence of fresh flux at the surface over a period of a year or more (see, for instance, Gaizauskas et al., 1983; Castenmiller, Zwaan, and van der Zalm, 1986).

The rapid disappearance of magnetic flux from the surface of the Sun has long been a puzzling observational fact. Parker (1984a) suggested that most of the emerging bipolar fields are pulled back beneath the surface, so that there is relatively little total azimuthal flux lost into space. This may be understood as a direct consequence of the convective downdraft in the cool shadow above each band of azimuthal field in the convective zone. However, there remains much observational work yet to establish that most of the magnetic flux disappears by retraction (see the review by S. F. Martin in this issue).

An interesting feature of the dynamo driven exclusively by horizontal variation of the angular velocity is that the period is readily adjusted to the observed 22 years. A long standing problem with dynamos based on a radial variation of the angular velocity (Parker, 1955, 1957) is that the theoretical period is much shorter than 22 years – usually 1–5 years – if one employs current estimates of the eddy diffusivity. To elaborate on this problem, suppose that the eddy diffusivity is as small as $1–4 \times 10^{11}$ cm^2 s^{-1}. Then the turbulent diffusion of field is slow and the theoretical dynamo period is as long as 22 years (Köhler, 1973; Yoshimura, 1975). But a straightforward application of the mixing-length theory, on which the model convective zone is based (Spruit, 1974), yields an eddy diffusivity of about 10^{12} cm^2 s^{-1}, which produces a dynamo period of only a few years (see discussion in Parker, 1979c, p. 762; Rai Choudhuri, 1984). That is to say if the convective helicity and radial shears are strong enough to overcome turbulent diffusion, the phase velocity of the dynamo wave is too fast.

The theoretical dynamo model based on horizontal shear avoids this problem. The migration toward the equator arises only because of the nonuniform distribution of the helicity and horizontal shear. The period depends quantitatively on that distribution and can in principle be arbitrarily long for uniform helicity, uniform boundaries, etc. Such a dynamo provides a ready answer for the well-known variability of the period of the magnetic cycle, being particularly sensitive to the distribution of cyclonic convection, meridional flow, and nonuniform rotation.

The combined horizontal shear, downward migration, and suppression of buoyancy by the cool shadow accounts in a natural way for negligible loss of azimuthal magnetic flux through the surface of the Sun (Parker, 1984b). As pointed out elsewhere (Parker, 1984a), the fields emerging through the surface pull back into the Sun rather than escape into space. The cool shadow plays an important role in the process (Parker, 1987a, 1988a), and we now have the migration of the dynamo wave adding to the suppression of escaping flux.

Note, however, that there is still the possibility that small flux bundles – magnetic debris – may continually escape upward from the lower half of the convective zone, as originally suggested by Golub *et al.* (1981) and Durney (1988), because the cool shadow is ineffective in suppressing the buoyancy of any flux bundle whose horizontal width is small compared to the local pressure scale height (see Parker, 1987a, Figure 4, 1988c). The downward phase velocity of the dynamo wave would be only a few meters s^{-1}, at most, while the rate of buoyant rise of a small flux bundle with a width of, say, 10^3 km, might be as fast as 10^2 m s^{-1}.

The overall picture, then, is one of broad bands of intense $(3-10 \times 10^3$ G) azimuthal magnetic field crawling toward the equator along the bottom of the convective zone. The local dynamics of such bands, with widths of the order of $2-4 \times 10^5$ km, sends thermal plumes to the surface, which carry the magnetic flux responsible for the normal active regions. The associated poloidal field is diffuse and not easily observed. Its most obvious manifestation is in the polar fields, which keep in step with the migrating alternating bands of azimuthal field.

Consider, then, the upper half of the convective zone, where the convective motions are not as robust and where the helicity is presumed to be negative in the northern hemisphere and positive in the southern, as a consequence of the expansion of the thermal plumes rising from below (Steenbeck and Krause, 1969). Again it is assumed that the corresponding subsidence of gas occurs around the periphery of the updrafts, so that its helicity is of opposite sign but so small in magnitude that it can be neglected. The net negative helicity in the northern hemisphere, in combination with the decline of Ω with increasing latitude, causes an upward migration of the dynamo waves.

Now the equipartition field declines from 3×10^3 G at mid level (a depth of 10^{10} cm) to 2×10^3 G at a depth of 0.4×10^{10} cm, to a few hundred gauss at the visible surface.

Hence, the cool shadow cannot effectively suppress the magnetic buoyancy of flux bundles above a depth of the order of 1.5×10^9 cm (Parker, 1988c). So there is no firm barrier for the dynamo wave to push against.

The fields are carried up and out by the combined magnetic buoyancy and migration of the dynamo wave. Consequently it appears that the dynamo action in the upper half of the convective, if it exists at all, can contribute no more than the network fields and the small bipolar regions (e.g., ephemeral active regions). About all that can be said is that these very small magnetic features are the result of the magnetic debris from the lower half of the convective zone, amplified by whatever dynamo action there may be in the upper half. This is, of course, the point of view suggested by Golub *et al.* (1981) many years ago. It is interesting that we are led to the idea by the simple dynamo based on horizontal shear and by a consideration of thermal shadows, which cannot suppress the magnetic buoyancy within about 1.5×10^9 cm of the surface of the Sun. Indeed, the observed behavior of the small flux bundles and the transitory ephemeral active regions are just what would be expected from the theory of cool shadows, thermal plumes, and the retraction of magnetic flux bundles at depths of the order of 2×10^9 cm (Parker, 1988b, c).

With this brief summary of ideas on the origin of magnetic fields, in the convective zone, and their intermittent appearance at the surface, consider some of the dynamical effects of the fields above the surface.

3. Magnetic Activity above the Surface

The structures erected above the surface of the Sun, by the magnetic field emerging through the surface, are subject to direct observation and are consequently well known in general form. We are all familiar with the fibril structure of the magnetic field at the photosphere (Muller, 1985, and references therein) and with the spontaneous clustering of the fibrils to form pores and sunspots (Zwaan, 1985, and references therein). The cause of the fibril structure has yet to be confirmed for lack of the necessary high-resolution observations. Several ideas have been proposed. First of all, the magnetic field energy of a fibril field is larger than a uniform field with the same mean intensity $\langle B \rangle$ by the factor $B_f / \langle B \rangle$ where $B_f = 1-2 \times 10^3$ G, is the field in the individual fibril. This ratio is of the order of 10^2 in quiet regions and 10 in active regions. On the other hand, the widely separated fibrils offer less impediment to the convective transport of heat than the same total flux spread uniformly. Hence, the heat transport is more effective, and the thermal and gravitational energies are reduced by the fibril state of the field. It can be shown that the minimum total energy occurs for fibrils with B_f of the order of $1-4 \times 10^3$ G (Parker, 1984c), in rough agreement with the values of $1-2 \times 10^3$ G inferred from observation. It is interesting, then, that the fibril field observed at the photosphere may represent an energy minimum in the surface layers of the Sun, but, of course, so general a statement does not disclose the physical mechanism that creates the individual fibril.

Nor can it be concluded that the fields are in an intense fibril state throughout the convective zone, with B_f considerably in excess of the equipartition value. Brants (1985) and Zwaan (1985) infer from observation that B_f is only as large as the equipartition value, of about 0.5×10^3 G, in the magnetic flux bundles freshly emerging through the surface of the Sun. This suggests that the intensification to a value several times the equipartition value is intrinsically a surface effect, along the lines suggested by Spruit. That is to say, the global energy minimum can be achieved only where the means to achieve it is available and that is only at the surface.

In general terms, the creation of the fibril at the surface of the Sun requires that the gas pressure p_i within the fibril be reduced to about half the ambient external pressure p_e. Even a very small cooling ΔT may achieve this if it extends over several scale heights, because the pressure scale height is correspondingly reduced and the gas slides down out of the fibril, evacuating the upper regions and compressing the field according to the approximate horizontal equilibrium condition

$$B_f^2 = 8\pi(p_e - p_i).$$

Thus, for instance, a small internal temperature reduction ΔT extending over n scale heights of an isothermal atmosphere reduces the pressure by Δp, where

$$\frac{\Delta p}{p} \cong n \frac{\Delta T}{T}.$$

Then, if $\Delta T/T$ has the modest value 0.1 extending over 5 scale heights (about 10^3 km), the result is $\Delta p/p \cong 0.5$ and $p_e - p_i \cong 0.5 p_e$, yielding $B_f^2 \cong 4\pi p_e \cong 1-2 \times 10^3$ G.

A modest downdraft within the fibril can easily achieve such a cooling to produce the observed B_f of $1-2 \times 10^3$ G (Parker, 1978a, 1979c, pp. 260–271). Spruit (1979) has pointed out that the gas pressure reduction $p_e - p_i$, caused by the magnetic pressure $B^2/8\pi$, decreases the opacity and permits a more transparent atmospheric state in which the photospheric surface ($\tau = 1$) lies at a lower level within the fibril. The effect seems to be sufficient to produce the observed concentration of field. Deinzer et al. (1984a, b) and Hasan (1985) have explored the idea in some detail. Their modeling shows a quasi-steady state and a vigorously oscillatory state, respectively, for the reduced photospheric gas pressure and compressed magnetic fibril. Detailed studies of line profiles (Stenflo and Harvey, 1985; Stenflo, 1985) suggest that the oscillatory state worked out by Hasan may be more realistic than the quasi-steady model. High-resolution observations, down to 0.1″ (75 km), are necessary to confirm or deny these theoretical ideas.

Priest (1982, pp. 280–324) provides a general description and review of the structure of sunspots. The clustering of magnetic fibrils to form pores and sunspots during (and only during) periods of flux emergence at the surface of the Sun is a puzzling observational fact (Zwaan, 1978, 1985). In as much as the magnetic fibrils expand to fill all the available volume, creating a continuum field, at heights of a few hundred km above the visible surface, the fibrils at the photosphere exert short range repulsive forces on their nearest neighbors. The clustering, in opposition to the magnetic repulsion, must be driven by powerful hydrodynamic forces. There are modest attractive forces arising from the motion of the individual fibrils through the ambient fluid (Parker, 1978b, 1979e), but in order to drive the main event, forming a sunspot with fields compressed to 3×10^3 G, the only possibility seems to be a strong converging flow at depths of several thousand km (Meyer *et al.*, 1974). We suggest that the converging flow feeds a downdraft along the field (Parker, 1979a, b) which sweeps away the thermal energy whose upward convective transport is blocked by the field.

It is an observed fact that magnetic fibrils cluster to form pores and sunspots only while fresh magnetic flux is emerging in the active region, i.e., only while there is azimuthal field being carried to the surface by the thermal eruption from the azimuthal band of field in the lower convective zone. If the idea that the formation of sunspots at the surface is a consequence of a hydrodynamic flow converging on the points of clustering, then it follows that, somehow, there is a tendency for converging flows to form unseen beneath the surface in those regions of thermal upwelling. It is not evident on theoretical grounds why this should occur. We expect (Parker, 1987a–d) a general downdraft, with a converging flow at its upper end, in the cool shadow above each band of azimuthal field. Is there some reason why this general convective flow should become so strong locally in the regions of flux emergence as to sweep the magnetic fibrils into the highly compressed state that forms a pore or sunspot?

We have suggested that the subsurface magnetic field of a sunspot is a loose assembly of intense ($\sim 10^4$ G) nearly vertical magnetic fibrils, with field-free fluid flowing in and down between the fibrils (Parker, 1979a–d). The fibrils flare out at the visible surface to fill the available space and to provide a nearly continuous umbral field. We suggest that the bright umbral dots are a result of the transitory upward intrusion of the field-free fluid (Parker, 1979e). High-resolution observations of sunspot umbrae appear to support this picture of the sunspot (Garcia de la Rosa, 1987). The influence of the subsurface magnetic field structure on the local p-mode oscillations holds promise for helioseismological probing of the subsurface structure (Thomas, Lites, and Nye, 1982; Thomas and Scheuer, 1982; Thomas *et al.*, 1987; Bogdan and Zweibel, 1987; Bogdan, 1987a, b; Abdelatif and Thomas, 1989). It is clear from observation that the sunspot has profound effects on the local oscillations (cf. Balthasar, Küveler, and Wiehr, 1987, and references therein).

Moving up into the atmosphere above the surface of the Sun there is the quiescent prominence, representing a quasi-stable condensation of gas suspended on the magnetic field protruding upward from the surface (see, for instance, Tandberg-Hanssen, 1974; Hirayama, 1985; Sakai, Colin, and Priest, 1987; Wu and Low, 1987; Ballester and Priest, 1987, and references therein).

The coronal mass ejections have become an active subject for researrch, since their discovery a few years ago. It appears that the coronal mass ejection arises when a bipolar magnetic arcade, or similar structure, anchored in the surface of the Sun, is sheared lengthwise beyond a critical amount so that there is no longer a closed magnetic equilibrium (Low, 1977a, b, 1981, 1984, 1985, 1986; Jockers, 1978; Birn, Goldstein, and Schindler, 1978; Parker, 1981a; Seehafer, 1985; Browning and Priest, 1986, and references therein; Biskamp and Welter, 1988). To put the matter in its simplest terms, the magnetic energy of a magnetic arcade confined to a long quonset hut (horizontal semi-circular cylinder) can be increased linearly without bound as the lengthwise shearing increases. On the other hand, an elemental magnetic flux tube can be extended radially from the surface of the Sun to infinity with a finite amount of energy. In particular, a field B_0 extending radially from the surface of the Sun ($r = R$) to infinity possesses a magnetic energy dE in the solid angle $d\omega$, given by

$$dE = \frac{B_0^2}{8\pi} R^3 \, d\omega \,.$$

The energy between R and r is proportional to $(1/R - 1/r)$, so it does not matter how the field is deflected at $r = \infty$ to other directions. It follows, therefore, that beyond a critical shear, the field can reduce its energy if it expands upward from the original arcade and extends to infinity. One imagines, then, that a magnetic arcade, or similar structure, at the surface of the Sun is progressively and slowly sheared by the massive motions of the photosphere. When the shearing reaches some critical value, the lowest energy state of the field involves an increasing radial extension, which reaches to infinity for finite shear. That is to say, quasi-static magnetic equilibria form discontinuous families in configuration space, so that continuous deformation beyond a given point introduces a discontinuous – and in this case infinite – jump in the equilibrium configuration. The equilibrium ceases to exist in the form of an arcade and is replaced by an equilibrium which extends far out into space. The coronal gas tied to the field is pitched outward with the expanding field, creating the spectacular coronal mass ejection. Once underway the ejection is no longer near quasi-static equilibrium because the velocity of expulsion is comparable to the Alfvén speed. The finite kinetic energy of the catapulted coronal gas permits at least some portion of the gas and field to move to infinity.

4. Spontaneous Formation of Current Sheets

Solar flares, microflares, and the active X-ray corona appear to be a consequence of magnetic neutral point reconnection. The essential point is that these extreme supra-thermal phenomena are the result of intense, localized dissipation of magnetic field energy, producing high speed jets ($10^2–10^3$ km s^{-1}), fast particles, and generally intense heating of the ambient gas. They occur in regions where the temperature is already quite high ($10^5–10^6$ K) and the resistivity of the gas is low. The classical resistive diffusion coefficient $\eta = c^2/4\pi\sigma$ is $10^3–10^5$ cm^2 s^{-1}, so that the characteristic resistive dissipation time over typical granule scales of 500 km is $10^{10}–10^{12}$ s ($3 \times 10^2–3 \times 10^5$ years). Dissipation on so long a time-scale is uninteresting.

The current ideas on wave heating of the X-ray corona appear to be ruled out by the observation (Rosner, Tucker, and Vaiana, 1978) that the surface brightness of the X-ray corona is essentially independent of the scale of the emitting region, from the normal active region at 2×10^5 km down to the smallest resolvable emitting regions of

$2-4 \times 10^3$ km. Waves with periods of 1–2 s or less would be required to accomplish the heating of the smallest regions. Waves with such power at such high frequencies would be a revelation in themselves. In any case, the large active regions would then be heated by all waves with lengths equal to or less than the dimensions of the active region, rendering the longer magnetic loops much brighter than the shorter loops.

The essential point is that the random displacement of the footpoints (the magnetic fibrils) by the photospheric convection deforms the bipolar magnetic fields that extend outward from the photosphere. Large-scale shears produce the abrupt outward expansion of the static equilibrium of the field, as described in the preceding section. The large-scale motion may also be responsible for the large current sheet believed to produce the hard component of the flare emission. The reader is referred to the general studies of flares by Švestka (1976), Sturrock (1980), Priest (1980), and de Jager and Švestka (1985), and to the specific ideas spelled out by Priest (1982, pp. 344–381) and Seehafer (1985, 1986).

More pertinent to the problem of suprathermal heating is the fine-scale winding and interweaving of the lines of force of a bipolar field as a consequence of the continuous random walk of the footpoints of the field carried about in the granules. The small-scale winding and interweaving leads to the spontaneous formation of current sheets (tangential discontinuities) in the field (Parker, 1972, 1979c, pp. 359–391, 1987c, 1988d, e; Tsinganos, 1982; Tsinganos, Distler, and Rosner, 1984, and references therein). The heat input responsible for the X-ray corona and a large part of the solar flare appears to be a consequence of the dissipation of these spontaneous current sheets (Parker, 1972, 1979c, pp. 359, 391, 1981a, b; 1983a, b, 1987f, g, 1988d). The dissipation arises from the rapid (neutral point) reconnection of the fields across each current sheet, which is essentially independent of the very large electrical conductivity of the medium.

Figure 1 is a sketch of the lines of force of a simple bipolar field of length L. Figure 2 is a convenient idealization of that field in which the overall curvature has been removed so that the initial unperturbed field extends uniformly with intensity B_0 from the 'photosphere' at $z = 0$ to the 'photosphere' at $z = L$. Suppose, then, that the footpoints of the field at $z = L$ are fixed while the footpoints at $z = 0$ are transported in a random continuous velocity field with r.m.s. velocity u, correlation length λ, and correlation time

Fig. 1.   A schematic drawing of the lines of force of a bipolar magnetic region above the surface of the Sun.

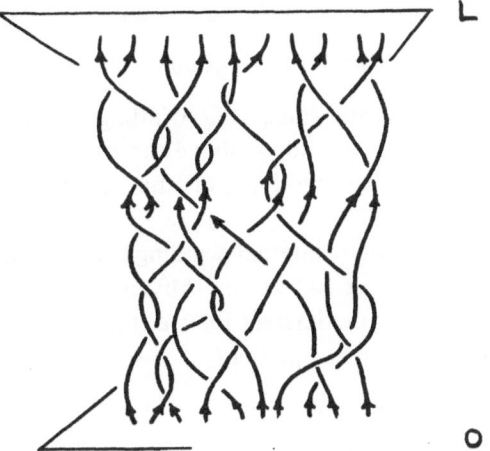

Fig. 2. A schematic drawing of the idealized form of a bipolar field.

$\tau \cong \lambda/u$ (which we identify with the granule motions for which $\lambda = 500$ km, $u = 0.5$ km s^{-1}). Thus, in a time t the typical footpoint travels a distance $s = ut$ along a random path composed of $n \cong t/\tau$ statistically independent steps of length λ. The lines of force trail out behind the moving footpoint at $z = 0$, as sketched in Figure 3, following the same general pattern throughout $0 < z < L$ as traced out by the moving footpoint at $z = 0$. Hence, the mean inclination θ of the line to the original z-direction is given by $\tan \theta = ut/L$. The mean value of the z-component of the field is unaffected by this winding and interweaving of the lines of force and remains equal to B_0, while the mean transverse component is $B_t = B_0 \tan \theta$. The tension in the field opposes the forward random march of the footpoint with the Maxwell stress $B_t B_0/4\pi$ so that the motion of

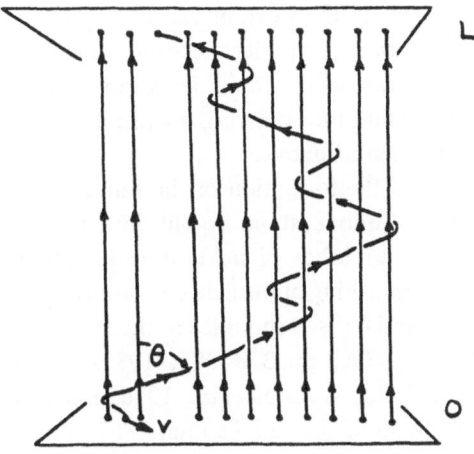

Fig. 3. A schematic drawing of the wandering of a single line of force among neighboring lines of force.

the footpoint does work on the field at the rate (Parker, 1983a)

$$W = uB_tB_0/4\pi = B_0^2u^2t/4\pi L \quad \text{ergs cm}^{-2}\text{ s}^{-1}.$$

The result is a progressive internal small-scale winding and interweaving of the lines of force of the re-entrant fields extending above the surface of the Sun. The rate of energy input by the convective motions of the footpoints is of the order of $uB_0^2 \tan \theta/4\pi$ ergs cm^{-2} s^{-1}.

Now the field, with its random internal winding and wrapping is in quasi-static equilibrium, with u of the order of 10^{-3} of the Alfvén speed V_A ($= 10^3$ km s^{-1} in the corona). The gas pressure is approximately 10^{-2} of the magnetic pressure, so that the field satisfies the familiar force-free equations

$$\nabla \times \mathbf{B} = \alpha \mathbf{B}, \quad \mathbf{B} \cdot \nabla \alpha = 0,$$

where α is the helicity of the field,

$$\alpha = \mathbf{B} \cdot \nabla \times \mathbf{B}/B^2.$$

Note, then, that α is rigorously constant along each line of force. On the other hand, the random field shown in Figure 2 involves a random sequence of right- and left-hand winding of the individual line of force about its neighbors. The helicity α cannot follow these changes in the geometrical helicity of the various independent windings along a given line of force. The field resolves the issue by developing internal tangential discontinuities, across which the magnitude of the field is continuous but the direction changes discontinuously (by an amount of the order of θ) so as to accommodate the successive right- and left-handed winding among the neighboring flux bundles (Parker, 1985, 1986a, b, 1988e). The tangential discontinuity occupies no volume and contains no magnetic flux (in the limiting case that the fluid has infinite electrical conductivity). Hence, the equilibrium equations are satisfied everywhere except on a set of points of measure (volume) zero.

The reader may wish to consult the work of van Ballegooijen (1988) who holds that tangential discontinuities do not occur under these circumstances. His numerical simulations provide a vivid illustration of the rapid development of current sheets in a field subject to successive winding and interweaving operations, which, however, he asserts is not the same effect as discussed above.

Another way to understand the formation of the tangential discontinuities (current sheets) is to consider a continuous static equilibrium field involving one or more separatrices between regions or lobes of field. It is readily demonstrated that any nonuniform deformation or squashing of the field (by motion of the footpoints or further twisting, etc.) produces tangential discontinuities at the separatrices (Syrovatskii, 1971, 1978, 1981; Parker, 1972, 1979c, pp. 378–383, 1982, 1983b, 1987c; Rosenbluth, Dagazian, and Rutherford, 1973; Waddell et al., 1976; Hu and Low, 1982; Low and Hu, 1983; Moffatt, 1987; Steinolfson and Tajima, 1987; Low, 1987, 1988; Low and Wolfson, 1988; Otani and Strauss, 1988; Strauss and Otani, 1988).

It must also be remembered that the footpoints of the coronal magnetic fields of the

Sun are separated into the individual fibrils, each of which may move independently of its nearest neighbors. This introduces discontinuities directly into the field above, without any consideration for the winding and interweaving of the flux bundles from the separate fibrils. Whole flux tubes may be displaced relative to the ambient field, producing tangential discontinuities along the entire length of the displaced flux bundle (Parker, 1981a, b).

The spontaneous appearance of tangential discontinuities through the re-entrant field of the solar atmosphere introduces dissipation, where none of any significance is otherwise expected. These current sheets are subject to resistive instabilities, leading to rapid reconnection of the fields across the current sheet. The high electric current densities in the discontinuity suggest the possibility of plasma turbulence and anomalous resistivity, although it must be remembered that this requires the electron conduction velocities to exceed at least the ion, if not the electron, thermal velocity.

Laboratory experience and numerical simulation indicates that the reconnection progresses only very slowly until the discontinuity θ exceeds some critical value, θ_c, whereupon there is a burst of reconnection, reducing θ to some fraction of θ_c (Rosenbluth, Dagazian, and Rutherford, 1973; Kadomtsev, 1975, 1984; Waddell et $al.$, 1976; Finn and Kaw, 1977; Montgomery, 1982; Seehafer, 1985; Lichtenburg, 1984; Dahlburg et $al.$, 1986).

Heyvaerts and Priest (1984) and Dixon, Browning, and Priest (1987) suggest that the reconnection proceeds in such a way as to satisfy Taylor's hypothesis (Taylor, 1974, 1986). That is to say, the total helicity $\int d^3r B^2/8\pi$ declines to the minimum allowed by the given total helicity. With this constraint the minimum energy occurs for uniform α. Seehafer (1985, 1986) has studied the evolution of force-free fields with uniform but time varying α, showing the various singular transitions in the topology of the field with a continuously increasing $\alpha(t)$. He suggests that major flares may be associated with these transitions.

The central point is then that, with the continual shuffling and intermixing of the footpoints of the field, one has the picture of the gradual increase of θ (over a period of hours in a normal active region with a scale $L = 10^{10}$ cm) until some critical value θ_c is exceed, whereupon the individual current sheets undergo random transient bursts of dissipation, and a statistically steady state is reached with the mean θ (over the field) of the same order as θ_c. Withbroe and Noyes (1977) estimate an energy input of the order of 10^7 ergs cm^{-2} s^{-1} to the brighter regions of the solar X-ray corona. With a typical field $B_0 = 10^2$ G and an assumed $u = 0.5$ km s^{-1}, the result is $\tan \theta \cong \frac{1}{4}(\theta \cong 14°)$. The energy of each individual burst is some fraction of the energy associated with the characteristic volume $\lambda^3 \cot \theta$ and transverse magnetic energy density $B_0^2 \tan^2 \theta/8\pi$. With $\lambda = 500$ km and $\tan \theta = \frac{1}{4}$, the characteristic energy is 10^{25} ergs. The characteristic burst, then, might involve 10^{24} ergs, which we have referred to as a $nanoflare$, it being about 10^{-3} of the typical 10^{27} ergs of the microflare (Parker, 1987g, 1988d).

It follows from these theoretical considerations that the individual X-ray coronal region represents a cloud of nanoflares (Parker, 1983a). Observations of the X-ray corona without high resolution show only the general glow of the individual X-ray loops.

But increasing resolution in space and in time progressively shows increasing spatial structure and rapid time variation. On scales of $1-2 \times 10^3$ km the X-ray corona is, in fact, a rapidly flickering spotty structure with correlation times 20–200 s (Golub, Krieger, and Vaiana, 1976a, b; Sheeley and Golub, 1979; Nolte, Solodyna, and Gerassimenko, 1979; Habbal and Withbroe, 1981; Brueckner, 1981; Brueckner and Bartoe, 1983; Lin *et al.*, 1984; Porter, Toomre, and Gebbie, 1984; Dere, Bartoe, and Brueckner, 1986; Porter *et al.*, 1987). The observations of Brueckner and Bartoe (1983) and Porter, Toomre, and Gebbie (1984) show that the individual nanoflare is of the general order of 10^{24} ergs, with a characteristic time of 10^2 sec. Thus, the theory appears to be borne out in detail by the observations (see discussion in Porter, Toomre, and Gebbie, 1984). The coronal emission rate of 10^7 ergs cm^{-2} s^{-1} requires about one nanoflare per square second of arc $(0.5 \times 10^{16}$ cm$^2)$ in progress at any given point in time.

We have suggested (Parker, 1987b) on the basis of the high-resolution observations of Machado *et al.* (1988a, b) that the typical solar flare represents a coordinated burst of nanoflares throughout a finite volume of field. Machado *et al.* point out that "The basic structure of a flare usually consists of an initiating closed bipole plus one or more adjacent closed bipoles impacted against it... The flare energy release begins either within the initiating bipole or at the interaction site between it and the impacted bipole. The initiating and impacted bipoles interact strongly in the impulsive phase of the flare [during which] most of the energy... is released inside the initiating bipole and/or inside one or more of the adjacent bipoles rather than at the interacting site between these." They point out that most of the hardest radiation may be emitted from the interaction site, but constitutes only a small fraction of the total X-ray energy.

It appears from this vivid description that the flare is a consequence of the large number of tangential discontinuities in the bipoles prior to the interaction, i.e., mutual deformation, of the bipoles. The flare occurs when the large-scale motion of the footpoints squashes two or more bipoles together, creating further current sheets and setting off a major portion of the pre-existing small-scale current sheets in a coordinated burst of nanoflares. The deformation of the bipoles may lead to rapid instabilities, violently shaking and squashing the bipoles along the lines described by Seehafer (1985, 1986). Further observational study of the initiating deformation is an essential step in understanding the diverse natures of individual flares. Indeed, the initial magnetic configuration, the intensity of the small-scale internal current sheets, and the particular mode of deformation are presumably the key to the nature of the flare.

It is interesting to note that Sturrock *et al.* (1984) pointed out several years earlier that the extremely spiky nature of the X-ray and microwave emission from solar flares suggests that the flare is a consequence of many very small reconnection events. They suggested that the individual magnetic fibrils rotate independently of each other, with the result that current sheets are produced at the contact surfaces between the individual flux bundles at coronal altitudes. Our own work has modified this concept by adding the spontaneous current sheets that arise throughout the field subject to any continuous random displacement of the footpoints at the photosphere. The observations of

Machado *et al.* (1988a, b) have placed the whole picture of the flare, as a coordinated burst of nanoflares, on a firm footing.

It is obvious that (a) further theoretical work needs to be done (see Parker, 1987f, and references therein) and (b) the final proof of the nature of the flare and the X-ray corona must come from detailed observation. Not only is it necessary to extend the present preliminary observational studies of individual microflares and nanoflares in the X-ray corona and in the flare itself, but much of the basic data on the structure and the motions of the individual magnetic fibrils (which are the footpoints of the coronal fields) have yet to be determined. The numbers employed in the present discussion ($\lambda = 500$ km, $u = 0.5$ km s^{-1}) are conjectures based on the observed granule motions. These observations will have to be done from an orbiting diffraction-limited mirror of 1 m diameter or more, or failing that, employing active optics on a large ground-based telescope.

The necessary optics have yet to be fully developed, although the technology holds great promise. The nature of the solar flare and the solar X-ray emission is fundamental to the terrestrial environment, to X-ray astronomy, and to the activity of all stars. It is not possible to extract much science from the extensive recording of the X-ray emission from the distant stars, until we have exploited the nearest star to understand what physical process is implied by the X-ray emission.

Finally, note that we have said nothing about the heating in coronal holes, which are responsible for the fast streams in the solar wind. The recent work by Withbroe (1988) indicates substantial energy deposition within the first few hundred thousand km. Current sheets do not form spontaneously in the open fields of the coronal hole. It remains to be shown what hydromagnetic wave spectrum might accomplish the heating close to the Sun where phase mixing is presumably not effective. It is essential that we take a hard headed attitude toward this problem, because for too many years now we have been without a real scientific understanding of the energy source for the solar wind. It is time that we recognize the problem and attack it on realistic terms.

References

Abdelatif, T. and Thomas, J. H.: 1989, *Astrophys. J. Suppl.* (May issue).
Anderson, B. N.: 1988, *Solar Phys.* **114**, 207.
Ballester, J. L. and Priest, E. R.: 1987, *Solar Phys.* **109**, 335.
Balthasar, H., Küveler, G., and Wiehr, E.: 1987, *Solar Phys.* **112**, 37.
Beckers, J. M.: 1979, *Oss. Astron. Catania Publ.*, No. 162, p. 166.
Birn, J., Goldstein, H., and Schindler, K.: 1978, *Solar Phys.* **57**, 81.
Biskamp, D. and Welter, H.: 1989, *Solar Phys.* **120**, 49.
Bogdan, T. J.: 1987a, *Astrophys. J.* **318**, 888.
Bogdan, T. J.: 1987b, *Astrophys. J.* **318**, 896.
Bogdan, T. J. and Zweibel, E. G.: 1987, *Astrophys. J.* **312**, 444.
Brants, J. .: 1985, *Solar Phys.* **98**, 197.
Browning, P. K. and Priest, E. R.: 1986, *Solar Phys.* **106**, 335.
Brueckner, G. E.: 1981, in F. Q. Orrall (ed.), *Skylab Solar Workshop on Active Regions*, Associated University Press, Boulder, Colorado, pp. 113–127.
Brueckner, G. E. and Bartoe, J. D. F.: 1983, *Astrophys. J.* **272**, 329.

Castenmiller, M. J. M., Zwaan, C., and van der Zalm, E. B. J.: 1986, *Solar Phys.* **105**, 237.
Dahlburg, J. P., Montgomery, D., Doolen, G. D., and Matthaeus, W. H.: 1986, *J. Plasma Phys.* **35**, 1.
De Jager, C. and Švestka, Z.: 1985, *Solar Phys.* **100**, 435.
Deinzer, W., Hensler, G., Schüssler, M., and Weishaar, E.: 1984a, *Astron. Astrophys.* **139**, 426.
Deinzer, W., Hensler, G., Schüssler, M., and Weishaar, E.: 1984b, *Astron. Astrophys.* **139**, 435.
Delucca, E. E. and Gilman, P. A.: 1986, *Geophys. Astrophys. Fluid Dyn.* **37**, 85.
Dere, K. P., Bartoe, J. D. F., and Brueckner, G. E.: 1986, *Astrophys. J.* **310**, 456.
Dixon, A. M., Browning, P. K., and Priest, E. R.: 1987, *Astron. Astrophys. Fluid Dyn.* **40**, 203.
Durney, B.: 1975, *Astrophys. J.* **199**, 761.
Durney, B.: 1989, *Astrophys. J.* (in press).
Duvall, T. L.: 1979, *Solar Phys.* **63**, 3.
Duvall, T. L., Harvey, J. W., and Pomerantz, M. A.: 1986, *Nature* **321**, 500.
Finn, J. M. and Kaw, P. K.: 1977, *Phys. Fluids* **22**, 2140.
Gaizauskas, V., Harvey, K. L., Harvey, J. W., and Zwaan, C.: 1983, *Astrophys. J.* **265**, 1056.
Galloway, D. J. and Weiss, N. O.: 1981, *Astrophys. J.* **243**, 945.
Garcia de la Rosa, J. I.: 1987, *Solar Phys.* **112**, 49.
Gilman, P. A.: 1983, *Astrophys. J. Suppl.* **53**, 243.
Gilman, P. A. and Howard, R.: 1984a, *Solar Phys.* **93**, 171.
Gilman, P. A. and Howard, R.: 1984b, *Astrophys. J.* **283**, 385.
Gilman, P. A. and Howard, R.: 1985, *Astrophys. J.* **295**, 233.
Gilman, P. A., Morrow, C. A., and DeLuca, E. E.: 1989, *Astrophys. J.* (in press).
Glatzmaier, G. A.: 1985, *Astrophys. J.* **291**, 300.
Golub, L., Krieger, A. S., and Vaiana, G. S.: 1976a, *Solar Phys.* **49**, 79.
Golub, L., Krieger, A. S., and Vaiana, G. S.: 1976b, *Solar Phys.* **50**, 311.
Golub, L., Rosner, R., Vaiana, G. S., and Weiss, N. O.: 1981, *Astrophys. J.* **243**, 309.
Habbal, S. R. and Withbroe, G. L.: 1981, *Solar Phys.* **69**, 77.
Hasan, S. S.: 1985, *Astron. Astrophys.* **143**, 39.
Heyvaerts, J. and Priest, E. R.: 1984, *Astron. Astrophys.* **137**, 63.
Hirayama, T.: 1985, *Solar Phys.* **100**, 415.
Howard, R.: 1979, *Astrophys. J.* **228**, L45.
Howard, R.: 1984, *Ann. Rev. Astron. Astrophys.* **22**, 131.
Howard, R. and Gilman, P. A.: 1986, *Astrophys. J.* **307**, 389.
Howard, R. and Harvey, J.: 1970, *Solar Phys.* **12**, 23.
Howard, R. and LaBonte, B.: 1980, *Astrophys. J.* **239**, L33.
Howard, R. and LaBonte, B.: 1981, *Solar Phys.* **74**, 131.
Howard, R., Gilman, P. A., and Gilman, P. I.: 1984, *Astrophys. J.* **283**, 373.
Hu, Y. Q. and Low, B. C.: 1982, *Solar Phys.* **81**, 107.
Jockers, K.: 1978, *Solar Phys.* **56**, 37.
Kadomtsev, B. B.: 1975, *Soviet J. Plasma Phys.* **1**, 389.
Kadomtsev, B. B.: 1984, *Plasma Phys. Contr. Fusion* **26**, 217.
Köhler, H.: 1973, *Astron. Astrophys.* **25**, 467.
Lerche, I. and Parker, E. N.: 1972, *Astrophys. J.* **176**, 213.
Lichtenburg, A. J.: 1984, *Nuclear Fusion* **24**, 1277.
Lin, R. P., Schwartz, R. A., Kane, S. R., Pelling, R. M., and Hurley, K. C.: 1984, *Astrophys. J.* **283**, 421.
Low, B. C.: 1977a, *Astrophys. J.* **212**, 234.
Low, B. C.: 1977b, *Astrophys. J.* **217**, 988.
Low, B. C.: 1981, *Astrophys. J.* **251**, 352.
Low, B. C.: 1984, *Astrophys. J.* **281**, 392.
Low, B. C.: 1985, *Solar Phys.* **100**, 309.
Low, B. C.: 1986, *Astrophys. J.* **307**, 205.
Low, B. C.: 1987, *Astrophys. J.* **323**, 358.
Low, B. C.: 1988, *Solar Phys.* **115**, 269.
Low, B. C. and Hu, Y. Q.: 1983, *Solar Phys.* **84**, 83.
Low, B. C. and Wolfson, R.: 1988, *Astrophys. J.* **324**, 574.
Machado, M. E., Moore, R. L., Hernandez, A. M., Rovira, M. G., Hagyard, M. J., and Smith, J. B.: 1988a, *Astrophys. J.* **326**, 425.

Machado, M. E., Xiao, Y. C., Wu, S. T., Prokakis, Th., and Dialetis, D.: 1988b, *Astrophys. J.* **326**, 451.

Mayer, F., Schmidt, H. U., Weiss, N. O., and Wilson, P. R.: 1974, *Monthly Notices Roy. Astron. Soc.* **169**, 35.

Moffatt, H. K.: 1987, in G. Comte-Bellot and J. Mathieu (eds.), *Advances in Turbulence*, Springer-Verlag, Berlin, pp. 240–241.

Montgomery, D.: 1982, *Phys. Scripta* **T2**, 83.

Muller, R.: 1985, *Solar Phys.* **100**, 237.

Newton, H. W. and Nunn, M. L.: 1955, *Monthly Notices Roy. Astron. Soc.* **111**, 413.

Nolte, J. T., Solodyna, C. V., and Gerassimenko, M.: 1979, *Solar Phys.* **63**, 113.

Otani, N. F. and Strauss, H. R.: 1988, *Astrophys. J.* **325**, 468.

Parker, E. N.: 1955, *Astrophys. J.* **122**, 293.

Parker, E. N.: 1957, *Proc. Natl. Acad. Sci.* **43**, 8.

Parker, E. N.: 1971, *Astrophys. J.* **166**, 295.

Parker, E. N.: 1972, *Astrophys. J.* **174**, 499.

Parker, E. N.: 1975, *Astrophys. J.* **198**, 205.

Parker, E. N.: 1978a, *Astrophys. J.* **221**, 368.

Parker, E. N.: 1978b, *Astrophys. J.* **222**, 357.

Parker, E. N.: 1979a, *Astrophys. J.* **230**, 905.

Parker, E. N.: 1979b, *Astrophys. J.* **232**, 291.

Parker, E. N.: 1979c, *Cosmical Magnetic Fields*, Clarendon Press, Oxford, pp. 659–681.

Parker, E. N.: 1979d, *Astrophys. J.* **234**, 333.

Parker, E. N.: 1979e, *Astrophys. J.* **231**, 270.

Parker, E. N.: 1981a, *Astrophys. J.* **244**, 631.

Parker, E. N.: 1981b, *Astrophys. J.* **244**, 644.

Parker, E. N.: 1982, *Geophys. Astrophys. Fluid Dyn.* **22**, 195.

Parker, E. N.: 1983a, *Astrophys. J.* **264**, 642.

Parker, E. N.: 1983b, *Geophys. Astrophys. Fluid Dyn.* **23**, 85.

Parker, E. N.: 1984a, *Astrophys. J.* **280**, 423.

Parker, E. N.: 1984b, *Astrophys. J.* **281**, 839.

Parker, E. N.: 1984c, *Astrophys. J.* **283**, 343.

Parker, E. N.: 1985, in M. Zeilik and D. M. Gibson (eds.), *Cool Stars, Stellar Systems and the Sun*, Springer-Verlag, Berlin, pp. 341–362.

Parker, E. N.: 1986a, *Geophys. Astrophys. Fluid Dyn.* **34**, 243.

Parker, E. N.: 1986b, *Geophys. Astrophys. Fluid Dyn.* **35**, 277.

Parker, E. N.: 1987a, *Astrophys. J.* **312**, 868.

Parker, E. N.: 1987b, *Solar Phys.* **110**, 11.

Parker, E. N.: 1987c, *Astrophys. J.* **318**, 876.

Parker, E. N.: 1987d, *Astrophys. J.* **321**, 984.

Parker, E. N.: 1987e, *Astrophys. J.* **321**, 1009.

Parker, E. N.: 1987f, in A. I. Poland (ed.), *Coronal and Prominence Plasmas*, NASA Goddard Space Flight Center, pp. 9–17.

Parker, E. N.: 1987g, in R. Altrock (ed.), *Workshop on Solar and Stellar Coronal Structure and Dynamics*, Sacramento Peak Observatory, August 1987.

Parker, E. N.: 1988a, *Astrophys. J.* **325**, 880.

Parker, E. N.: 1988b, *Astrophys. J.* **326**, 395.

Parker, E. N.: 1988c, *Astrophys. J.* **326**, 407.

Parker, E. N.: 1988d, *Astrophys. J.* **330**, 474.

Parker, E. N.: 1988e, *Geophys. Astrophys. Fluid Dyn.* (in press).

Perez Garde, M., Vazquez, M., Schwan, H., and Wöhl, H.: 1981, *Astron. Astrophys.* **93**, 67.

Porter, J. G., Toomre, J., and Gebbie, K. B.: 1984, *Astrophys. J.* **283**, 879.

Porter, J. G., Moore, R. L., Reichmann, E. J., Engvold, O., and Harvey, K. L.: 1987, *Astrophys. J.* **323**, 380.

Priest, E. R. (ed.): 1981, *Solar Flare Magnetohydrodynamics*, Gordon and Breach Science Publishers, New York.

Priest, E. R.: 1982, *Solar Magnetohydrodynamics*, D. Reidel Publ. Co., Dordrecht, Holland.

Rai Choudhuri, A.: 1984, *Astrophys. J.* **281**, 846.

Rosenbluth, M. N., Dagazian, R. Y., and Rutherford, P. H.: 1973, *Phys. Fluids* **16**, 1894.

Rosner, R., Tucker, W. H., and Vaiana, G. S.: 1978, *Astrophys. J.* **220**, 643.

Sakai, J., Colin, A., and Priest, E. R.: 1987, *Solar Phys.* **114**, 253.
Schmitt, D.: 1987, *Solar Phys.* **174**, 281.
Schmitt, J. H. M. M. and Rosner, R.: 1983, *Astrophys. J.* **265**, 901.
Schröter, E. H.: 1985, *Solar Phys.* **100**, 141.
Schwarzschild, M., 1958, *Structure and Evolution of the Stars*, Princeton University Press, Princeton.
Seehafer, N.: 1985, *Solar Phys.* **96**, 307.
Seehafer, N.: 1986, *Solar Phys.* **105**, 223.
Sheeley, N. R. and Golub, L.: 1979, *Solar Phys.* **63**, 119.
Spruit, H. C.: 1974, *Solar Phys.* **34**, 277.
Spruit, H. C.: 1979, *Solar Phys.* **61**, 363.
Spruit, H. C. and van Ballegooijen, A. A.: 1982, *Astron. Astrophys.* **106**, 58.
Steenbeck, M. and Krause, F.: 1969, *Astron. Nachr.* **291**, 49, 271.
Steinolfson, R. S. and Tajima, T.: 1987, *Astrophys. J.* **322**, 503.
Stenflo, J. O.: 1985, *Solar Phys.* **100**, 189.
Stenflo, J. O. and Harvey, J. W.: 1985, *Solar Phys.* **95**, 99.
Strauss, H. E. and Otani, N. F.: 1988, *Astrophys. J.* **326**, 418.
Sturrock, P. A.: 1980, *Solar Flares*, Colorado Associated University Press, Boulder.
Sturrock, P. A., Kaufman, P., Moore, R. L., and Smith, D. F.: 1984, *Solar Phys.* **94**, 341.
Švestka, Z.: 1976, *Solar Flares*, D. Reidel Publ. Co., Dordrecht, Holland.
Syrovatskii, S. I.: 1971, *Soviet Phys. JETP* **33**, 933.
Syrovatskii, S. I.: 1978, *Solar Phys.* **58**, 89.
Syrovatskii, S. I.: 1981, *Ann. Rev. Astron. Astrophys.* **19**, 163.
Tandberg-Hanssen, E.: 1974, *Solar Prominences*, D. Reidel Publ. Co., Dordrecht, Holland.
Taylor, J. B.: 1974, *Phys. Rev. Letters* **33**, 1139.
Taylor, J. B.: 1986, *Rev. Mod. Phys.* **58**, 741.
Thomas, J. H. and Scheuer, M. A.: 1982, *Solar Phys.* **79**, 19.
Thomas, J. H., Lites, B. W., and Nye, A. H.: 1982, *Nature* **297**, 485.
Thomas, J. H., Lites, B. W., Gurman, J. B., and Ladd, E. F.: 1987, *Astrophys. J.* **312**, 457.
Tsinganos, K. C.: 1982, *Astrophys. J.* **259**, 832.
Tsinganos, K. C., Distler, J. and Rosner, R.: 1984, *Astrophys. J.* **278**, 409.
van Ballegooijen, A. A.: 1982, *Astron. Astrophys.* **113**, 99.
van Ballegooijen, A. A.: 1988, *Geophys. Astrophys. Fluid Dyn.* **41**, 181.
Waddell, B. V., Rosenbluth, M. N., Monticello, D. A., and White, R. B.: 1976, *Nucl. Fusion* **16**, 528.
Ward, F.: 1965, *Astrophys. J.* **141**, 534.
Withbroe, G. L.: 1988, *Astrophys. J.* **325**, 442.
Withbroe, G. L. and Noyes, R. W.: 1977, *Ann. Rev. Astron. Astrophys.* **15**, 363.
Wu, F. and Low, B. C.: 1987, *Astrophys. J.* **312**, 431.
Yoshimura, H.: 1971, *Solar Phys.* **18**, 417.
Yoshimura, H.: 1972, *Solar Phys.* **22**, 20.
Yoshimura, H.: 1975, *Astrophys. J. Suppl.* **29**, 467.
Yoshimura, H.: 1981, *Astrophys. J.* **247**, 1102.
Zwaan, C.: 1978, *Solar Phys.* **60**, 213.
Zwaan, C.: 1985, *Solar Phys.* **100**, 397.

HYDRODYNAMIC MODELS OF SOLAR AND STELLAR FLARES

GIOVANNI PERES

Osservatorio Astronomico di Palermo, Piazza del Parlamento 1, I-90134 Palermo, Italy

Abstract. This paper discusses the hydrodynamic modeling of flaring plasma confined in magnetic loops and its objectives within the broader scope of flare physics. In particular, the Palermo–Harvard model is discussed along with its applications to the detailed fitting of X-ray light curves of solar flares and to the simulation of high-resolution Ca XIX spectra in the impulsive phase. These two approaches provide complementary constraints on the relevant features of solar flares. The extension to the stellar case, with the fitting of the light curve of an X-ray flare which occurred on Proxima Centauri, demonstrates the feasibility of using this kind of model for stars too. Although the stellar observations do not provide the wealth of details available for the Sun, and, therefore, constrain the model more loosely, there are strong motivations to pursue this line of research: the wider range of physical parameters in stellar flares and the possibility of studying further the solar-stellar connection.

1. Introduction

Hydrodynamic loop models of flares aim at calculating the evolution of confined plasma during the thermal phase of flares, under simplifying hypotheses. They achieve a reasonable level of detail and of realism by describing the simultaneous influence of the effects dominating the thermal phase of flares (e.g., thermal conduction, compressible hydrodynamics, radiative losses), in the fully non-linear regime and in the loop-like geometry shown by observations.

Their scope covers only some aspects of flares but, on the other hand, a complete theory of flares, describing all the phenomena and their intricate interaction, has never been developed. Most of the phenomena consider to be the primary cause of the flare, like magnetic field dissipation, MHD instabilities, electron acceleration etc, have been the object of a considerable amount of theoretical work but there are too few direct data on them. Most of the secondary effects, like plasma motion, hard X-rays, soft X-ray light curves and line profiles, are just a consequence of the primary effects; yet they have been observed with a much higher level of detail.

Hydrodynamic models of flares can provide a link between the well-observed thermal phase, characterized by the secondary effects, and the primary effects. After (and if) the model has provided a satisfactory reproduction of the observations, one can impose constraints on the primary mechanisms using the parameters of the model yielding the best agreement. For instance the localization and extension of the heating function used in the model, as well as the timing required to fit a flare, might give information on the primary energy release mechanism. Analogously the computed maximum plasma pressure provides a lower limit on the magnetic field needed to confine the plasma. Of course some diagnostic power lies also in the possibility of proving the ineffectiveness of some specific mechanism. This approach evidences the diagnostic possibilities of hydrodynamic models, an interesting expansion of their use as theoretical tools.

Solar Physics **121** (1989) 289–298.
© 1989 *by Kluwer Academic Publishers.*

Several groups have developed models of this kind and used them to tackle several of the problems existing in solar physics (see references in Peres and Serio, 1984; in Kopp *et al.*, 1986; and in Peres *et al.*, 1987).

Although each of the different models has unique features, either in the treatment of the physical effects or in the numerical technique used, they share many similarities. In order to illustrate their possibilities in more detail, I discuss below the features and some results on solar and stellar flares, of the Palermo–Harvard model, the one developed by my group.

2. The Palermo–Harvard Model

Space observations have shown the magnetic structuring of the whole solar atmosphere (for a review, see Vaiana and Rosner, 1978) and, more recently, the evidence of magnetic confinement of coronae in late-type stars (for a review, see Rosner, Golub, and Vaiana, 1985). Therefore, the thermal conduction and any fluid motion of the atmospheric plasma is channelled along the field lines and any closed magnetic loop can be modelled as a structure dynamically and thermodynamically independent from the neighbouring structures. Except for high-speed electrons in the transition region (see Section 5), the particle mean free path is shorter than the loop dimensions by orders of magnitudes and this permits a fluid treatment. Thermal flares have been shown to occur in closed loops (Pallavicini *et al.*, 1975; Pallavicini, Serio, and Vaiana, 1977) whose morphology, for the case of *compact flares*, does not change significantly. *Two-ribbon flares*, instead, show evident changes in the loop morphology and, therefore, have a qualitatively different evolution.

Our model describes directly only the thermal phase of compact loop flares. It is based on the single fluid differential equations of mass, momentum and energy conservation, taking into account the effect of ionization, gravity, viscosity, heating, optically thin plasma radiative losses and thermal conduction along with the equation of state of the gas (Peres *et al.*, 1982). The loop is assumed to be semi-circular, of constant cross section, to be infinitely rigid and anchored to the photosphere. The magnetic field plays no explicit role other than confining the plasma inside the coronal loop. The model is symmetric with respect to the loop apex and the initial atmosphere is static.

The heating function consists of two components: a steady heating, needed to maintain the atmosphere in static conditions by balancing the energy losses, and an impulsive heating which drives the flare by overheating the confined plasma. We have chosen two formulations of the impulsive heating. The first is a separable function of time and field-line coordinate:

$$Q(s, t) = E_H \times f(t) \times g(s), \tag{1}$$

where E_H is the maximum intensity of the heating, $g(s)$ is a Gaussian whose center and spread can be chosen at will, and $f(t)$ yields the chosen evolution of the heating. The alternative formulation describes the effect of electron beams, precipitating from the loop apex to the loop footpoints, with a power-law energy distribution, of index δ, abruptly truncated at the low-energy cutoff E_c, and of a given intensity and temporal evolution. In this approach we have followed the treatment of Nagai and Emslie (1984).

The equations are nonlinear and too complex to be solved analytically and, therefore, have to be solved numerically, by translating the differential equations to difference equations which in turn are solved by computer.

3. Applications to the Solar Flares

The fitting of the observations of the X-Ray Polychromator (XRP) on board the Solar Maximum Mission (SMM) satellite first showed the diagnostic possibilities of the model. This instrument measures the flux of seven X-ray lines (O VIII at 18.97 Å, Ne IX at 13.45 Å, Mg XI at 9.17 Å, Si XIII at 6.65 Å, S XV at 5.04 Å, Ca XIX at 3.17 Å, Fe XXV at 1.85 Å) whose temperatures of maximum emissivity cover the range of temperature of the coronal flaring plasma (Acton et al., 1980).

Our approach was to synthesize the plasma emission from the evolution of density, temperature and velocity computed with the model, and then to compare it with the observations. We checked the general agreement of the model with the observations (Pallavicini et al., 1983), and found a few general results, such as the importance of the Fe XXV line, among the observed ones, to trace the timing of the heating. We then turned to the detailed fitting of a specific flare observed in high level of detail by SMM and by Earth-based instruments (Peres et al., 1987a). We made the calculations using two models of local thermal heating (either localized at the loop apex, or at the base of the corona) and two models of heating by electron beams (either with $E_c = 10$ keV or with $E_c = 25$ keV, and both with $\delta = 8$). We performed several simulations for all heating models.

As an example of the results obtained, Figure 1 reports the best fitting cases for the heating with electron beams for $E_c = 10$ keV and Figure 2 for $E_c = 25$ keV. The fitting is satisfactory for the first case but much less for the second. The implications are that in this flare, although high-energy electrons are present, as shown by the hard X-ray observations (MacNeice et al., 1985), they do not seem to drive the soft X-ray flare.

A rather dynamic evolution, with impulsive heating of the plasma, rise of plasma temperature, and chromospheric evaporation, characterizes the rise phase of the thermal flares. Diagnostics of this phase can be obtained with the high-resolution spectra of X-ray lines such as that of Ca XIX around 3.17 Å, taken by the Bent Crystal Spectrometer (BCS) on board SMM. Antonucci et al. (1982) proved that, during the early phase of flares, the main component of this line typically broadens and a blue-shifted component, taken as evidence of plasma moving toward the observer at hundreds of kilometers per seconds, becomes prominent. The analysis of spectra at different times provides the evolution of plasma velocity and temperature (Antonucci et al., 1982).

We are synthesizing Ca XIX spectra from loop flare simulations under a wide range of hypotheses for the impulsive heating and for the pre-flare atmosphere, to understand their effect on the spectral features of typical flares observed with the BCS. So far we have completed the analysis of cases of thermal heating. Figure 3 shows two spectra,

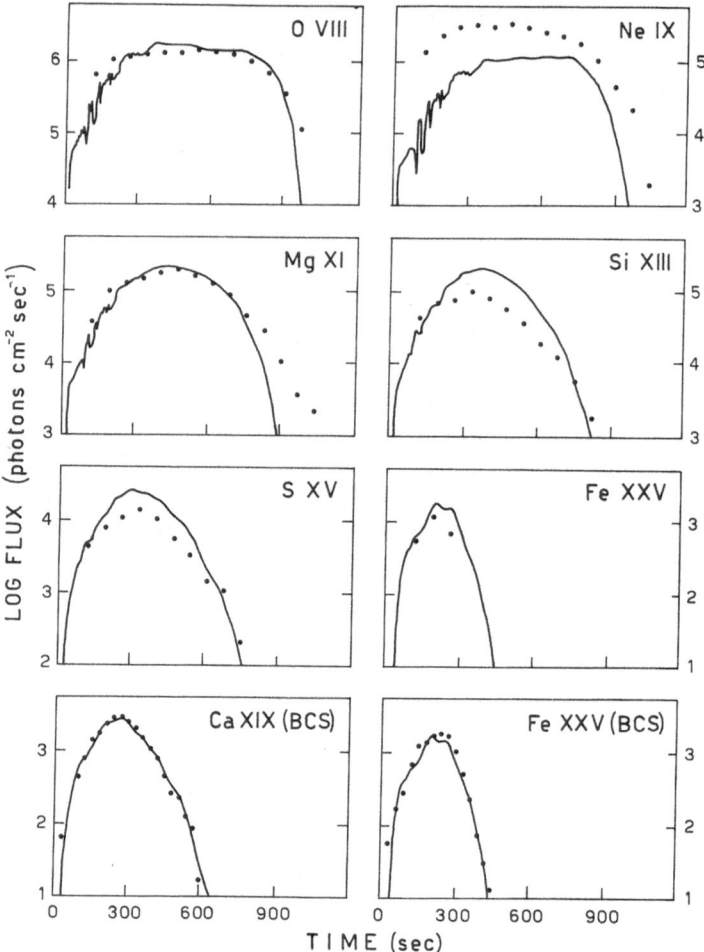

Fig. 1.   Modelling of the solar flare of November 12, 1980 at 17:00 UT: observed (dots) and computed (solid lines) light curves in the seven XRP lines. The flux is in photon s^{-1} cm^{-2} at Earth. In the simulation we assume heating by beams of non-thermal electrons with E_c = 10 keV and δ = 8. (From Peres *et al.*, 1988).

from Antonucci *et al.* (1987), computed with the Palermo–Harvard code for local thermal heating at the loop apex (Figure 3(a)), and for heating at the loop base (Figure 3(b)) in otherwise identical model loops. The case of heating at the footpoints reproduces features, like blue shifts, observed in at least 40% of the cases and, therefore, we can infer that impulsive heating in these flares most likely occurs near the loop footpoints rather than at the loop apex. On the other hand signatures for heating at the loop apex typically evolve rapidly and occur mostly at the beginning of the flare, when the emission is still rather weak. Since, at the beginning of the flare, present day instrumentation has to integrate over an extended time, any possible signature of heating at the loop apex is cancelled.

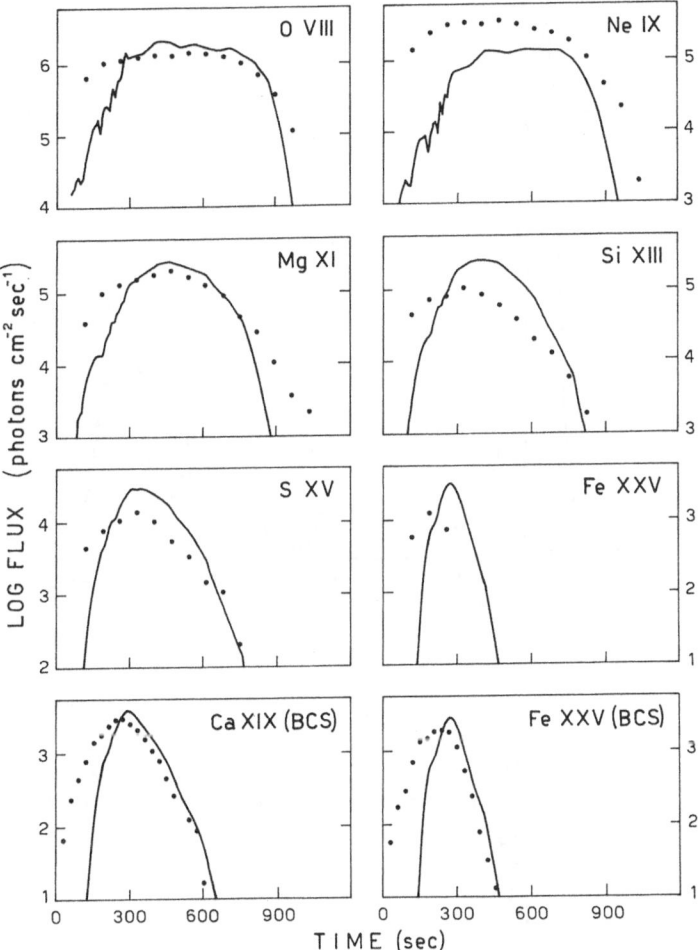

Fig. 2.   As in Figure 1, but for $E_c = 25$ keV. (From Peres *et al.*, 1987.)

4. The Extension to Stellar Flares

The satisfactory results obtained for the Sun have encouraged us to extend this technique to the stellar case to show similarities or differences between solar and stellar flares. Further interest lies in the wider range of physical parameters of the stellar environment but, on the other hand, the observation of events on other stars necessarily lack the fine details available for the Sun. Therefore, this novel approach aims also at constraining the dimensions of the flaring structure, and possibly to explore the dependence of magnetic field, of heating and of effective temperature on stellar characteristics.

Reale *et al.* (1988) have tackled the simulation of a flare which occurred on Proxima Centauri and was observed by the Einstein telescope (Giacconi *et al.*, 1979) on August 20, 1980 (Haisch *et al.*, 1983). The detector in the focal plane of the telescope was the Imaging Proportional Counter. Figure 4(a) reports the light curve as seen by the Imaging

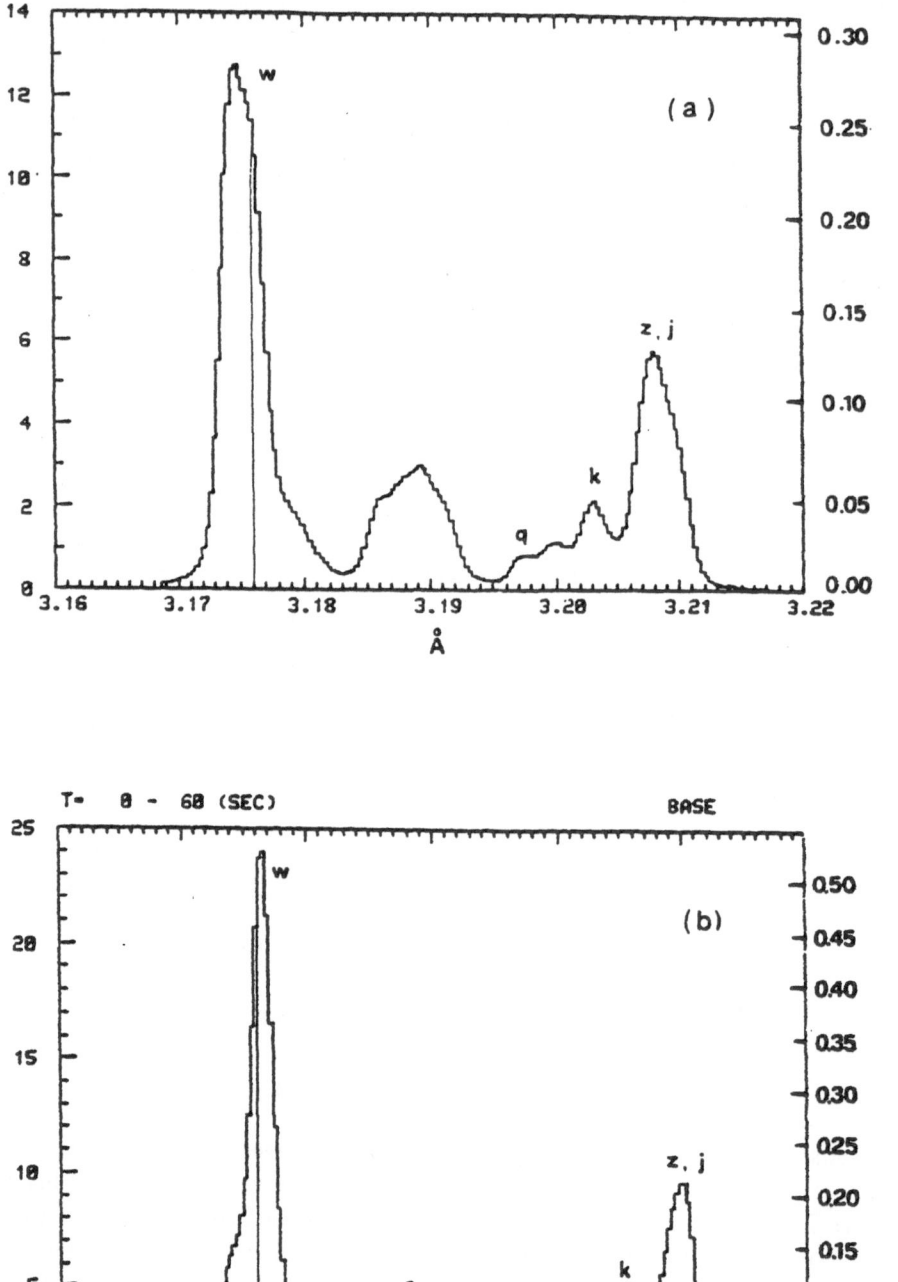

Fig. 3.　(a) Ca XIX spectra syntesized from a flaring loop heated at the loop apex; the emission is integrated over the first 60 seconds from the beginning of the heating. (b) As in (a) but for heating at the base of the corona. (From Antonucci *et al.*, 1987.)

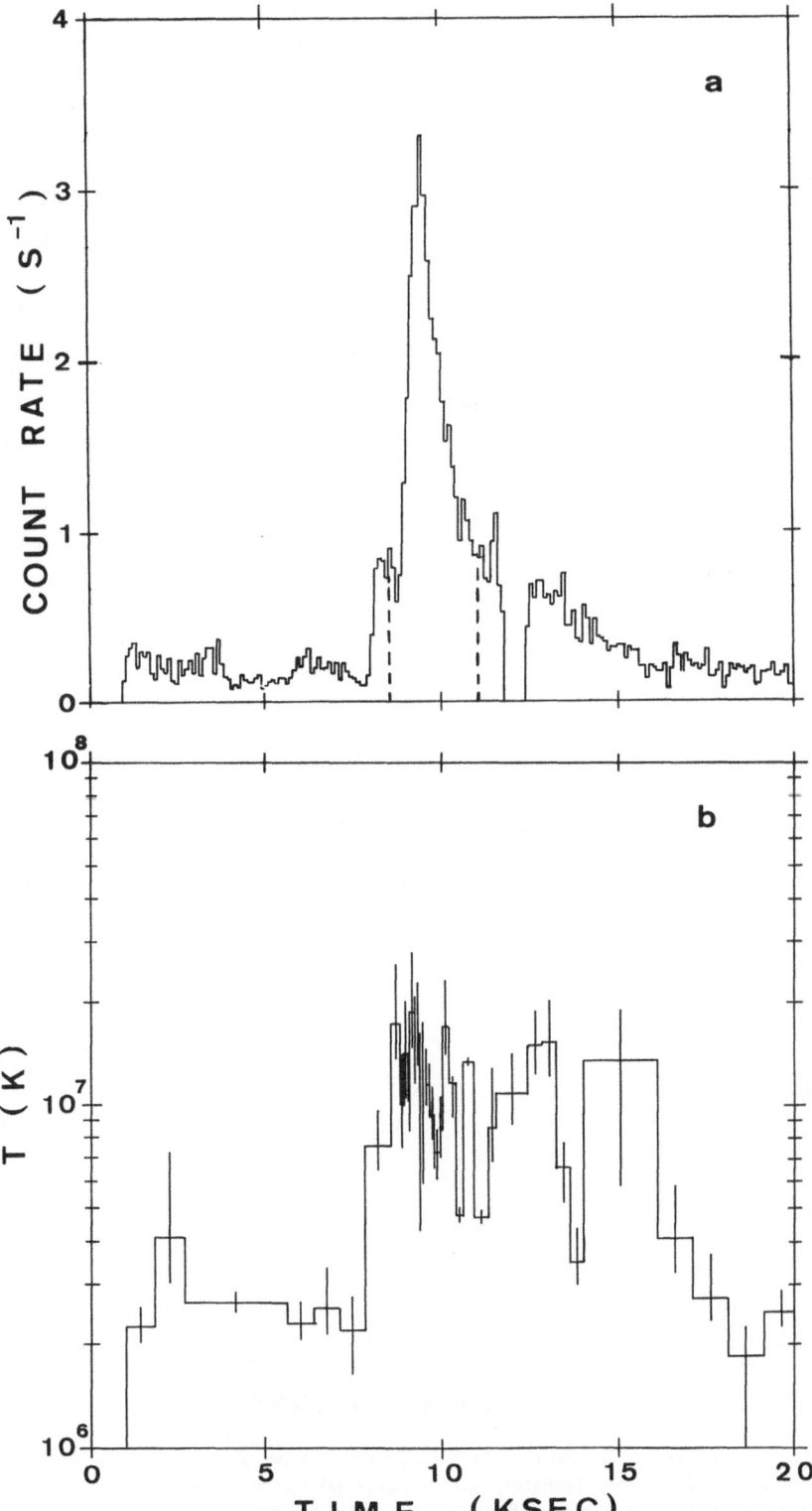

Fig. 4.   (a) Observed light curve of the flare on Proxima Centauri detected in the IPC X-ray band with the Einstein satellite telescope (from Haisch *et al.* , 1983). The dashed vertical lines bound the time range covered by the hydrodynamic calculations. (b) Evolution of the single-component temperature, derived from the data. (From Reale *et al.*, 1988.)

Proportional Counter and Figure 4(b) reports the evolution of the plasma temperature derived from the data. The effective plasma temperature increases close to the flare maximum and then gradually decreases.

Observations of stellar flares lack all the information coming from spatial resolution; moreover, since the IPC is a wide band detector, the information from tracers of the heating, like the Fe XXV line at 1.85 Å, so useful for the solar simulations (Pallavicini *et al.*, 1983, Peres *et al.*, 1987), is missing. Part of our initial effort, therefore, has been to estimate from the observed features, and to extrapolate from the solar experience, a plausible set of pre-flare conditions, as well as the features of the heating. We have concentrated on local thermal heating, because the data of the Monitor Proportional Counter of the Einstein telescope do not show any clear evidence of a nonthermal hard X-ray burst, a tracer of nonthermal beams of energetic electrons.

We have simulated the flare assuming three different values of loop semi-length L: 10^{10} cm, 7×10^9 cm and 5×10^9 cm. Figure 5 shows the light curve synthesized from the three simulations and evidences that the one with $L = 7 \times 10^9$ cm provides the best

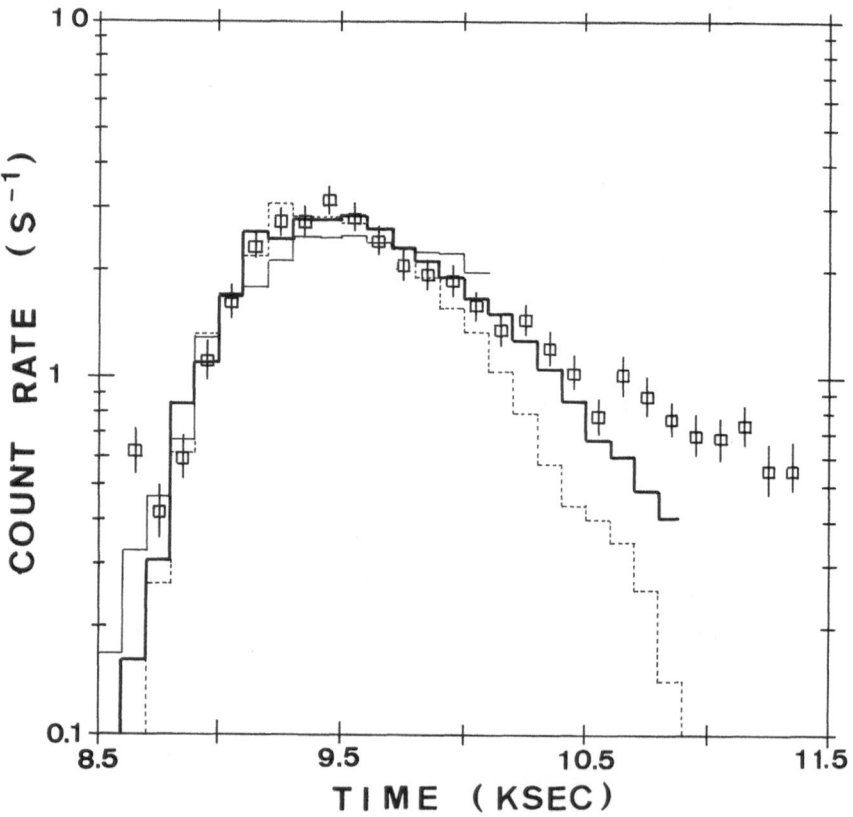

Fig. 5. Comparison between the Proxima Centauri flare light curve observed with the Einstein satellite telescope in the IPC X-ray band (squares) and the computed light curves for flare loops of semi-length $L = 10^{10}$ cm (thin solid line), $L = 7 \times 10^9$ cm (heavy), and $L = 5 \times 10^9$ cm (dashed) and subjected to a heating flux pulse of $F = 1.25 \times 10^{10}$ erg cm^{-2} s^{-1} lasting 700 s. (From Reale *et al.*, 1988.)

fit to the observations. We have assumed that the impulsive heating is active at a constant intensity for the first 700 s of the flare and inactive afterwards. Under this assumption the decay phase allows the diagnostics of the loop length. The sensitivity of the results to the adopted value of L, evident in Figure 5, provides a means of establishing the accuracy of the value derived. The simulation with $L = 10^{10}$ cm was interrupted at a time earlier than the other two because the computed decay time was evidently too long relative to the observed one. For further details see Reale *et al.* (1988).

Our approach shows that the model, used in conjunction with observations, can help to constrain physical quantities, like linear dimensions and coronal magnetic fields, not directly observable. For example, in the case of the Proxima Centauri flare, the hypothesis of confinement yields the constraint $B \geq 100$ G.

5. Conclusions

The Palermo–Harvard model has proved to describe fairly well the gradual phase of solar flares by means of the detailed simulation of the light curves in the XRP lines (Pallavicini *et al.*, 1983; Peres *et al.*, 1987); it is also providing the fine diagnostics of the impulsive phase through the detailed simulation of high resolution spectra of the Ca XIX (3.17 Å) line (Antonucci *et al.*, 1987).

Stellar studies are a relatively recent extension. The stellar observations cannot provide as stringent constraints as in the solar case; however, the extension is important because the range of physical parameters involved in the stellar flares is much wider than for solar flares alone. Stellar simulations allow physical insight and constraints on the dimensions and physical parameters not directly observable.

All these simulations show that the hydrodynamic and thermodynamic evolution of the confined plasma is a fundamental physical phenomenon of the gradual phase of compact loop flares. The fine tuning of the models to reproduce the observations typically provides the diagnostics of other physical aspects of the flare.

Many improvements have been proposed for these models. For instance they should use, in the transition region, the appropriate resolution of the numerical grid and, more importantly, a better treatment of the physics. In particular thermal conduction determines the steepness of the transition region and the structure of the whole atmosphere, thus imposing the requirements on the spacing of the numerical grid. However, the Spitzer (1962) formulation of heat flux, commonly used in these models, is not always appropriate because the mean free path of high speed electrons is often too long, with respect to the scale heights of temperature and density. A non-local treatment of thermal conduction would, instead, be more appropriate (Smith, 1986; Peres, Serio, and Rosner, 1987; Karpen and DeVore, 1987). In any case the computed evolution of the flare does not appear to be affected by the poor resolution (few tens of km) in the transition region (e.g., Reale *et al.*, 1988). An even more radical improvement would be to provide for a changing geometry, in order to model non-compact flares.

The interest in improvements shows the liveliness of this field. In fact hydrodynamical models of loop are now established theoretical and diagnostic tools in solar research and they promise to be useful also in stellar research.

Acknowledgements

I thank S. Serio and G. S. Vaiana for stimulating discussions, F. Reale for helpful criticism, and the organizers of the meeting, B. Haisch and M. Rodonò, for their kind invitation. This work was supported by Ministero della Pubblica Istruzione and by Piano Spaziale Nazionale.

References

Acton, L. W., Culhane, J. L., Gabriel, A. H., and 21 co-authors: 1980, *Solar Phys.* **65**, 53.

Antonucci, E., Gabriel, A. H., and 7 co-authors: 1982, *Solar Phys.* **78**, 107.

Antonucci, E., Dodero, M. A., Peres, G., Serio, S., and Rosner, R.: 1987, *Astrophys. J.* **322**, 522.

Giacconi, R. and 30 co-authors: 1979, *Astrophys. J.* **230**, 540.

Haisch, B. M., Linsky, J. L., Bornmann, P. L., Stencel, R. E., Antiochos, S. K., Golub, L., and Vaiana, G. S.: 1983, *Astrophys. J.* **267**, 280.

Karpen, J. T. and DeVore, C. R.: 1987, *Astrophys. J.* **320**, 904.

Kopp, R. A., Fisher, G. H., MacNeice, P., McWhirter, R. W. P., and Peres, G.: 1986, in M. R. Kundu and Woodgate (eds.), *Energetic Phenomena on the Sun*, NASA Conference Publication 2439.

MacNeice, P., Pallavicini, R., Mason, H. E., Simnett, G. M., Antonnucci, E., Shine, R. A., Rust, D. M., Jordan, C., and Dennis, B. R.: 1985, *Solar Phys.* **99**, 167.

Nagai, F. and Emslie, A. G.: 1984, *Astrophys J.* **279**, 896.

Pallavicini, R., Serio, S., and Vaiana, G. S.: 1977, *Astrophys. J.* **216**, 108.

Pallavicini, R. and Vaiana, G. S., Kahler, S. W., and Krieger, A. S.: 1975, *Solar Phys.* **45**, 411.

Pallavicini, R., Peres, G., Serio, S., Vaiana, G. S., Acton, L., Leibacher, J., and Rosner, R.: 1983, *Astrophys. J.* **270**, 270.

Peres, G., Rosner, R., Serio, S., and Vaiana, G. S.: 1982, *Astrophys. J.* **252**, 791.

Peres, G., and Serio, S.: 1984, *Mem. Soc. Astron. Ital.* **55**. 749.

Peres, G., Serio, S., and Rosner, R.: 1987, *Nuovo Cimento* **B99**, 29.

Peres, G., Reale, F., Serio, S., and Pallavicini, R.: 1987, *Astrophys. J.* **312**, 895.

Reale, F., Peres, G., Serio, S., Rosner, R., and Schmitt, J. H. M. M.: 1988, *Astrophys. J.* **328**, 256.

Rosner, R., Golub, L., and Vaiana, G. S.: 1985, *Ann. Rev. Astron. Astrophys.* **23**, 413.

Smith, D. F.: 1986, *Astrophys. J.* **302**, 836.

Spitzer, L.: 1962, *The Physics of Fully Ionized Gases*, Interscience, New York.

Vaiana, G. S. and Rosner, R.: 1978, *Ann. Rev. Astron. Astrophys.* **16**, 393.

A REVIEW OF STELLAR FLARES AND THEIR CHARACTERISTICS

B. R. PETTERSEN

Institute of Theoretical Astrophysics, University of Oslo, P.O. Box 1029 Blindern, N-0315 Oslo 3, Norway

Abstract. We review the flaring activity of stars across the HR-diagram. Brightenings have been reported along the entire Main Sequence and in many stars off the Main Sequence. Some stars are decidedly young, others are in advanced stages of stellar evolution. Flares are common on stars with outer convection zones and outbursts have been reported also on other types of stars, although confirmations are needed for some of them.

Analyses of flare occurrence sometimes find flares to be randomly distributed in time, and sometimes indicate a tendency for flares to come in groups. Preferred active longitudes have been suggested. Recent solar results, where the occurrence rate for flares is found to exhibit a periodicity of 152 days, suggest that stellar flare data should be reanalyzed over long time baselines to see if the present confusing situation can be resolved.

The radiation from stellar flares is dominated by continuum emission and about equal amounts of energy have been recorded in the optical, UV, and X-ray regions of the spectrum. In solar flares strong continuum emission is rarely recorded and a large collection of bright emission lines takes prominence. Small flares occur more frequently than large ones and the latter have longer time-scales. Flare energies can exceed 10^{37} erg. The most productive flare stars are those where the convective envelopes occupy large volumes. Slow stellar rotation rates are believed to reduce the level when the star has been braked significantly from its young rotation rate.

1. Introduction

The term 'flare star' has been used to designate dwarf K and M stars with transient optical brightenings (UV Ceti stars). Objects known to be members of young clusters and star formation complexes were called flash stars or sometimes cluster flare stars. They are mostly of late spectral types, sometimes with affinities to T Tauri stars and related objects. Reports of flaring in a few early spectral type stars gave rise to the term non-classical flare stars (Kunkel, 1975).

In recent years improved detector performance and access to new spectral regions have led to unexpected detections of flare activity in many other kinds of stars. Often the only observation available is a light curve that resembles those seen in classical flare stars. There is only indirect evidence that stellar flares are localized eruptions in magnetic fields as in the Sun, and for some of the flaring objects it is possible that we are seeing atmospheric responses to mass transfer or other dynamical processes. The discovery observations alone never demonstrate beyond doubt that the cause of the observed licht curve is the same as in the Sun. It could be that merely the symptoms are similar.

For these reasons we shall simply apply the principle of recognition when scanning the literature in search of flare stars. Any observation that reminds me, even in an extreme form, of something seen in solar flares or in classical stellar flares is good enough

Solar Physics **121**: 299–312, 1989.
© 1989 *Kluwer Academic Publishers*.

to count the star among those that flare. In practice, since we have no spatial resolution when observing stars, I will try to recognize a light curve from its shape, or look for new appearances or enhancements of emission lines on time-scales from minutes to hours. The solar experience allows me to explore many spectral windows: X-rays, UV, optical, and radio. I shall have a relaxed attitude towards the range of some parameters since solar flare analogs on other stars will reflect local conditions there.

In addition to guidance provided by solar flares, a sideview to flaring dMe stars may also prove useful as a background for selection criteria of flaring in other stars. Large stellar flares on red dwarfs last long enough to be time-resolved both spectroscopically and photometrically. In some cases one has succeeded in recording the flare in several spectral windows from X-rays to radio. Several characteristic flare properties have been noted. H I Balmer emission lines are often enhanced over quiescent values before the continuum flux begins to rise in the flare. Emission lines reach their peak flux after the continuum peak. Ca II and Mg II are more delayed than H I Balmer lines. He I follows the higher Balmer lines, while He II emission is very shortlived. During maximum flare emission the lines contribute negligibly to the total flare luminosity in the optical region, beginning at 5% of the U and B band fluxes near flare peak and increasing to 20% during the tail of the flare decay. Only when the continuum component has almost died out does the line flux begin to match that of the continuum, i.e., in the very last stages of the flare. Flare colours show individual changes during the maximum phase, but decay remarkably slowly after flare peak. Stellar flares on red dwarfs are, therefore, predominantly continuum events in terms of energy, but emission lines are enhanced before and well after the continuum component is gone. Contrast effects on the Sun help to mask out the continuum and optical solar flares are studied predominantly in lines.

Estimates from few and scattered stellar flare data in other spectral regions suggest that roughly equal amounts of flare energy are emitted in the optical, ultraviolet, and X-ray windows. *UBVRI* photometry usually gives $3E_U \leq E_{opt} \leq 5E_U$ (in erg), so $18E_U \leq E_{bol} \leq 30E_U$ (Pettersen, 1988).

2. Which Stars Flare?

A literature search reveals flaring in many types of stars across the HR diagram (see Figure 1). Identifications and references are given in Table I. Flaring has been reported in a few T Tau stars and in star formation complexes, aggregates, and young open clusters with ages from 3×10^5 years to 6×10^8 years. Several hundred flaring objects have been cataloged in Orion and the Pleiades. Many have been observed to flare more than once. Most young flare stars were detected through multiple exposures with Schmidt cameras, but some were X-ray recordings by Einstein and Exosat. Flares have also been seen on stars close to the Main Sequence and in post-T Tauri stars.

Thus, flare activity starts at an early phase in the evolution of stars, and flaring is a common phenomenon in young stars.

On the Main Sequence, flaring has been reported in all spectral classes from Wolf–Rayet stars and B stars to K and M dwarfs. The largest number of flare stars have

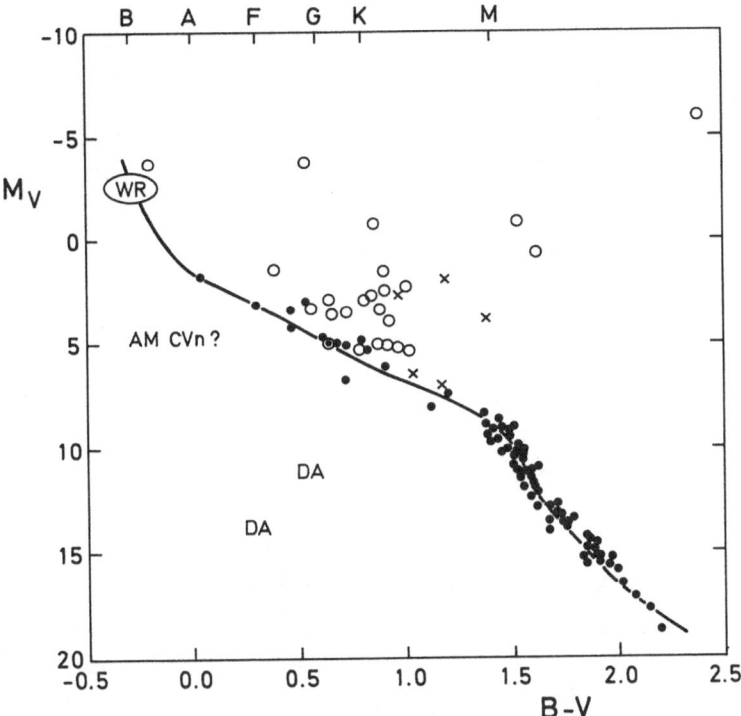

Fig. 1.   The positions of some flaring objects in the HR-diagram. Dots are Main-Sequence stars, crosses are young stars and circles are (sub)giants and supergiants at various stages of stellar evolution.

been found among the classical UV Ceti flare stars of spectral types dKe and dMe. Recent years have seen extensions into spectral types G, F, and even A, both in optical and X-ray observations. (Although a couple of the F- and G-type binaries are often referred to as RS CVn systems in the literature (σ^2CrB, SV Cam, XY UMa), we list them here as Main-Sequence stars since both components have luminosity class V. RS CVn stars, where at least one component is a (sub)giant, are listed separately below.) The earliest spectral types on the Main Sequence have flare reports in the optical only, and were listed by Kunkel (1975) as non-classical flare stars.

Thus, flaring has been reported in all spectral classes along the Main Sequence. Flares occur on single stars and on members of multiple systems.

Among the evolved stars and systems the most well-known class of flaring objects are the RS CVn binaries, where at least one component is a subgiant. The presence of chromospheres and coronae have been established by satellite observations, and photospheric starspots are commonly detected by optical photometry. The occurrence of flares in all spectral windows adds to this solar-like picture. In the same region of the HR-diagram we also find stars that are members of Algol and W UMa systems. Both of these are characterized by short orbital periods and small distances between components, so that mass transfer takes place when one component overflows its Roche lobe. FK Com may be an advanced result of this process where one component has

TABLE I

Stars with observed flare activity

Star type	X-ray	UV	Optical	Radio	References
Young stars					
T Tau stars	AS 205, DG Tau		DI Cep		7, 21, 27, 68
ρ Oph cloud	several		4 stars		38
Orion (3–10 $\times 10^5$ yr)			482 stars		36
Tau–Aur ($<10^6$ yr)	several		102 stars		36
NGC 7000 (3×10^6 yr)			67 stars		36
NGC 2264 (10^5–10^7 yr)			42 stars		36
Post-T Tau stars	HD 560 B, AB Dor	FK Ser	FK Ser, V4046 Sgr		9, 16, 17, 65, 66
Pleiades (8×10^7 yr)			546 stars		36
Praesepe (3×10^8 yr)			54 stars		36
Coma Ber (3×10^8 yr)			4 stars		36
UMa stream (3×10^8 yr)	π^1 UMa				31
Hyades (6×10^8 yr)	HD 27130		H II 2411, vA 351		45, 53, 64
Main-Sequence stars					
Wolf–Rayet stars			CQ Cep		26
B-type stars			66 Oph, BD + 31°1048, HD 160202		2, 42
					4
A-type stars	α Gem		SS 199 II, HD 12211		43, 48, 54
F-type dwarfs	σ^2 CrB		5 Ser, o Aql, HD 137050		1, 5, 41, 64
G-type dwarfs	Sun	Sun	Sun, SV Cam, XY UMa	Sun	44, 74
			κ Cet, RZ Psc, HD 97766		50, 52, 75
K-type dwarfs	BY Dra	BY Dra	BY Dra, EQ Vir+++		46, 47
M-type dwarfs	EQ Peg, YY Gem	AD Leo, V654 Cen	UV Cet, EV Lac++	AD Leo, YZ CMi	46, 47
Evolved stars and systems					
RS CVn systems	V711 Tau, HD 8357, HD 101379, II Peg, DM UMa	V711 Tau, UX Ari, AR Lac	V711 Tau, SZ Psc, II Peg, AR Lac, VY Ari, EZ Peg	V711 Tau, UX Ari, AR Lac	10, 11, 14, 15, 18, 22, 23, 24, 33, 49, 51, 55, 57, 60, 69
Algols	β Per	(BH CVn)	DI Peg	(BH CVn)	12, 13, (33), 71
W UMa stars	VW Cep		44i Boo, U Peg, W UMa, VW Cep, CN And	VW Cep	19, 20, 25, 30
FK Com stars			FK Com		66, 72
Giants			BW Vul, HD 129246/7, α Tau, V654 Her, IU Ori, HD 282773	α Cet	14, 39, 67, 8, 20, 28, 29, 32, 34
Supergiants			ε Aur, μ Cep		37
Systems containing white dwarfs		AY Cet, (λ And)	V471 Tau, G44-32, Case 1, VY Scl, AM CVn		3, 40, 6, 35, 56, 58, 59, 65, 70, 73

References to Table I:

(1) Agrawal, P. C. *et al.*: 1986, *Monthly Notices Roy. Astron. Soc.* **219**, 777.
(2) Andrews, A. D.: 1964, *Irish Astron. J.* **6**, 212.
(3) Arsenijevic, J.: 1985, *Astron. Astrophys.* **145**, 430.
(4) Bakos, G. A.: 1969, in *Non-Per. Phen. in Var. Stars*, p. 159.
(5) Bakos, G. A.: 1983, *Astron. J.* **88**, 674.
(6) Baliunas, S. L. *et al.*: 1984, *Astrophys. J.* **282**, 733.
(7) Bastian, U. and Mundt, R.: 1979, *Astron. Astrophys.* **78**, 181.
(8) Boice, D. C. *et al.*: 1981, *Astrophys. J.* **245**, L71.
(9) Busko, I. C. and Torres, C. A. O.: 1978, *Astron. Astrophys.* **64**, 153.
(10) Buzasi, D. L. *et al.*: 1987, *Astrophys. J.* **322**, 353.
(11) Charles, P. A.: 1983, in *Activity in Red Dwarf Stars*, p. 415.
(12) Chaubey, U. S.: 1980, *Inf. Bull. Var. Stars*, No. 1739.
(13) Chaubey, U. S.: 1982, *Astrophys. Space Sci.* **81**, 283.
(14) Chugainov, P. F.: 1976, *Izv. Krymsk. Astrofiz. Obs.* **55**, 85.
(15) Collier, A. C.: 1983, in *Activity in Red Dwarf Stars*, p. 254.
(16) Collier Cameron, A. *et al.*: 1988, *Monthly Notices Roy. Astron. Soc.* **231**, 131.
(17) Darius, J.: 1978, *Inf. Bull. Var. Stars*, No. 1429.
(18) Dorren, J. D. *et al.*: 1981, *Astron. J.* **86**, 572.
(19) Egge, K. E. and Pettersen, B. R.: 1983, in *Activity in Red Dwarf Stars*, p. 481.
(20) Eggen, O. J.: 1948, *Astrophys. J.* **108**, 15.
(21) Feigelson, E. D. and De Campli, W. M.: 1981, *Astrophys. J.* **243**, L89.
(22) Feldman, P. A. *et al.*: 1978, *Astron. J.* **83**, 1471.
(23) Garcia, M. *et al.*: 1980, *Astrophys. J.* **240**, L107.
(24) Hobbs, L. *et al.*: 1978, *Astron. J.* **83**, 1525.
(25) Huruhata, M.: 1952, *Publ. Astron. Soc. Pacific* **64**, 200.
(26) Kartashova, T. A.: 1970, *Inf. Bull. Var. Stars*, No. 473.
Kartashova, T. A.: 1972, *Per. Zvez.* **18**, 459.
(27) Kelemen, J.: 1985, *Inf. Bull. Var. Stars*, No. 2744.
(28) Kovalchuk, G. U.: 1986, in *Flare Stars and Related Obj.*, p. 16.
(29) Kovalchuk, G. U. and Pugach, A. F.: 1984, *Inf. Bull. Var. Stars*, No. 2557.
(30) Kuhi, L.: 1964, *Publ. Astron. Soc. Pacific* **76**, 430.
(31) Landini, M. *et al.*: 1986, *Astron. Astrophys.* **157**, 217.
(32) Liller, W.: 1966, *Astron. J.* **71**, 862.
(33) Little-Marenin, I. R. *et al.*: 1986, *Astrophys. J.* **303**, 780.
(34) Ludendorff, H. and Eberhard, G.: 1905, *Astron. Nachr.* **170**, 165.
(35) Marar, T. M. K. *et al.*: 1988, *Astron. Astrophys.* **189**, 119.
(36) Mirzoyan, L. V. and Oganyan, G. B.: 1986, in *FS and Rel. Obj.*, p. 68.
(37) Moffett, T. J. and VandenBout, P.: 1973, *Inf. Bull. Var. Stars*, No. 833.
(38) Montmerle, T. *et al.*: 1983, *Astrophys. J.* **269**, 182.
(39) Morris, S. and Milone, E.: 1983, *Publ. Astron. Soc. Pacific* **95**, 376.
(40) Nha, I.-S. and Lee, S. J.: 1983, *Inf. Bull. Var. Stars*, No. 2405.
(41) Olson, E. C.: 1980, *Inf. Bull. Var. Stars*, No. 1825.
(42) Page, A. A. and Page, B.: 1970, *Publ. Astron. Soc. Australia* **1**, 324.
(43) Pallavicini, R.: 1988, in *Solar and Stellar Physics*, p. 98.
(44) Patkos, L.: 1981, *Astrophys. Letters* **22**, 1.
(45) Pesch, P.: 1972, *Astrophys. J.* **178**, 203.
(46) Pettersen, B. R.: 1976, *Catalogue of Flare Star Data*.
(47) Pettersen, B. R.: 1988, (in prep.).
(48) Philip, A. G. D.: 1968, *Publ. Astron. Soc. Pacific* **80**, 171.
(49) Pope, S. K.: 1983, *Inf. Bull. Var. Stars*, No. 2388.
(50) Pugach, A. F.: 1976, in *Var. Stars and Stellar Evol.*, p. 144.
(51) Ramsey, L. and Nations, H.: 1981, *Publ. Astron. Soc. Pacific* **93**, 732.
(52) Robinson, C. R. and Bopp, B. W.: 1988, *Cool Stars 1987*, p. 509.
(53) Rodono, M.: 1974, *Astron. Astrophys.* **32**, 337.
(54) Rovithis, P. *et al.*: 1988, *10th European IAU Meeting*, p. 107.
(55) Schwartz, D. A. *et al.*: 1981, *Monthly Notices Roy. Astron. Soc.* **196**, 95.
(56) Shugarov, S.: 1984, *Inf. Bull. Var. Stars*, No. 2612.
(57) Simon, T. *et al.*: 1980, *Astrophys. J.* **239**, 911.
(58) Simon, T. *et al.*: 1985, *Astrophys. J.* **295**, 153.
(59) Simon, T. and Sonneborn, G.: 1987, *Astron. J.* **94**, 1657.
(60) Srivastava, R. K.: 1983, *Inf. Bull. Var. Stars*, No. 2450.
(61) Stern, R.: 1983, *Astrophys. J.* **264**, L55.
(62) Stienon, F. M.: 1971, *Inf. Bull. Var. Stars*, No. 545.
(63) Tagliaferri, G. *et al.*: 1988, *Astrophys. J. Letters* (in press).
(64) van den Oord, G.: 1987, *Cool Star 1987*, p. 494.
(65) van Genderen, A. M.: 1973, *Inf. Bull. Var. Stars*, No. 815.
(66) Vilhu, O. *et al.*: 1988, *Astrophys. J.* **330**, 922.
(67) Walter, F. and Basri, G. S.: 1982, *Astrophys. J.* **260**, 735.
(68) Walter, F. and Kuhi, L. V.: 1984, *Astrophys. J.* **284**, 194.
(69) Walter, F. *et al.*: 1987, *Astron. Astrophys.* **186**, 241.
(70) Warner, B. *et al.*: 1970, *Nature* **226**, 67.
(71) White, N. *et al.*: 1985, *Exosat preprint*, No. 10.
(72) Yang, Y.-L. and Liu, Q.-Y.: 1985, *Inf. Bull. Var. Stars*, No. 2705.
(73) Young, A. *et al.*: 1983, *Astrophys. J.* **267**, 655.
(74) Zeilik, M. *et al.*: 1983, *Astron. J.* **88**, 532.
(75) Zhang, Z. *et al.*: 1987, *Inf. Bull. Var. Stars*, No. 3050.

transferred almost all of its mass to the other star. It is not clear whether the flares, seen in X-ray, optical and radio, are directly linked to the mass transfer. Satellite observations again demonstrate the presence of very active chromospheres and coronae, which may be stellar or may be associated with an accretion disk.

A number of late-type giants and supergiants have been reported to flare in the optical and radio. Any event detected on such stars would have to be very powerful (corresponding to flare energies of 10^{34}–10^{39} erg), and none of these stars has been reported to flare more than once.

Other evolved systems are those that contain white dwarfs. The flaring primary star can be a G giant as in AY Cet (λ And could perhaps belong to this class?), or an active K or M dwarf, perhaps with a previous history of mass transfer to the white dwarf companion. From the point of view of atmospheric activity, these stars may be reclassified as (sub)giants or Main-Sequence dwarfs, respectively. The disturbing example is AM CVn, a binary consisting of two helium white dwarfs. It is an open question whether the flares reported have a solar-like nature.

This literature search has demonstrated flaring in young objects contracting towards the Main Sequence, on Main-Sequence stars, and in various types of evolved stars. If we regard reports of single flares as interesting for follow-up studies, but restrict confirmed flare stars to be those where repeated flares have been reported (irrespective of spectral window), the group would be somewhat reduced. Surviving the scrutiny are various age groups among the young stars (T Tau and post-T Tau stars, members of the Orion and Tau–Aur complexes, NGC 7000, NGC 2264, the Pleiades, Praesepe, and the Hyades). On the Main Sequence, 66 Oph (B2e; mass loss instability of disk?) and HD 12211 (A5V + G0V) are early-type stars that have flared more than once. Repeated flaring has been seen in σ^2CrB (F6V + G0V), SV Cam (G3V – K4V), the Sun (G2), HD 97766 (G5), and in a number of dKe and dMe classical flare stars. Among the evolved stars, repeated flaring has been seen in RS CVn systems, Algols, W UMa stars, and FK Com stars, and perhaps in the helium white dwarf binary AM CVn.

A comparison of the HR-diagrams of confirmed flare stars (Figure 2; stars with more than one flare) in clusters of different ages (Orion $\sim 10^6$ years; Pleiades 5×10^7 years; Hyades 6×10^8 years) to that of field stars (Figure 1) reveals that the majority of flare stars in clusters are on or near the Main Sequence. In Orion they cover the spectral classes K2–M2, in the Pleiades K3–M6, and the Hyades F8 – M4. In the Orion and Pleiades diagrams a few stars appear below the Main Sequence. They are either background stars or grossly subluminous. Foreground stars in the direction of nearby clusters can be identified from proper motion studies. None of the clusters show evolved or early-type flare stars. In the Hyades, stars of $M > 2$–$3\,M_\odot$ have begun their post-Main-Sequence evolution, while in the Pleiades the turnoff point is near $6\,M_\odot$. Since the observations are predominantly photographic (with low time resolution) this may preclude detection of energetic, but low amplitude flare-ups in early type and other intrinsically bright stars.

Thus flaring is confirmed to take place in stars with outer convection zones, on or above the Main Sequence. This includes both young and evolved stars, singles and

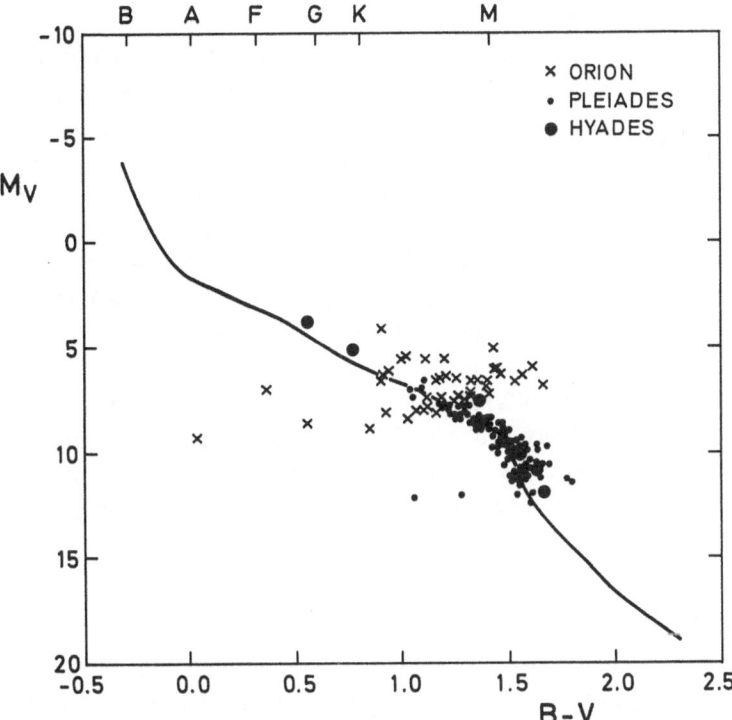

Fig. 2.   The HR-diagram for confirmed flare stars in three clusters.

members of binaries. Flaring may take place in other stars as well, but this should be confirmed by further observations.

3. Flare Characteristics

Many stars have been seen to flare only once so typical characteristics can hardly be determined. A similar problem arises for some confirmed flare stars where one flare was seen in optical and another in X-rays or radio. Extensive flare recording has been done for solar neighbourhood dKe and dMe stars, and flare stars in various young clusters and aggregates. These observations show that individual stars span considerable ranges in flare energy, amplitudes, and time-scales. There are notable differences also between some stars. Due to technicalities such as different detection thresholds, limited time resolution, etc., the dominating selection effect in existing observations is that only powerful flares are recorded in intrinsically bright stars. The weak photospheres of intrinsically faint stars allow much smaller flares to be detected, and they occur in much larger numbers than large flares. In what follows we will discuss flare properties in well-observed flare stars, i.e., we consider data for dMe stars, flare stars in clusters, and individual data points for a few evolved stars. The young stars were observed by photographic techniques with Schmidt cameras and suffer from bad time resolution.

3.1. The time distribution of flares

Statistical analyses of flare occurrence for several stars have concluded that flares are randomly (Poisson) distributed in time (Oskanyan and Terebizh, 1971; Lacy, Moffett, and Evans, 1976; Pettersen *et al.*, 1986; Pettersen, Coleman, and Evans, 1984). In contrast to this there have been repeated claims by observers that flares tend to come in groups, and some have found indications of periodic behaviour from time to time. Some data sets, when subjected to statistical analysis, give the conflicting result that flares are not randomly distributed in time (Pazzani and Rodono, 1981; Melikian and Grandpierre, 1984; Pettersen, 1988; unpublished result for UV Cet). Doyle (1987) even suggested a preferred active longitude for flare production on EV Lac over a certain length of time.

Viewed against recent results for the Sun, where the occurrence rate of flares has been found to exhibit a periodicity of about 152 days in many spectral windows (see Bai and Sturrock, 1987, and references therein), it might prove useful to re-analyze stellar flare data over very long time baselines to see if the present confusing situation can be resolved.

3.2. Flare time-scales

More than 30 years ago, Haro and Chavira (1955) noted that flare duration correlated with the spectral type of the flaring star in a cluster, in that short duration flares occurred preferentially on stars of late spectral type. This was eventually confirmed for dMe stars in the solar vicinity, where longer duration flares occurred mostly on the more luminous stars (Kunkel, 1969, 1974, 1975; Pettersen *et al.*, 1984). The functional relationship between $t_{0.5}$, the average decay time from maximum luminosity to half that level, and the absolute magnitude of the star is such that $t_{0.5} \sim g^{-2}$, where g is the surface gravity of the star. This would imply that the average decay time is proportional to the square of the pressure scale height of the atmosphere, so it follows that the star imposes conditions on the flare region in accordance with its size and mass. Gershberg and Shakhovskaya (1973) considered flare brightness and decay rates, and arrived at the alternative conclusion that large flares last longer than small flares, and only large flares are seen on luminous stars due to contrast effects while small flares dominate in number on faint stars.

Sufficient U-filter flare data are now available at each flare energy level to consider this problem. We have arranged the data in order of flare energy, considering windows of width 1 decade in $\log E_u$ (erg). Flares fainter than 10^{27} erg were seen on the intrinsically faintest stars, while the brightest dKe stars showed flares up to 10^{34} erg. Cluster flare stars and evolved stars showed flares up to 10^{37} erg. Stars with less than about 50 recorded flares showed a range in flare energy of 2–3 orders of magnitude, while stars with about 200 observed flares spanned 4–5 orders of magnitude in flare energy. For each window we determined the average flare time-scale for rise and decay $(t_{0.5})$ in units of seconds. The result is shown in Figure 3, which reveals the same basic properties for rise and decay:

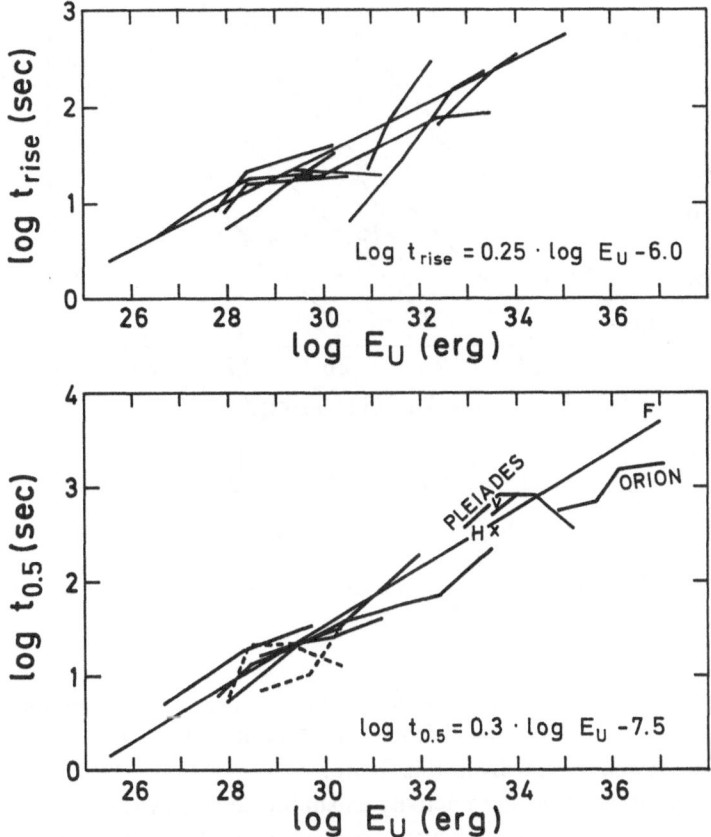

Fig. 3.   Relationships between flare time-scales and flare energy. Typical error bars are $\pm^{0.3}_{0.7}$ in $\log t_{\text{rise}}$ and $\log t_{0.5}$.

(1) Energetic flares are seen predominantly on bright stars due to selection effects.
(2) Large flares tend to last longer than small flares.

It appears that the time-scales of flares are characterizing the flare region rather than the host star. Within each energy window the time-scales of individual flares may span 1–2 orders of magnitude, and the trend is revealed here by considering average time-scales and a large range in energy. Lack of time resolution for cluster flare stars allow us to include them only for the decay time-scale. A few observations of evolved stars do not contradict the trend in Figure 3.

3.3. FLARE LUMINOSITIES

Photometric U-filter monitoring has resulted in numerous flare light curves on the best observed stars. Integration under the flare portions of such time series measurements and subsequent absolute calibration, yields a time average of the amount of radiative energy emitted during the flare, here referred to as the flare luminosity of a star, $L_f(U)$, in erg s^{-1}. Reliable estimates of $L_f(U)$ require a reasonable number of observed flares

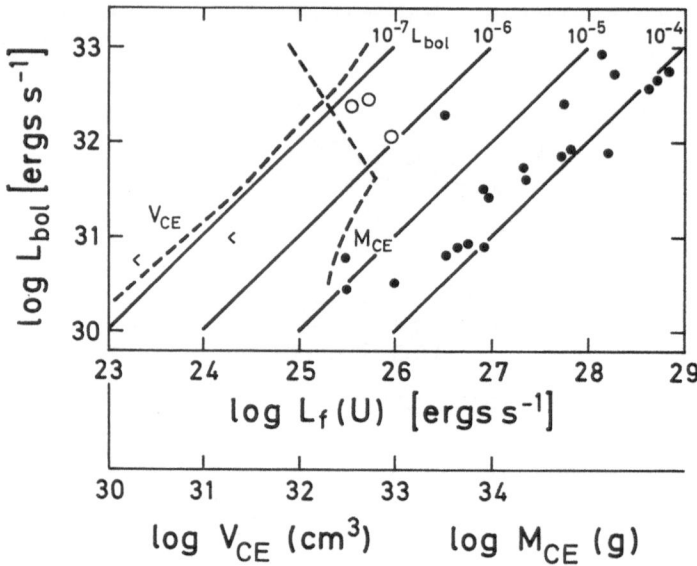

Fig. 4.   *U*-filter flare luminosity of red dwarfs versus their bolometric quiescent luminosities. Also indicated
is the run of convection zone mass ($M_{\rm CE}$) and volume ($V_{\rm CE}$) for such stars. See text for details.

for each star, say 50. In Figure 4 we have plotted $\log L_f(U)$ for a number of dKe and
dMe stars, characterized by their bolometric luminosity $L_{\rm bol}$. Two important aspects
of flare activity are apparent from this figure:

(1) A saturation level for flare activity exists for all stars with $7 \leq M_V \leq 17$. In the
U-filter the maximum amount of flare production is about 10^{-4} of the bolometric
radiative output of a star. We estimate that this corresponds to a bolometric flare
luminosity of $0.003 L_{\rm bol}$ (Pettersen, 1988).

(2) Many stars are less active than the saturation level. The non-emission line dM
stars are 500–1000 times less active than their saturated dMe counterparts. Reduced
activity may be a result of age or slow rotation.

It has been shown (Pettersen, 1988) that a change in flare luminosity is due to a change
in flare frequency at each energy level, $L_f(U) \sim (N/T)$. It has not been decided if active
stars produce larger flares than less active stars. It appears that both classes of stars
have the same basic properties, but active stars flare more often at every flare energy
level. It, therefore, requires great perseverance (or luck!) to observe a giant flare in a low
activity star.

The stars in Figure 4 represent two kinds of structures. Those with $\log L_{\rm bol} > 31.5$
have radiative cores and outer convection zones. Those fainter than $\log L_{\rm bol} = 31.5$ are
fully convective. It has been suggested that different types of dynamos are responsible
for the generation of magnetic fields on these two types of stars (Rosner, 1980). Flare
activity is a result of these magnetic fields, and we note that both bright and faint stars
in Figure 4 are capable of producing flares at the saturation level. This suggests that
dynamo efficiencies are not very different in the two types of stars, and the smooth decay

of the saturation level with bolometric luminosity indicates that the same mechanism is at work in both types of stars. The bolometric luminosity of solar-type stars is proportional to the volume of the star, so it follows from Figure 4 that the saturation level for stellar activity is in some way connected to the size of the star. The often quoted link between atmospheric activity and surface magnetic fields and the convection zone of the star, invites a search for empirical relationships between convection zone parameters and stellar activity parameters.

3.4. ATMOSPHERIC ACTIVITY AND STELLAR CONVECTIVE ENVELOPES

According to Main-Sequence models by Copeland, Jensen, and Jørgensen (1970) the mass of the convection zone increases from less than $0.01\,M_\odot$ for the Sun to a maximum for a fully convective star of $0.3\,M_\odot$. The extent of the radiative core is $0.8\,R_\odot$ in the Sun, decreasing to $0.2\,R_\odot$ (50%) for a $0.4\,M_\odot$ star. The radiative core is nonexistent in stars smaller than $0.3\,M_\odot$. As a result the average density of the convection zone is nearly a linear function of stellar mass and can be approximated by

$$\log\bar{\rho}_{CE}(\mathrm{g\ cm}^{-3}) = 2.4-4(M/M_\odot)\,.$$

Thus, there is no correspondence between the mass of the convection zone and the flare luminosity. The run $\bar{\rho}_{CE}$ is roughly parallel to the run of $L_f(U)$, but in the opposite direction. The most active dMe flare stars have average convection zone densities smaller than those at the lower end of the Main Sequence.

An almost perfect correspondence is found between the volume of the convective envelope, V_{CE}, and $L_f(U)$. The former parameter was computed from Table 2 in Copeland, Jensen, and Jørgensen (1970), using

$$V_{CE} = \tfrac{4}{3}\pi(R^3 - R_{rc}^3)\,,$$

where R is the radius of the star and R_{rc} is the radius of the radiative core. The latter becomes zero for stars smaller than $0.3\,M_\odot$. When V_{CE} is plotted against the bolometric luminosity of the stars, using the same scales on the axes as in Figure 4, an almost straight line with slope unity results, namely

$$\log V_{CE}(\mathrm{cm}^3) = \log L_{bol}(\mathrm{erg\ s}^{-1}) - 0.23$$

for stars between $0.1\,M_\odot$ and $1\,M_\odot$. It appears that the saturated (maximum) flare activity level of a dwarf star is directly proportional to the volume of its convection zone.

Since some stars do not meet such high levels of flare activity, other parameters must also be important in determining the activity levels of individual stars. One such parameter is stellar rotation, since it would influence dynamo action through the component of differential rotation. If stellar rotation slows down with the ageing of stars, this would immediately explain age effects in flare activity.

Figure 5 shows the empirical relationship between flare luminosity and the volume of the convection zone for a number of solar neighbourhood flare stars.

Radii were taken from Pettersen (1980) or relationships derived therefrom, and the sizes of radiative cores were interpolated in the tables of Copeland, Jensen, and

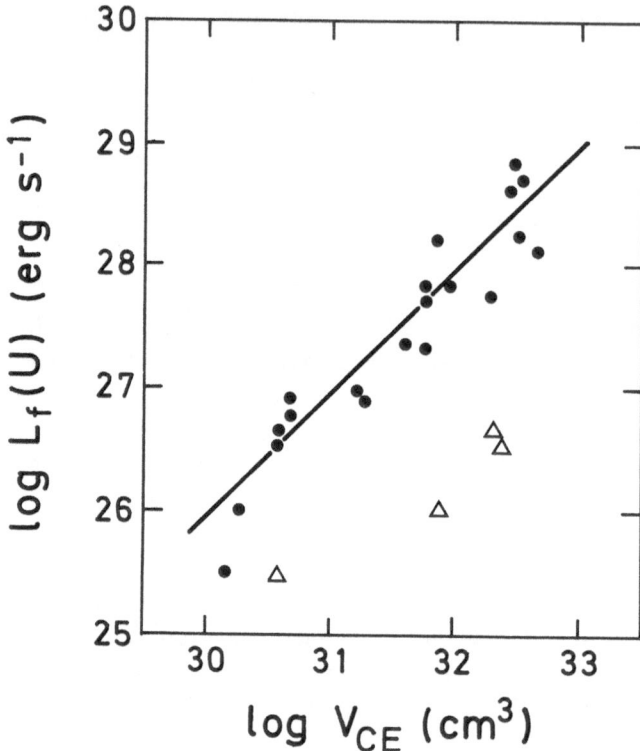

Fig. 5.   Relationship between flare luminosity and the volume of the convective envelope, for very active
dKe and dMe flare stars. Less active dM stars are indicated by triangles.

Jørgensen (1970). The same kind of relationship can be produced for coronal X-ray
luminosity and chromospheric Hα luminosity, outside the flares. It appears that all the
common proxies of stellar activity are proportional to the volume of the convection zone
for the most active stars. (Similar plots would result if V_{CE} were replaced by the volume
of the entire star because the volume of the radiative core is $\leq V_{CE}$ for stars smaller than
the Sun, as pointed out by J. H. M. M. Schmitt during the discussion session.) In Figure
6 we have plotted L_x versus V_{CE} for solar neighbourhood flare stars as well as RS CVn
stars and one T Tau star. The latter has been assumed to be fully convective, while it
is very difficult to estimate the depth of the convection zones in the evolved stars since
their ages are not well determined. We have considered their position in the HR-diagram
and have interpolated in tables by Novotny (1973). The uncertainties in $\log V_{CE}$ must
be considerable since even radii are only crudely estimated. There is no doubt that the
trend continues to hold for these very active stars, even if the positions in Figure 6 are
only approximate. This scenario explains the high level of activity in young pre-
Main-Sequence stars since they are rapid rotators, fully convective, and have large
volumes since they are still contracting towards their Main-Sequence sizes. On the
post-Main-Sequence end of stellar evolution one would expect large convection zones

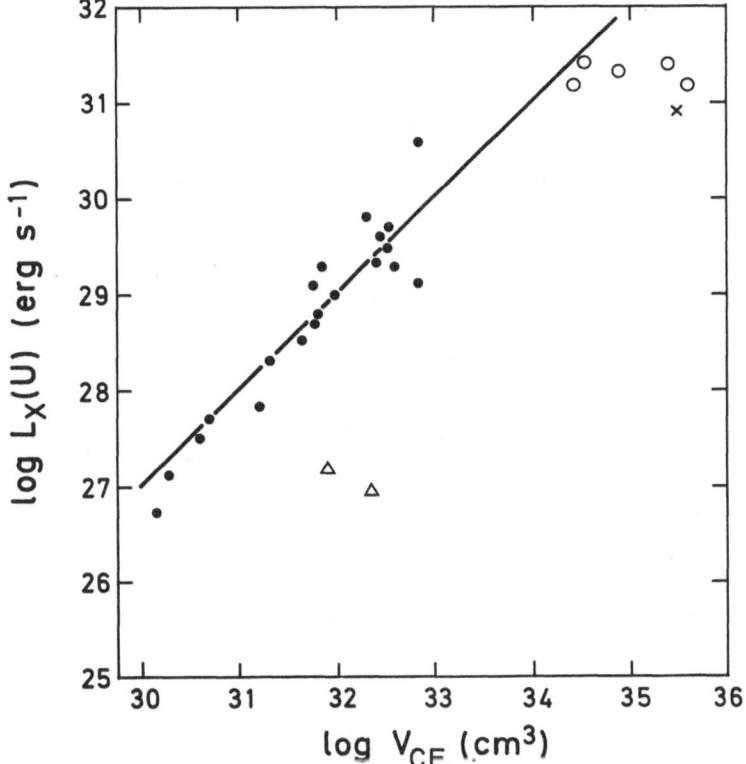

Fig. 6.   Relationship between coronal X-ray flux and the volume of the convective envelope. Dwarfs are indicated as in Figure 5. Open circles are RS CVn stars and the cross is a T Tau star.

in giants and supergiants, but volume expansion from Main-Sequence dimensions would greatly reduce stellar rotation rates. Single evolved stars are therefore not expected to show high levels of activity, but members of binary systems that are close enough for synchronization effects to maintain rapid rotation should be very active. One example is the RS CVn class.

4. Conclusions

The main conclusions of this review are:

  – Frequent flaring occurs on stars with outer convection zones, on or above the Main Sequence. This includes young as well as evolved stars; singles and members of binaries.

  – Flare-like events have also been seen elsewhere in the HR-diagram.

  – The time-scales of energetic flares are longer than for small flares, irrespective of host star.

  – Studies of flare occurrence as a function of time show contradicting results, sometimes favouring random (Poisson) distributions and sometimes indicating periodic components. Further studies are needed.

– Atmospheric activity, including flaring, grows with increasing volume of the convective envelope of the host star.

References

Bai, T. and Sturrock, P. A.: 1987, *Nature* **327**, 601.
Copeland, H., Jensen, J. O., and Jørgensen, H. E.: 1970, *Astron. Astrophys.* **5**, 12.
Doyle, J. G.: 1987, *Astron. Astrophys.* **177**, 201.
Gershberg, R. E. and Shakhovskaya, N. I.: 1973, *Nature Phys. Sci.* **242**, 85.
Haro, G. and Chavira, E.: 1955, *Bol. Obs. Tonantzintla Tacubaya*, No. 12.
Kunkel, W. E.: 1969, in S. S. Kumar (ed.), *Low Luminosity Stars*, p. 195.
Kunkel, W. E.: 1974, *Nature Phys. Sci.* **248**, 571.
Kunkel, W. E.: 1975, in V. E. Sherwood and L. Plaut (eds.), 'Variable Stars and Stellar Evolution', *IAU Symp.* **67**, 15.
Lacy, C. H., Moffett, T. J., and Evans, D. S.: 1976, *Astrophys. J. Suppl.* **30**, 85.
Melikian, N. D. and Grandpierre, A.: 1984, *Inf. Bull. Var. Stars*, No. 2683.
Novotny, E.: 1973, *Introduction to Stellar Atmospheres and Interiors*, Oxford University Press, Oxford, p. 373ff.
Oskanyan, V. S. and Terebizh, V. Yu.: 1971, *Astrofizika* **7**, 83.
Pazzani, V. and Rodono, M.: 1981, *Astrophys. Space Sci.* **77**, 347.
Pettersen, R. R.: 1980, *Astron. Astrophys.* **82**, 53.
Pettersen, B. R.: 1988, in O. Havnes *et al.* (eds.), *Activity in Cool Star Envelopes*, Kluwer Academic Publishers, Dordrecht, Holland, p. 49.
Pettersen, B. R., Coleman, L. A., and Evans, D. S.: 1984, *Astrophys. J. Suppl.* **54**, 375.
Pettersen, B. R., Panov, K. P., Sandmann, W. H., and Ivanova, M. S.: 1986, *Astron. Astrophys. Suppl.* **66**, 235.
Rosner, R.: 1980, in A. K. Dupree (ed.), *Cool Stars, Stellar Systems and the Sun*, SAO Special Report No. 389, p. 79.

LONG-DURATION SOLAR AND STELLAR FLARES

GIANNINA POLETTO

Osservatorio Astrofisico di Arcetri, Firenze, Italy

Abstract. According to one of the most popular classifications, solar flares may be assigned either to the category of small short-lived events, or to the category of large, long-duration two-ribbon (2-R) flares. Even if such a broad division oversimplifies the flare phenomenon, our knowledge of the characteristics of stellar flares is so poor, that it is worthwhile to investigate the possibility of adopting this classification scheme for stellar flares as well. In particular we will analyze Einstein observations of a long duration flare on EQ Peg to establish whether it might be considered as a stellar analogy of 2-R solar events. To this end we apply to EQ Peg data a reconnection model, developed originally for solar 2-R flares, and conclude that stellar observations are consistent with model predictions, although additional information is required to identify uniquely the physical parameters of the flare region. Application of the model to integrated observations of a 2-R solar flare, for which high spatial resolution data are also available, shows, however, that future integrated observations may allow us to solve the ambiguities of the model and use it as a diagnostic tool for a better understanding of stellar flares.

1. Introduction

Solar flares may differ widely in a number of properties such as the size of the flaring region, the duration of the event, and the energy released. Small events, for instance, last only a few minutes and release about 10^{29} ergs, i.e., about 10^{-3} less energy than is released in the larger events which span a time interval on the order of a few hours. Nevertheless, several authors tried, in the past, to classify solar flares according to their 'basic' characteristics. As a result, different classifications have been proposed (Pallavicini, 1977; Tanaka, 1983; Švestka, 1986) which oversimplify the flare phenomenon, the more they are concise, but still provide an overall description of the more fundamental aspects of a flare. In the well-known two-group classification which we will adopt in the following, small events are interpreted as involving one or a few loops and are referred to as compact or single-loop flares, while large events, which occur in an arcade of loops and are characterized at the chromospheric level by the appearance of the two bright Hα ribbons from which they take their name, are considered as typical representative of the second class of flares.

The different behavior of flares pertaining to the two classes is usually ascribed to the different physical mechanisms involved in the event. A short-lived energy input and a local magnetic field disruption account for compact flares; a continuous energy release and a global magnetic field disruption are invoked to explain two-ribbon events. In these latter flares the growing system of loops which rises in the corona and the increasing ribbon separation at the chromospheric level are interpreted as different manifestations of a reconnection process which closes back to a lower energy state an arcade of loops which had been torn open beforehand by an unspecified eruptive phenomenon.

Solar Physics **121** (1989) 313–322.

The energy released from stellar flares on M dwarf stars, although much larger than the energy released in solar flares, is delivered, as in the Sun, on time-scales ranging from a few minutes to a few hours. This behavior may be taken as an indication that flares on M stars imply the same mechanisms invoked in the solar case and possibly may be classified either as compact or as two-ribbon flares. If this were the case, it would be possible to apply to stellar flares the methods which proved successfull in modeling solar flares.

Attempts along this line are quite scanty. A hydrodynamic numerical code which describes the behavior of plasma confined in a rigid closed loop and heated by a strong heat pulse has been used by Reale *et al.* (1988) to model an X-ray flare on Prox Cen. This hydrodynamical model has been able to reproduce fairly well the temporal behavior of the X-ray emission from compact solar flares. Fisher and Hawley (1989) applied a flare loop evolution model, valid in the limit of weak evaporation, to reproduce the rise phase and the initial decay of a flare on AD Leo. This work is based on an attempt to extend to the stellar case the 'scaling laws', widely used in modeling solar loops. Poletto, Pallavicini, and Kopp (1988) and Van den Oord and Mewe (1988) interpreted a few stellar flares on the basis of a model (Kopp and Poletto, 1984) which had been originally developed for solar two-ribbon flares.

Clearly any attempt to extrapolate solar models to the stellar case is risky: integrated stellar observations do not provide all the information which would be needed for a safe extension to stellar phenomena of models originally developed to interpret events observed on the Sun with high spatial and temporal resolution. This, however, represents a workable technique for a better use of stellar data, provided one keeps in mind the uncertainties implied in such an extension. With this respect it is important to establish if, and how, a model capable of describing high-resolution data can be successfully applied to spatially-integrated data.

The present paper will focus on long-duration flares, in an attempt to establish whether long-lasting stellar flares can be interpreted as stellar counterparts of solar two-ribbon flares. To this end we apply a reconnection model, successfully applied to 2-R solar flares, to a flare observed on EQ Peg (Section 2), and check whether the profile of the magnetic energy release, as predicted by model, matches the X-ray light curve, as observed by EXOSAT. Extending previous works (Poletto, Pallavicini, and Kopp, 1988; Van den Oord and Mewe, 1988), we show (Section 3) that it is possible to reach an agreement between the data and the analytical predictions throughout the whole flare lifetime (after the impulsive phase). However, even if consistency between model predictions and observations is easily reached, the model is unable to choose, among the different scenarios which allow for consistency, the one which better approximates the real situation. To overcome this severe limitation, we apply (Section 4) the model to integrated observations of a solar 2-R flare, whose characteristics are well known from high-resolution data, and compare the parameters predicted by the model on the basis of integrated observations with results obtained from high-resolution data. This comparison shows that the correct parameters of the flareing region can be uniquely determined whenever its density is known from observations. Future integrated observations of stellar flare spectra may provide this data and we conclude that the 2-R model, complemented by this further information, may be used as a diagnostic tool to infer unidentified flare parameters and define uniquely the flare scenario.

2. The Reconnection Model

As mentioned above, the reconnection model hypothesizes that an open magnetic field, created by an eruptive phenomenon, closes back to a lower energy state. In this process open field lines are driven toward the neutral sheet by an unbalanced Lorentz force, and reconnect at progressively higher levels in the corona. The excess magnetic energy released by reconnection shows up as thermal energy of the bright X-ray loops (Hirayama, 1974; Kopp and Pneuman, 1976).

These basic ideas have been worked out subsequently by a few authors (Pneuman, 1981, 1982; Cargill and Priest, 1982) to provide a comprehensive description of the relationships between the various phenomena observed in two-ribbon flares. When the magnetic field configuration is simple enough, a two-dimensional model may be adequate to describe the flare region topology, provided is understood that such a simplified description has, by necessity, to overlook some characteristics – e.g., magnetic shears – which may be relevant when a more detailed representation of the event is required. Such a simplified model has been developed analytically by Kopp and Poletto (1984), who assumed the initial open configuration to be radial and the reconnected one to be potential between the solar surface ($r = R_0$) and a spherical equipotential source surface at $r = r_1$ (neutral point height), and non-potential and still radial above this height. With these assumptions Kopp and Poletto solved Laplace equation for the scalar potential ψ and expressed its solution in terms of a single Legendre polynomial of degree n. Once the scalar potential is known, its partial derivatives, with respect to spatial coordinates, give the magnetic field components at any position r. The effect of reconnection progressing to higher and higher altitudes is mimicked by allowing r_1 to take increasingly higher values and constructing a sequence of field configurations with successively higher closed field lines. These, if the function $r_1 = r_1(t)$ is known, may be directly compared with the configuration observed at time t. Conversely, when the function $r_1(t)$ is unknown, its profile can be determined a posteriori, by matching the calculated magnetic topology to the observed one.

From the analytical expression of the magnetic field it is trivial to derive the total magnetostatic energy $E = \int (B^2/8\pi)\, dV$ as a function of the height $r_1 = r_1(t)$. Since the field is potential up to r_1 and non-potential above, the stored energy decreases as r_1 rises to higher altitudes, and at any time t the liberated energy is given by

$$dE/dt = (1/8\pi)2n(n + 1)(2n + 1)^2 R_0^3 B_0^2 \{I_{12}(n)/P_n^2(\vartheta_{12})\} \times$$

$$\times \{(r_1/R_0)^{2n}[(r_1/R_0)^{2n + 1} - 1]/[n + (n + 1) \times$$

$$\times (r_1/R_0)^{2n + 1}]^3\}\, d/dt(r_1/R_0).$$

$$(1)$$

An exhaustive description of the reconnection model can be found in the above quoted reference. Here it will suffice to clarify the relation between individual parameters in (1) and observations. The degree n of the Legendre polynomial is a free parameter which is adjusted to fit the latitudinal width of the flaring region, smaller n corresponding to larger regions. B_0 is a parameter related to the magnetic field strength in the region: larger B_0's imply a larger magnetic energy content. The function $r_1 = r_1(t)$ can be derived from high spatial resolution observations by measuring, at different times t during the flare lifetime, the height of the most recently formed X-ray loop, generally assumed to lie immediately below the reconnection altitude. Finally, R_0 is the stellar radius, and $I_{12}(n) = \int P_n^2 \, d(\cos \theta)$, with the integration extended over the lobe width.

3. Modeling EQ Peg Flare

When modeling solar flares, the choice of n, B_0, $r_1 = r_1(t)$, is not free: observations of the size of the region, of the magnetic field strength in the flaring area, and of the rise of X-ray loops in the corona provide all the information necessary to determine these parameters and evaluate uniquely the temporal profile of the magnetic energy release. Stellar flare integrated observations provide none of the previous information. Therefore, in order to apply the reconnection model, one has to devise some criteria to establish the behavior of these quantities.

The value n of the degree of the Legendre polynomial can be parametrized to represent different flaring region sizes. There is evidence that stellar active regions may be larger than solar active regions. Therefore, besides $n = 17, 35$ (latitudinal width, respectively, 5, 10 deg) we considered also smaller n values ($n = 5, 9$; latitudinal width, respectively, 33 and 20 deg). The function $r_1 = r_1(t)$ can be determined only by resorting to the solar case. From the observations of a number of two-ribbon solar events, it appears that the upward expansion of the hot loop system can be described by an exponential law of the form

$$r_1(t) = 1 + (H/R_0) \left[1 - \exp(-t/t_0) \right] , \qquad (2)$$

where H is the maximum height reached by the X-ray loops (on the order of the width of the flaring region), and t_0 is a constant which depends on the duration of the event. In the following we assume this law to hold also in the stellar case. We notice that if we make the further assumption that the longitudinal extent of the region is about twice its latitudinal width (as in the Sun), (2) allows us to evaluate the volume of the flaring region at any time t. The magnetic field strength in the region can be either chosen *a priori* to be large enough to provide for the flare energy requirement (see, for instance, Poletto, Pallavicini, and Kopp, 1988), or determined by assuming that the loop plasma is magnetically confined. In this case the gas pressure, at any time t, is derived from the temperature and emission measure given by X-ray observations and the volume calculated from the model – as described above – and the magnetic field strength is set to the value required by a balance between gas and magnetic pressures.

These criteria allow us to calculate, through (1), the magnetic energy release rate and to check whether the calculated energy release matches the X-ray light curve. To this end we used EXOSAT observations of the flare which occurred on EQ Peg on 6 August, 1985. Figure 1 shows the four curves which give the rate of magnetic energy release for $n = 5, 9, 17, 35$ superposed to the light curve of the flare as derived from the Medium Energy experiment data (≈ 2–10 keV) (dots). Each curve provides a different scenario for the flaring region. Low n values correspond to large regions, with large volumes (high H values), and, since densities are evaluated from the emission measure, densities lower than in regions characterized by larger n. As a consequence, also the gas pressure and the magnetic field strength are smaller for low n, even if, due to the large volume, the magnetic energy content is higher for lower n. In spite of these differences each curve may be scaled down to match quite closely the observed X-ray light curve. While the reconnection model can represent the data, its inability to make a choice among sets of different parameters appear as a major shortcoming in all practical applications of the model. Next section shows how to overcome this difficulty.

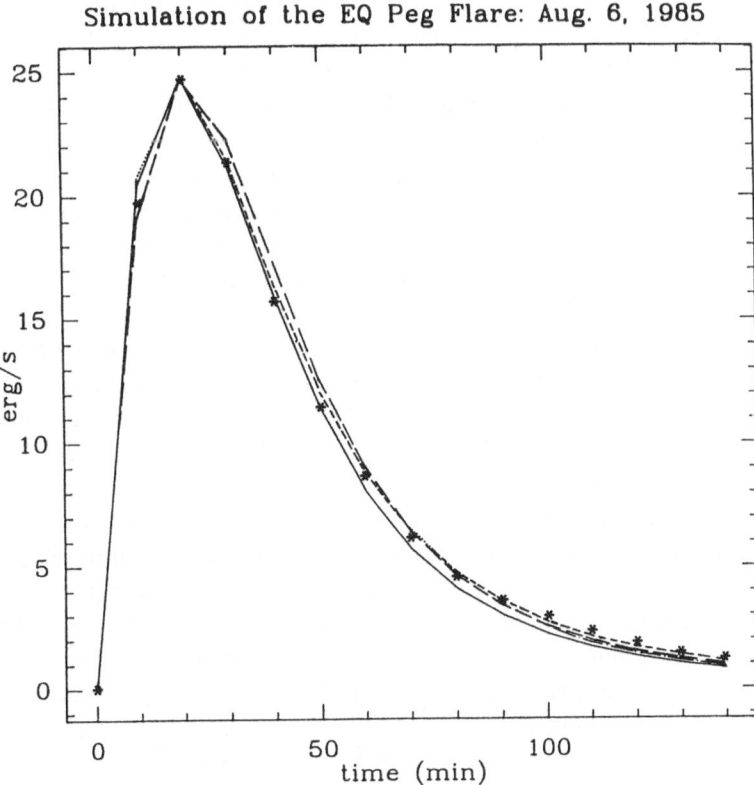

Fig. 1.   The rate of energy release during the flare observed on EQ Peg on 6 August, 1985 is shown here as derived from EXOSAT Medium Energy observations in the band 2–10 keV (dots) and as predicted by the reconnection model simulations (solid, dotted, long-dash, short-dash curves). The analytical curves have been calculated through the relationships (1) and (2) (see text): t_0 has been set to 8000 s; the constant H is set equal to the region width and the parameter B_0 is derived from the balance between gas and magnetic pressure.

4. Modeling Integrated Solar Flare Data

In spite of the pessimistic conclusion of the previous paragraph, we notice that the specification of an additional parameter would be sufficient to identify a realistic scenario. For instance, assuming recent observations of magnetic field in the dM star AD Leo (Saar and Linsky, 1985) to be representative of typical magnetic field strength in M stars, we will be drawn to consider, as the most plausible scenario for the EQ Peg flare region, a width on the order of 10–20 deg, a magnetic field strength on the order of 2000–3000 G, and an electron density which from 3.6–9.7×10^{12} cm^{-3} at peak X-ray intensity, decreases to 3–8×10^{11} cm^{-3} by the end of the flare.

Observations of stellar magnetic field are, however, much too scanty to rely heavily on them. Also the size of the active region, which determines n, and, therefore, points to a unique model, is obviously not provided by X-ray integrated stellar observations. We apparently lack the means to verify whether the model is realistic. However, applying the model to integrated observations of solar flares, for which high-resolution observations are also available, we have the possibility of checking if and how well the real scenario is approximated, and, possibly, to find out how integrated observations can be supplemented to provide the scenario known, from high-resolution data, to be realistic. Such an opportunity is offered by the SOLRAD satellites, which made integrated solar observations at the same time that Skylab high-resolution data were gathered.

The solar two-ribbon flare which occurred on 29 July, 1973, had been observed both by Skylab and SOLRAD, and showed such a simplified and regular behavior as to be the very prototype of 2-R flares (Martin, 1979; Nolte *et al.*, 1979; Petrasso *et al.*, 1979; Moore *et al.*, 1980; Švestka *et al.*, 1982). The size of the active region where the flare occurred, its magnetic field strength distribution, the growth of the hot X-ray loops with time, have all been observed in great detail. Only SOLRAD-type integrated data would be available, however, if the Sun had been observed as a star. Assuming this is the case, we repeated the procedure used in the EQ Peg flare simulation, and matched the observed light curve with the magnetic energy release curve predicted by the reconnection model. Results of this simulation are shown in Figure 2. Analytical curves calculated with the same parameters and assumptions adopted when modeling the EQ Peg flare (namely, same values for the n parameter; maximum height of the X-ray loops equal to the width of the active region; balance between magnetic and gas pressure) are superposed to integrated data in the 1–8 Å band from SOLRAD 9 observations. Once more, independently of the choice of the parameters, the analytical curves fit the data quite well, even if, as shown in Figure 3, the gas pressure, evaluated from SOLRAD data and volumes predicted by the model, strongly varies as a function of n. Nevertheless, we are unable to select the more realistic scenario.

Skylab high-resolution observations, however, provide a means to evaluate the pressure of the loop plasma. Densities in hot loops have been derived from the time delay an X-ray loop takes to cool down to chromospheric temperatures (Švestka *et al.*, 1982). From these densities and temperatures determined from the ratio between fluxes in two SOLRAD channels, the gas pressure may be calculated (solid line in Figure 3). Clearly

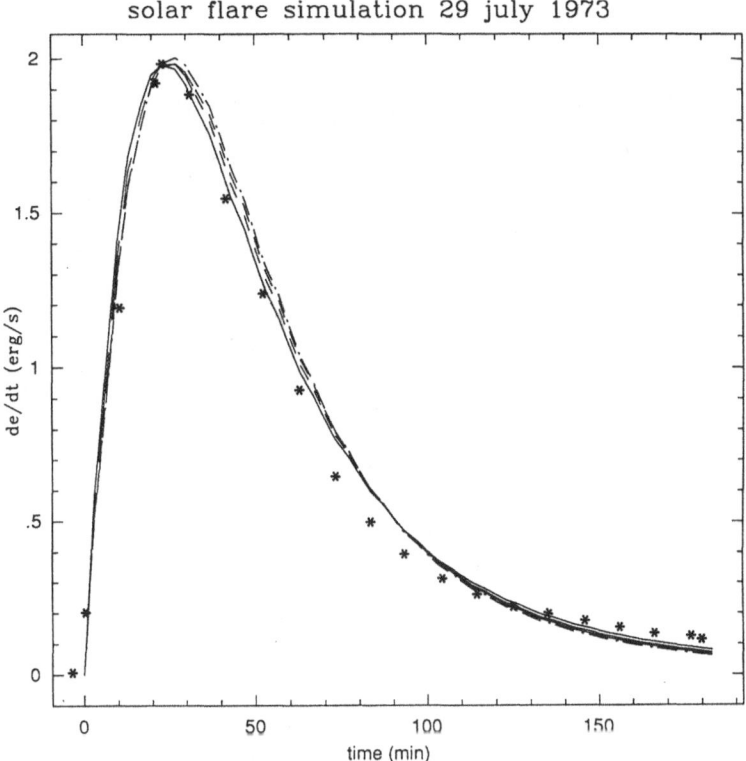

Fig. 2.   The rate of energy release during the solar flare observed on 29 July, 1973 is shown here as derived from SOLRAD 9 data in the 1–8 Å band, and as predicted by the reconnection model simulations (solid, dotted, long-dash, short-dash). As in Figure 1, simulations with different n are indistinguishable when scaled down to fit the observed energy release rate.

the model with $n = 17$ approximates the physical conditions of the loop plasma better than the others.

In fact high-resolution observations allow us to establish that the model with $n = 17$ gives a realistic description of the flaring region. Both its volume (Moore *et al.*, 1980) and the value of the average magnetic field at the photospheric level (100 G) (Michalitsanos and Kupferman, 1974), are identified correctly. At higher levels a comparison between the values given by the model for the magnetic pressure at the heights new loops are formed and the values of the gas pressure in these loops, calculated from X-ray observations, shows that, between 7000 and 90 000 km, the magnetic field strength in the potential approximation adopted by the model is within 40% of the value required for a magnetic confinement of the loop plasma. We conclude that the model with $n = 17$ is realistic and may be selected from the set of different simulations whenever densities in the flaring region are known.

Although unavailable at present, this requirement does not appear impossible to meet. Future X-ray missions, like AXAF and XMM, will obtain spectra with high enough resolution to separate a number of density-sensitive lines (Linsky, 1987; Barr *et al.*,

solar flare simulation 29 july 1973

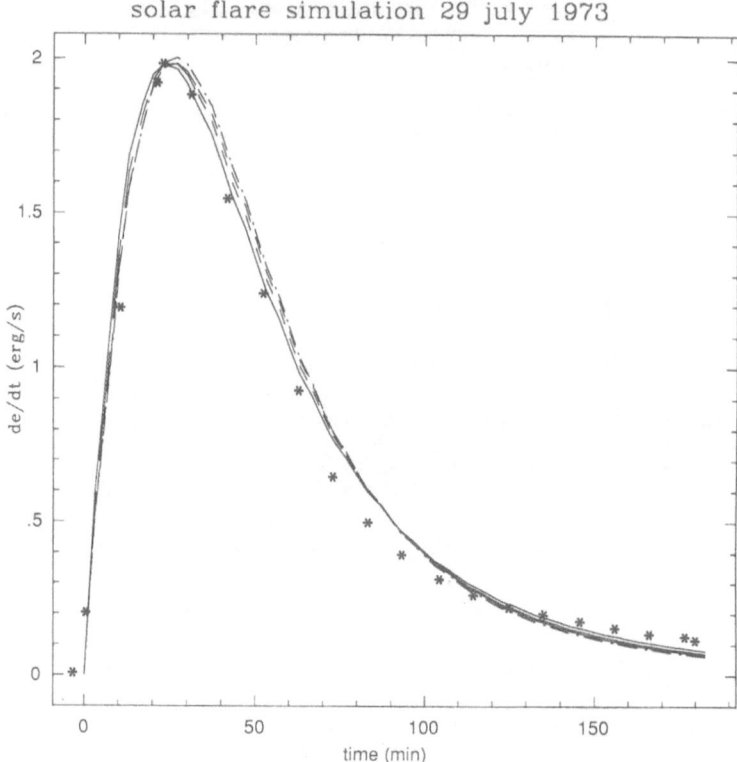

Fig. 3.   Gas pressure vs time in the solar flare of 29 July, 1973. The solid curve shows the behavior of the empirical gas pressure at different times as evaluated from temperatures derived from the ratio of SOLRAD 9 fluxes in the bands 1–8 and 8–20 Å, and densities derived from the time a loop takes to cool down from X-ray to chromospheric temperatures. At late times, only lower limits to the density values can be given; the corresponding gas pressures are marked with arrows. The four curves labelled by the value of the parameter n give the profile of the gas pressure vs time as evaluated from temperatures and emission measures derived from SOLRAD 9 data and volumes predicted by the reconnection model. Smaller n imply larger volumes and, as a consequence, smaller densities and gas pressures.

1988). Integrated observations will, therefore, supply the value of density in stellar flares, thus giving the reconnection model the capability of providing crucial information about the stellar flaring regions: their size at the photospheric level (via the value of n), their volume at different times during the flare (via the law describing the rise of hot X-ray loops in the corona), and the average magnetic field strength (via magnetic confinement of the loop plasma) at different heights in the region.

5. Concluding Remarks

Before drawing any conclusion from these results, a number of points need to be discussed. If the energy released by reconnection has to account for the total energy released in the gradual phase of the flare, we have to make sure not only that the profile of the magnetic energy release vs time reproduces the actual temporal profile of the rate of flare energy release, but also that the energy released by reconnection meets quantitatively the total energy output. Therefore, two questions need to be asked: can we safely assume the X-ray light curve to be representative of the temporal profile of the overall flare energy release? And how does the total energy released by the flare compare with the energy released in the X-ray band?

We refer the reader to Poletto, Pallavicini, and Kopp (1988) for a more thorough discussion of these points. Briefly, we recall here, as to the first point, that, resorting to solar observations, we can consider the profile of the energy released in the X-ray band as representative of the overall profile of the radiative flare energy. Flare light curves at different wavelengths, after the precursor and impulsive phase, do not show relevant differences, except for the microwave range, energetically unimportant (e.g., Kane, 1974; Lin, 1975). Whether this is representative of the profile of the overall flare energy output, is a more difficult question to answer: information about the temporal profile of the mechanical energy associated with the flare is scanty, at best.

As to our second question, following Canfield et al. (1980) and resorting once more to the solar analogy, we may assume the total energy radiated by the flare to be about ten times larger than the energy released in the X-ray bands we considered. The magnetic energy supply has to be at least as large, and possibly larger, if it has also to provide for other flare energy losses.

A solution to these problems will allow us to define more precisely the constraints the reconnection model has to satisfy. Still, such a model, besides accounting for solar two-ribbon flares, is capable of reproducing the X-ray light curve of stellar long-duration flares, while providing for their energy requirement (the magnetic energy released in the EQ Peg flare simulations is larger than the X-ray energy losses by a factor ranging between 14 and 35, depending on the value of n). Lacking further observational evidence about the characteristics of stellar flares, rather than claim that two-ribbon stellar flares have been identified, it may be safer to claim that the predictions from the reconnection model are consistent with observations of long duration stellar flares. We recall, however, that experimental evidence for a temporary increase in the column absorption density along the line-of-sight at the time EINSTEIN observed a long lasting event on the dM star Prox Cen was fundamental to an earlier suggestion for the presence of two-ribbon flares on stars (Haisch et al., 1988). In fact we have shown that the X-ray light curve of the Prox Cen flare can be simulated by the reconnection model (Poletto, Pallavicini, and Kopp, 1988). This, however, is not the only interpretation supported by the data (Peres, 1989).

Besides establishing, with the limitations mentioned above, that there may be counterparts to two-ribbon solar flares, it has to be stressed that the reconnection model, when supplemented by integrated observations of density-sensitive lines, has the capability of determining a number of parameters which provide a complete description of the flaring region. Its determination of the magnetic field strength and size of stellar flaring regions may be compared with independent determinations of the same quantities from other techniques and provide a cross-check on the derivation of stellar flaring region parameters. In a field where the unavailability of high resolution observations sets severe limits to the amount of information that can be drawn from the data, it is especially relevant to establish alternative means to achieve a reliable description of phenomena which are unobservable in detail. This seems the more promising area of applicability of the model.

References

Barr, P. and 15 co-authors: 1988, *The High-Throughput X-Ray Spectroscopy Mission*, ESA SP-1097.

Canfield, R. C., Cheng, C.-C., Dere, K. P., Dulk, G. A., Mc Lean, D. J., Robinson, R. D. Jr., Schmahl, E., Jr., and Schoolman, S. A.: 1980, in P. A. Sturrock (ed.), *Solar Flares*, Colo. Assoc. Univ. Press, Boulder, p. 451.

Cargill, P. J. and Priest, E. R.: 1982, *Solar Phys.* **76**, 357.

Fisher, G. H. and Hawley, S. L.: 1989, in B. M. Haisch and M. Rodonò (eds.), IAU Colloq. 104, *Solar and Stellar Flares*, Poster Papers, Publ. Catania Astrophys. Obs., Special Volume, p. 353.

Haisch, B. M., Linsky, J. L., Bornmann, P. L., Stencel, R. E., Antiochos, S. K., Golub, L., and Vaiana, G. S.: 1983, *Astrophys. J.* **267**, 280.

Hirayama, T.: 1974, *Solar Phys.* **34**, 323.

Kane, S. R.: 1974, in G. Newkirk, Jr. (ed.), 'Coronal Disturbances', *IAU Symp.* **57**, 105.

Kopp, R. A. and Pneuman, G. W.: 1976, *Solar Phys.* **50**, 85.

Kopp, R. A. and Poletto, G.: 1984, *Solar Phys.* **93**, 351.

Lin, R. P.: 1975, in S. R. Kane (ed.), 'Solar Gamma-, X- and EUV Radiation', *IAU Symp.* **68**, 385.

Linsky, J. L.: 1987, *Astrophys. Letters Com.* **26**, 385.

Martin, S. F.: 1979, *Solar Phys.* **64**, 175.

Michalitsanos, A. G. and Kupferman, P.: 1974, *Solar Phys.* **36**, 304.

Moore, R. L. and 15 co-authors: 1980, in P. A. Sturrock (ed.), *Solar Flares*, Colo. Assoc. Univ. Press, Boulder, p. 341.

Nolte, J. T., Gerassimenko, M., Krieger, A. S., Petrasso, R. D., and Švestka, Z.: 1979, *Solar Phys.* **62**, 123.

Pallavicini, R., Serio, S., and Vaiana, G. S.: 1977, *Astrophys. J.* **216**, 108.

Peres, G.: 1989, *Solar Phys.* **121**, 289 (this issue).

Petrasso, R. D., Nolte, J. T., Gerassimenko, M., Krieger, A. S., Krogstad, R., Seguin, F. H., and Švestka, Z.: 1979, *Solar Phys.* **62**, 133.

Pneuman, G. W.: 1981, in E. R. Priest (ed.), *Solar Flare Magnetohydrodynamics*, Gordon and Breach, New York, p. 379.

Pneuman, G. W.: 1982, *Solar Phys.* **78**, 229.

Poletto, G., Pallavicini, R., and Kopp, R. A.: 1988, *Astron. Astrophys.* **201**, 93.

Reale, F., Peres, G., Serio, S., Rosner, R., and Schmitt, J. H. M. M.: 1988, *Astrophys. J.* **328**, 256.

Saar, S. A. and Linsky, J. L.: 1985, *Astrophys. J.* **299**, L47.

Švestka, Z.: 1987, in D. Neidig (ed.), *The Lower Atmosphere of Solar Flares*, NSO/Sac Peak, p. 332.

Švestka, Z., Dodson-Prince, H. W., Martin, S. F., Mohler, O. C., Moore, R. L., Nolte, J. T., and Petrasso, R. D.: 1982, *Solar Phys.* **78**, 271.

Tanaka, K.: 1983, in P. B. Byrne and M. Rodonò (eds.), 'Activity in Red Dwarfs', *IAU Colloq.* **71**, 307.

Van den Oord, G. H. J., and Mewe, R.: 1988, *Astron. Astrophys.* (in press).

Webb, D. F., Cheng, C.-C., Dulk, G. A., Edberg, S. J., Martin, S. F., McKenna-Lawlor, S., and McLean, D. J.: 1980, in P. A. Sturrock (ed.), *Solar Flares*, Colo. Assoc. Univ. Press, Boulder, p. 471.

SOLAR FLARES: HIGH-ENERGY RADIATION AND PARTICLES

ERICH RIEGER

Max-Planck-Institut für Physik und Astrophysik, Institut für extraterrestrische Physik, 8046 Garching, F.R.G.

Abstract. Due to the Sun's proximity flares can be investigated in the gamma-ray regime and flare generated particles can be measured in space and related to particular events. In this review paper we focus on the problem of particle acceleration by using as observational ingredients: the fluxes and spectra of particles inferred from gamma-ray measurements and observed in interplanetary space, the temporal characteristics of flares at high-energy X- and gamma-rays and the distribution of gamma-ray flares over the solar disc.

1. Introduction

The storage and sudden explosive release of energy in sheared magnetic fields involving the acceleration of particles is a common phenomenon occurring in plasmas throughout the Universe from a place as close as the Earth's magnetosphere to objects at cosmological distances such as quasars. The physical understanding of these processes is of basic importance to astrophysics and solar terrestrial physics. Over this vast distance scale the Sun is of crucial importance. Due to its proximity, flares can be investigated in the gamma-ray regime and flare-generated particles can be measured in space and related to particular events, which provides unique information about the acceleration of particles. Furthermore, flares can be localized on the disc, allowing us to study them under different aspect angles. Observations of center-to-limb variations of properties of the high-energy radiation place constraints on flare models.

We begin this article with a brief summary of the mechanisms which produce gamma-rays and neutrons and then comment on a correlation between the continuum and line radiation during solar flares. We follow with a comparison of particle fluxes measured in interplanetary space and particle number inferred from the gamma-ray measurements. After a discussion of the energy spectra of the primary accelerated particles and of the temporal characteristics of gamma-ray flares relevant particle acceleration mechanisms will be briefly reviewed. Finally, the implications on the particle beaming deduced from the nonuniform distribution of gamma-ray flares on the solar disc will be discussed.

Solar elemental abundance determinations, the anomalous ^{3}He content during certain flares and periodicity studies carried out with energetic events are not covered in this paper. The reader is referred to reviews by Chupp (1984, 1987), Ramaty and Murphy (1987), and Kocharov (1987).

Solar Physics **121** (1989) 323–345.
© 1989 *by Kluwer Academic Publishers.*

2. Production of Gamma-Rays and Neutrons

The gamma-ray and neutron production mechanisms, most relevant for the Sun, are shown in Figure 1. They can be classified according to the species of the parent particle (electron or proton).

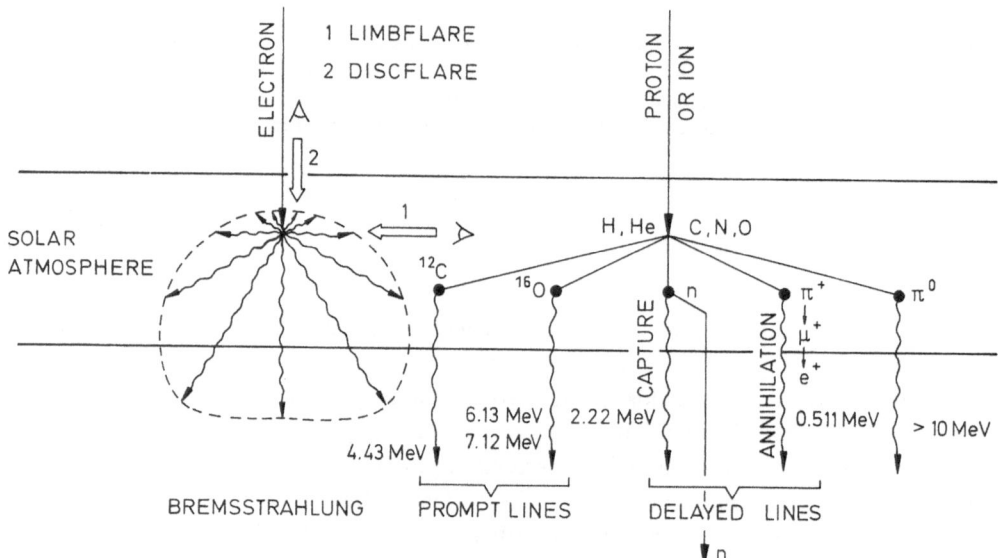

Fig. 1. Schematic drawing of the gamma-ray and neutron production mechanisms. The dashed oval illustrates the anisotropic emission of photons resulting from electron bremsstrahlung. Energy of incident electron $\sim$ 150 keV; photon energy $\sim$ 50 keV (Elwert and Haug, 1971). The direction given to the arrows symbolizing the gamma rays and neutrons produced by ion interactions does not imply a directivity of the radiation.

2.1. ELECTRON BREMSSTRAHLUNG

From the interaction of electrons with matter a continuous spectrum results extending from the energy of the most energetic electrons to almost zero. Using the cross section given by Koch and Motz (1959) for electron-proton bremsstrahlung and the cross section for electron-electron bremsstrahlung given by Haug (1975), Bai (1977) has calculated the photon spectrum resulting from accelerated electrons interacting with a fully-ionized hydrogen plasma. Assuming an electron power-law spectrum, the photon spectrum is also a power law between 10 and 80 keV, hardening at higher energies. Above $\sim$ 400 keV the spectrum is again a power law.

2.2. ION INTERACTIONS

Neutral radiation from ion interactions can be divided into three classes, whose relative importance depends upon the energetic particle energy.

2.2.1. *Nuclear De-excitation Lines*

As opposed to the continuum spectrum of electron bremsstrahlung, narrow gamma-ray lines result from the interaction of accelerated protons and α-particles with He and heavier nuclei in the solar atmosphere. The cross sections for the various nuclear de-excitation lines have been discussed by Ramaty, Kozlovsky, and Lingenfelter (1979). The strongest lines result from the de-excitation of ^{12}C at 4.438 MeV and ^{16}O at 6.129 MeV. The proton energies most relevant for the production of these lines are 10–30 MeV. Because the lifetimes of the excited states are 10^{-12} s or shorter, the lines are emitted without a measurable delay, and are thus called prompt lines.

2.2.2. *Neutron Production*

Free neutrons are produced when accelerated ions interact with the ambient atmosphere. The most prolific neutron producing reaction is that of protons on ^4He with a threshold of about 30 MeV (Ramaty, Kozlovsky, and Lingenfelter, 1975). These neutrons can have different fates:

– They escape from the Sun and those reaching the Earth before decaying, can be detected with suitable detectors (Chupp *et al.*, 1982; Debrunner *et al.*, 1983; Kocharov, 1983; Chupp *et al.*, 1987). Neutrons leaving the Sun which do not survive can be identified through their energetic decay protons which are trapped by the interplanetary magnetic field (Evenson, Meyer, and Pyle, 1983; Evenson, Kroeger, and Meyer, 1985).

– Neutrons which remain at the Sun can be captured by nuclei before decaying. The reaction $^3He(n, p)\,^3H$ proceeds without the emission of radiation, whereas the reaction $^1H(n, \gamma)\,^2H$ produces the 2.223 MeV line. Because this line is produced in the photosphere it is strongly attenuated in limb flares (Wang and Ramaty, 1974; Hua and Lingenfelter, 1987a). For flares off the limb it is by far the strongest one. As opposed to the de-excitation lines, the 2.223 MeV line is emitted with a delay originating from the finite capture time of neutrons on protons. This is evident in Figure 2 which shows the dynamic spectrum of the flare of 24 April, 1984 in the energy range from 0.3 to 9 MeV. The decay of the line can be followed for about 20 min. Because the neutron capture time depends on the proton and ^3He densities, the study of the time history of the 2.223 MeV line can give information on the abundance of ^3He and the depth in the atmosphere where these reactions take place. Analysis of the decay of the line shows that the capture occurs mainly in the photosphere at hydrogen densities of $\sim 1.3 \times 10^{17}$ cm^{-3} (Kanbach *et al.*, 1981; Prince *et al.*, 1983). The $^3He/H$ ratio derived by Hua and Lingenfelter (1987b) from the 2.223 MeV time history of the flare of 3 June, 1982 is $(2.3 \pm 1.2) \times 10^{-5}$. This value is close to that obtained by Yang *et al.* (1982) who assume that turbulent mixing of the solar interior does not significantly alter the ^3He abundance from that of primordial nucleosynthesis.

2.2.3. *Gamma-Rays from Pion Decay*

Pion production, which requires proton energies of > 100 MeV leads to photons with energies > 10 MeV, either through the prompt decay of neutral pions or from the decay

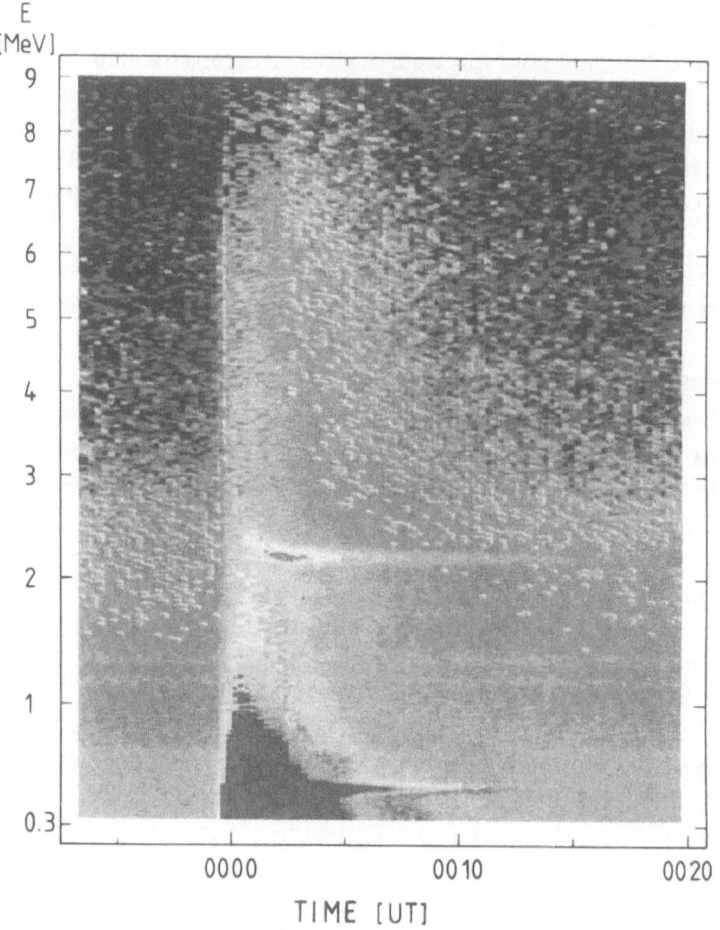

Fig. 2. Temporal evolution of the spectrum of the 24 April, 1984 flare in the energy range 0.3–9 MeV recorded by the gamma-ray spectrometer on SMM. Note the 'afterglow' in the delayed lines at 0.511 and 2.223 MeV. The flare is shown without background subtraction.

of charged pions to positrons and electrons. The positrons and electrons slow down mostly due to Coulomb collisions, bremsstrahlung, and synchrotron losses. After thermalization the positrons annihilate with ambient electrons either directly to produce two 0.511 MeV gamma rays per positron or via positronium, where the triplet state (75%) decays into 3 photons with energies less than 0.511 MeV (see, e.g., Murphy, Dermer, and Ramaty, 1987). Pion decay positrons have energies of a few tens of MeV, whereas positrons originating from the decay of radioactive nuclei have energies of ~1 MeV (Kozlovsky, Lingenfelter and Ramaty, 1987). Due to the high density of the interaction region the slowing down time of both populations of positrons is short compared to the radioactive decay times, so that the delay seen in Figure 2 reflects the half lives of the parent radioactive nuclei (Murphy and Ramaty, 1984).

3. Gamma-Ray Line – Bremsstrahlung Correlation

It is evident from the foregoing that the solar flare gamma-ray spectrum will appear as a superposition of electron bremsstrahlung, line radiation, and pion decay emission. As pointed out by Ibragimov and Kocharov (1977) and by Ramaty, Kozlovsky, and Suri (1977), above about 1 MeV line radiation will dominate the spectrum. This important fact facilitates the separation of the ionic from the electronic component. To illustrate this, the time-integrated count spectrum of the flare of 7 December, 1982 observed with the Solar Maximum Mission (SMM) Gamma-Ray Spectrometer (GRS) is shown in Figure 3 in the energy range from 0.28 to 9 MeV. Between 0.28 MeV (the threshold of the detector) and 1 MeV, where strong nuclear lines are absent, the emission is a continuum resembling a power law in spectral shape. Above about 1 MeV the contribution of line radiation, modified by the response of the detector, is apparent. The absense of intense lines above about 7.5 MeV (Crannell, Crannell, and Ramaty, 1979) shows up as a cutoff in the count spectrum. Above 10 MeV (not shown in the figure) a contribution of pion decay gamma-rays and of electron bremsstrahlung has to be considered. To separate the ionic from the electronic component the best fit power-law obtained between 0.28 and 1 MeV is extrapolated to higher energies (dashed line). The

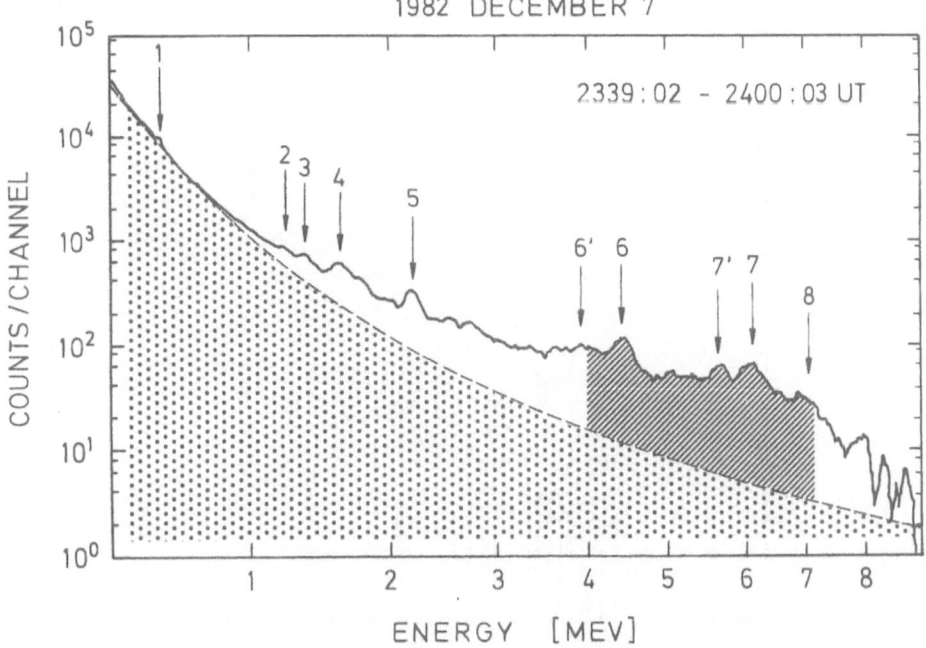

Fig. 3.   Time integrated and background subtracted count spectrum of the 7 December, 1982 flare in the energy range 0.28–9 MeV. Significant lines and their origin are: (1) 0.511 MeV (annihilation line); (2) 1.24 MeV (^{56}Fe); (3) 1.38 MeV (^{24}Mg); (4) 1.63 MeV (^{20}Ne); (5) 2.223 MeV (neutron capture line, attenuated due to limb proximity of the flare); (6) 4.43 MeV (^{12}C); (7) 6.13 MeV (^{16}O), and (8) 7.12 (^{16}O). (6′) and (7′) are the instrumentally produced escape peaks of the lines at 4.43 and 6.13 MeV, respectively. The dashed and dotted areas are explained in the text.

excess above this extrapolation, which is prominent especially between 4 and 7 MeV (dashed area), is ascribed to nuclear interactions.

Since the launch of the SMM satellite in February 1980 until 1987, the GRS has recorded 150 flares with an excess of the time-integrated flux (fluence) at the threshold of the detector (0.28 MeV); 90 out of these 150 flares allow spectral analysis. Of these events 50% show an excess between 4 and 7 MeV. In Figure 4 this excess fluence (dashed area of Figure 3) is plotted versus the fluence above 280 keV (dotted area of Figure 3) attributed to electron bremsstrahlung. For flares without nuclear line excess, the upper limits are indicated. No distinction is made between short impulsive (< 1 min) and long duration events (~ 30 min). The striking feature of Figure 4 is the close correlation between the nuclear line excess, and the > 280 keV continuum fluence, from which the following conclusions are made.

– The smaller number of flares showing observable evidence for nuclear gamma-ray lines is due only to a lower instrument sensitivity for nuclear emissions (~ 1 phot cm^{-2}).

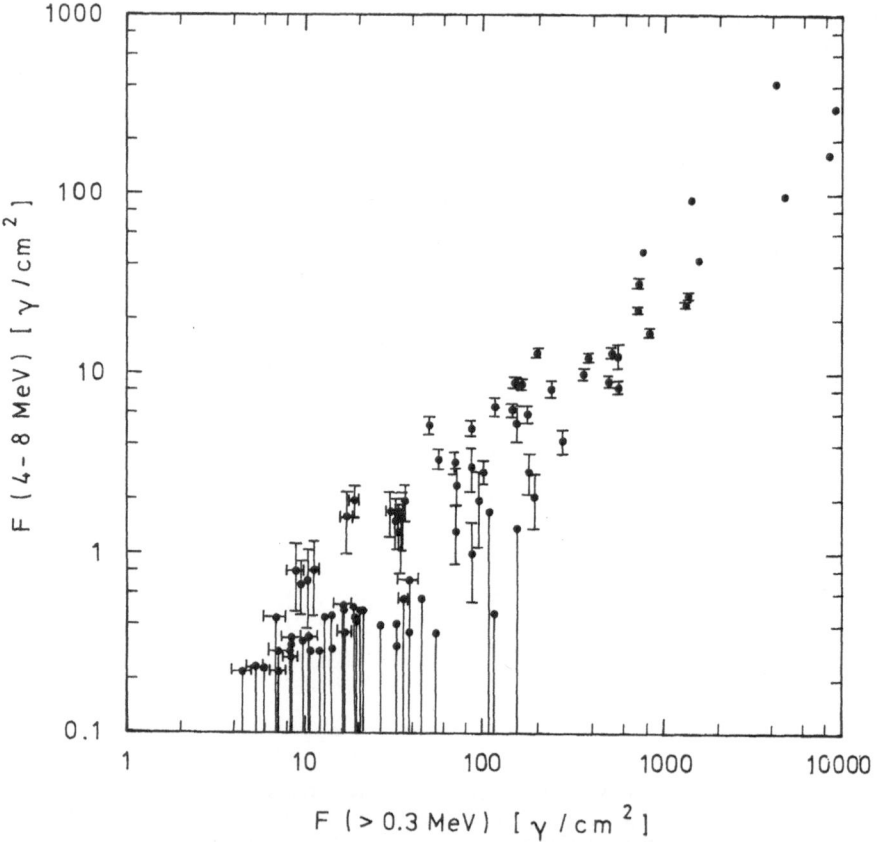

Fig. 4.   4–8 MeV line fluence versus > 0.3 MeV continuum fluence for 90 flares observed by the GRS on SMM (Forrest, private communication). The energy interval 4–8 MeV instead of 4–7 MeV is used to include the ^{16}O line at 7.12 MeV.

– If there is a threshold for the flare size, below which only electrons and no ions can be accelerated, it is below the instrument's sensitivity for nuclear emissions.

– There are no pronounced proton or electron events (missing points in the upper left and lower right of the figure), and

– short impulsive and long-duration events do not appear as separate classes in the diagram. This requires an acceleration mechanism producing high-energy electrons and ions in close proportion independent of the flare duration (Forrest, 1983).

4. Solar Flare Gamma-Rays and Solar Energetic Particles (SEP)

When gamma-ray measurements of solar flares became available, it was of interest to examine the relationship between the charged particles producing the gamma-ray emissions at the Sun and those observed in interplanetary space after these flares. In order to minimize the influence of coronal and interplanetary transport effects, it has been proposed to consider only those events which were magnetically well connnected to the observer. Under these conditions the peak flux is a reliable indicator of the particle fluence injected into space (van Hollebeke, Ma Sung, and McDonald, 1975; van Hollebeke, 1979). As the nuclear line fluence is a measure of the number of protons precipitated into the solar atmosphere (Murphy and Ramaty, 1984), a correlation with the peak flux of 10 MeV protons, would be expected.

In Figure 5 the 10 MeV peak proton flux measured with detectors on IMP-8, ISEE-3 and Helios is plotted for 45 well-connected events versus the gamma-ray line fluence obtained directly from the GRS on SMM or inferred from the hard X-ray experiment on ISEE-3 (Cliver et al., 1987). It is evident from the missing points in the lower right hand corner of the figure that large gamma-ray line events also produce large particle fluxes. The converse, however, is not true. There are a number of large particle events for which no nuclear line fluence is measured.

A correlative study of particle events observed with the Helios 1 satellite and the GRS gives a similar result (Kallenrode et al., 1987). This lack of correlation was also recognized in previous investigations (Chambon et al., 1981; von Rosenvinge, Ramaty, and Reames, 1981; Pesses et al., 1981; Yoshimori and Watanabe, 1985). It was pointed out by Cane, McGuire, and von Rosenvinge (1986) and Bai (1986) that the duration of a flare measured at low X-ray energy is an ordering parameter in the sense that long-duration events are more prolific in producing interplanetary protons than short ones. The high correlation of big SEP-events with type II radio emission is explained by a shock accelerating the protons high in the corona, where they have ample access to open field lines, whereas in short-duration events the shock acts in low-lying, mostly closed magnetic flux tubes (Pallavicini, Serio, and Vaiana, 1977; Cliver et al., 1987). Coronal shocks, therefore, seem to accelerate predominantly protons (Kallenrode et al., 1987). This conclusion is also substantiated by the facts, that

– for short impulsive events the number of protons interacting in the solar atmosphere is much higher than the number of protons escaping into interplanetary space, whereas

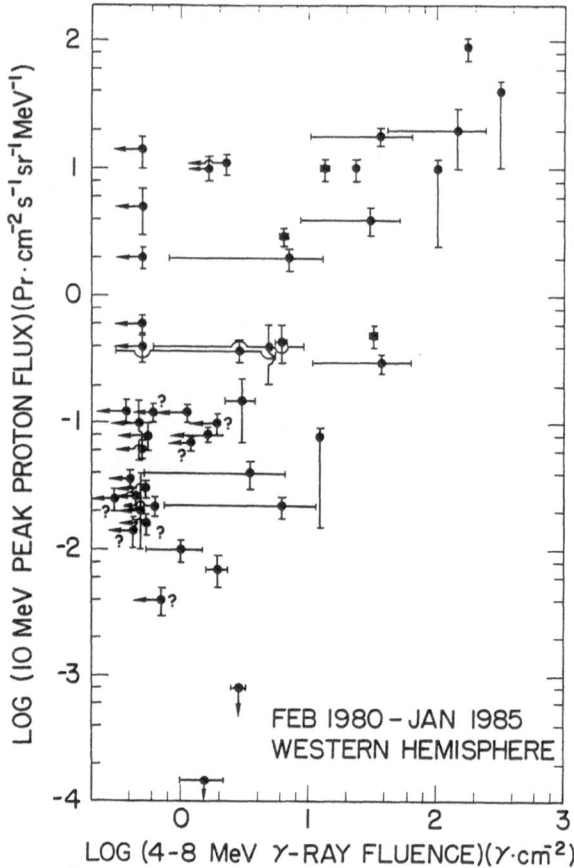

Fig. 5.   Peak ~10 MeV proton flux versus 4–8 MeV gamma-ray line fluence for 45 magnetically well-connected flares (from Cliver *et al.*, 1987).

for long-duration events more particles escape than are precipitated (Ramaty and Murphy, 1987; Hua and Lingenfelter, 1987a), and

– the electron to proton ratio of the escaping particles tends to be higher for flares emitting gamma-rays than for flares which only produce interplanetary particles (Evenson *et al.*, 1984; Cane, McGuire, and von Rosenvinge, 1986).

5. Proton Energy Spectra

As already mentioned in Section 2 the relative importance of the emissions caused by nuclear interactions, namely the prompt lines of ^{12}C and ^{16}O, the 2.223 MeV capture line and the pion decay radiation, depends upon the energy of the accelerated nuclei. It is, therefore, possible to deduce spectral parameters of the primary particles. Murphy and Ramaty (1984) have calculated the yields of the ^{12}C and ^{16}O de-excitation lines, neutrons, pions, and positrons as a function of the primary proton spectrum under the

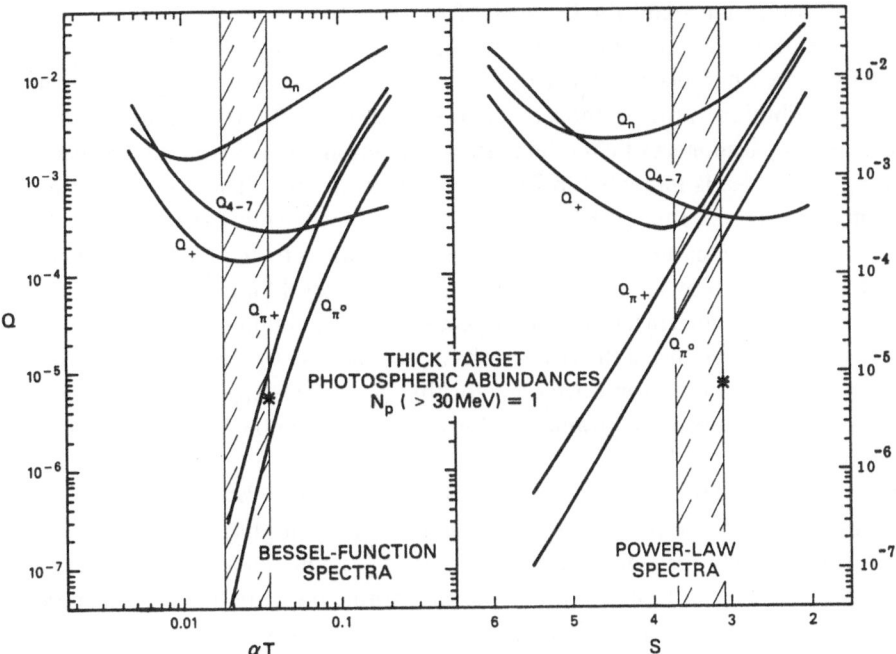

Fig. 6. Neutron, 4–7 MeV nuclear gamma-ray, positron, π^+ and π^0 production versus the spectral parameter of a Bessel function spectrum and power-law spectrum for the primary protons (from Murphy and Ramaty, 1984). Indicated by the dashed area is the range of the spectral parameters deduced from observations. Asterisk: π^0-decay fluence of the 3 June, 1982 flare.

assumption that the interacting energetic protons have an isotropic distribution and that they loose all their energy in the ambient medium (thick target case). The results are shown in Figure 6, a reproduction of their Figure 1. The yields are normalized to one proton with an energy > 30 MeV. The calculations were performed for a hard and a soft spectrum. On the right-hand side the protons have a power-law spectrum characterized by the exponent s. On the left-hand side the proton spectrum has the form of a modified Bessel function of second-order where a high αT indicates a hard spectrum (Ramaty, 1979; Forman, Ramaty, and Zweibel, 1986). The slope of this function increases with increasing energy.

From Figure 6 we see, that the yield of the ^{12}C and ^{16}O lines (here the yield is calculated for the energy band 4–7 MeV, which includes both lines) does not depend very much on the spectral parameters if we exclude very soft proton spectra, whereas the neutron and especially the pion yield increases as the spectrum hardens. To get the 2.223 MeV line fluence, the neutron yield has to be multiplied by the neutron-to-2.223 MeV photon conversion factor. This factor depends upon the spectrum and the angular distribution of the protons and the position of the flare at the Sun (Murphy and Ramaty, 1984; Hua and Lingenfelter, 1987a). The ratio of the 2.223 MeV and 4–7 MeV fluence is a measure of the steepness of the proton spectrum predominantly in the energy range 10–100 MeV. It is important to note that because of the different time characteristics of the prompt lines and the delayed 2.223 MeV line the fluences (time integrated fluxes) rather than the fluxes must be compared. Up to now the fluence ratio for 12 gamma-ray flares has been published (see Hua and Lingenfelter, 1987a). Assuming isotropic distribution of the protons, one obtains $0.018 < \alpha T < 0.034$ and $3.1 < s < 3.7$ for the spectral parameters of a modified Bessel function and a power law, respectively. This range of values is indicated in Figure 6 by dashed lines. There is at present only one event, the flare of 3 June, 1982, for which also the fluence due to the decay of neutral pions is available. This neutral pion fluence, normalized to the 4–7 MeV fluence, is inserted into Figure 6 by an asterisk, showing that a Bessel function spectral shape of the protons is favoured. This, however, applies only to the first phase of the event (about the spectral hardening during the second phase, see Section 6). There is in fact no flare, for which an unbroken proton power law for the spectrum is suggested by the gamma-ray measurements. Even the ground level event of 7 December, 1982 is no exception (Rieger et al., 1987).

If we try to deduce proton-energy spectra from the particle observations in space, we have to bear in mind that particles originating from different phases of a flare, implying different acceleration mechanisms, can be mixed together in the signal (Neustock, Wibberenz, and Iwers, 1985). In this respect photon measurements are superior to particle measurements, because different phases can generally be distinguished by their temporal characteristics (see Figure 8). For charged particles the influence of coronal propagation and interplanetary transport must also be considered. As already mentioned, these effects can be minimized by using particle measurements of magnetically well-connected flares. McGuire and von Rosenvinge (1984) have reviewed the measurements of such well-connected events and found that the spectra can be best approximated with a Bessel function. For protons the αT-values are within the same range as those obtained from the gamma-ray data. These observations suggest, that for most flares a common mechanism could accelerate both particle populations. It must, however, be pointed out that there are only 4 flares for which spectral information both from gamma-ray (GRS) and from particle measurements exists. These are the events of 7 June, 1980 (McGuire and von Rosenvinge, 1984); 21 June, 1980; 3 June, 1982 (McDonald and van Holleke, 1985); and 7 December, 1982 (Rieger et al., 1987). A Bessel function can be fit to the particle data of the flares of 7 June, 1980 and 21 June, 1980, in accord with the results of the gamma-ray measurements. The particle spectra of the flares of 3 June, 1982 and 7 December, 1982, however, are too hard to be fit with a Bessel function. They could, therefore, result from a different acceleration mechanism operating during a second phase (see Ramaty and Murphy, 1987).

Shortly after the reports of ground-level cosmic-ray enhancements (Ground Level Events) following solar flares, Biermann, Haxel, and Schlüter (1951) conjectured that the GeV protons at the Sun could produce neutrons observable at Earth. Neutrons were detected by the GRS and ground based neutron monitors after the flares of 21 June, 1980 and 3 June, 1982. Because these detectors do not measure the neutron energy directly, it must be derived from the time of flight between the Sun and Earth. If the duration of a flare, determined by the emission time history in the energy band 4–7 MeV and/or at > 25 MeV is short compared to the flight time between Sun and Earth of the fastest neutrons, the neutron production can be approximated by a delta function in time, coincident with the peak of the photon emission. Applying this criterion to the neutron flare observations of 21 June, 1980 and 3 June, 1982, Murphy and Ramaty (1984) and Hua and Lingenfelter (1987c) show that a good fit of the calculated neutron flux time history with the data is achieved when a Bessel function spectral type instead of a power law is assumed for the primary protons. The spectral parameters αT are in good agreement with the values obtained from the ratio of the 2.223 MeV line – and 4–7 MeV fluence during the impulsive phase. They point out that the bulk of the neutrons is a product of the impulsive phase. On the other hand Chupp *et al.* (1987), by analyzing the GRS *and* neutron monitor data, find that the majority of the neutrons is produced after the impulsive phase. This discrepancy is further discussed by Ramaty and Murphy (1987).

6. Temporal Characteristics of Flares

In addition to the energy spectra of the primary particles, deduced from gamma-ray line, neutron, and charged particle measurements, the time history of a flare in different energy bands is an important observational ingredient setting constraints on particle acceleration mechanisms.

Before gamma-ray measurements were available, Wild, Smerd, and Weiss (1963) used radio data in their search for the origin of the flare-energy release. From the observation that impulsive meter-wave type III bursts were followed after several minutes by type II meter-wave emissions, indicative of a shock wave moving through the solar corona, they developed a concept of two acceleration phases. According to this idea, electrons should be accelerated impulsively during phase one to energies of ~ 100 keV. The second more gradual phase, which needs phase one as a trigger, then accelerates electrons to relativistic energies and creates high-energy ions producing ground level events. This concept of two phases or two steps (Bai and Ramaty, 1979) of particle acceleration during solar flares was basically accepted until the early 1980's, when flare measurements with high time resolution in the X- and gamma-ray regime became available through the SMM- and Hinotori-detectors. Among the many flares with gamma radiation, events were observed with impulsive emission occurring simultaneously in X- and gamma-rays within the 1–2 s time resolution of the instruments (Rieger, 1982; Forrest and Chupp, 1983; Rieger *et al.*, 1983; Yoshimori *et al.*, 1983; Kane *et al.*, 1986). To stress this important observational point, the time history of the flare of 7 May, 1983 is shown in Figure 7. This intense event consists mainly of one impulsive burst of about 1 min duration visible in three decades of energy. The emission seen in panel 1 (time resolution 1 s) is bremsstrahlung of subrelativistic electrons. The emission in panel 2 (time resolution 2 s) which appears more bursty than that at lower energies, originates mainly from the nuclear de-excitation lines of ^{12}C and ^{16}O at 4.4 and 6.1 MeV, respectively (see Section 2). Above 10 MeV primary electron bremsstrahlung is again the dominant radiation mechanism, unless the primary proton

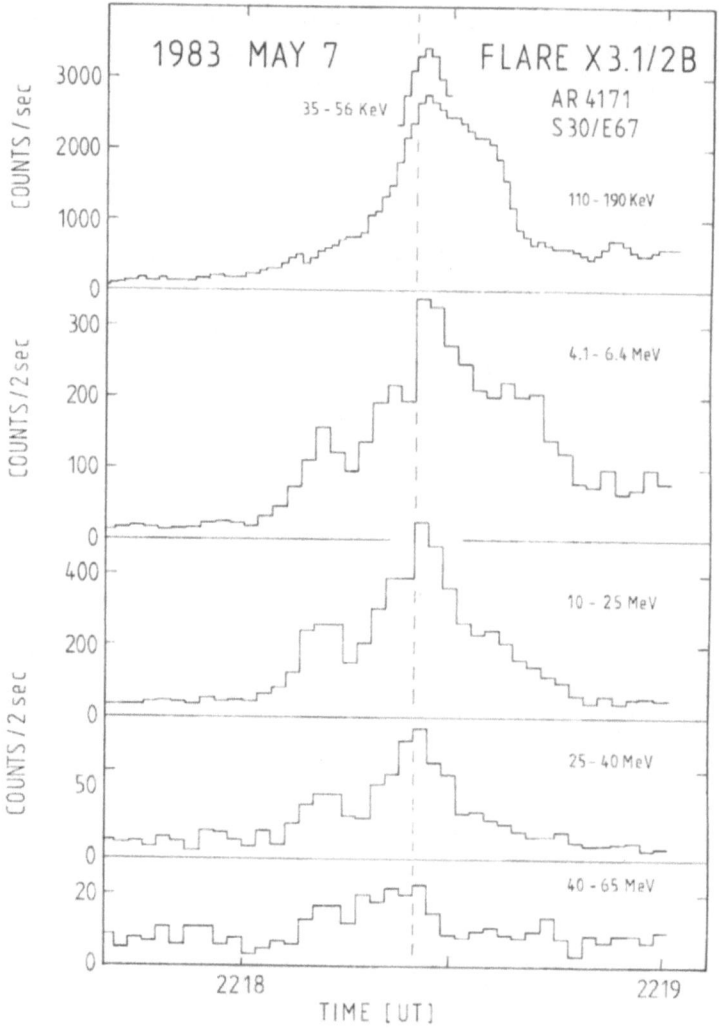

Fig. 7.   Time history of the 7 May, 1983 flare in different energy bands (SMM GRS).

spectrum is very hard (Rieger *et al.*, 1983; Ramaty *et al.*, 1983). It is evident from Figure 7 that the peak of the emission at gamma-ray energies occurs simultaneously within the limit of the time resolution with respect to the peak of the medium energy X-rays, usually taken as the reference time of the impulsive phase (Bai and Ramaty, 1979). Strangely enough, there is a precursor preceding the main peak, which is prominent only in gamma-rays, but not in X-rays. We will, however, not refer to this curiosity. These observations show, that ions and electrons can be accelerated impulsively (~ 1 s) to energies of > 10 MeV and ~ 100 MeV, respectively. If we take into account that the emission versus time is a complicated superposition of acceleration-, transport-, and energy-loss processes and that the last two have a tendency to smooth out quick temporal changes, the acceleration by itself may even be more bursty in time.

Under rare conditions ions (protons) can be accelerated to greater energies during the impulsive phase. This is evident from the analysis of the flare of 3 June, 1982 (Figure 8). This very intense event had a distinctive 2-phase nature at energies > 10 MeV: an impulsive emission of $\lesssim 1$ min duration was followed by a more gradual one lasting several minutes. The primary peak spectrum shows a flattening above 40 MeV which has been attributed to neutral pion decay photons (Forrest *et al.*, 1985, 1986). Measurements from the Jungfraujoch neutron monitor show that high energy neutrons (~ 1 GeV) were produced already during the impulsive phase (Debrunner *et al.*, 1983; Chupp *et al.*, 1987). These observations imply that during the first or impulsive phase protons can be accelerated to GeV-energies.

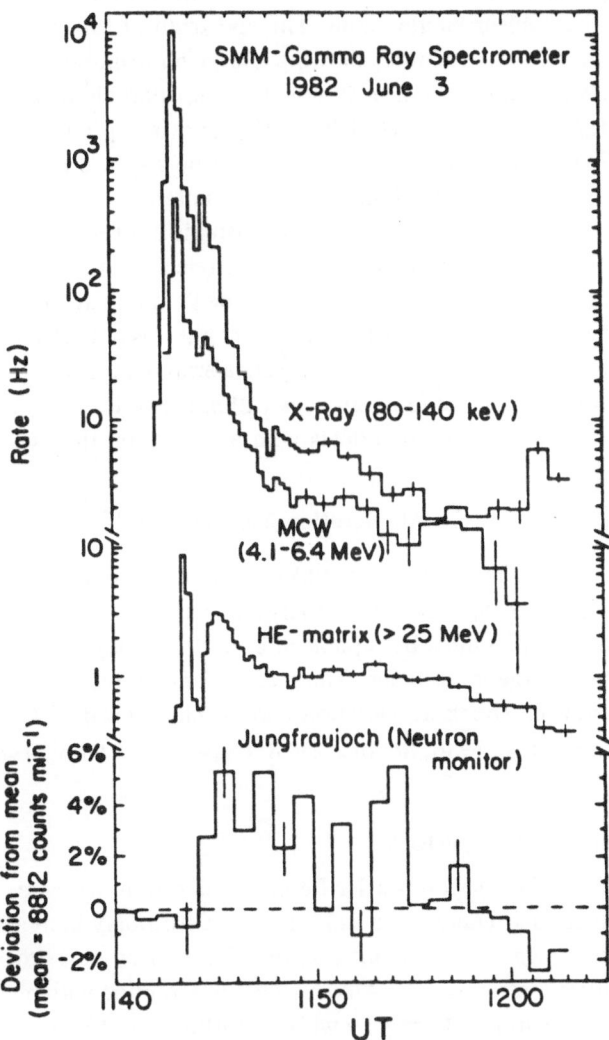

Fig. 8.   Time history for several data channels of the SMM GRS and of the Jungfraujoch neutron monitor count rate for the 3 June, 1982 flare (from Chupp *et al.*, 1985).

The SMM and Hinotori spectrometers and detectors on earlier missions recorded numerous other flares, where a delay between the emissions at lower (~ 50 keV) and higher energies was observed (see, e.g., Vlahos et al., 1986). While these delays have been interpreted as evidence for a second step (or second stage) acceleration (Bai and Ramaty, 1979; Bai, 1982; Bai et al., 1983a, b; Bai and Dennis, 1985), a trap plus precipitation model (Kane, 1974; Melrose and Brown, 1976; Bai and Ramaty, 1979; Vilmer, Kane, and Trottet, 1982; MacKinnon et al., 1983; Ryan, 1986) and a trap plus turbulent propagation model (Bespalov, Zaitsev, and Stepanov, 1987), a more direct observational proof for a second phase acceleration comes from the gamma-ray measurements of the 3 June, 1982 flare. As already mentioned a gradual peak (> 10 MeV) appears after the first intense burst, which does not stand out at X-ray energies and in the nuclear energy band. The spectrum of this second peak is harder than that of the first impulsive burst, showing a large contribution of π^0-decay photons (Forrest et al., 1985). This second peak together with particle measurements in space after the flare (McDonald and van Hollebeke, 1985) are interpreted by Murphy, Dermer, and Ramaty (1987) as a second phase acceleration by a shock wave going through the solar corona. Another event with similar temporal characteristics is the flare of 24 April, 1984, for which a detailed evaluation of the spectrum is not yet available. It is of interest to note that these two flares are by far the most intense events observed by a detector sensitive to energies above 10 MeV. Thus, from the gamma-ray measurements the two phase or two-step acceleration process during solar flares is not ruled out. The important new finding, however, is that electrons can be accelerated impulsively (~ 1 s) in a single step to ~ 100 MeV and ions to GeV energies within seconds. This 'inconvenience' has to be taken into account, when one deals with acceleration theories.

7. Acceleration Mechanisms

Any viable acceleration mechanism should be able to account for the time-scales of electron and proton acceleration, their energy spectra, the total number or ratio of accelerated particles and it must be capable of imparting a large fraction of the available flare energy to the energetic particles. The mechanisms, most widely investigated, are stochastic acceleration, shock acceleration and acceleration by DC electric fields (for a review see Forman et al., 1986; Scholer, 1988; Cargill and Vlahos, 1989; and Sturrock, 1989).

7.1. STOCHASTIC ACCELERATION

This process is essentially second-order Fermi acceleration, where particles in a turbulent plasma change their energy in a random way with many increases and decreases in energy. Various wave-particle interactions have been considered, for instance resonant pitch angle scattering from Alfvén waves, interaction with magnetosonic waves (Forman et al., 1986) and scattering by whistler turbulence (Miller and Ramaty, 1987). By assuming that the acceleration efficiency α and the average time T during which the particles experience acceleration are energy independent, it was shown, that the particle energy spectrum can be represented by a modified second-order Bessel function in the nonrelativistic regime, and by a power law in the ultra relativistic regime (Ramaty, 1979). In both cases the product of the parameters α and T characterizes the shape of the spectra in the sense that a larger value of αT corresponds to a harder spectrum.

For the nonrelativistic case ($E \ll m_0 c^2$) and at energies $E \gg 3.26(\alpha T)^2$ MeV nucl.$^{-1}$ the formula

$$\frac{\mathrm{d}J}{\mathrm{d}E} = E^{3/8} \exp(-(E/(3.26(\alpha T)^2))^{1/4}), \tag{1}$$

where $\mathrm{d}J/\mathrm{d}E$ is the differential particle energy spectrum, is a good approximation to the modified Bessel function (Forman et al., 1986). In the ultrarelativistic case the power-law exponent is given by

$$\gamma = \tfrac{1}{2}(9 + 12/\alpha T)^{1/2} - \tfrac{1}{2}. \tag{2}$$

Using this model, energy spectra of protons and α-particles observed in space were successfully fitted (Ramaty, 1979). As already mentioned in Section 5 the spectral parameter αT deduced from gamma-ray measurements varies in narrow limits between 0.018 and 0.034. If these values are used to fit the electron spectra above 1 MeV by using Equation (2), we get a slope which is much steeper than observed (e.g., Ramaty, 1979; Evenson, Kroeger, and Meyer, 1985; Dröge et al., 1989).

The acceleration time in stochastic acceleration was investigated by Miller, Ramaty, and Murphy (1987). It is shown that the time to accelerate ions from ambient energies to 5, 30, 400, and 6000 MeV nucl.$^{-1}$ is 1, 2, 6, and 16 s, respectively. This is not in serious contradiction to what can be deduced from the temporal characteristics of gamma-ray flares. A problem, however, arises with the time-scale implied by the observations for electron acceleration. To accelerate them to tens of MeV in the same acceleration region as the ions takes about 100 s, which is much too slow (Ramaty and Murphy, 1987). The time-scales are shorter if electron acceleration due to whistler turbulence is taken into account. Miller and Ramaty (1987) showed that in this case electrons can be accelerated to 100 MeV in about 10 s. Similar time-scales are obtained by Dröge and Schlickeiser (1986) who investigated the combined effects of first- and second-order Fermi acceleration. But in view of the simultaneous peaking (1–2 s) of the emission at ~ 50 keV and ~ 50 MeV observed in some flares (see Section 6) there still remain doubts if stochastic acceleration is a mechanism suitable for the impulsive phase.

7.2. DIFFUSIVE SHOCK ACCELERATION

In this case the basic acceleration mechanism is the first order Fermi process with a turbulent medium on both sides of a quasi parallel shock. The particles are scattered by the inhomogeneities so that they experience multiple accelerations at the shock. When losses are due only to convection of the particles away from the shock downstream and the shock size is large with respect to particle diffusion lengths (planar shock), the particle spectrum deduced in this case is a power law in momentum space with an exponent determined by the compression ratio r of the shock (Blandford and Ostriker, 1978). The corresponding differential particle intensity, $(dJ/dE)_0$, in units of particles cm^{-2} s^{-1} sr^{-1} MeV^{-1} is given by (Ellison and Ramaty, 1985)

$$(dJ/dE)_0 \sim (E^2 + 2Em_0c^2)^{-\gamma}, \tag{3}$$

where m_0c^2 is the rest mass energy and $\gamma = \frac{1}{2}(r + 2)/(r - 1)$. Several effects can truncate this power-law behaviour at high energies, for instance shock life times comparable to particle acceleration times (Forman, 1981) and shock sizes comparable to particle diffusion length (e.g., Ellison, 1984). If one incorporates these effects, then Equation (3) is replaced by

$$dJ/dE \sim (dJ/dE)_0 \exp(-E/E_0), \tag{4}$$

where E_0 is called the turnover kinetic energy, above which the particle spectrum steepens considerably. Using this two-parameter equation Ellison and Ramaty (1985) were able to fit particle energy spectra of several flares. It is important to note, that for flares where electron and proton measurements are available, a satisfactory fit to the spectra of both types of particles is obtained by using the same values for the parameters r and E_0. This formula was also applied successfully to ground level events. The differential proton energy spectrum measured after the 3 June, 1982 flare (McDonald and van Hollebeke, 1985) was fitted by Equation (4) with $\gamma = 1.7$ and $E_0 > 300$ MeV. Using these values and the time history of the 0.511 MeV annihilation line (Share et al., 1983) it is possible to model the high-energy emission during the second phase of this event (Ramaty, Murphy, and Dermer, 1987).

Ellison and Ramaty (1985) point out that shock acceleration not only can explain the low e/p ratio observed at high energies, but also the short acceleration times for electrons and protons implied by the gamma-ray measurements. If the background turbulence is strong enough to produce scattering meanfree paths that are less than about 100 times the gyroradius, electrons and protons can be accelerated to 100 MeV in about 1 s. With the appropriate choice of the shock compression ratio r and the turnover kinetic energy E_0 shock acceleration could thus be a candidate also for the impulsive phase provided the time to generate the shock, or the shocks, is short (< 1 s).

Papadopoulos et al. (1985) and Decker and Vlahos (1986) have investigated the acceleration of charged particles by oblique turbulent shocks through numerical simulations. In their model a sudden energy release heats the plasma in some volume and a shock is formed in a few microseconds. The ions in the tail of the heated plasma serve as injected particles, which are then accelerated most effectively by quasiperpendicular shocks to energies of 50 MeV in less than milliseconds. In this treatment, however, the acceleration of electrons has not been addressed.

Quasi-perpendicular fast magnetosonic shock waves have been invoked by Ohsawa and Sakai (1987) and Sakai and Ohsawa (1987) as an acceleration mechanism for ions and electrons to relativistic energies in a short enough time to explain the gamma-ray observations of the 3 June, 1982 flare. The authors point out that in loops with strong magnetic fields (low-lying loops) magnetosonic waves could lead to impulsive ion and electron acceleration (first or impulsive phase). As the shock propagates out to weak magnetic field regions, strong electron acceleration ceases, while ion acceleration continues (second or gradual phase). But if we apply this model to the temporal characteristics of the flare (see Figure 8) one would expect to see a smooth transition above 25 MeV from phase 1 to phase 2. The difference in the temporal and spectral characteristics of both peaks, however, makes it tempting to assume that two physically different acceleration mechanisms were in operation causing an impulsive and a gradual phase of the flare (see Ramaty and Murphy, 1987). It is of interest to note that Kocharov *et al.* (1988) interpret the time history by a single acceleration process occurring during the impulsive peak. To explain the second maximum at > 25 MeV a fast rise of plasma turbulence is assumed to occur at the end of the impulsive peak, leading to precipitation due to pitch angle scattering of impulsively created high-energy protons.

7.3. ELECTRIC FIELDS

In addition to stochastic processes there is also the possibility to accelerate particles in direct electric fields (see, e.g., Forman, Ramaty, and Zweibel, 1986). Two approaches invoking DC electric fields, which have their routes in the investigation of auroral phenomena, shall be briefly considered here. It has been shown with ion cloud experiments carried out in the auroral zone (Haerendel *et al.*, 1976; Wescott *et al.*, 1976) and by measurements made with the S3-3 satellite (Mozer *et al.*, 1980; Temerin *et al.*, 1982) that under certain conditions electrostatic potential drops along the magnetic field lines can develop, by which auroral particles are accelerated. A necessary condition for this process to operate is the existence of intense field aligned currents exceeding a critical value (e.g., Spicer, 1982).

Assuming, that this magnetic field parallel potential drop is located over a very short parallel distance (supported by laboratory measurements) Carlquist (1986) has shown that the maximum potential drop attained by this 'double layer' is a function of the total current rather than the current density. In the solar atmosphere, where currents of the order 10^{11}–10^{12} A are known to exist (Anzer and Tandberg-Hanssen, 1970) he obtains $\sim 10^5$ MV for the maximum potential drop of the double layer, which would be enough to account for even the highest energy particles observed. The time to accelerate the particles is a fraction of a second provided the double layer attains the full power in a short time-scale.

In the model of Haerendel (1987) the flare is explained by a strong twisting and subsequent untwisting of a large number ($\sim 10^5$) of narrow flux tubes, which are thought to be embedded in an active region. The sudden release of the flare energy which is triggered by an unidentified cause (e.g., new flux emergence, reconnection in the corona, etc.) leads to a partial reinvestment of the flux tubes which can leave behind even more highly twisted ones. The corresponding currents create the low temperature flare in the lower chromosphere by classical Joule heating. In a small minority of flux tubes the current reaches values that makes them unstable and cause strong field aligned potential drops. One GV and more can be attained. In sufficiently strong primary fields (~ 2000 G) the 'reverse current' problem (e.g., Colgate, 1978) does not arise in this model, because the currents are assumed to be contained in small scale structures with random orientation. Thus, the net current can be many orders of magnitude below the sum of all the current filaments. It is pointed out, however, that the model rests upon the ability of the flux tubes to be stable against a twisting of many revolutions over the length of the tube.

8. Evidence for Beaming of the High-Energy Particles?

Acceleration by DC electric fields leads to particle beams directed along the magnetic field lines, whereas stochastic acceleration processes have a tendency to create isotropic particle distributions. Investigation of anisotropies in the velocity distribution of flare generated particles can, therefore, provide important clues about particle acceleration and about the particle transport in the flaring loops. The directionality of electrons can be studied, because bremsstrahlung photons are emitted anisotropically in the direction of motion of the electrons (see Figure 1), increasing in anisotropy with increasing electron energy (Elwert and Haug, 1971). If the electrons move predominantly downward towards the Sun (pencil beam) or tangential to the solar surface due to mirroring in the converging magnetic fields (fan beam), a flare observed near the limb will appear brighter than a flare observed at the disc. Weak flares are, therefore, detectable only if they occur close to the limb. But all efforts to show this 'limb brightening' by using X-ray flares as a proxy did not lead to a positive result (Kane, 1974; Datlove et al., 1977). This, however, does not rule out the existence of beams, because electrons of the solar atmosphere scatter the photons back (Compton backscattering), thereby smearing out any anisotropy of the primary radiation (Bai and Ramaty, 1978). A limb brightening begins to show up if we mark all the SMM GRS flares with known position at the solar disc. The number of flares close to the limb is higher than is expected for isotropically emitting flares (Vestrand et al., 1987). The finding that the best fit power-law electron continuum between 0.28 and 1 MeV is harder for flares near the limb than for flares closer to the disc center can be interpreted as an additional indication of a beaming of the electrons. Spectral hardening in the energy range 50–300 keV towards the limb was also found in a sample of flares observed by the Venera 13 spacecraft (Bogovalov et al., 1985). Evidence for a directivity is also given by Kane et al. (1988) in their analysis of flares observed stereoscopically by the ISEE-3 and Pioneer–Venus orbiter in the energy range 0.1–1 MeV. A third method is used by Bai (1988). Instead of counting the number of limb and disc flares he plots the fluence of the GRS flares after normalization with respect to the total counts measured by the hard X-ray burst spectrometer (HXRBS) on SMM versus the heliocentric angle and finds a systematic increase of the luminosity of the GRS flares towards the limb. The most favourable energy band to investigate the directivity of the electrons is between 8 and 40 MeV, where the flare radiation is predominantly electron bremsstrahlung. At these high energies the Compton backscatter is negligible and the bremsstrahlung cross-section is highly anisotropic. Figure 9 shows the Hα-position at the Sun of the flares with photon emission above 10 MeV. A concentration towards the limb is clearly visible (Rieger et al., 1983; Canfield et al.,

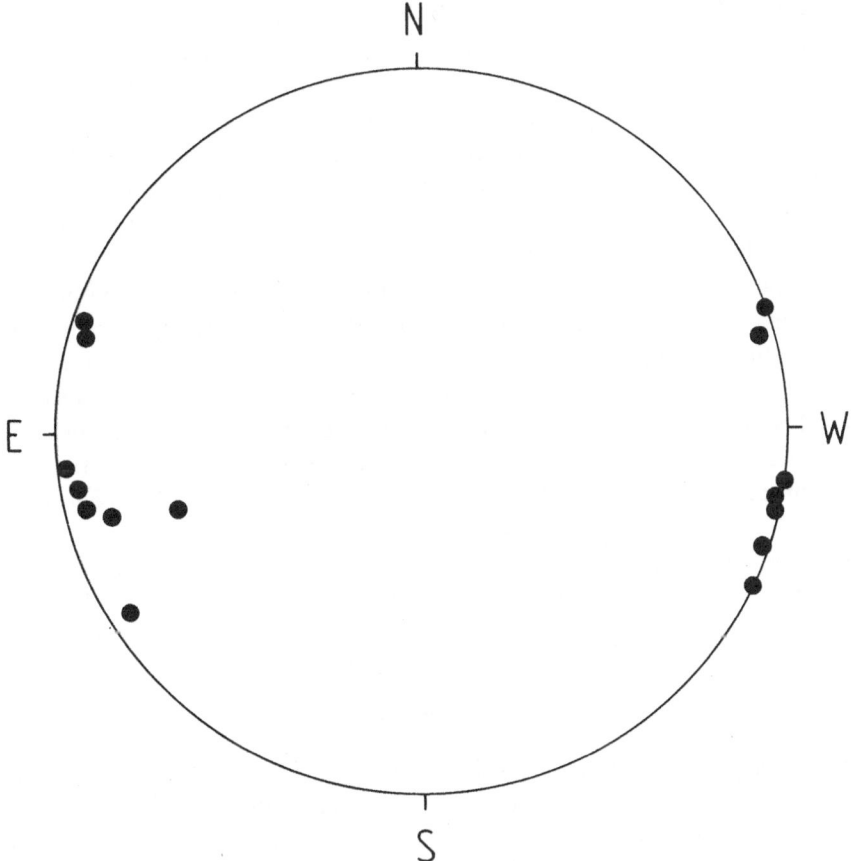

Fig. 9.   Position of flares with emission above 10 MeV at the solar disc observed by the GRS on SMM.

1986). Only two out of 15 events are at heliocentric angles $< 64°$. Although the number of flares with emission > 10 MeV is still small, this observational fact has led to theoretical investigations about the directionality of the primary high-energy electrons.

Naively interpreted one could imagine a configuration as shown in Figure 1, where electrons move vertically downward to the denser solar atmosphere. A motion tangential to the solar surface would also lead to a limb brightening (Canfield *et al.*, 1986; Dermer, 1987). Petrosian (1985) taking into account the transport of the electrons in the solar atmosphere concluded, that a downward isotropic distribution can explain the concentration of events towards the limb and that mirroring of the electrons is unimportant, if the magnetic fields do not depend strongly on altitude. On the other hand Kocharov *et al.* (1987, 1988) and Ramaty *et al.* (1988) assuming a more realistic magnetic field geometry, have shown that electrons injected isotropically at the top of a magnetic loop loose their energy preferentially when they mirror in the converging magnetic field. In their model a beaming of the primary particles along the magnetic field is not a necessary condition to explain the observational fact that flares with photon emission above

10 MeV are concentrated towards the limb. Thus, the implication of the limb brightening on the directionality of the primary particles depend to a large degree on the geometry of the magnetic field, which may differ from flare to flare (MacKinnon and Brown, 1989). It was, therefore, proposed to observe flares stereoscopically in the energy bands 0.28–1 MeV and > 10 MeV from different directions with identical detectors (Canfield *et al.*, 1986; Vestrand and Ghosh, 1987). The optimal orbital configuration for stereo-scopic observations is obtained if one spacecraft is in an earth orbit and the other in an ecliptic orbit 90° away in heliocentric longitude.

Information about the angular distribution of the protons cannot be obtained from the observation of the 4–7 MeV fluence, because this emission is only weakly dependent on the proton anisotropy (Ramaty, 1986). The flux of the escaping neutrons, however, depends on the angular distribution of the primary particles. It was shown by Hua and Lingenfelter (1987c) that the neutron flux observed after the flares of 21 June, 1980 and 3 June, 1982, when normalized to the 4–7 MeV emission is consistent with a fan beam or an isotropic distribution of the protons whereas a downward beamed distribution seems to be ruled out from the observations.

It must be noted, however, that this conclusion is based upon the measurements of only two flares. More observations are needed, to determine the angular distribution of the interacting particles.

9. Conclusion

The view about the acceleration of particles during solar flares has changed considerably since the early 1980's, when the Sun was monitored in the gamma-ray regime by the SMM and Hinotori detectors. This was to be expected, because in an environment, where *in situ* measurements are precluded and the motion of charged particles is constrained by strong and complex magnetic fields, the measurement of neutral radiations provides the closest approach to the flare phenomenon.

The good correlation between the continuum- and gamma-ray line emission, independent of the flare duration and flare size, may be a hint that the primary acceleration of particles proceeds under rather similar conditions from flare to flare.

Evidence for a two-phase acceleration mechanism is obtained by the observation of a distinct temporal development of the high-energy gamma-ray emission during two very intense solar flares. Whereas the second phase may be explained by a shock wave moving through the corona, a candidate which accelerates the particles during the first phase is not yet firmly established. The rapid acceleration to very high energies in one step and the abrupt switching off, inferred from certain gamma-ray measurements (Forrest and Chupp, 1983; Kane *et al.*, 1986) are challenges for any theorist dealing with acceleration mechanisms. In this respect the capabilities of stochastic processes may be strained. Therefore, acceleration by DC-electric fields, which is known to operate in the Earth's magnetosphere, has been investigated as an alternative.

The observation that flares emitting high-energy gamma-rays are concentrated to-wards the solar limb, suggests that the radiating electrons are anisotropic. It is still a matter of debate whether this detection can help to choose between different acceleration mechanisms.

Acknowledgements

The major contributors to the SMM GRS solar flare data analysis are E. L. Chupp, D. J. Forrest, G. Kanbach, C. Reppin, G. H. Share, and W. T. Vestrand, to whom the author is indebted. This work was partly supported by the Bundesministerium für Forschung und Technologie under 010K017–ZA/WS/WRK0275:4 in West Germany.

References

Anzer, U. and Tandberg-Hanssen, E.: 1970, *Solar Phys.* **11**, 61.
Bai, T.: 1977, 'Studies of Solar Hard X-Rays and Gamma Rays: Compton Backscatter Anisotropy, Polarization and Evidence for Two Phases of Aceleration', Ph.D. thesis, University of Maryland.
Bai, T.: 1982, in L. E. Lingenfelter, H. S. Hudson, and D. M. Worrall (eds.), AIP, New York, p. 409.
Bai, T.: 1986, *Astrophys. J.* **308**, 912.
Bai, T.: 1988, *Astrophys. J.* **334**, 1049.
Bai, T. and Dennis, B. R.: 1985, *Astrophys. J.* **292**,699.
Bai, T. and Ramaty, R.: 1978, *Astrophys. J.* **219**, 705.
Bai, T. and Ramaty, R.: 1979, *Astrophys. J.* **227**, 1072.
Bai, T., Dennis, B. R., Kiplinger, A. L., Orwig, L. E., and Frost, K. J.: 1983a, *Solar Phys.* **86**, 409.
Bai, T. Hudson, H. S., Pelling, R. M., Lin, R. P., Schwartz, R. A., and von Rosenvinge, T. T.: 1983b, *Astrophys. J.* **267**, 433.
Bespalov, P. A., Zaitsev, V. V., and Stepanov, A. V.: 1987, *Solar Phys.* **114**, 127.
Biermann, L., Haxel, O., and Schlüter, A.: 1951, *Z. Naturforsch.* **6a**, 47.
Blandford, R. P. and Ostriker, J. P.: 1978, *Astrophys. J.* **221**, L29.
Bogovalov, S. V., Kel'ner, S. R., Kotov, Yu. D., Zenchenko, V. M., Vedrenne, G., Barat, C., Chambon, G., and Talon, R.: 1985, *Soviet Astron. Letters* **11**, 322.
Cane, H. V., McGuire, R. E., and von Rosenvinge, T. T.: 1986, *Astrophys. J.* **301**, 448.
Canfield, R. C. *et al.*: 1986, *Energetic Phenomena on the Sun*, NASA CP-2439, p. 3–1.
Cargill, P. and Vlahos, L.: 1989, in B. M. Haisch and M. Rodonò (eds.), *IAU Colloq.* **104**, *Solar and Stellar Flares*, Poster Papers, Publ. Catania Astrophys. Obs., Special Volume, p. 325.
Carlquist, P.: 1986, *IEEE Trans. Plasma Sci.* **PS-14**, 794.
Chambon, G., Hurley, K., Niel, M., Talon, R., and Vedrenne, G.: 1981, *Solar Phys.* **69**, 147.
Chupp, E. L.: 1984, *Ann. Rev. Aston. Astrophys.* **22**, 359.
Chupp, E. L.: 1987, *Phys. Scripta* **T18**, 5.
Chupp, E. L., Forrest, D. J., Ryan, J. M., Heslin, J., Reppin, C., Pinkau, K., Kanbach, G., Rieger, E., and Share, G. H.: 1982, *Astrophys. J.* **263**, L95.
Chupp, E. L., Forrest, D. J., Vestrand, W. T., Debrunner, H., Flückiger, E., Cooper, J. F., Kanbach, G., Reppin, C., and Share, G. H.: 1985, *19th Int. Cosmic Ray Conf. Papers* **4**, 126.
Chupp, E. L., Debrunner, H., Flückiger, E., Forrest, D. J., Golliez, F., Kanbach, G., Vestrand, W. T., Cooper, J., and Share, G. H.: 1987, *Astrophys. J.* **318**, 913.
Cliver, E. W., Forrest, D. J., McGuire, R. E., von Rosenvinge, T. T., Reames, D. V., Cane, H. V., and Kane, S. R.: 1987, *20th Int. Cosmic Ray Conf. Papers* **3**, 61.
Colgate, S. A.: 1978, *Astrophys. J.* **221**, 1068.
Crannell, C. J., Crannell, H., and Ramaty, R.: 1979, *Astrophys. J.* **229**, 762.
Datlowe, D. W., O'Dell, S. L., Peterson, L. E., and Elcan, M. J.: 1977, *Astrophys. J.* **212**, 561.
Debrunner, H., Flückiger, E., Chupp, E. L., and Forrest, D. J.: 1983, *18th Int. Cosmic Ray Conf. Papers* **4**, 75.
Decker, R. B. and Vlahos, L.: 1986, *Astrophys. J.* **306**, 710.
Dermer, C. D.: 1987, *Astrophys. J.* **323**, 795.

Dröge, W. and Schlickeiser, R.: 1986, *Astrophys. J.* **305**, 909.

Dröge, W., Meyer, P., Evenson, P., and Moses, D.: 1989, *Solar Phys.* **121**, 95 (this issue).

Ellison, D. C.: 1984, *J. Geophys. Res.* **90**, 29.

Ellison, D. C. and Ramaty, R.: 1985, *Astrophys. J.* **298**, 400.

Elwert, G. and Haug, E.: 1971, *Solar Phys.* **20**, 413.

Evenson, P., Meyer, P., and Pyle, K. R.: 1983, *Astrophys. J.* **274**, 875.

Evenson, P., Kroeger, R., and Meyer, P.: 1985, *19th Int. Cosmic Ray Conf. Papers* **4**, 130.

Evenson, P., Meyer, P., Yanagita, S., and Forrest, D. J.: 1984, *Astrophys. J.* **283**, 439.

Evenson, P., Hovestadt, D., Meyer, P., and Moses, D.: 1985, *19th Int. Cosmic Ray Conf. Papers* **4**, 74.

Forman, M. A.: 1981, *Adv. Space Res.* **1**, 41.

Forman, M. A., Ramaty, R., and Zweibel, E. G.: 1986, in P. A. Sturrock, T. E. Holzer, D. Mihalas, and R. K. Ulrich (eds.), *The Physics of the Sun*, Vol. II, Chap. 13, D. Reidel Publ. Co., Dordrecht, Holland, p. 249.

Forrest, D. J.: 1983, in M. L. Burns, A. K. Harding, and R. Ramaty (eds.), *Positron-Electron Pairs in Astrophysics*, AIP, New York, p. 3.

Forrest, D. J. and Chupp, E. L.: 1983, *Nature* **305**, 291.

Forrest, D. J., Vestrand, W. T., Chupp, E. L., Rieger, E., Cooper, J. F., and Share, G. H.: 1985, *19th Int. Cosmic Ray Conf. Papers* **4**, 146.

Forrest, D. J., Vestrand, W. T., Chupp, E. L., Rieger, E., Cooper, J. F., and Share, G. H.: 1986, *Adv. Space Res.* **6**, No. 6, 115.

Haerendel, G.: 1987, *ESA* **SP-275**, 205.

Haerendel, G., Rieger, E., Valenzuela, A., Föppl, H., Stenbaek-Nielsen, H. C., and Wescott, E. M.: 1976, *ESA* **SP-115**, 203.

Haug, E.: 1975, *Z. Naturforsch.* **30a**, 1099.

Hua, X.-M. and Lingenfelter, R. E.: 1987a, *Solar Phys.* **107**, 351.

Hua, X.-M. and Lingenfelter, R. E.: 1987b, *Solar Phys.* **113**, 229.

Hua, X.-M. and Lingenfelter, R. E.: 1987c, *Astrophys. J.* **323**, 779.

Ibragimov, J. A. and Kocharov, G. E.: 1977, *Soviet Astron. Letters* **3**, 211.

Kallenrode, M.-B., Rieger, E., Wibberenz, G., and Forrest, D. J.: 1987, *20th Int. Cosmic Ray Conf. Papers* **3**, 70.

Kanbach, G., Pinkau, K., Reppin, C., Rieger, E., Chupp, E. L., Forrest, D. J., Ryan, J. M., Share, G. H., and Kinzer, R. L.: 1981, *17th Int. Cosmic Ray Conf. Papers* **10**, 9.

Kane, S. R.: 1974, in G. Newkirk Jr. (ed.), 'Coronal Disturbances', *IAU Symp.* **57**, 105.

Kane, S. R., Chupp, E. L., Forrest, D. J., Share, G. H., and Rieger, E.: 1986, *Astrophys. J.* **300**, L95.

Kane, S. R., Fenimore, E. E., Klebesadel, R. W., and Laros, J. G.: 1988, *Astrophys. J.* **326**, 1017.

Koch, H. W. and Motz, J. W.: 1959, *Rev. Mod. Phys.* **31**, 920.

Kocharov, G. E.: 1983, *Invited Talks 18th Eur. Cosmic Ray Symp.*, Bologna, p. 51.

Kocharov, G. E.: 1987, in R. A. Syunyaev (ed.), *Soviet Sci. Rev. Section E. Astrophys. Space Phys.* Vol. 6, Harwood Acad. Publ. G.m.b.H., U.K., p. 155.

Kocharov, G. E., Kovaltsov, G. A., Mandzhavidze, N. Z., and Semukhin, P. E.: 1987, *20th Int. Cosmic Ray Conf. Papers* **3**, 74.

Kocharov, G. E., Kocharov, L. G., Kovaltsov, G. A., and Mandzhavidze, N. Z.: 1988, Preprint 1258, Academy of Sciences of the USSR, A. F. Ioffe Physico-Technical Institute, Leningrad.

Kozlovski, B., Lingenfelter, R. E., and Ramaty, R.: 1987, *Astrophys. J.* **316**, 801.

MacKinnon, A. L. and Brown, J. C.: 1989, unpublished contribution to *IAU Colloq.* 104.

MacKinnon, A. L., Brown, J. C., Trottet, G., and Vilmer, N.: 1983, *Astron. Astrophys.* **119**, 297.

McDonald, F. G. and van Hollebeke, M. A. I.: 1985, *Astrophys. J.* **290**, L67.

McGuire, R. E. and von Rosenvinge, T. T.: 1984, *Adv. Space Res.* **4**, Nos. 2–3, 117.

Melrose, D. B. and Brown, J. C.: 1976, *Monthly Notices Roy. Astron. Soc.* **176**, 15.

Miller, J. A. and Ramaty, R.: 1987, *Solar Phys.* **113**, 195.

Miller, J. A., Ramaty, R., and Murphy, R. J.: 1987, *20th Int. Cosmic Ray Conf. Papers* **3**, 33.

Mozer, F. S., Cattell, C. A., Hudson, M. K., Lysak, R. L., Temerin, M., and Torbert, R. B.: 1980, *Space Sci. Rev.* **27**, 155.

Murphy, R. J. and Ramaty, R.: 1984, *Adv. Space Res.* **4**, No. 7, 127.

Murphy, R. J., Dermer, C. D., and Ramaty, R.: 1987, *Astrophys. J. Suppl.* **63**, 721.

Neustock, H.-H., Wibberenz, G., and Iwers, B.: 1985, *19th Int. Cosmic Ray Conf. Papers* **4**, 102.

Ohsawa, Y. and Sakai, J.-I.: 1987, *Astrophys. J.* **313**, 440.

Pallavicini, R., Serio, S., and Vaiana, G. S.: 1977, *Astrophys. J.* **216**, 108.

Papadopoulos, K., Goodrich, C. C., Cargill, P., and Vlahos, L.: 1985, *EOS Trans. Am. Geophys. Union* **66**, 331.

Pesses, M. E., Klecker, B., Glöckler, G., and Hovestadt, D.: 1981, *17th Int. Cosmic Ray Conf. Papers* **3**, 36.

Petrosian, V.: 1985, *Astrophys. J.* **299**, 987.

Prince, T. A., Forrest, D. J., Chupp, E. L., Kanbach, G., and Share, G. H.: 1983, *18th Int. Cosmic Ray Conf. Papers* **4**, 79.

Ramaty, R.: 1979, in J. Aarons, C. Max, and C. McKee (eds.), *Particle Acceleration Mechanisms in Astrophysics*, AIP, New York, p. 135.

Ramaty, R.: 1986, in P. A. Sturrock, T. E. Holzer, D. Mihalas, and R. K. Ulrich (eds.), *The Physics of the Sun*, Vol. II, Cap. 14, D. Reidel Publ. Co., Dordrecht, Holland, p. 291.

Ramaty, R. and Murphy, R. J.: 1987, *Space Sci. Rev.* **45**, 213.

Ramaty, R., Kozlovsky, B., and Lingenfelter, R. E.: 1975, *Space Sci. Rev.* **18**, 341.

Ramaty, R., Kozlovsky, B., and Suri, A. N.: 1977, *Astrophys. J.* **214**, 617.

Ramaty, R., Kozlovsky, B., and Lingenfelter, R. F.: 1979, *Astrophys. J. Suppl.* **40**, 487.

Ramaty, R., Murphy R. J., and Dermer, C. D.: 1987, *Astrophys. J.* **316**, L41.

Ramaty, R., Murphy, R. J., Kozlovsky, B., and Lingenfelter, R. E.: 1983, *Solar Phys.* **86**, 395.

Ramaty, R., Miller, J. A., Hua, X.-M., and Lingenfelter, R. E.: 1988, in G. H. Share and N. Gehrels (eds.), *Nuclear Spectroscopy of Astrophysical Sources*, AIP, New York (in press).

Rieger, E.: 1982, in *Hinotori Symp. on Solar Flares* (Tokyo Inst. Space Astronautical Sci.), p. 246.

Rieger, E., Reppin, C., Kanbach, G., Forrest, D. J., Chupp, E. L., and Share, G. H.: 1983, *18th Int. Cosmic Ray Conf. Papers* **10**, 338.

Rieger, E., Forrest, D. J., Bazilevskaya, G., Chupp, E. L., Kanbach, G., Reppin, C., and Share, G. H.: 1987, *20th Int. Cosmic Ray Conf. Papers* **3**, 65.

Ryan, J. M.: 1986, *Solar Phys.* **105**, 365.

Sakai, J.-I. and Ohsawa, Y.: 1987, *Space Sci. Rev.* **46**, 113.

Scholer, M.: 1988, in O. Havnes *et al.* (eds.), *Activity in Cool Star Envelopes*, Kluwer Academic Publishers, Dordrecht, Holland, p. 195.

Share, G. H., Chupp, E. L., Forrest, D. J., and Rieger, E.: 1983, in M. L. Burns, A. K. Harding, and R. Ramaty (eds.), *Positron-Electron Pairs in Astrophysics*, AIP, New York, p. 15.

Spicer, D. S.: 1982, *Space Sci. Rev.* **31**, 351.

Sturrock, P.: 1989, *Solar Phys.* **121**, 387 (this issue).

Temerin, R. B., Cattell, C. A., Mozer, F. S., and Meng, C.-I.: 1982, *Phys. Rev. Letters* **48**, 1175.

van Hollebeke, M. A. I.: 1979, *Rev. Geophys. Space Phys.* **17**, 545.

van Hollebeke, M. A. I., Ma Sung, L. S., and McDonald, F. B.: 1975, *Solar Phys.* **41**, 189.

Vestrand, W. T. and Ghosgh, A.: 1987, *20th Int. Cosmic Ray Conf. Papers* **3**, 57.

Vestrand, W. T., Forrest, D. J., Chupp, E. L., Rieger, E., and Share, G. H.: 1987, *Astrophys. J.* **322**, 1010.

Vilmer, N., Kane, S. R., and Trottet, G.: 1982, *Astron. Astrophys.* **108**, 306.

Vlahos, L. *et al.*: 1986, *Energetic Phenomena on the Sun*, NASA, CP-2439, p. 2–1.

von Rosenvinge, T. T., Ramaty, R., and Reames, D. V.: 1981, *17th Int. Cosmic Ray Conf. Papers* **3**, 28.

Wang, H. T. and Ramaty, R.: 1974, *Solar Phys.* **36**, 129.

Wescott, E. M., Stenbaek-Nielsen, H. C., Hallinan, T. J., Davis, T. N., and Peek, H. M.: 1976, *J. Geophys. Res.* **81**, 4495.

Wild, J. P., Smerd, S. F., and Weiss, A. A.: 1963, *Ann. Rev. Astron. Astrophys.* **1**, 291.

Yang, J., Turner, M. S., Steigman, G., Schramm, D. N., and Olive, K. A.: 1982, *Astrophys. J.* **281**, 493.

Yoshimori, M. and Watanabe, H.: 1985, *19th Int. Cosmic Ray Conf. Papers* **4**, 90.

Yoshimori, M., Okudaira, K., Hirasima, Y., and Kondo, I.: 1983, *Solar Phys.* **86**, 375.

MAGNETIC EQUILIBRIA AND INSTABILITIES

TAKASHI SAKURAI

National Astronomical Observatory, Mitaka, Tokyo, Japan

Abstract. Solar flares are understood as a process of explosive liberation of magnetic energy, coming after a slow phase of energy build-up. The slow evolution of magnetic equilibria may end up with (a) the termination of an equilibrium sequence, or (b) an instability. The distinction between the two can be made by drawing schematic potential curves. Case (a) has been extensively studied in two-dimensional models. The appearance of multiple solutions, or disappearance of a solution takes place as the system evolves away from the current-free configuration. Case (b) can be discussed in terms of ideal MHD or resistive MHD instabilities. A possible route to explosive energy release is suggested by combining these two cases.

1. Introduction

This review addresses theoretical aspects of energy build-up and energy liberation processes in solar flares. Solar flares take place in the solar corona above active regions. The heating of the corona from its quiescent 2 MK state to a 20–30 MK flare state indicates that the density of energy which produces flares must greatly exceed the thermal energy density of the quiescent coronal plasma. The most dominant source of energy in the corona is the magnetic field. The current consensus is that solar flares represent a process of liberating the magnetic energy stored in the corona.

The existence of the magnetic field in the corona does not necessarily mean that the magnetic energy is available there to heat the plasma. The electric currents that create the magnetic field are the ultimate source of energy. When no electric currents exist in the corona, the magnetic field in the corona is totally due to the currents flowing in the lower layers of the solar atmosphere (i.e., in the photosphere and below). Such magnetic fields are said to be free from distortion and have no energetic contribution to the phenomena in the corona. When the currents are induced in the corona, the magnetic field there is distorted, and contains the excess energy that can be liberated as a flare.

When an active region is born, there might be transient processes of energy liberation associated with the emergence of magnetic flux. After the initial relaxation is over, flare-productive active regions continue to grow and develop a highly distorted magnetic field configuration. This energy build-up phase will be regarded as a slow evolution of magnetic equilibria. This phase is followed by an explosive liberation of the stored magnetic energy, namely a flare.

The magnetic field in the corona is anchored in the dense photosphere and is controlled (passively moved around) by the flow of gas in the photosphere. Such a situation may be modeled by taking the photosphere as the lower boundary on which the boundary conditions can be specified. Therefore, the slow build-up of magnetic energy in the corona reflects the slow change in the boundary conditions near the photosphere. The energy release will be related to some sort of instability in the distorted magnetic field configuration. Although there can be other possibilities concerning the basic scenario of the flare process, in this review we will look into this canonical view in detail.

2. Instability vs Non-Equilibrium

Stability of a magnetic equilibrium can be most easily discussed by using an analogy with the equilibrium of a ball placed on a slope (Figure 1). The shape of the slope reflects an environment of the magnetic configuration, and may deform according to the change in the boundary condition for example.

The state O is a stable equilibrium. As it evolves to A_1, the system becomes marginally stable. At A_2, the curve becomes slightly convex and the system is now unstable. This instability is 'weak', because the system will evolve into a neighboring stable equilibrium, without any drastic process.

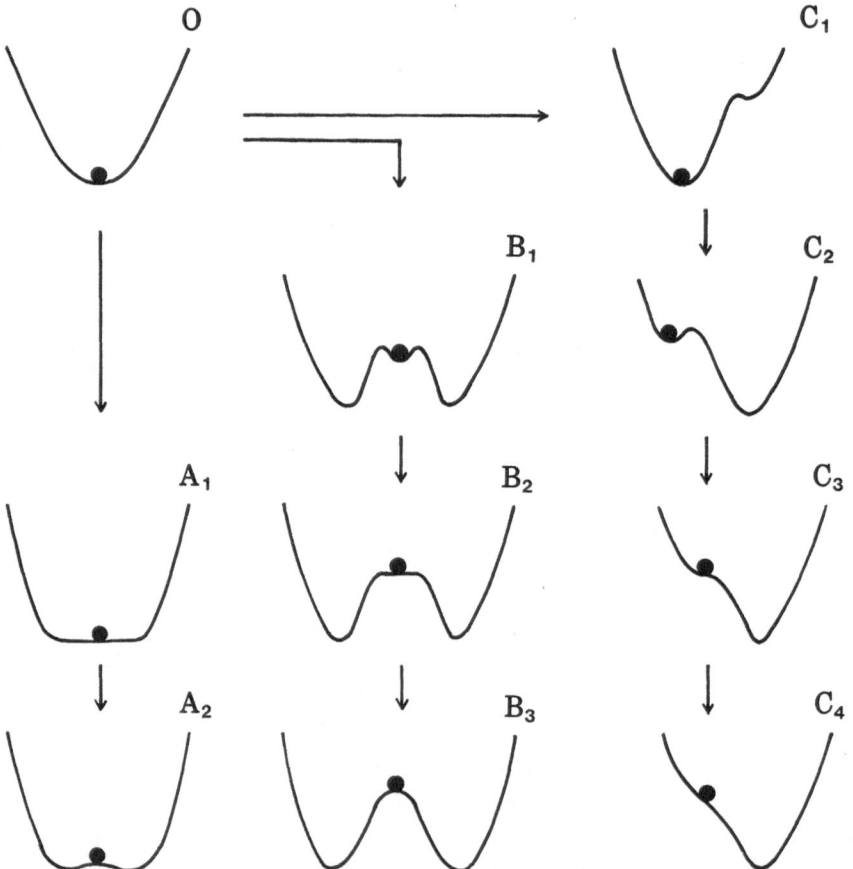

Fig. 1.   Schematic representation of equilibrium sequences. The state O is the initial stable equilibrium. Sequence A leads to a weak instability, and sequence B ends up with an explosive instability. Sequence C leads to the loss of equilibrium.

The state B_1 is called a meta-stable equilibrium, because it is stable against small amplitude perturbations but is unstable to large amplitude perturbations (Sturrock, 1966). At B_2 the system is marginally stable. The state B_3 is 'explosively unstable', because there is no stable equilibrium in the neighborhood of the original (unstable) equilibrium and the system undergoes a jump of finite amplitude.

The state C_2 is meta-stable and C_3 is marginally stable. As the system evolves to C_4, the equilibrium (i.e., the plateau in C_3) is lost and the system exhibits a finite-amplitude jump toward a new equilibrium.

One notable difference between the explosive instability (sequence B) and the loss of equilibrium (sequence C) is that the resulting perturbation that grows is unidirectional in sequence C. In sequence B the direction in which the system moves depends on the initial (infinitesimal) perturbation.

3. Two-Dimensional Equilibria

3.1. BASIC EQUATION

In order to study instabilities and disequilibrium of magnetic field configurations discussed above, we will next consider two-dimensional equilibria as an example (Birn and Schindler, 1981; Low, 1982). The x-coordinate is supposed to be an ignorable coordinate, and the magnetic field $\mathbf{B}$ is a function of y and z in the form

$$\mathbf{B} = \left(B_x, \ \frac{\partial A}{\partial z}, \ -\frac{\partial A}{\partial y} \right), \tag{1}$$

where A is a function of y and z, and B_x is a function of A. If the gas is stratified isothermally, the gas pressure p must be written as

$$p = p_0(A) \exp(-z/H), \tag{2}$$

where H is the (constant) scale height. The vector potential A is determined by the so-called Grad–Shafranov equation

$$\nabla^2 A + \frac{1}{2} \frac{\partial}{\partial A} \{B_x^2 + 8\pi p_0 \exp(-z/H)\} = 0. \tag{3}$$

The current-free state is represented by $\nabla^2 A = 0$, so that the second term reflects the distortion in the magnetic field. Therefore, we write

$$\nabla^2 A + \lambda F(A, z) = 0 \tag{4}$$

and the field configuration is increasingly distorted as λ is varied from zero to larger values.

The boundary condition for A is given at the $z = 0$ plane (the photosphere) as $A = A_b$. By defining $A' = A - \overline{A}$ with

$$\nabla^2 \overline{A} = 0, \qquad \overline{A} = A_b \quad \text{at} \quad z = 0, \tag{5}$$

the equation is reduced to a canonical form

$$-\nabla^2 A' = \lambda F(\overline{A} + A', z), \qquad A' = 0 \quad \text{at} \quad z = 0. \tag{6}$$

3.2. MULTIPLICITY IN THE SOLUTIONS

At this point it is useful to compare this equation with the equation describing the equilibrium of a membrane, namely,

$$-\tau \nabla^2 \xi = \sigma g. \tag{7}$$

Here $\tau(\xi)$ is the tension, σ is the surface density, g is the gravitational acceleration, and ξ is the displacement in the membrane which vanishes on the boundary. (We have slightly deviated from the linearization assumption that led to Equation (7) and introduced the ξ-dependence in τ.) We find the following correspondence:

$$\xi \leftrightarrow A', \qquad \tau \leftrightarrow 1/F, \qquad \sigma g \leftrightarrow \lambda.$$

We can see that larger λ corresponds to heavier mass loading in the membrane. The larger F is, the more fragile the membrane is.

We may roughly set $-\nabla^2 \rightarrow 1/L^2$, where L stands for the size of the system. Then Equations (4) and (7) are simplified as

$$\lambda F = A'/L^2 \tag{8}$$

and

$$\sigma g/\tau = \xi/L^2, \tag{9}$$

respectively, and the multiplicity in the solutions can be argued graphically.

Figure 2(a) shows that the solution exists and is unique when $\partial F/\partial A < 0$. On the other hand Figure 2(b) reveals that if $F > 0$, $\partial F/\partial A > 0$, and $\partial^2 F/\partial A^2 > 0$, no solution exists for $\lambda > \lambda_*$ but two (or more) solutions exist for $\lambda < \lambda_*$. (Similar argument holds if $F < 0$, $\partial F/\partial A > 0$, and $\partial^2 F/\partial A^2 < 0$.)

Figure 2(c) is equivalent to Figure 2(b) but in a different format. The curve in Figure 2(c) represents the restoring force in the membrane as a function of the displacement ξ. The solution with a smaller magnitude of ξ (designated as the branch I solution) is stable. The branch II solution will be unstable because slightly larger displacement ξ leads to smaller restoring force. The solution at $\lambda = \lambda_*$ represents the critical equilibrium and is marginally stable. These stability properties are for the cases in which the perturbations given to the system are also two-dimensional. If three-dimensional perturbations are considered, the branch I solutions may be unstable.

If the function F satisfies $\partial F/\partial A < 0$, the solution formally exists for an arbitrary value of λ. However, the solution sequence terminates physically at some λ, beyond which the

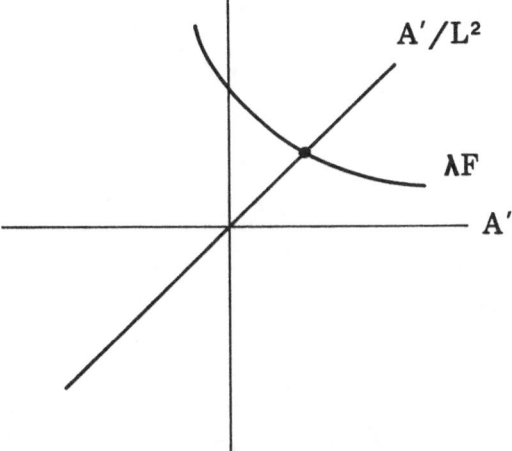

Fig. 2a.   Schematic diagram showing the uniqueness in the solutions to Equation (8) when $\partial F/\partial A < 0$.

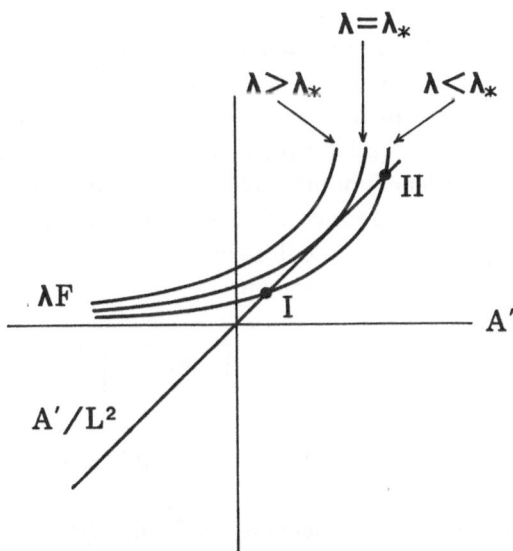

Fig. 2b.   Schematic diagram showing the multiplicity in the solutions to Equation (8) when $\partial F/\partial A > 0$.

quantity $B_x^2 + 8\pi p_0 \exp(-z/H)$ goes to negative. In the following we will concentrate on the cases with multiple solutions, because of their relevance to the flare models.

3.3. LOSS OF EQUILIBRIUM BY INCREASING SHEAR

First we will discuss the case with $p = 0$, so that λF measures the effect of B_x. The presence of B_x makes the field lines tilted from the yz-plane. The separation between the two footpoints (Δx) is called the shear. The variation of Δx as a function of λ is as in Figure 3. Solutions may be lost if λ is driven beyond λ_*.

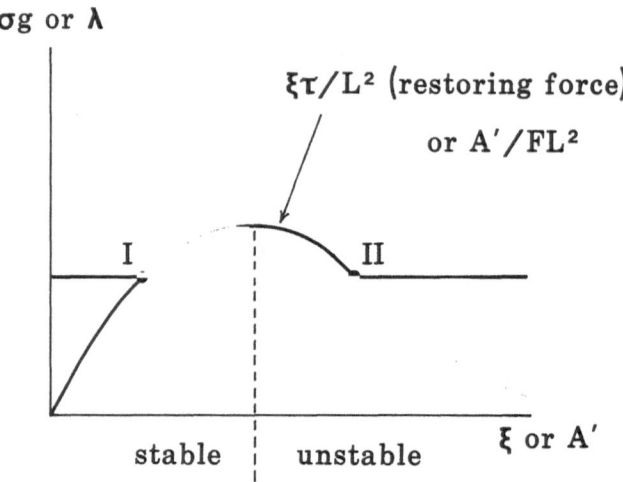

Fig. 2c.   Same as Figure 2(b) but in a different format. The ordinate corresponds to the equivalent restoring force.

However, it is not B_x but is Δx that directly represents the effect of boundary motions deforming the magnetic configuration. If Δx is taken as the control parameter, then λ varies from zero to λ_* and then decreases along the branch II as Δx increases. This behavior is explained by Jockers (1976) as follows. The shear Δx is proportional to $B_x \times L_F$, where L_F is the length of the field line. Along branch I solutions, L_F is nearly constant and B_x is proportional to Δx. On branch II the increase in Δx is accounted for by the increase in L_F, in spite of the decrease in B_x or λ.

Aly (1985) derived the integral inequality

$$\int B_x(A)\Delta x(A)\,\mathrm{d}A \leqq \left[\int B_z^2\,\mathrm{d}y \int y^2 B_z^2\,\mathrm{d}y \right]^{1/2} \tag{10}$$

and found that $B_x \to 0$ as $\Delta x \to \infty$. He interpreted this as the asymptotic formation of current sheets as Δx becomes very large.

The implication of the two arguments above is that the equilibrium will not be lost as the shear Δx increases. The solutions with very large shear may, however, belong to the branch II and will be unstable against two-dimensional perturbations. The stability of the equilibria with respect to three-dimensional perturbations will be discussed later.

3.4. LOSS OF EQUILIBRIUM BY INCREASING PRESSURE

Next we will look into the case with $B_x = 0$, so that λF is solely due to the gas pressure. Then λ (namely pressure) is a natural control parameter, and there would be no reason to exclude the cases with $\lambda > \lambda_*$. Therefore, the equilibrium will be lost when λ is driven beyond λ_*. Low (1981) argued that such disequilibrium may occur in two ways. One is when the prescribed pressure is too high to be magnetically confined. The effect of gravity is not essential in this case. The other case is due to the magnetic buoyancy.

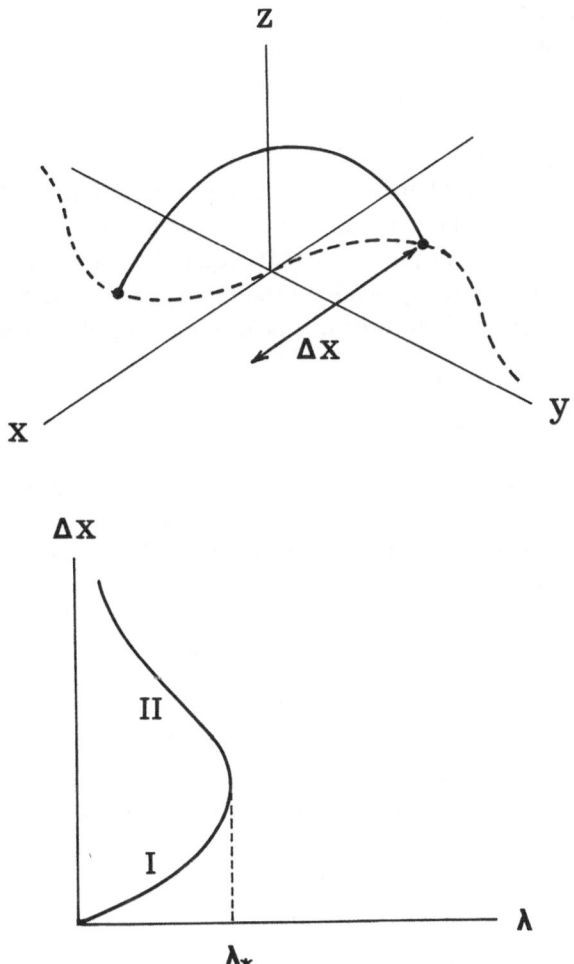

Fig. 3.   Definition of the shear Δx (*top*) and the behavior of Δx as a function of the parameter λ (*bottom*).

Zwingmann (1987) performed the numerical simulations including both the shear and the pressure effects, and found that the catastrophe requires finite pressure but can take place without shear. Therefore, the presence of finite pressure is necessary for the loss of equilibrium. However, shearless configurations may well be susceptible to the interchange instability (Section 4.2) even if $\lambda < \lambda_*$, when three-dimensional perturbations are considered. It is, therefore, debatable whether pressure-dominated equilibria can reach the critical state $\lambda = \lambda_*$ before getting unstable.

4. Ideal MHD Instabilities

When the effect of resistivity is negligible (ideal MHD), instabilities that can take place may be divided into two classes. One is the kink instability, driven by the Lorentz force.

This instability may occur even with the absence of the pressure. The other is the interchange instability, driven by the pressure force.

4.1. KINK INSTABILITY OF LOOPS

Kink instability appears in cylindrical loops as a growing helical perturbation. Anzer (1968) showed that infinitely long loops are always unstable to the kink instability. In actuality the loop has a finite length, and its footpoints are rooted in the dense photosphere. Therefore, the displacements are required to vanish at two footpoint cross sections. This effect of 'line-tying' tends to stabilize the kink instability (Raadu, 1972). For example the constant-pitch force-free field in cylindrical coordinates (r, θ, z),

$$B_z = B_0/[1 + r^2/b^2]^{-1},$$

$$B_\theta = (r/b)B_z \quad (b = \text{const.}),$$

$$(11)$$

is stable if the length of the loop L is such that $L/b \lesssim 2.5\pi$. That is, this loop is stable if the twist over the length of the loop does not exceed 1.25 turn (Hood and Priest, 1981).

The nonlinear development of the kink instability will be described by the equation for the mode amplitude ξ as

$$\frac{\partial^2 \xi}{\partial t^2} = \gamma^2 \xi + \delta \xi^3 + \cdots.$$

$$(12)$$

Even when the system is linearly unstable ($\gamma^2 > 0$), nonlinear effects stabilize the instability if $\delta < 0$. On the other hand if $\delta > 0$, the nonlinear effects enhance the instability, leading to the explosive instability. For example, in the case of a sharp boundary pinch (plasma in $r < r_1$, vacuum in $r_1 < r < r_2$, with a current sheet at $r = r_1$ and a conducting wall at r_2), δ is found to be

$$\delta \sim \frac{(r_2/r_1)^4 - 2}{(r_2/r_1)^4 - 1},$$

$$(13)$$

so that the kink instability is explosive if $r_2 \gtrsim 1.2 r_1$ (Pao, 1978). The nonlinear development of the kink instability was studied by Sakurai (1976).

4.2. STABILITY OF ARCADES

Roughly speaking, the magnetic arcade is equivalent to a straight magnetic loop sliced in half along the loop axis and placed on the photosphere. Helical perturbations that characterize the kink instability are prohibited in magnetic arcades due to the line-tying condition. If the effect of pressure is large enough, the arcade can be unstable to the interchange instability. Hood (1986) derived the sufficient condition for instability as

$$\frac{B^2}{r^2} + \frac{\pi^2}{4} B_x^2 \left(\frac{1}{q}\frac{dq}{dr}\right)^2 + \frac{2}{r}\frac{dp}{dr} + \frac{16\pi\Gamma p B_\theta^2}{r^2(4\pi\Gamma p + B^2)} < 0$$

$$(14)$$

$$(q = rB_x/B_\theta).$$

Here Γ denotes the ratio of specific heats, and the cylindrical coordinates (r, θ) around the x-axis are taken as in Figure 4(a). The equilibrium quantities are functions of r. The instability arises when the plasma confined by the magnetic field $(dp/dr < 0)$ escapes from the confinement by making thin sheets intruding through the magnetic field. Stabilizing effects come from varying pitch of the field lines (the second term) and from the line tying (the first and the fourth terms). For force-free arcades ($p = 0$) this equation does not suggest instability, and a higher-order analysis is necessary to determine the stability. Hood and Anzer (1987) studied the stability of several force-free arcade configurations and could not find any unstable modes. However, if the configuration contains closed field lines detached from the photosphere (the magnetic island, Figure 4(b)), which might be related to dark filaments on the Sun (Anzer and Tandberg-Hanssen, 1970), the situation is similar to a loop surrounded by an external magnetic field and the rigid boundary (i.e., the photosphere). Hood and Priest (1980) showed that if the height of the closed loop above the photosphere is sufficiently large, the loop may be unstable against the kink instability.

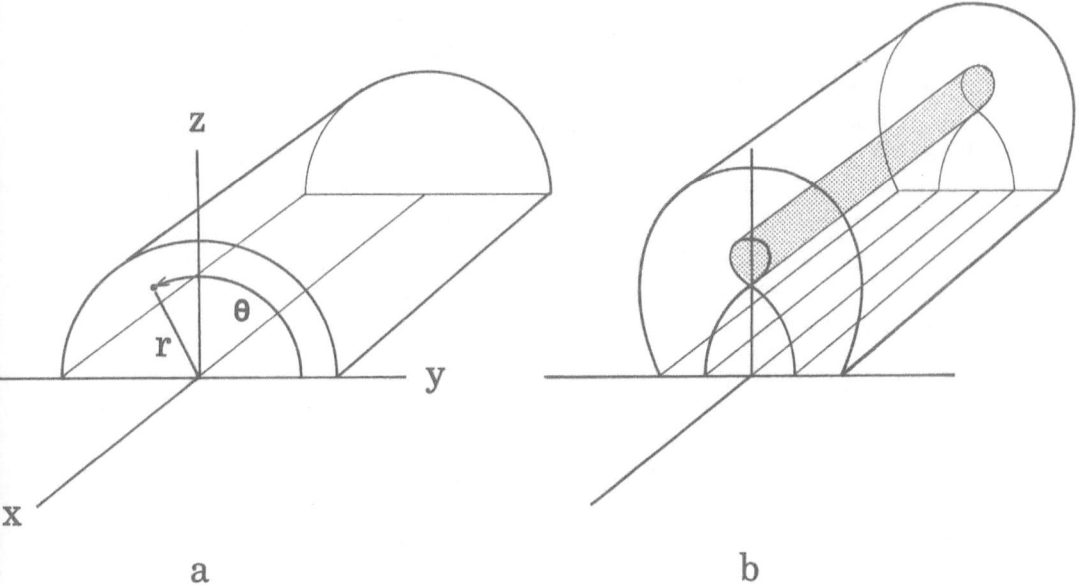

Fig. 4.   (a) The magnetic arcade configuration. (b) The arcade with a magnetic island or 'a filament' (shaded).

5. Non-Ideal Effects

Deviation from the ideal MHD situation can be found in various ways. The effect of non-zero resistivity is to violate the frozen-in condition of lines of force, and to modify the energy equation by Joule heating. Non-adiabatic effects may also come from the heat conduction and radiative processes.

5.1. FINITE RESISTIVITY

A well-known instability in the presence of a finite resistivity is the tearing instability
(Furth, Killeen, and Rosenbluth, 1963). Let us denote the dynamical and resistive
time-scales by τ_A and τ_R, respectively. They are defined as

$$\tau_A = L/V_A, \qquad \tau_R = L^2/\eta, \tag{15}$$

where L is the size of the system, V_A is the Alfvén speed, and η is the resistivity. The
magnetic Reynolds number S is defined by $S = \tau_R/\tau_A$, which is much larger than unity
in the solar atmosphere. The growth rate of the tearing instability is expressed as

$$\gamma \sim S^\alpha/\tau_R, \tag{16}$$

where α ranges from $\frac{2}{5}$ to $\frac{2}{3}$ (Steinolfson and Van Hoven, 1983).

   Another important process is the magnetic reconnection. Original analytic but simplified treatment (Petschek, 1964) assumed a steady state, but recent progress has been
made by time-dependent numerical simulations. An important implication derived from
them is that the rate of reconnection can be as fast as one likes, if the flow field is set
up which forces the reconnection to occur (Sato and Hayashi, 1979). Then the question
is how such a favorable flow is set up.

   In this respect it is adequate to mention the so-called coalescence instability (Finn
and Kaw, 1977). This instability takes place, for example, when two loops with parallel
electric currents attract with each other. Until two loops come in contact and merge,
the process is described essentially as an ideal MHD process. The explosive nature of
this instability is demonstrated by using a self-similar scaling (Sakai and Tajima, 1986).
In one-dimensional coalescence in the x-direction, they found the velocity v, the density
ρ, and the magnetic field B to scale as

$$v = \frac{1}{a}\frac{da}{dt}\, x, \tag{17}$$

$$\rho = \frac{\rho_0}{a}, \tag{18}$$

and

$$B = B_0\, \frac{x}{x_0}\, \frac{1}{a^2}. \tag{19}$$

Here ρ_0, B_0, and x_0 are constants, and the scale factor a varies as $a \sim (t_0 - t)^{2/3}$. The
time t_0 represents the epoch when two current systems collide.

   It may well be that not only the coalescence instability but also other MHD instabilities are able to provide a driving force which promotes a rapid reconnection. For
example the tearing instability shows faster growth (α in Equation (16) is close to $\frac{2}{3}$) when
it is driven externally (Spicer and Brown, 1981).

5.2. THERMAL INSTABILITIES

Thermal instabilities may arise in several ways. If the radiative cooling rate decreases as the temperature increases, there will be a thermal runaway (Field, 1965). The time-scale of this radiative instability is

$$\tau_{rad} = \frac{3nkT}{n^2 Q_{rad}(T)} , \tag{20}$$

where n is the number density of the plasma and $Q_{rad}(T)$ is the radiative output power per unit number density at temperature T.

Similarly if the Joule heating rate increases as the temperature increases, the plasma is unstable to the so-called Joule heating mode (Heyvaerts, 1974). This instability arises because

$$T\uparrow \to \eta \sim T^{-3/2}\downarrow \to j\uparrow (E \sim \text{const.}) \to \text{heating}\uparrow \to T\uparrow .$$

For the electric field E to stay constant, the time-scale of this sequence must be as long as τ_R. For faster time-scales the decrease in η will lead to the decrease in E while the current j is kept constant, and the heating is reduced instead.

In the solar corona the ordering $\tau_A < \tau_{rad} < \tau_R$ holds. The interplay between thermal and tearing instabilities was studied, e.g., by Steinolfson (1983). When the radiation is unstable, there are cases in which the radiative instability grows faster than the tearing instability. The radiative instability in such a case involves the reconnection of magnetic fields as well. Joule heating mode does not appear unless the radiative energy loss is artificially suppressed.

Thermal instabilities can play a role in setting up a pre-flare current filamentation or in triggering other more energetic instabilities. It seems unlikely, however, that they serve as the primary mechanism of flare energy release.

5.3. THERMAL STABILITY OF CORONAL LOOPS

Thermal stability of coronal loops is more complicated than the argument given above, due to the coupling between the coronal loop and the chromosphere/photosphere at its footpoints. The treatment of the mass flow into or out of the loop has yielded contradictory results. McClymont and Craig (1985) claimed that the coronal loops are stable, while Antiochos et al. (1985) found instabilities. Martens and Kuin (1983) suggested that the loops we see are not static but are exhibiting a limit cycle behavior. Recent analysis (Klimchuk, Antiochos, and Mariska, 1987) indicated that the stability of coronal loops critically depends on the height or the length of loops. Apparently more study is needed to resolve this important issue.

6. Summary

A possible sequence from a current-free (non-distorted) magnetic field configuration to an explosive energy release is shown in Figure 5.

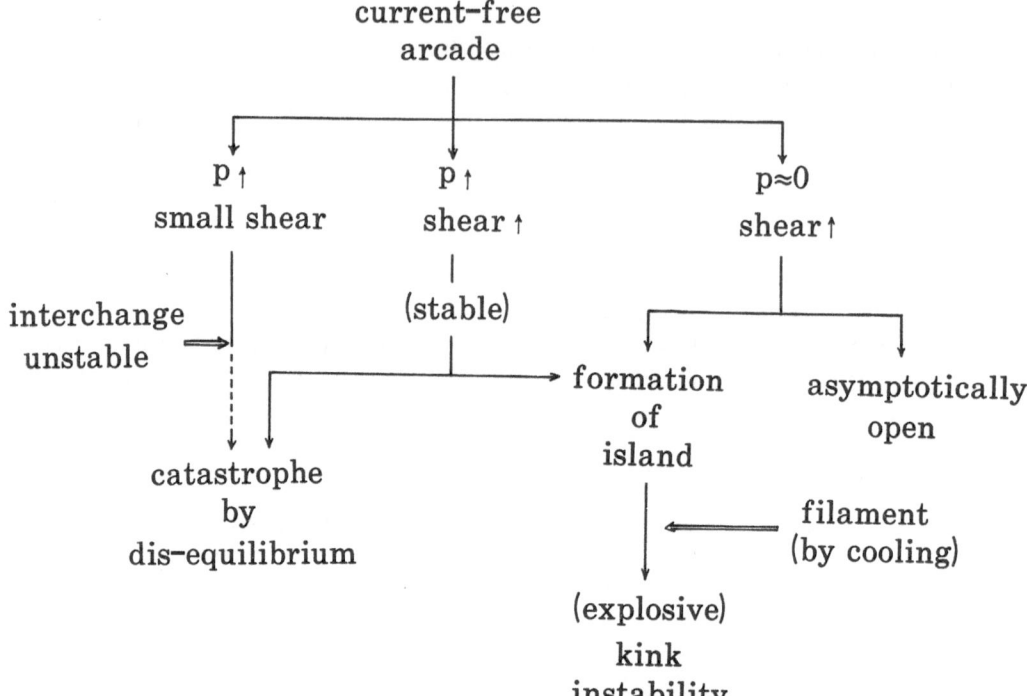

Fig. 5.   Various possibilities which may lead to catastrophic energy release, starting from an initial stable
equilibrium.

(a) If the pressure increases in an arcade with small magnetic shear, 2-D equilibrium
sequence models predict a catastrophe by disequilibrium. However, it is likely that the
system may become unstable to the interchange instability before the catastrophe. [The
interchange instability is here assumed to be a local, non-catastrophic instability. This
may not be true if the magnetic reconnection is involved (Parker, 1973; Uchida and
Sakurai, 1977).]

(b) If the pressure increases in a sheared arcade, the system may remain stable
against the interchange instability until the system loses the equilibrium. However, the
requirement of a large pressure in cases (a) and (b) is unsatisfactory, in considering the
magnetically dominated condition in the corona. Therefore, the case (c) below is
concluded to be the most promising.

(c) If the shear increases in a nearly force-free arcade, there is evidence that the
equilibrium always exists and the magnetic configuration asymptotically approaches the
open magnetic field. Force-free arcades are highly stable, with the absence of a magnetic
island. The configuration with the magnetic island can become unstable against the kink
instability.

It is premature to say that any force-free arcades without magnetic islands are stable.
Nevertheless it is suggestive that the presence of the magnetic island in an arcade could
be crucial in considering the stability of the system. Physically speaking, the magnetic

island may correspond to a filament if cool material is present there. In some pressure-dominated cases (Melville, Hood, and Priest, 1987), the island forms before the catastrophic point is reached. In all the force-free arcade models available up to now, the island only forms at or after the critical point is reached (Low, 1977; Jockers, 1978). It is not clear whether this is a general rule or because only a small number of examples were studied. Therefore, the important issues raised here are (i) to thoroughly study the stability of force-free arcades, and (ii) to clarify whether the magnetic island can form in stable force-free equilibrium sequences.

In the models developed so far, the island either emerges from the bottom boundary (Low, 1977; Jockers, 1978; Melville, Hood, and Priest, 1987) or forms at some height in the volume $z > 0$ (Zwingmann, 1987). The latter case physically requires the magnetic reconnection (Anzer and Priest, 1985). When the magnetic island (or a magnetic loop) is formed in an arcade, it may undergo explosive kink instability. This is in accord with the fact that flares are often associated with the activation of filaments. The role played by the cool material which comprises the filament is not clear, however. It might be that an invisible filament, namely a magnetic island not filled with cool material, can play the same role.

The scenario described above may correspond to the so-called two-ribbon flares, which are in contrast with the simple loop flares (Priest, 1981). For the latter class of flares, they could be due to the kink-unstable loops which may form simply by twisting the existing coronal loops. The magnetic reconnection (or the tearing instability) driven by explosive MHD instabilities may, therefore, be a possible explanation for both types of flares. This conjecture does not exclude the possibility of other important classes of flare models such as the emerging flux model (Heyvaerts, Priest, and Rust, 1977).

References

Aly, J. J.: 1985, *Astron. Astrophys.* **143**, 19.
Antiochos, S. K., Shoub, E. C., An, C.-H., and Emslie, A. G.: 1985, *Astrophys. J.* **298**, 876.
Anzer, U.: 1968, *Solar Phys.* **3**, 298.
Anzer, U. and Priest, E. R.: 1985, *Solar Phys.* **95**, 263.
Anzer, U. and Tandberg-Hanssen, E.: 1970, *Solar Phys.* **11**, 61.
Birn, J. and Schindler, K.: 1981, in E. R. Priest (ed.), *Solar Flare Magnetohydrodynamics*, Gordon and Breach, London, p. 337.
Field, G. B.: 1965, *Astrophys. J.* **169**, 379.
Finn, J. M. and Kaw, P. K.: 1977, *Phys. Fluids* **20**, 72.
Furth, H. P., Killeen, J., and Rosenbluth, M. N.: 1963, *Phys. Fluids* **6**, 459.
Heyvaerts, J.: 1974, *Astron. Astrophys.* **37**, 65.
Heyvaerts, J., Priest, E. R., and Rust, D. M.: 1977, *Astrophys. J.* **216**, 123.
Hood, A. W. and Anzer, U.: 1987, *Solar Phys.* **111**, 333.
Hood, A. W. and Priest, E. R.: 1980, *Solar Phys.* **66**, 113.
Hood, A. W. and Priest, E. R.: 1981, *Geophys. Astrophys. Fluid Dyn.* **17**, 297.
Hood, A. W.: 1986, *Solar Phys.* **103**, 329.
Jockers, K.: 1976, *Solar Phys.* **50**, 405.
Jockers, K.: 1978, *Solar Phys.* **56**, 37.
Klimchuk, J. A., Antiochos, S. K., and Mariska, J. T.: 1987, *Astrophys. J.* **320**, 409.
Low, B. C.: 1977, *Astrophys. J.* **212**, 234.
Low, B. C.: 1981, *Astrophys. J.* **251**, 352.

Low, B. C.: 1982, *Rev. Geophys. Space Phys.* **20**, 145.

Martens, P. C. H. and Kuin, N. P. M.: 1983, *Astron. Astrophys.* **123**, 216.

McClymont, A. N. and Craig, I. J. D.: 1985, *Astrophys. J.* **289**, 820.

Melville, J. P., Hood, A. W., and Priest, E. R.: 1987, *Geophys. Astrophys. Fluid Dyn.* **39**, 83.

Pao, Y.: 1978, *Phys. Fluids* **21**, 765.

Parker, E. N.: 1973, *Astrophys. J.* **180**, 247.

Petschek, H. E.: 1964, in W. N. Hess (ed.), *NASA Symposium on the Physics of Solar Flares*, p. 425.

Priest, E. R.: 1981, in E. R. Priest (ed.), *Solar Flare Magnetohydrodynamics*, Gordon and Breach, London, p. 2.

Raadu, M. A.: 1972, *Solar Phys.* **22**, 425.

Sakai, J. and Tajima, T.: 1986, *ESA SP-251*, 77.

Sakurai, T.: 1976, *Publ. Astron. Soc. Japan* **28**, 177.

Sato, T. and Hayashi, T.: 1979, *Phys. Fluids* **22**, 1189.

Spicer, D. S. and Brown, J. C.: 1981, in S. Jordan (ed.), *The Sun as a Star*, NASA SP-450, p. 413.

Steinolfson, R. S.: 1983, *Phys. Fluids* **26**, 2590.

Steinolfson, R. S. and Van Hoven, G.: 1983, *Phys. Fluids* **26**, 117.

Sturrock, P. A.: 1966, *Phys. Rev. Letters* **16**, 270.

Uchida, Y. and Sakurai, T.: 1977, *Solar Phys.* **51**, 413.

Zwingmann, W.: 1987, *Solar Phys.* **111**, 309.

A SOLAR FLARE OBSERVED WITH THE SMM AND *EINSTEIN* SATELLITES

J. H. M. M. SCHMITT

Max-Planck-Institut für extraterrestrische Physik, 8046 Garching, F.R.G.

J. R. LEMEN

Lockheed Palo Alto Research Laboratories, U.S.A.

and

D. ZARRO

Applied Research Corporation at NASA/GSFC, U.S.A.

Abstract. We present X-ray observations of the 21 July, 1980 flare which was observed both with the *Einstein* Observatory Imaging Proportional Counter (IPC) and the X-Ray Polychromator (XRP) and Gamma-Ray Spectrometer onboard the SMM satellite. The *Einstein* observations were obtained in scattered X-ray light, i.e., in X-rays scattered off the Earth's atmosphere. In this way it is possible to obtain spatially unresolved X-ray data of a solar flare with the same instrument that observed many X-ray flares on other stars. This paper juxtaposes the results and implications of the 'stellar interpretation' to those obtained from the far more detailed SMM observations. The result of this 'calibration' observation is that the basic properties of the flaring plasma can be reliably determined from the 'stellar' data, however, the basic physics issues can only be studied through models.

1. Introduction

Imaging soft X-ray telescopes such as flown onboard the *Einstein* and EXOSAT satellites have provided a large number of observational examples of stellar X-ray flares (Haisch, 1983; Pallavicini, 1987; Collura, Pasquini, and Schmitt, 1988; Pallavicini and Schmitt, 1988). Flares are observed not only on the previously known (optical) flare stars, i.e., on late-type M dwarf stars, but also on RS CVn systems, solar-like stars and possibly even on A-type stars (Pallavicini *et al.*, 1989). On the Sun the occurrence of flares is, of course, quite a common phenomenon and flares have been the object of study of a concert of dedicated instruments flown onboard Skylab and SMM. In particular, the SMM satellite allowed simultaneous monitoring of solar flares from γ-rays to soft X-rays with very high time resolution. At soft X-ray energies, spectral resolution of $\lambda/\Delta\lambda \sim 10\,000$ could be accomplished, allowing the measurement of line profiles and Doppler shifts; further, spatial resolution of ~ 15 arc sec was possible up to photon energies of ~ 10 keV. The wealth and detail of the SMM observations greatly advanced our understanding of the physical processes occurring in solar flares.

The data quality of stellar X-ray observations is of course far inferior to that obtained with state-of-the-art observations of the Sun in every respect. First of all, spatial resolution is totally absent and, hence, no direct information on the size of the emitting regions is available. Second, the stellar X-ray observations are typically carried out at lower energy (in the pass band 0.2–4.0 keV with the *Einstein* Observatory and

Solar Physics **121**: 361–373, 1989.
© 1989 *Kluwer Academic Publishers.*

0.04–2.0 keV with the EXOSAT LE), and little or no sensitivity remains at higher energies; in particular, no information is available on non-thermal X-ray emission and further, the energy resolution of the soft X-ray data is very limited ($\Delta E/E \sim 1$) for the *Einstein* Observatory IPC or absent (for the EXOSAT LE). Lastly, the time resolution of stellar flare observations is far worse than that of solar observations due to counting statistics; although the X-ray observations are typically carried out in photon counting mode with the arrival times of individual photons measured with an accuracy of a few milliseconds or better, the count rates of the observed stellar flare events are typically rather low (say, 0.1–1 counts s^{-1}), and, therefore, it is difficult to detect variability on time scales less than ~ 1 min in a statistically significant way.

Consequently, the basic X-ray observations of a stellar flare consist of a light curve (possibly in various energy bands) and thus an estimate of a decay time, a flux (implying an emission measure) and possibly a rather coarse estimate of X-ray temperature obtained from fitting theoretical model spectra to broad-band X-ray data (cf. Pallavicini *et al.*, 1987). The inference of the physical parameters of the flaring plasma (such as temperature, size, and density of the flaring region) on a star must, therefore, be rather indirect and in particular involves assumptions on the relevant cooling time scales. Using such indirect arguments one typically finds stellar flares to be far more energetic than their solar counterparts; however, the sizes of the flaring regions are almost always found to be small compared to the stellar radius, implying the presence of rather dense and hot material in the coronae of stars.

Schmitt, Harnden, and Fink (1987) report X-ray observations of solar flares with the *Einstein* Observatory. While this telescope was designed exclusively for non-solar cosmic X-ray observations and could, therefore, not be pointed directly at the Sun, solar X-rays scattered off the Earth's atmosphere were recorded while the Sun-lit Earth intercepted the telescope's field-of-view. By modelling the propagation of solar X-rays in the Earth's atmosphere (see Fink, Schmitt, and Harnden, 1988 for details) it is possible to disentangle variations in the observed scattered X-ray flux due to the continuously changing viewing geometry from intrinsic variations in the incident solar X-ray flux. Thus an X-ray light curve of a solar flare can be obtained which has – since obtained with the same instrument – all the observational attributes of a stellar flare observation. The purpose of this paper is to present and discuss the SMM observations of the same flare, to compare the parameters derived from the stellar modelling of the July 21, 1980 flare to the direct measurements obtained with the SMM satellite, and to investigate how much of the physics of stellar flares can be deduced from the available soft X-ray observations.

2. The *Einstein* Observations of the Flare on July 21, 1980

A detailed description of the *Einstein* observations of scattered solar X-rays was given by Schmitt, Harnden, and Fink (1987; hereafter abbreviated as SHF) and Fink, Schmitt, and Harnden (1988; abbreviated as FSH); here we only summarize the most important issues and refer those readers interested in the detailed modelling to the above papers.

When the X-ray telescope is pointed at the Sun-lit Earth, solar X-ray photons scatter off the atoms and molecules of the Earth's atmosphere, enter the telescope aperture and form a diffuse 'image' of the Sun in the detector plane. Depending on the photon energy, either Thomson scattering or fluorescent scattering predominates, however in the so-called carbon window between 0.2 and 0.28 keV where the IPC is most sensitive all photons are Thomson scattered. In both cases, however, the ratio between scattering and absorption cross section is small, so that for a theoretical description of photon propagation at X-ray energies in the Earth's atmosphere a single-scattering description suffices. Due to the inherent three-dimensional nature of the problem this simplifies the radiative transfer calculations enormously; in planar geometry which approximately applies under certain viewing conditions the problem can be solved analytically.

FSH showed that the X-ray light curves of *Einstein* Observatory bright Earth passes can be satisfactorily modelled under a variety of viewing conditions assuming the incident solar flux to be constant. SHF reported three bright Earth passes which could not be modelled under the constant flux assumption and argued that the modelling procedure failed because of the occurrence of solar flares. The event which was by far the strongest among those reported by SHF was the flare on July 21, 1980, commencing at 2 : 56 UT.

In the upper panel of Figure 1 we show the observed IPC light curve during the bright Earth pass; the smooth curve shows the best model fit to the data and the lower

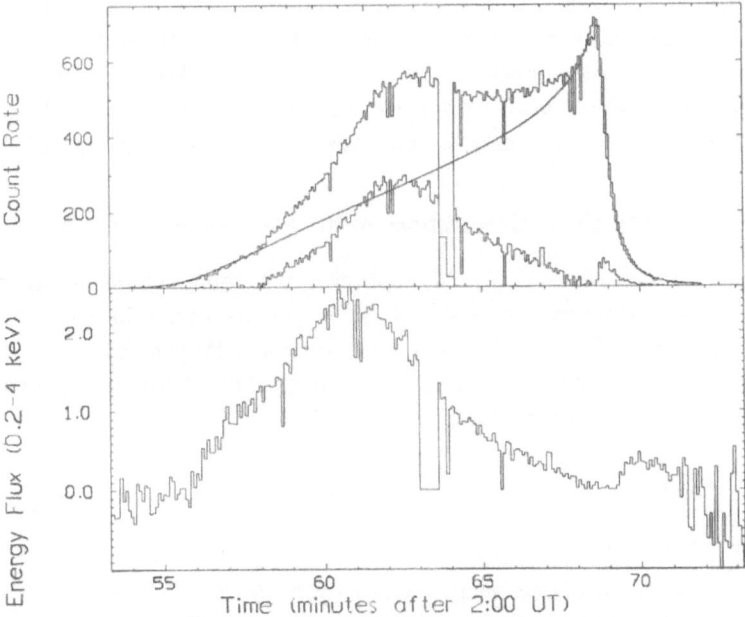

Fig. 1a–b. (a) Observed IPC light curve (upper histogram) as function of time; the smooth curve represents the best fit assuming constant incident flux, the bottom curve is the residual attributed to the flare, containing both intrinsic as well as geometric variations. (b) Inferred energy flux in the IPC flare as a function of time.

histogram shows the excess count rate attributed to the flare. Note that this histogram still contains contributions from geometric variability as well as intrinsic flux variations.

Our model fits to the data have significantly improved compared to the method used by SHF and FSH. In particular, instead of a monochromatic calculation we now use an empirical solar X-ray spectrum for the quiet and flaring Sun (extracted from Manson, 1976) to compute the scattered solar X-ray flux. This procedure removes the difficulty encountered by SHF and FSH of having to fix the flux scattered through Thomson and fluorescent scattering and allows an immediate computation of the energy flux measured by direct observation of the Sun. This inferred energy flux in the IPC pass band is plotted in the lower panel of Figure 1. Flare onset occurs at about $2:55:40$ UT with the peak count rate occurring at about $3:00:40$ UT. The rise to maximum is not linear but has an inflection point at $2:58:20$ UT. At flare maximum a flux of ~ 2.4 erg s^{-1} cm^{-2} is inferred; it should, however, be kept in mind that the energy fluxes inferred from the broad band data are probably no better than $\sim 50\%$. (Note that an IPC count rate of $\sim 3 \times 10^{11}$ counts s^{-1} would have been expected had the instrument been pointed at the Sun!).

In Table I, extracted from SHF, we summarise the physical parameters of the flare of July 21, 1980 as derived from the IPC observations. Clearly, the bright earth data offer an excellent opportunity to compare full disk IPC observations of solar flares to full disk IPC observations of stellar flares. The IPC measurements clearly show the soft X-ray emitting flare plasma heat and cool; further, the derived parameters are consistent with the properties of compact loop flares, the most commonly type of flare found on the Sun. Specifically, the IPC data indicate peak temperatures of the flare plasma of $\log T \sim 7.25$; and second, characteristic length scales of $\sim 3 \times 10^9$ cm. We re-emphasise that both of these findings were obtained with broad band data having no spatial resolution. In the following we want to examine to what extent the SMM data support this picture.

3. The SMM Observations of the Flare on July 21, 1980

The Solar Maximum Mission (SMM) spacecraft was launched on February 14, 1980, being operational through November 1980 before its repair in April 1984. SMM observed the flare on July 21, 1980 with both the X-ray Polychromator (XRP) and the Gamma-Ray Spectrometer (GRS); HXIS imaging data at higher energies are unfortunately not available for this flare.

3.1. THE XRP DATA

3.1.1. *The Instrument*

A detailed description of the XRP is given by Acton *et al.* (1980). Briefly the XRP consists of a Flat Crystal Spectrometer (FCS) and a Bent Crystal Spectrometer (BCS). The BCS can be used for high resolution ($\lambda/\Delta\lambda \sim 10\,000$) X-ray spectroscopy in the wavelength range between ~ 1.8 and 2.5 Å, in particular it can study the dielectronic satellite spectra of the helium-like ions Ca XIX and Fe XXV which are rich in diagnostic applications; the BCS collimator field of view is 6 arc min with no imaging capability. The crystals of the FCS are rotatable and, therefore, cover a rather wide spectral range between ~ 1.4–22.4 Å, in addition, by rastering spectroheliograms with an angular resolution of ~ 14 arc sec can be obtained in the resonance lines of O VIII, Ne IX, Mg XI, Si XIII, Su XV, and Fe XXV.

For the flare on July 21, 1980 a complete BCS light curve in the lines of Ca XIX and Fe XXV is available. The FCS was used for rastering in the rise time of the flare (i.e., from 2:54 to 03:02 UT); afterwards, i.e., from 3:02 to 3:14 UT an X-ray light curve of the brightest spot in the previously obtained X-ray image in the lines of O VIII, Ne IX, Mg XI, Si XIII, Su XV, and Fe XXV is available. In the following sections we shall describe these data in detail.

3.1.2. *The BCS Observations*

In Figure 2 the Ca XIX and Fe XXV light curves as recorded by the BCS are shown. Flare onset occurs slightly before 2:55 UT, with the count rate first rather smoothly increasing, followed by a rapid increase between 2:56–2:57:30 UT. The Ca XIX count rate stays approximately constant, and the Fe XXV count rate decreases slightly before another rapid increase in count rate starts at 2:59 UT. Flare maximum is reached at 3:00 UT followed by a swift decay in count rate to preflare levels in about 5 min.

A comparison between the BCS light curves in Ca XIX and Fe XXV (see Figure 2) and the IPC light curve (Figure 1) reveals a striking similarity. Flare onset in the IPC occurs a little later; note, however, that the beginning of the flare depends sensitively on the subtraction procedure, hence, the difference is not significant. A break in the rising part

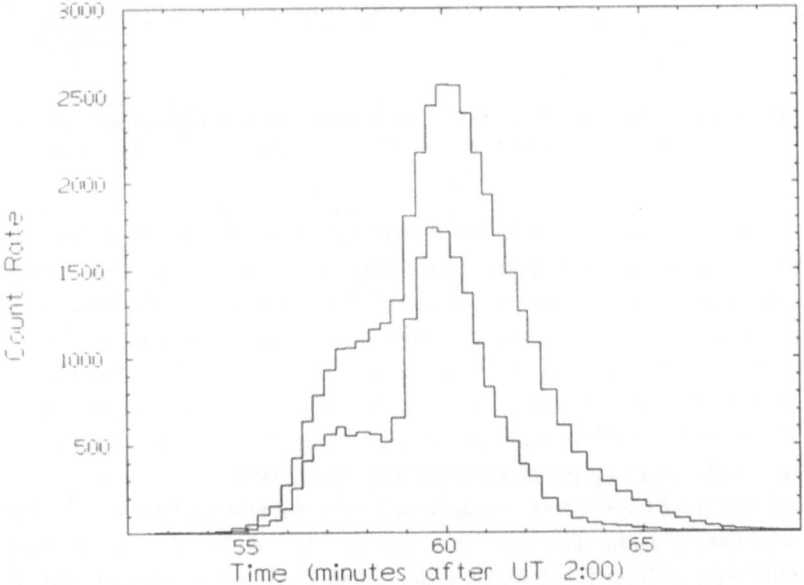

Fig. 2.   BCS light curves in the Ca XIX (upper curve) and Fe XXV (lower curve) lines.

of the flare at 2:57:30 UT is visible; the flare maximum is reached at the same time. The decay time-scale of the IPC light curve seems to be somewhat larger; this is of course not surprising since the IPC is more sensitive to lower temperature material and has little or no sensitivity at the CaXIX and FeXXV wavelengths, respectively. At any rate, the correctness of the interpretation of the IPC data obtained by SHF without knowledge of the SMM data is beyond any doubt.

For purposes of comparison to the broadband fluxes measured with the IPC one needs to know the run of temperature and emission measure with time, which are plotted in Figures 3(a) and 3(b) as derived from the CaXIX and FeXXV data respectively. The

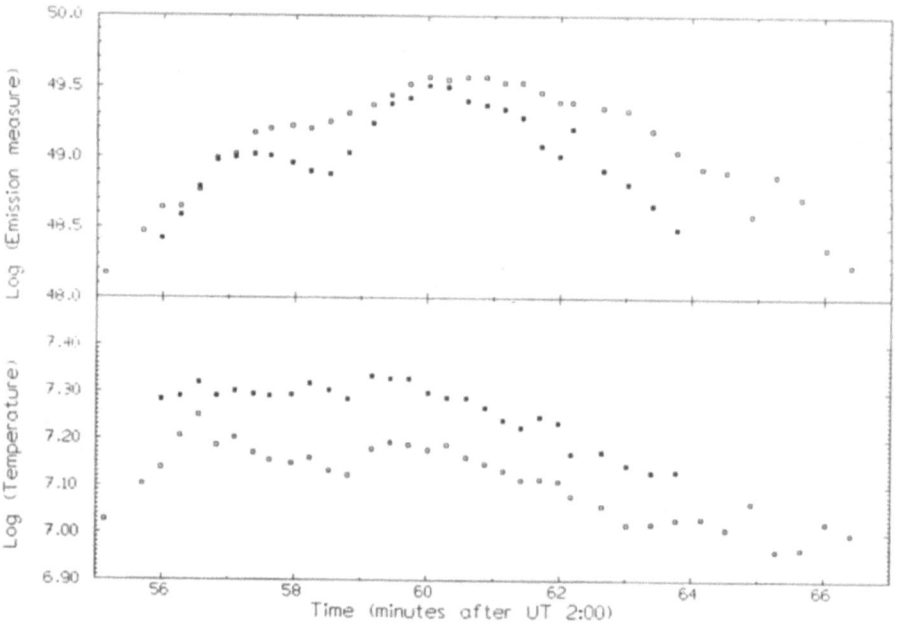

Fig. 3.   (Logarithmic) temperature (*lower panel*) and emission measure (*upper panel*) derived from the CaXIX (open squares) and FeXXV (solid squares) data as a function of time.

theory of dielectronic recombination spectra and the problems of derivations of physical parameters from such spectra are extensively discussed in the literature (see, for example, Gabriel, 1972; Antonucci *et al.*, 1982; Lemen *et al.*, 1984). Note that the BCS emission measures have not been corrected for the collimator response which amounts to a factor of ~ 2 since the flare occurred close to the edge of the field of view as shown by the FCS raster images. As is evident from Figure 3 temperatures and emission measures derived from different lines do not agree, an effect attributed to the heterothermal nature of the emitting plasma (see Strong *et al.*, 1986).

In summary, the BCS spectra and light curve show a temperature stratification of the hot thermal flare plasma. The emission measure derived for the CaXIX and FeXXV evolve differently with time, the FeXXV light curve hints at a second episode of energy input after an initial energy release.

3.1.3. *The FCS Observations*

FCS images of the active region 2562 were taken on July 21, 1980 starting at 2 : 53 UT, i.e., in the rising phase of the flare, at 3 : 13 UT, i.e., immediately after the end of the flare, and at 3 : 34 UT. As an example we show (in Figure 4) a contour plot of the O VIII raster image taken between 2 : 53 and 3 : 02 UT. In this line (as well as in the images

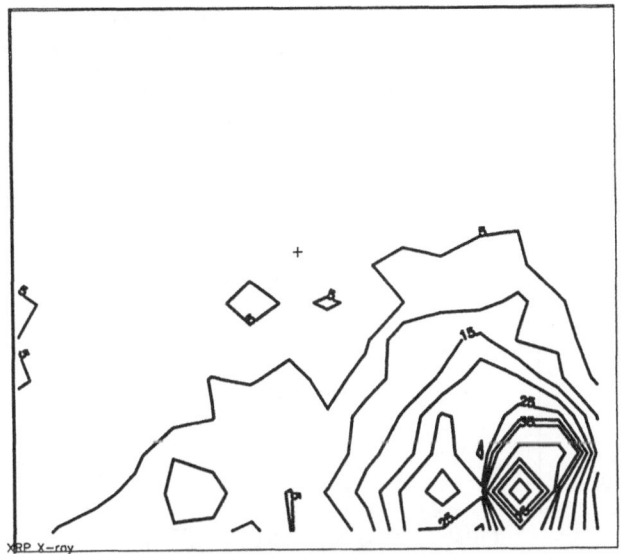

Fig. 4. Contour plot of the O VIII FCS raster image taken between 02 : 54 and 03 : 02 on July 21, 1980; note the diffuse emission from active region 2562 as well as the flare in the lower right hand part.

in the lines of Ne IX and Mg XI) a region of diffuse emission is definitely visible during and after the flare, presumably part of the loop system of the active region 2562. The flare itself is clearly visible in all FCS images taken in the time interval 2 : 53–3 : 02 UT. At higher energies the flaring region becomes less pronounced, and in Si XIII and S XV virtually no excess emission is seen later than 3 : 13 UT.

In the lines of Si XIII and S XV the flare is unresolved, whereas in the lower temperature lines the flare may be resolved; this is, however, difficult to ascertain since the flare is located on top of an area with diffuse emission (cf. Figure 4). We, therefore, adopt as angular size of the flaring region $\theta \leq 14''$, the pixel size of the FCS, which translates into $L \leq 10^9$ cm. Therefore, the FCS images demonstrate directly the compact nature of the event on July 21, 1980.

Between 3 : 02 and 3 : 13 UT a light curve of the brightest spot in the FCS image is available in each spectral line. This brightest spot is likely to contain most of the flaring region; problems would however arise if the flaring region moved with respect to the instrument during FCS data taking. Figure 5 shows the FCS light curves in the lines of O VIII, Ne IX, Mg XI, Si XIII, and S XV, respectively. At higher temperatures no significant emission is found later than 3 : 06 UT, very much in line with the BCS Ca XIX

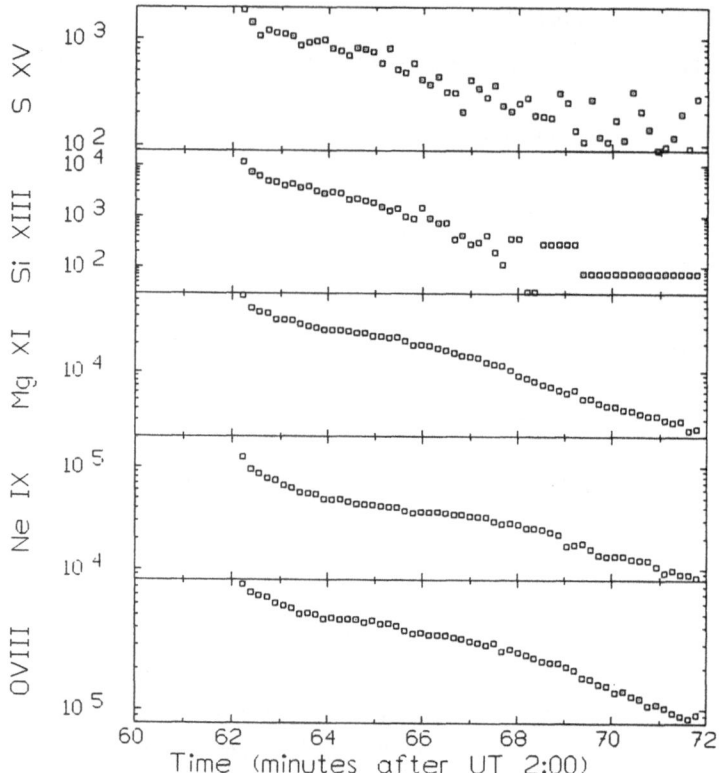

Fig. 5.   FCS light curves in the O VIII, Ne IX, Mg XI, Si XIII, and S XV lines respectively; the fluxes (in
phot cm^{-2} s^{-1}) refer to that pixel in the FCS raster with peak intensity.

and Fe XXV light curves. At lower temperatures significant time-variable emission is
found right until the beginning of the subsequent raster image.

In summary, the FCS raster image directly demonstrates the compact nature of the
flaring event, while the FCS light curves show the longer decay time-scales for the lower
temperature lines.

3.2. THE GRS DATA

3.2.1. *The Instrument*

The SMM Gamma Ray Spectrometer (GRS) consists of seven Na I (Tl) scintillation
spectrometers sensitive to γ-rays between 0.28 and 9.28 MeV; pulse height analysis is
provided in 476 digital pulse height channels with an energy resolution of $\sim 7\%$ at
0.628 MeV. An auxiliary system of two Na I X-ray detectors gives simultaneous
coverage in the hard X-ray band between 10 and 140 keV. The two X-ray detectors are
identical except for the entrance filters. Here we consider only data from the detector
with an Al–Fe filter giving 50% transmission at ~ 30.8 keV. Pulse height analysis is
provided in four energy channels, yielding the energy bands 14–28, 28–56, 56–114, and

114–119 keV. The GRS provides no spatial resolution, and in fact views almost half the sky at any given time. A detailed description of the GRS is given by Forrest *et al.* (1980).

3.2.2. *The Hard X-Ray and γ-Ray Light Curves*

In Figure 6 (left panel) we show the measured light curves in the four energy channels of the hard X-ray detector. The light curves show an interesting double peak structure. In all bands flare onset occurs at 2:55:40 UT, but some 'precursor activity' is definitely present. Towards higher energies the relative strength of the second (later) maximum decreases rapidly and becomes virtually absent at energies above ~ 50 keV; at energies higher than ~ 50 keV only one hard X-ray burst centered on 2:56:10 UT is observed. In the right panel of Figure 6 we show the light curves as measured by the γ-ray spectrometer. The GRS light curves continue the trend observed in the hard X-ray light curves. At γ-ray energies only one burst centered on 2:56:10 UT is visible; at the time of soft X-ray flare maximum, i.e., near 3:00 UT, no substantial γ-ray emission is produced. While we show only the γ-ray light curves up to 2 MeV the γ-ray emission can be detected up to energies of ~ 4 MeV when the signal disappears in the noise.

In summary, the GRS observations demonstrate the very hard spectrum of the first burst, extending into the MeV regime; however, they also show very clearly the existence of a second, presumably purely thermal energy release with a much softer spectrum, that is probably triggered by the first flare event. This complicated double peak structure then explains the slight dip in the Fe xxv emission measure observed at 2:58 UT; while the temperature derived from the Ca xix and Fe xxv data remains almost unchanged, the second energy release leads to yet another increase in emission measure at temperatures above 10^7 K.

4. The Solar-Stellar Connection: Comparison of the SMM and IPC Observations and Conclusions

With the SMM data we can immediately verify (or falsify) the 'stellar' results as summarised in Section 2. The FCS raster scans can be used directly to measure the extent of the flaring region. Inspection of the FCS images shows that the flare occurred in the active region visible in the lower temperature lines. While it is difficult to tell whether the actual flare emission is extended or not in the lower temperature lines, the flare appears spatially unresolved in the higher temperature lines. Consequently we can only state that the characteristic size scale of the flaring plasma should be less than 14 arc sec, the FCS pixel size, i.e., less than $\sim 10^9$ cm. Thus the IPC measurements overpredict the actual length scale by a factor ~ 3 (or possibly more). Such a disagreement does not come totally unexpectedly considering the multitude of assumptions (as detailed by SHF) that go into a 'stellar' flare size estimates. Also it is important to note that the size estimate depends on the assumed emission measure EM; in order to determine the emission measure SFH assumed the flare plasma to be isothermal at temperatures of ~ 7.2. If low temperature plasma contributes significantly

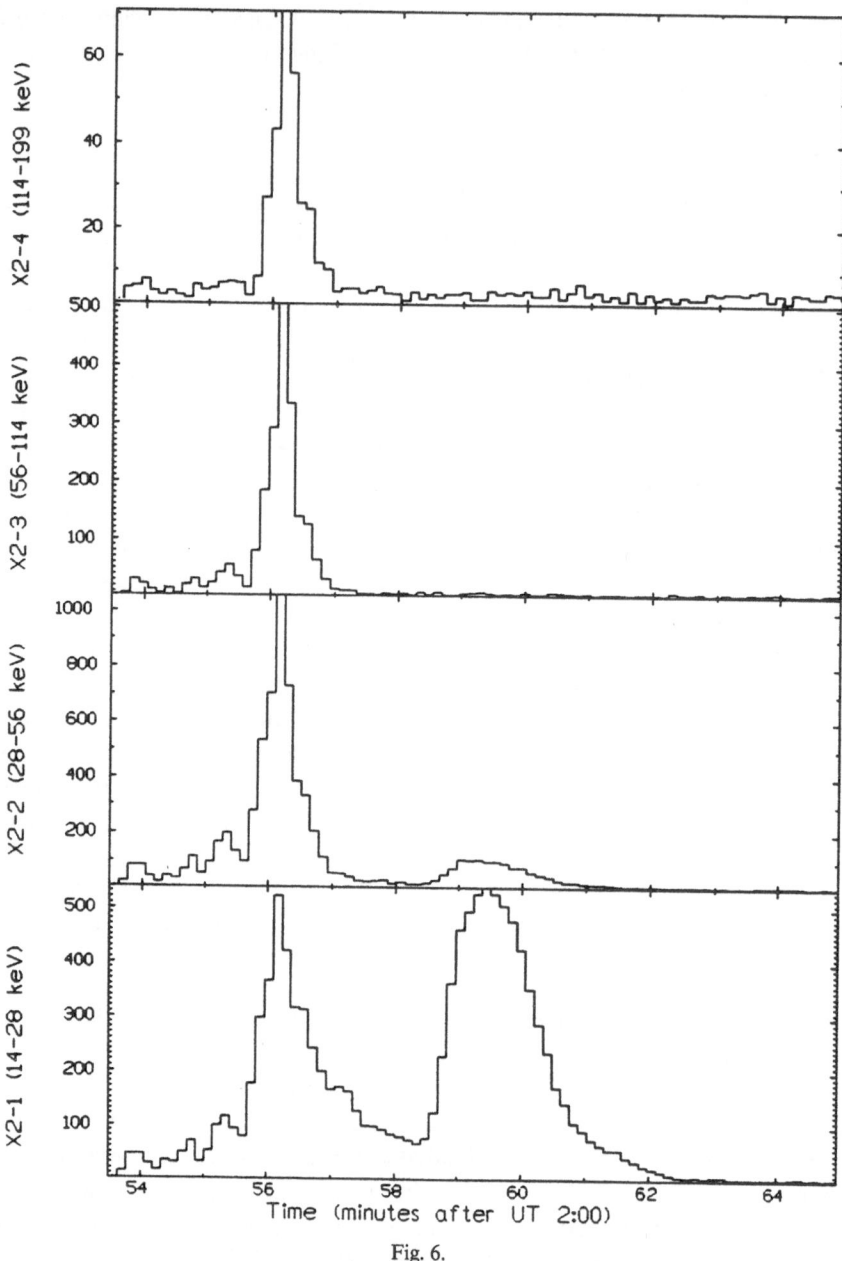

Fig. 6.

to the observed IPC count rate, the emission measure and length scales as derived by SFH will be overestimated as appears to be the case.

A simultaneous FCS and BCS light curve exists in the time interval between 3:02 UT and 3:05. We note, however, that the BCS light curve refers to a 6 arc min total field of view whereas the FCS light curve refers only to the 14 arc sec pixel with

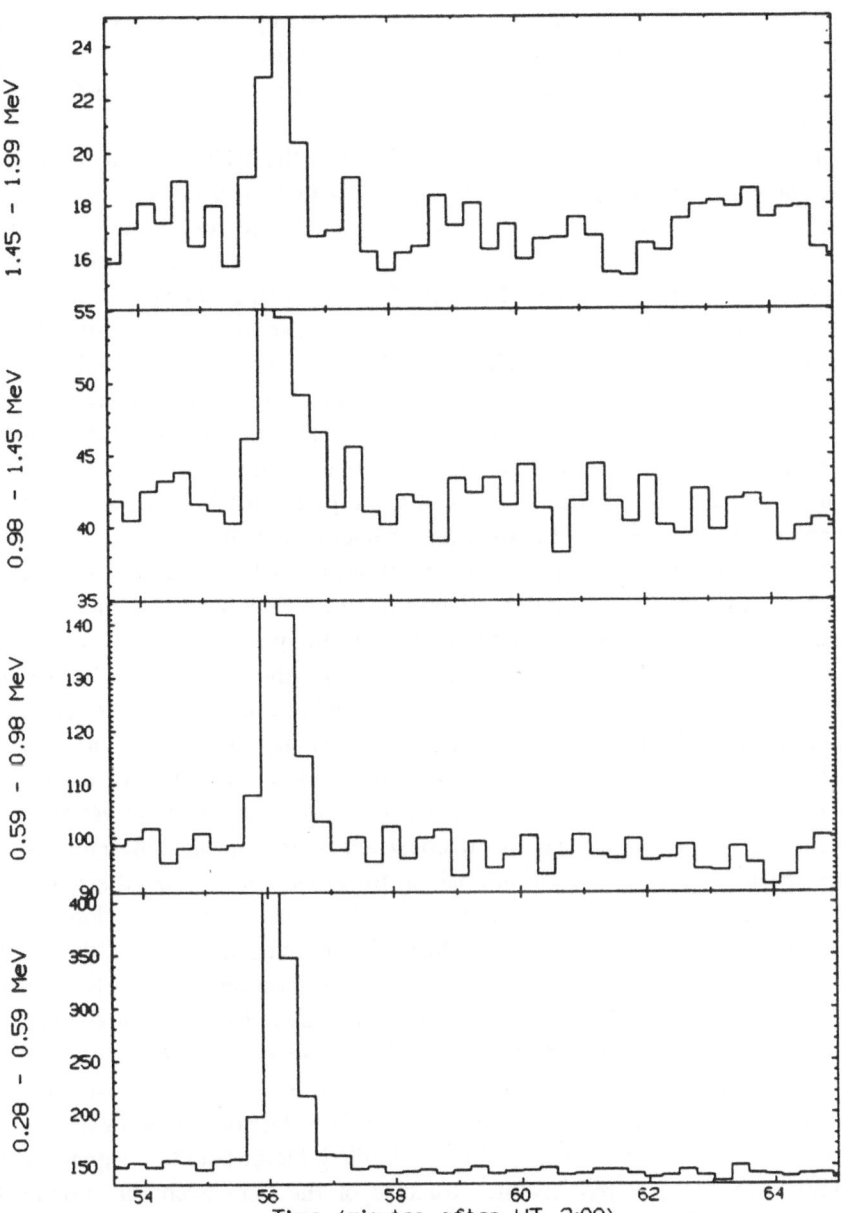

Fig. 6.   GRS hard X-ray light curves between 0.14–114 keV (*left panel*) and γ-ray light curves between 0.28 and 2 MeV (*right panel*).

the peak intensity in the previously obtained raster image. A comparison between the raster image taken at 2:56 UT and 3:10 UT shows that the pixel containing the peak emission has moved; hence, the FCS light curves are likely to have missed a significant portion of the flare flux. In fact, because of the cross calibration problem between FCS and BCS fields of view, we were unable to perform a consistent differential emission measure analysis. However, we find that the FCS low-temperature line fluxes in the first raster image (obtained during the rise phase of the flare) can be reasonably fit by an isothermal source with a mean electron temperature of 5×10^6 K and a volume emission measure of 3×10^{48} cm^{-3}.

Given an isothermal plasma model derived from the BCS spectra (over the full 6 arc min field of view) as well as the FCS model as derived from the raster scan data in the rise phase of the flare we can proceed to estimate how much of the broad band IPC emission can be accounted for by the emission seen with the XRP and compare it to the observed values (cf. Table II). Table II reveals that in the IPC band pass most of the flux comes from the BCS component in the assumed incident spectrum; however, the total predicted IPC count rate falls below the observed IPC count rate in all cases. This fact is of course not unexpected since the emission measure derived from FCS raster scan in the *rising* phase of the flare is likely to be too small for the time immediately after the flare peak. On the other hand, in order to account for *all* of the observed IPC emission with the FCS and BCS components, the FCS emission measure would have to be increased by a factor of 5–10. The same problem is encountered when the emission measures derived from the FCS light curves are used which – as discussed above – are likely to have missed a substantial fraction of the total FCS flare flux. In view of these uncertainties we are not able to determine whether the plasma model as derived from the SMM BCS and FCS data completely accounts for the observed IPC count rate or whether the IPC with its rather soft band pass sees additional low temperature material not visible in the harder FCS and BCS instruments. At any rate, because of the presence of low temperature material (i.e., material cooler than 10^7 K) it is now clear why the emission measures and hence size scales derived from the IPC observations are too large; however, because of our inability to construct an appropriate differential emission measure distribution we are not in a position to provide an improved size estimate.

What can then be learnt from spatially unresolved, low spectral resolution data on stellar flares? We have shown that one of the basic physical parameters of the soft X-ray emitting flaring plasma, i.e., the temperature, can be well estimated. Clearly, the detailed temperature stratification, the dynamics, and the detailed physics of the flaring plasma cannot be addressed observationally in the stellar context. The rather coarse estimates of density and length scale based on the IPC observations are shown to be approximately correct; however, it should be noted that also the XRP data do not allow direct density diagnostics, and therefore in principle the FCS filling factor could be significantly less than unity. The complicated heating structure of the flare event, its double burst structure with an initial nonthermal burst followed by another thermal burst producing extremely hot plasma, is at best hinted at by the soft X-ray data, but can only be revealed by the hard X-ray data; the lack of hard X-ray data is certainly a severe deficiency of the existing body of stellar X-ray flare data.

As far as solar flares are concerned hard X-ray observations are thought to be the key towards an understanding of the flare energetics. In the stellar context hard X-ray observations of flares have never been reported. Although a number of events observed in all-sky surveys of X-ray transients have been identified as huge flares on nearby active

M-dwarfs and somewhat more distant RS CVn systems (cf., Pye and McHardy, 1983; Connors, 1988), the observed emission is very likely to have thermal origin. Scaling solar hard X-ray and γ-ray flares to stellar distances results in mind-boggling observational requirements to detect such events. On the other hand, stellar (soft) X-ray flares are far more energetic than their solar counterparts; the same probably applies also to the impulsive phases of these stellar flares, but it is difficult to extrapolate from the soft X-ray to the hard X-ray and γ-ray regime. Under these circumstances it becomes very important to look for proxy indicators of the impulsive phases of stellar flares and associated hard X-ray and γ-ray emission. On the Sun, microwave bursts are known to correlate extremely well with hard X-ray bursts (cf., Wiehl *et al.*, 1985), and we, therefore, propose that simultaneous microwave and soft X-ray observations may provide (indirect) evidence for the existence of impulsive phases in stellar flares and concurring hard X-ray and γ-ray emission.

Acknowledgements

We thank Dr G. Kanbach for his help with the analysis and interpretation of the GRS data and Dr S. Snowden for help with analysis of scattered solar X-ray radiation.

References

Acton, L. W., Culhane, J. L., Gabriel, A. H., and 21 co-authors: 1980, *Solar Phys.* **65**, 53.
Antonucci, E., Gabriel, A. H., and 7 co-authors: 1982, *Solar Phys.* **78**, 107.
Collura, A., Pasquini, L., and Schmitt, J. H. M. M.: 1988, *Astron. Astrophys.* **205**, 197.
Connors, A.: 1988, Ph.D. Thesis, Univ. of Maryland, NASA report 88–014.
Fink, H. H., Schmitt, J. H. M. M., and Harnden, F. R., Jr.: 1988, *Astron. Astrophys.* **193**, 345.
Forrest, O. J., Chupp, E. L., Ryan, J. M., and co-authors: 1980, *Solar Phys.* **65**, 15.
Gabriel, A. H.: 1972, *Monthly Notices Roy. Astron. Soc.* **160**, 99.
Haisch, B. M.: 1983, in P. B. Byrne and M. Rodonò (eds.), *Activity in Red Dwarf Stars*, Astrophysics and Space Science Library, Vol. 102, p. 255.
Lemen, J. R., Philipps, K. J. H., Cowan, R. D., and Grant, I. P.: 1984, *Astron. Astrophys.* **135**, 313.
Manson, J. E.: 1976, in O. R. White (ed.), *The Solar Output and Its Variation*, Colorado Associated University Press, Boulder.
Pallavicini, R., Monsignori-Fossi, B. C., Landini, M., and Schmitt, J. H. M. M.: 1987, *Astron. Astrophys.* **191**, 109.
Pallavicini, R.: 1987, in E.-H. Schröter and M. Schüssler (eds.), 'Solar and Stellar Flares', *Lecture Notes in Physics* **292**, 98.
Pallavicini, R. and Schmitt, J. H. M. M.: 1988, in W. Hermsen (ed.), *Contribution to COSPAR Advances and Perspectives of X-ray and γ-ray Astronomy*, Helsinki.
Pallavicini, R. *et al.*: 1989, *Astron. Astrophys.* (submitted).
Pye, J. P. and McHardy, I. M.: 1983, *Monthly Notices Roy. Astron. Soc.* **205**, 875.
Schmitt, J. H. M. M., Harnden, F. R., Jr., and Fink, H. H.: 1987, *Astrophys. J.* **322**, 1034.
Strong, K. H. *et al.*: 1986, in *Energetic Phenomena on the Sun*, SMM Flare Workshop, Nasa Conference Publication 2439.
Wiehl, H. J., Batchelor, D. A., Crannell, C. J., Dennis, B. R., Price, P. N., and Magun, A.: 1985, *Solar Phys.* **96**, 339.

STELLAR FLARE STATISTICS – PHYSICAL CONSEQUENCES

N. I. SHAKHOVSKAYA

Crimean Astrophysical Observatory, 334413 p/o Nauchny, Crimea, U.S.S.R.

Abstract. The observational data permit us to establish clear statistical correlations between different parameters of stellar flare activity and the characteristics of quiet stars. These relations are:
  (i) between energies and frequencies of flares on stars of different luminosities;
  (ii) between total radiation energies of flares and quiet stars both in X-ray and Balmer emission lines;
  (iii) between flare decay rates just after the maxima and flare luminosities at maxima.

1. Introduction

In the vicinity of the Sun, flare activity similar to that of the Sun has been detected on more than 80 red dwarf stars of spectral classes dK0–dM8 (Gershberg, 1978; Page, 1988). About three dozen of these stars have been studied in detail: more than 2000 optical flares were registered during several thousands of hours of photoelectric patrol observations; the luminosities of quiet chromospheres, transition regions and coronal emission in soft X-rays have been estimated.

The analysis of these data permits us to establish some statistical correlations between activity parameters of stars of different spectral classes. Some of these correlations have clear physical consequences; others we need to understand and take into consideration while developing a general theory of flare activity.

2. Flare Energy Spectra

The energy and temporal parameters of stellar flares show an emormous variety in their duration and intensity. As an example, the data of the EV Lac photoelectric patrol observations carried out at Crimea Observatory in 1986–1987 are shown in Figure 1. Indeed, flares detected on one star have time-scales from several seconds (11 September, 1986, $23^h 25^m$) to many hours (10 September, 1987, $18^h 50^m$–$21^h 20^m$), covering a range of about 3–4 orders of magnitude of the total radiation output. There exists a wide variety of flare light curves: from strongly asymmetric with fast rise and slower decay (11 September, 1986, $19^h 20^m$–$20^h 00^m$) to almost perfectly sinusoidal form (12 September, 1986, $17^h 50^m$–$18^h 20^m$).

The variety of flare forms compels us to use statistics to describe the general properties of flare activity on different stars. 'Mean' characteristics do not fit our goal: these are strongly influenced by observational selection effects – only the most powerful flares can be detected on absolutely brighter stars. We must study the distribution of these characteristics instead of the mean parameters in order to eliminate selection effects. Kunkel (1968, 1975) found one of the first important results for statistical properties of

Solar Physics **121** (1989) 375–386.

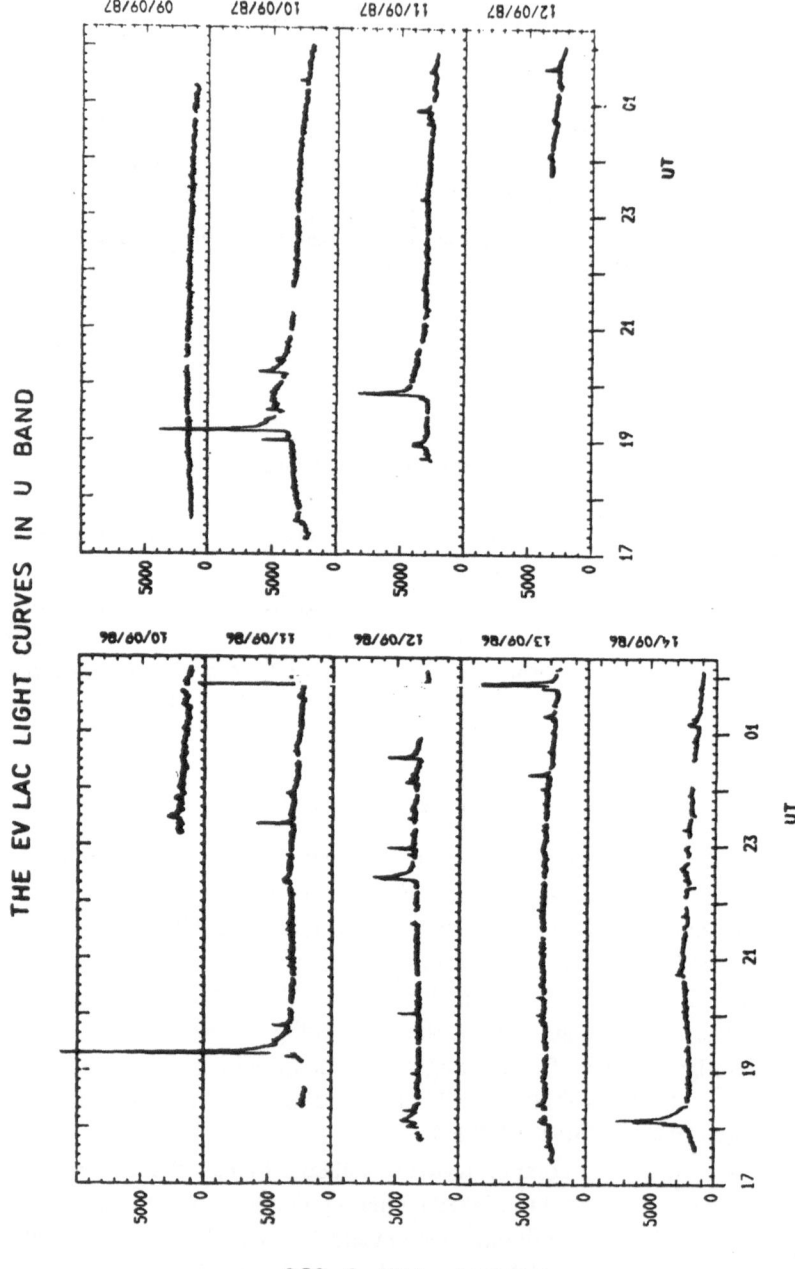

Fig. 1.   The photoelectric photometric monitoring of EV Lac in the *U*-band.

optical flares on UV Cet-type red dwarf stars. He showed that the mean frequency of stellar flares whose luminosity at maximum exceeds the brightness corresponding to the stellar magnitude m, can be given by the simple relation

$$R(m) = 10^{a(m - m_0)},$$

where m_0 is the magnitude of the brightest flare that takes place on the star during a given time interval and a is a coefficient close to a unity.

Later, the time-integrated flare energy at optical wavelengths was considered as the main characteristic of the flare (Lacy, Moffett, and Evans, 1976; Gershberg, 1972) instead of amplitude. In Figure 2, taken from Gershberg and Shakhovskaya (1983), the

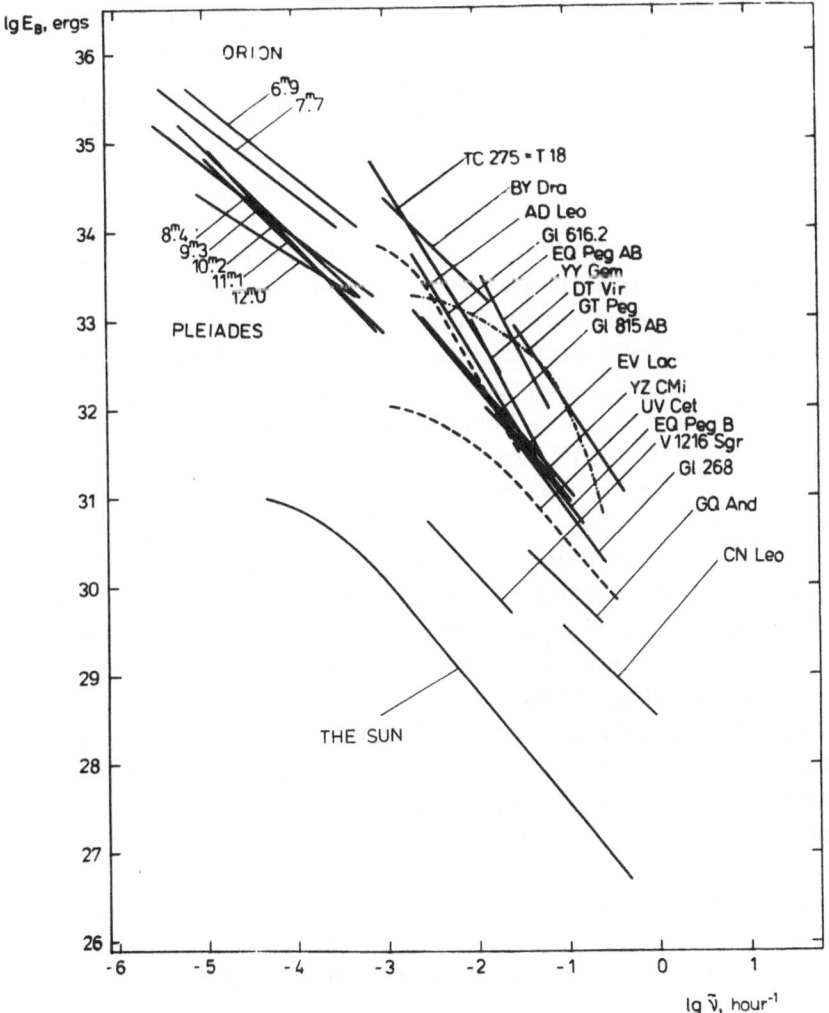

Fig. 2. Energy spectra of flares on red dwarf stars and the Sun. Total energy in the B-band flare radiation, E, is plotted versus frequency $\tilde{\nu}$ of flares with energy exceeding E.

energy spectra of flares of 23 red dwarf stars in the solar vicinity and several groups of flare stars in clusters are given: the total energy of B-band flare radiation – E (ergs) – is plotted versus frequency – $\tilde{\nu}$ (hr^{-1}) – for flares with energy exceeding E. The magnitudes of the quiet stars and their trigonometric parallaxes from Gliese (1969) were used for conversion of the observed counts of flares into flare energies. In Figure 2, the energy spectrum of solar flares is taken from Gershberg, Mogilevskij, and Obridko (1987) and constructed on the basis of 15 500 flares. Only the upper part of the energy spectra of flares beyond the photometric flare detection threshold – $E_{\lim}$ – are presented in Figure 2 (the brighter the star, the higher $E_{\lim}$).

As follows from Figure 2, the observed part of energy spectra of flares of different flare stars and the Sun (overlapping several orders of magnitudes in energy) can be represented by a power function with the spectral index β corresponding to the slope of the straight lines in Figure 2. The spectral indices in the energy spectra of flares have a rather narrow range of values: from 0.4 to 1.4.

The flare energy spectra form a rather narrow band in Figure 2. The upper limit of this band seems to be real and determined by the highest attainable efficiency of optical power in a star of a particular type. The reason for this must be explained in any theory of the origin of flares. The lower boundary of the band is the result of observational selection: flare stars with the greatest activity are those usually observed.

In Figure 3, the values of spectral indices β for flare stars of different absolute stellar magnitude (M_V) are given according to Gershberg (1988). If the binary components were not resolved by photoelectrical observations (e.g., UV Cet, V577 Mon, YY Gem, V1054 Oph, AT Mic, Gl 815 AB) then the value of M_v and the parameters of the flare activity refer to the stellar system. For flare stars in the solar vicinity, there exists a slight ($r = 0.60 \pm 0.14$) correlation between β and M_V: the fainter the star, the larger the β. For the flare stars in clusters, Figure 3 shows that β is an age-dependent parameter; the younger the cluster, the larger the β.

Although the physical meaning of the power-law representation of energy spectra is not yet known, its analytical presentation allows us to estimate some important characteristics of flare activity: namely the maximum energy of the most powerful stellar and solar flares, the intrinsic flare frequency, and the total radiation of optical flares of different stars (Gershberg, 1988).

Pustilnik (1988) has recently proposed that the observed energy spectra are determined by the structure of the photospheric magnetic field in the active regions in a regime of turbulent convection. He derived values of the index of the power law β closely matching those observed, assuming that the energy spectrum reflects the dimensional distribution of turbulent elements.

As Figure 3 indicates, this depends more on the age of the star than on its mass. Possibly, a strong influence of stellar rotation on the structure of turbulent convection leads to this dependence. This problem requires further theoretical and observational investigations.

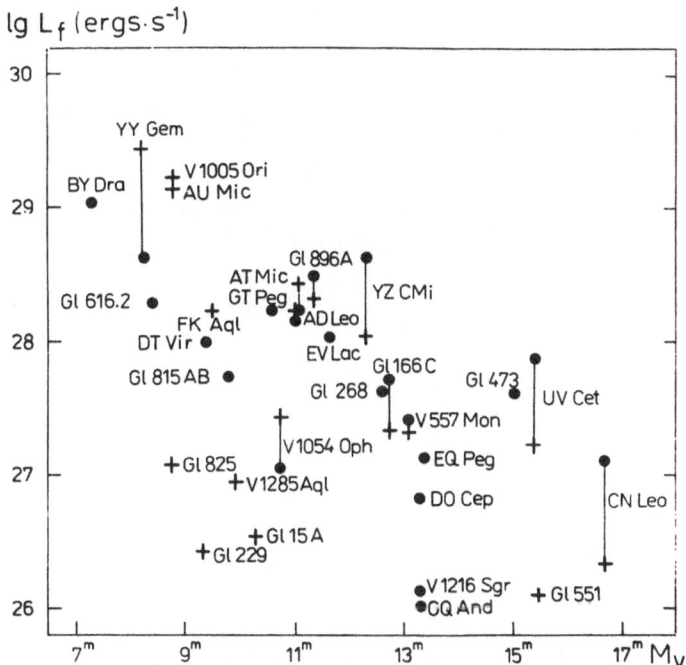

Fig. 4.   Time-averaged power of optical radiation of flares, L_f. Circles represent the data by Gershberg (1988), crosses by Doyle and Buttler (1985).

Figure 4 shows that the highest time-averaged optical flare power occurs on the intrinsically bright stars (spectral classes dK5–dM0). According to Gershberg (1985), the intrinsic frequency of flares is also higher for the intrinsically brighter stars, although flares are detected more frequently among fainter stars.

Another important characteristic of a stellar flare activity level is the ratio of L_f to stellar bolometric luminosity: L_f/L_{bol}. In Figure 5, the same symbols as in Figure 4 indicate computed ratios L_f/L_{bol} for the flare stars in the solar vicinity. Values of L_{bol} are taken from Pettersen (1980), when available, otherwise they were computed from M_V using the correlation presented by Agrawal, Rao, and Sreekantan (1986).

Thus, the upper boundary of absolute flare activity level increases from the late M stars to the K5 stars, although a portion of the total stellar energy that is released in flare activity is independent of the absolute magnitude of stars in the range from 7^m6 to 16^m7 and reaches 0.1% for the most active stars.

4. Flare Activity and Chromospheric and Coronal Emission

Stellar chromospheres manifest themselves in the optical range by strong emission in Balmer lines. The energy released in these lines for the flare stars, according to Linsky *et al.* (1982), exceeds the total radiation of all other observed chromospheric lines. However, according to Grinin and Katysheva (1980), for conditions approximating

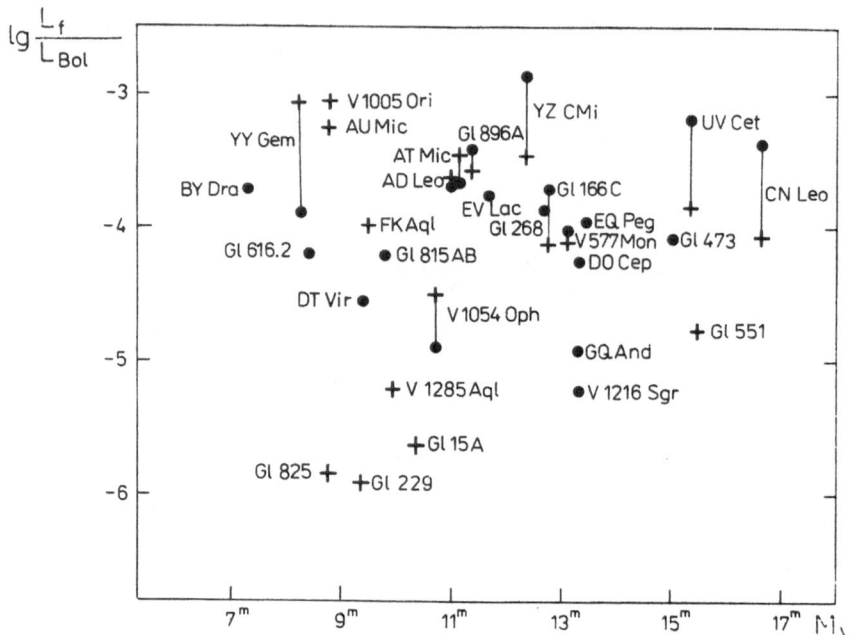

Fig. 5.   Ratios of time-averaged powers of the flare optical radiations of flare stars bolometric luminosity.
The symbols are the same as in Figure 4.

those of the chromosphere, the Lyman emission may be 1–2 orders of magnitude greater than in the Balmer lines.

For the flare stars, Gershberg and Shakhovskaya (1983) estimated the ratio of energy released in these lines by a stationary stellar chromosphere to the total radiated energy of the star, L_{Bal}/L_{bol}, and noted that this ratio is close to the value of L_f/L_{bol} and independent of M_V.

We next examine the correlation between L_f and L_{Bal}. In Figure 6, the values of L_f are plotted versus L_{Bal} on a log-log scale for 18 flare stars within the range of M_V, $7\overset{m}{.}6$–$15\overset{m}{.}3$. The symbols are the same as in Figures 4 and 5. The uncertainty of the value $\log L_{Bal}$ is about ± 0.3 because the flare stars show variability in the intensity of Balmer emission lines, even when photometrically quiescent (Bopp, 1974, Shakhovskaya, 1974b). The symbol $\odot$ corresponds to solar data: the result of Athay (1966) used for the definition of $\log L_{Bal}$ and L_f has been calculated from the energy spectra of solar flares (Figure 2). The following linear correlation has been found for 18 stars:

$$\log L_f = -6.98 + (1.24 \pm 0.28) \log L_{Bal},   \qquad (1)$$

with correlation coefficient of $r = 0.80$.

The corresponding straight line is shown in Figure 6. The scatter of any star from the line does not exceed the possible observational errors, but the deviation of the Sun from this correlation is significant.

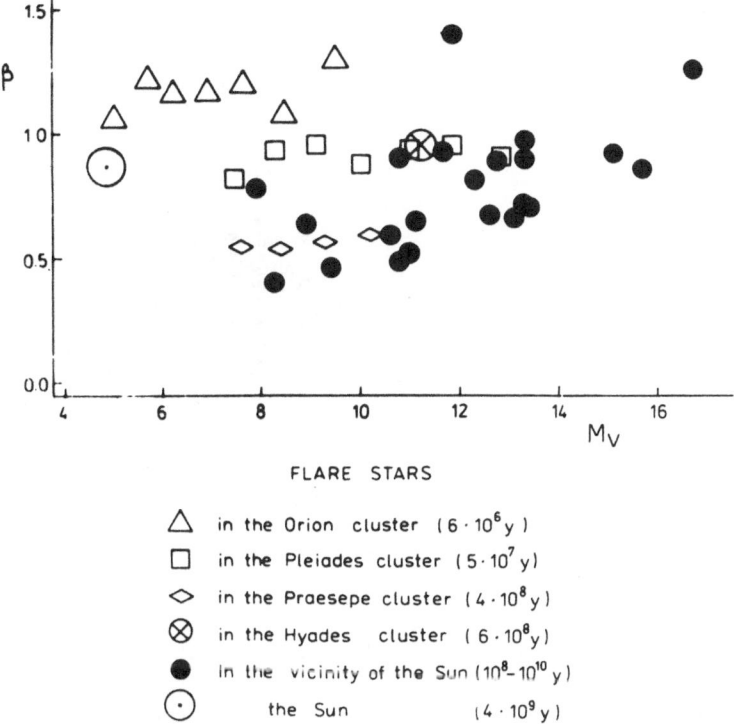

FLARE STARS

△ in the Orion cluster ($6 \cdot 10^6$ y)
□ in the Pleiades cluster ($5 \cdot 10^7$ y)
◇ in the Praesepe cluster ($4 \cdot 10^8$ y)
⊗ in the Hyades cluster ($6 \cdot 10^8$ y)
● In the vicinity of the Sun (10^8– 10^{10} y)
☉ the Sun ($4 \cdot 10^9$ y)

Fig. 3. Spectral indices of energy spectra of flares on red dwarf stars in the solar vicinity and in clusters.

3. Total Radiation of Flares

The time-averaged power of optical radiation of stellar flares L_f(erg s^{-1}) has been computed by the integration of their energy spectra and by conversion of the flare energy recorded in B or U bands into total optical radiation (Gershberg and Shakhovskaya, 1983). If the values of E for the most powerful and weakest flares actually registered in each flare star are used as limiting magnitudes, the integration will yield only the lower limit of L_f. However, if any flare star has a flare energy spectrum from $E = 3 \times 10^{35}$ erg (that is, the strongest flares registered in clusters) to $E = 2 \times 10^{27}$ erg (the weakest flares on the faintest star CN Leo), the real values L_f may be one or two orders of magnitude larger than the lower limits.

In Figure 4, the lower limits of L_f, computed for the stars in the solar vicinity, are indicated by circles together with the time-averaged flares observed by Doyle and Butler (1985) which are indicated by crosses. The symbols corresponding to one star have been connected by straight lines. The lengths of these lines characterize the uncertainty of the L_f values. As in Figure 2, the upper boundary of a region occupied in Figure 4 corresponds to the most active flare stars, while the lower left (unpopulated) corner corresponds to absolutely bright flare stars of low activity, with high thresholds for detection of flares.

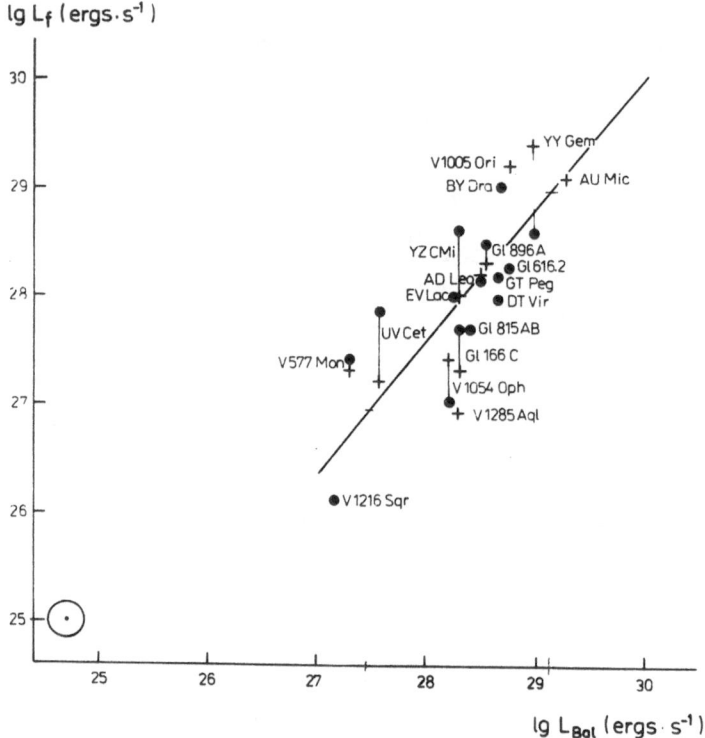

Fig. 6.   A plot of the time-averaged powers of optical radiation of flares, L_f, versus the total radiation in Balmer lines of quiet chromospheres, L_{Bal}. The symbols are the same as in Figure 4.

Gershberg and Shakhovskaya (1983) showed, that the ratio of coronal soft X-ray emission L_X (erg s^{-1}) to L_{bol} is close to L_f/L_{bol} and $L_{\mathrm{Bal}}/L_{\mathrm{bol}}$ and also independent of M_V. Later, Doyle and Butler (1985) found a linear correlation between $\log L_f$ and $\log L_X$. In Figure 7, this correlation is presented for a larger body of data than that of Doyle and Butler (1985): the estimates of L_X by Agrawal, Rao, and Sreekantan (1986) and L_f values have been added. The symbols are the same as in Figures 4–6. The L_f value for the Sun has been adopted from Ambruster, Sciortino, and Golub (1987). The following linear correlation has been found for 23 flare stars:

$$\log L_f = 4.8 + (0.80 \pm 0.17) \log L_X, \tag{2}$$

the correlation coefficient being $r = 0.81$. The corresponding straight line is shown in Figure 7.

The deviation of any star from this line does not exceed possible observational errors, although the Sun deviates somewhat from this correlation.

The slopes of lines (1) and (2) are close to unity, so the ratios L_f/L_{Bal} and L_f/L_X are constant within the range of M_V under consideration. This fact was also noticed by Doyle and Butler (1985) for correlation (2), found on the basis of a smaller data sample. To compare the energy losses by flares, Balmer emission of the chromospheres and soft

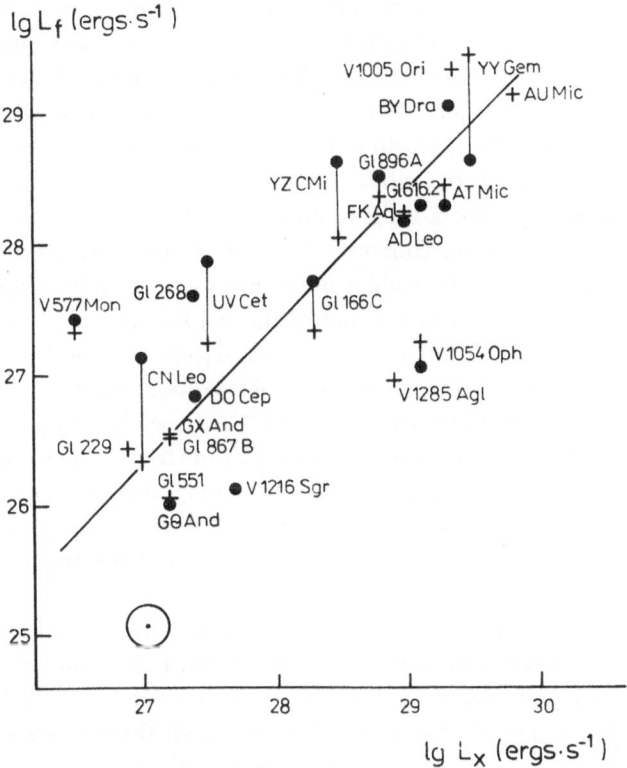

Fig. 7. A plot of the time-averaged powers of optical radiation of flares, L_f, versus the 'quiescent' X-ray flux, L_X. The symbols are the same as in Figure 4.

X-ray coronal emission, the following mean ratios have been calculated:

$$\log \frac{L_f}{L_{\mathrm{Bal}}} = -0.4 \pm 0.1 \quad \text{(for 18 stars)},$$

$$\log \frac{L_f}{L_X} = -0.6 \pm 0.2 \quad \text{(for 23 stars)}, \tag{3}$$

$$\log \frac{L_{\mathrm{Bal}}}{L_X} = -0.2 \pm 0.2 \quad \text{(for 14 stars)}.$$

The corresponding ratios for the Sun: $\log L_f/L_{\mathrm{Bal}} = -1.7$; $\log L_f/L_X = -2$ differ from (3) significantly. The ratios (3) are compatible with the ones estimated earlier by Gershberg and Shakhovskaya (1983) on the basis of a smaller data sample.

Doyle and Butler (1985) and later Butler *et al.* (1986) proposed that the tight linear correlation of L_f and L_X suggested that stellar microflares might be the dominant heating mechanism of late-type stellar coronae. However, Beskin *et al.* (1988) and Ambruster, Sciortino, and Golub (1987) present arguments against that supposition. The energy contribution of microflares can be calculated by integrating the flare frequency power law of the energy spectra of flares down to very low energies. Assuming that the true value E_{min} is about 10^{25} erg, then the microflare energy contribution to the observed L_f value is significant (about 50%) only for stars later than M5, where $\beta = 1.3-1.4$ (Figure 3). For the brighter stars with $\beta = 0.7$, the contribution of microflares to L_f is less than 10%. According to the ratios (3), it will not be enough to heat the coronae. Furthermore, following Figure 6, a tight linear correlation exists between L_f and L_{Bal} indicating that the energy of microflares is sufficient to heat the chromosphere. But if the contribution of microflares is significant for faint stars, one might expect some decay of the L_f/L_X for bright stars, which is not confirmed by the observations.

On the other hand, Beskin et $al.$ (1988) compared the values of the total radiation of microflares and individual flares using direct observations of flare stars with a time resolution of 3×10^{-7} s carried out at the 6-m telescope. The authors selected small patches of light curves, free from any trace of individual flares, and estimated the upper limit of the variable component, assuming that this component has a gradual fluctuation with characteristic time 1–10 s. Compared with the L_f value found for individual flares, the upper limit of the microflare power is markedly lower. Thus, the heating of coronae by stellar microflares is unlikely to be effective.

One might add to the correlation (1) and (2) the tight linear correlations between L_X and the transition region line luminosity, as well as between L_X and the luminosities of chromospheric Mg II lines found by Agrawal, Rao, and Sreekantan (1986). Evidently, all these correlations suggest that in the flare stars, all these kinds of radiation are physically related and that there exists a common heating mechanism in which the portions of energy released in these radiations are dependent neither upon the total energy of this mechanism nor upon the absolute magnitude of the star.

The quantitative similarity between luminosities of optical flares, chromospheres and coronae, proposed by Gershberg and Shakhovskaya (1983), remains to be explained.

5. Flare Decay Rates

Since flares occur on stars of different absolute magnitudes, we have the opportunity to study the correlation between stellar atmospheres and the rate of flare physical processes. The first result in this direction was obtained by Haro and Chavira (1955). They showed that the duration of flares on stars in clusters depends on the spectral class of the star: the later the type of the star, the shorter the flare. Later, Kunkel (1969a, b) confirmed this correlation for stars in the solar vicinity and supposed that this is a consequence of the physical dependence of flare decay rates on physical conditions in the stellar atmosphere. However, Gershberg and Shakhovskaya (1973) and Shakhovskaya (1974a) showed that there exists a strong observational selection effect on these correlations: flares of lower luminosities have been observed on less luminous stars only, and weaker flares are shorter.

In Figure 8, the absolute luminosities of flares at maxima L^{max} (erg s^{-1}) are compared with the decay rates just after flare maximum ($-dL/dt$) (erg s^{-2}). Photoelectric observations of flare stars in the U band have been used. To the 28 flares studied by Shakhovskaya (1974) we have added 66 flares observed with time resolution of 3×10^{-7} s by Beskin *et al.* (1988) and 37 flares with time resolution of about 10 s by Bruevich *et al.* (1980) and Iljin (1987). Figure 8 shows the symbols for different stars as a function of their absolute magnitudes M_V and the numbers of flares utilized. Only the observation with time resolution 3×10^{-7} s, marked by the letter R, have been used for the analysis of the correlation between L_U^{max} and $(-dL_U/dt)$. For each of two stars with the most numerous flares (V577 Mon, Wolf 424 = FL Vir), a linear correlation between $\log L_U^{max}$ and $\log(-dL_U/dt)$ was sought. The difference in regression equations obtained was random with a probability of more than 90%. The flares on the other stars, UV Cet and CN Leo, are consistent with these equations. The regression equation for all the 66 flares of these four stars is

$$\log \frac{dL_U}{dt} = 3.22 + (0.84 \pm 0.06) \log L_U^{max} \tag{4}$$

with $r = 0.77$. The corresponding straight line (4) is represented in Figure 8.

Individual flares, observed with lower time resolution are not inconsistent with this correlation (4), but a thorough analysis shows that there exists a systematic shift of these

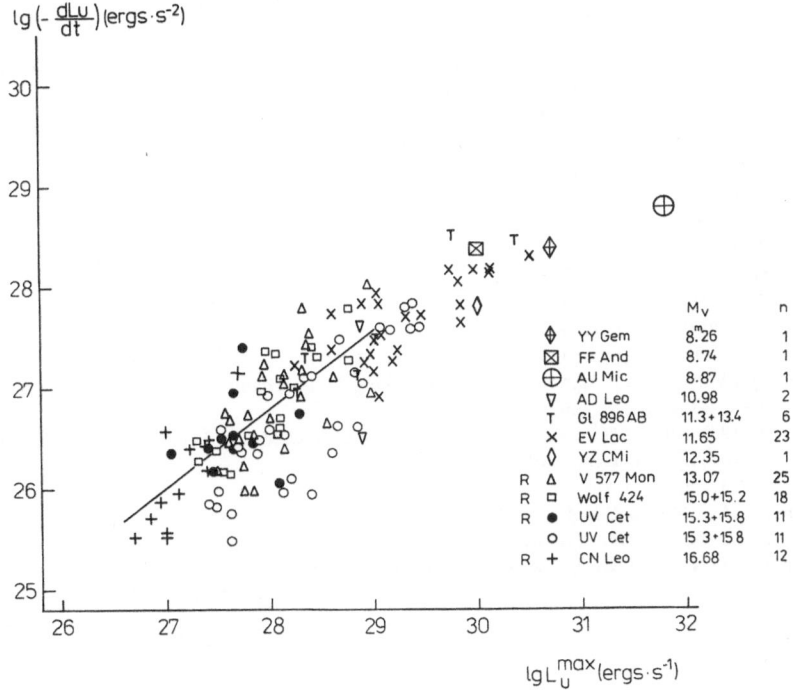

Fig. 8. The comparison of the absolute luminosities of flares at maximum, L_U^{max}, and decay rates just after the maximum, $(-dL_U/dt)$, in the U-band.

flares relative to Equation (4) having the value $\log(-\mathrm{d}L_U/\mathrm{d}t) = -0.15$. This spurious displacement is connected by the difference in time resolution.

We conclude that statistical relations between flare luminosities at maxima and flare decay rates exist covering the range of absolute magnitudes of flare stars from $M_V = 8\overset{m}{.}3$ up to $M_V = 16\overset{m}{.}7$. In other words, the flare decay rate is independent of stellar luminosity. The physical implications of this correlation are not yet clear.

References

Agrawal, P. C., Rao, A. R., and Sreekantan, B. V.: 1986, *Monthly Notices Roy. Astron. Soc.* **219**, 225.

Athay, R. G.: 1966, *Astrophys. J.* **146**, 223.

Ambruster, C. W., Sciortino, S., and Golub, L.: 1987, *Astrophys. J. Suppl.* **65**, 273.

Beskin, G. M., Gershberg, R. E., Neizvestnyi, S. I., Plakhotnichenko, V. L., Pustilnik, L. A., and Shvartsman, V. F.: 1988, *Izv. Krymsk. Astrofiz. Obs.* **79**, (in press).

Bopp, B. W.: 1974, *Monthly Notices Astron. Soc.* **168**, 255.

Bruevich, V. V., Burnashov, V. I., Grinin, V. P., Kiljachkov, N. N., Kotyshev, V. V., Shakhovskaya, N. I., and Shevchenko, V. S.: 1980, *Izv. Krymsk. Astrofiz. Obs.* **61**, 90.

Butler, C. J., Rodono, M., Foing, B. H., and Haisch, B. M.: 1986, *Nature* **321**, 679.

Doyle, J. G. and Butler, C. J.: 1985, *Nature* **313**, 378.

Gershberg, R. E.: 1972, *Astrophys. Space Sci.* **19**, 75.

Gershberg, R. E.: 1978, *Low Mass Flare Stars*, Nauka, Moscow.

Gershberg, R. E.: 1985, *Astrofizika* **22**, 531.

Gershberg, R. E.: 1988, 'XXXII Congresso della Soc. Astron. Ital.', Catania, 4–7 May, 1988, Catania Preprint No. 9.

Gershberg, R. E., Shakhovskaya, N. I.: 1973, *Nature, Phys. Sci.* **242**, 85.

Gershberg, R. E. and Shakhovskaya, N. I.: 1983, *Astrophys. Space Sci.* **95**, 235.

Gershberg, R. E., Mogilevskij, E. I., and Obridko, V. N.: 1987, *Kinematika i Fizika Nebesnykh Tel.* **5**, 3 (IZMIRAN Preprint No. 41 (655), Moscow, 1986).

Gliese, W.: 1969, *Veröff. Astron. Rechen-Institut*, No. 29, Heidelberg.

Grinin, V. P. and Katysheva, N. A.: 1980, *Izv. Krymsk. Astrofiz. Obs.* **62**, 66.

Haro, G. and Chavira, E.: 1955, *Bol. Obs. Tonantzintla Tacubaya* **12**, 3.

Ilyin, I. V.: 1987, *Inf. Bull. Var. Stars*, No. 2985.

Kunkel, W. E.: 1968, *Inf. Bull. Var. Stars*, No. 315.

Kunkel, W. E.: 1969a, in S. S. Kumar (ed.), *Low Luminosity Stars*, Gordon and Breach, New York, p. 195.

Kunkel, W. E.: 1969b, *Nature* **222**, 1129.

Kunkel, W. E.: 1975, in V. E. Sherwood and L. Plaut (eds.), *Variable Stars and Stellar Evolution*, D. Reidel Publ. Co., Dordrecht, Holland, p. 15.

Lacy, C. H., Moffett, T. J., and Evans, D. S.: 1976, *Astrophys. J. Suppl. Ser.* **30**, 85.

Linsky, J. L., Bornmann, P. L., Carpenter, K. G., Wing, R. F., Giampapa, M. S., and Worden, S. P., and Hege, E. K.: 1982, *Astrophys. J.* **260**, 670.

Page, A. A.: 1988, *Atlas of Flare Stars in the Solar Neighbourhood*, Mount Tamborine Observatory Publication No. 3, University of Queensland.

Pettersen, B. R.: 1980, *Astron. Astrophys.* **82**, 53.

Pustilnik, L. A.: 1988, *USSR Astron. J. Letters* (in press).

Shakhovskaya, N. I.: 1974a, *Izv. Krymsk. Astrofiz. Obs.* **50**, 84.

Shakhovskaya, N. I.: 1974b, *Izv. Krymsk. Astrofiz. Obs.* **51**, 92.

THE ROLE OF ERUPTION IN SOLAR FLARES

PETER A. STURROCK

Center for Space Science and Astrophysics, Stanford University, Stanford, CA 94305, U.S.A.

Abstract. This article focuses on two problems involved in the development of models of solar flares. The first concerns the mechanism responsible for eruptions, such as erupting filaments or coronal mass ejections, that are sometimes involved in the flare process. The concept of 'loss of equilibrium' is considered and it is argued that the concept typically arises in thought-experiments that do not represent acceptable physical behavior of the solar atmosphere. It is proposed instead that such eruptions are probably caused by an instability of a plasma configuration. The instability may be purely MHD, or it may combine both MHD and resistive processes. The second problem concerns the mechanism of energy release of the impulsive (or gradual) phase. It is proposed that this phase of flares may be due to current interruption, as was originally proposed by Alfvén and Carlqvist. However, in order for this process to be viable, it seems necessary to change one's ideas about the heating and structure of the corona in ways that are outlined briefly.

1. Introduction

There appear to be several types of solar flares (Bai and Sturrock, 1989), ranging from simple X-ray and Hα brightenings to the complexity of large two-ribbon flares. It appears that flares with high-energy manifestations, such as gamma-ray emission, tend to involve eruptive mass motion (see, for instance, Bai, 1986a). It has been argued (Kahler, 1982) that such an association may be due simply to the 'big flare syndrome', namely, that everything that can happen in a flare will happen in a big flare. However, the association between eruptive events and high-energy events in flares does raise legitimate questions, of which we here consider only two.

(1) What is the mechanism that leads to an eruption such as a coronal mass ejection? And

(2) Is there any physical reason why an eruption should lead to a high-energy event? We consider these questions in the next two sections.

2. Loss of Equilibrium

In recent years, it has been proposed by several authors that eruptions in solar active regions, that might give rise to CMEs, may be ascribed to a concept termed 'loss of equilibrium'. This concept arises in the study of force-free magnetic-field configurations of translational symmetry, that may be described in terms of a 'generating function'.

Consider a magnetic-field configuration that is uniform in the z-direction of Cartesian coordinates z, x, z. Since $\nabla \cdot B = 0$, we see that the magnetic field may be expressed as

$$\mathbf{B} = \left(\frac{\partial A}{\partial y}, \ -\frac{\partial A}{\partial x}, \ B_z \right). \tag{2.1}$$

Solar Physics **121**: 387–397, 1989.

We then find that the condition

$$\mathbf{j} \times \mathbf{B} = 0 \tag{2.2}$$

for the field to be force-free is satisfied if

$$\nabla B_z \times \nabla A = 0 , \tag{2.3}$$

which implies that B_z is expressible as a function of A, and if

$$\nabla^2 A + B_z \frac{dB_z}{dA} = 0 . \tag{2.4}$$

So far, the treatment is quite general, except that we are requiring translational symmetry.

At this point, we note that it is possible to define a *family* of solutions of Equation (2.4) by assuming that

$$B_z = \lambda F(A) . \tag{2.5}$$

Then Equation (2.4) becomes

$$\nabla^2 A = -\lambda^2 f(A) , \tag{2.6}$$

where

$$f(A) = F(A)F'(A) . \tag{2.7}$$

Equation (2.6) describes a family of force-free magnetic-field configurations of translational symmetry, corresponding to varying values of the parameter λ^2, for a given form of the generating function $f(A)$.

As a specific example, we consider the case discussed originally by Low (1977a) and later by Birn, Goldstein, and Schindler (1978) and Priest and Milne (1980). In our notation, the generating function is given by

$$f(A) = -k^2 \exp(-2A) . \tag{2.8}$$

The magnetic flux B_y at the plane $y = 0$, that represents the photosphere, is given by

$$A(x, 0) = \ln(1 + k^2 x^2) . \tag{2.9}$$

Low shows that Equation (2.6) and the boundary condition (2.9) are satisfied by the function

$$A(x, y) = \ln\left[1 + k^2 x^2 + 2 \left(\frac{1 - \mu^2}{1 + \mu^2} \right) ky + k^2 y^2 \right] , \tag{2.10}$$

where λ and μ are related by

$$\lambda = \frac{4\mu}{1 + \mu^2} . \tag{2.11}$$

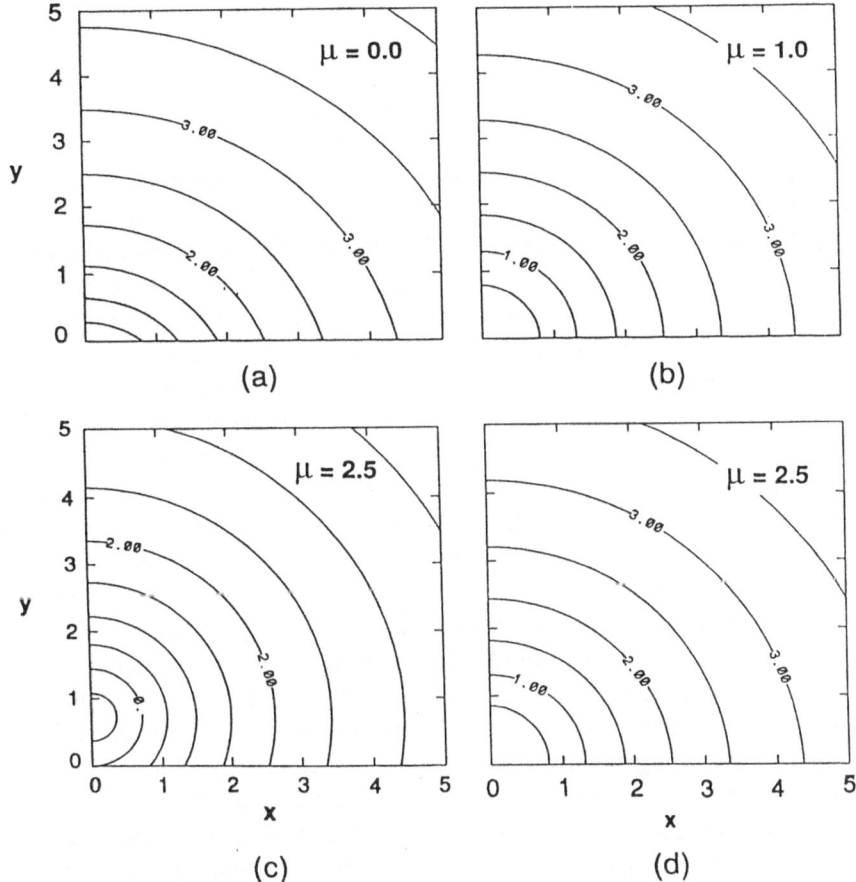

Fig. 1.   End-on view of magnetic field lines for the model defined by Equation (2.10). (a), (b), and (c) are taken from the generating-function model for the values $\mu = 0$, $\mu = 1$, and $\mu = 2.5$. It is seen from (c) that for $\mu > 1$ this model involves floating flux. (d) shows the magnetic-field configuration that arises from numerical solution of the force-free-field equations with the same boundary conditions as case (c). It is seen that it is indeed possible to meet those boundary conditions without the introduction of floating flux.

Low considers the sequence of magnetic-field configurations formed by allowing μ to increase from 0 to ∞. Then λ increases from 0 to a maximum of 2 (when $\mu = 1$) and then decreases back down to 0. For $0 < \mu \leq 1$, the magnetic-field configuration is that of a simple arcade (see Figures 1(a) and 1(b)) in which the distribution of footpoints on the photosphere is given by

$$kz = \frac{2\mu}{1 + \mu^2} \ \text{arc sin} \left\{ \left[\left(\frac{1 - \mu^2}{1 + \mu^2} \right)^2 + k^2 x^2 \right]^{-1/2} kx \right\}. \qquad (2.12)$$

For $\mu > 1$, the magnetic-field configuration is no longer that of a simple arcade. It contains a flux tube that runs above and parallel to the photosphere, that may be termed 'floating flux' (see Figure 1(c)). The distribution of footpoints is now given by

$$kz = \frac{2\mu}{1 + \mu^2} \left(\pi \frac{x}{|x|} - \arcsin \left\{ \left[\left(\frac{1 - \mu^2}{1 + \mu^2} \right)^2 + k^2 x^2 \right]^{-1/2} kx \right\} \right). \quad (2.13)$$

Low takes the position that the creation of floating flux is forbidden by the assumption of infinite electrical conductivity, and concludes that 'field configurations with $\mu > 1$ are not available to the evolving magnetic field'. He suggests that the quasi-steady evolution of the force-free field ceases at $\mu = 1$, whereupon explosive events take over.

The question we need to address is whether the physical problem represented by this mathematical model is relevant to processes that can occur in the Sun's atmosphere. For the range of parameters $\mu \leq 1$, the problem is equivalent to that of moving a prescribed distribution of magnetic flux according to the displacement given by Equation (2.12). This is a physically acceptable thought experiment.

On the other hand, for $\mu > 1$, the evolution of the sequence, that requires the global constraint of the functional relationship between B_z and A described by Equations (2.5), (2.7), and (2.8), is not equivalent simply to the footpoint motions described by Equation (2.13). As Low points out, it requires also the introduction of flux not connected to the photosphere. Klimchuk and Sturrock (1989) therefore argue that, for $\mu > 1$, the generating function problem defined by Low does not specify a physically acceptable thought experiment. This being the case, consequences of that thought experiment are not relevant to the actual behavior of the Sun's magnetic field.

Indeed, using the magneto-frictional method developed by Yang, Sturrock, and Antiochos (1986), we have been able to calculate force-free magnetic-field configurations corresponding to the continued evolution of the footpoints through Equation (2.12) for $0 < \mu \leq 1$, and then through Equation (2.13) for $\mu > 1$. We find that the magnetic field develops in a well-behaved manner and shows no evidence of catastrophic behavior (see Figure 1(d)).

As a result, we have concluded that the concept of 'loss of equilibrium' is an artifact of the specification of physically unacceptable thought experiments. Hence the concept cannot be invoked as an explanation of magnetic-field eruptions related to flares and CMEs.

In the above calculations, we have considered strictly force-free magnetic-field configurations. In reality, the coronal magnetic field is always interacting with plasma of non-zero density and pressure, so that the field is never exactly force-free. In articles following the one previously discussed (Low, 1977a), Low (1977b, 1980) has considered the implications of non-zero plasma pressure upon the evolution of a sequence of magnetic-field configurations defined by a particular choice of generating function. By using the Bernstein integral to test for MHD stability (Bernstein, 1973a, b), Low finds, in terms of his parameter λ, that the system is stable for $\lambda < \lambda^*$, where λ^* is the critical value of λ, but that it is only marginally stable for $\lambda = \lambda^*$. Hence Low conjectures that, in a system with non-zero gas pressure, an approach to the 'loss-of-equilibrium' state (as previously defined) is also an approach towards MHD instability. If this conjecture

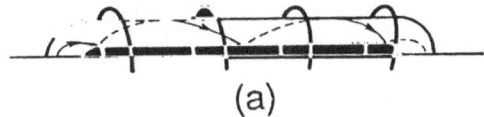

(a)

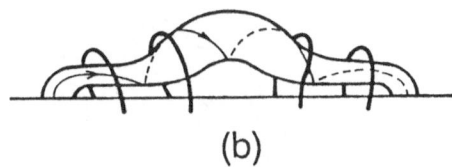

(b)

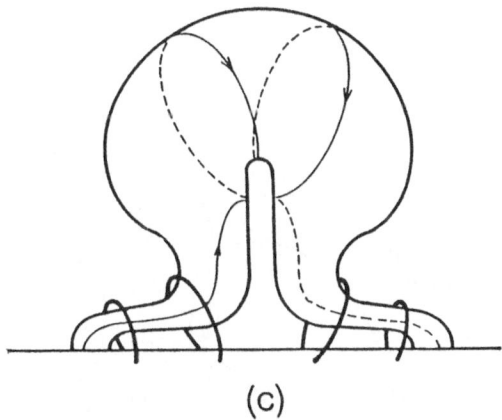

(c)

Fig. 2.   A possible MHD interpretation of solar eruptions. (a) shows a twisted flux tube anchored at both ends and held down by an overlying magnetic arcade. (b) indicates how the flux may begin to emerge by displacing selected field lines of the arcade. (c) shows that the flux tube would unwind in the region held down by the arcade, transferring most of the twist to the part that is erupting. If the original flux tube is long enough and sufficiently twisted, the initial state will have higher magnetic energy than the final state of a completely open configuration.

can be shown to have general validity, then the concept of loss of equilibrium would acquire general physical significance. However, this general connection between loss of equilibrium and MHD instability has so far not been demonstrated.

If magnetic-field eruptions are not due to 'loss of equilibrium', this leaves us with the problem of finding an alternative interpretation. The most likely explanation is that the eruption is due to some type of instability.

One possibility is that the instability is purely MHD in nature. For instance, consider the configuration shown schematically in Figure 2, comprising the horizontal flux tube rooted in the photosphere at its ends and held down by an overlying magnetic arcade.

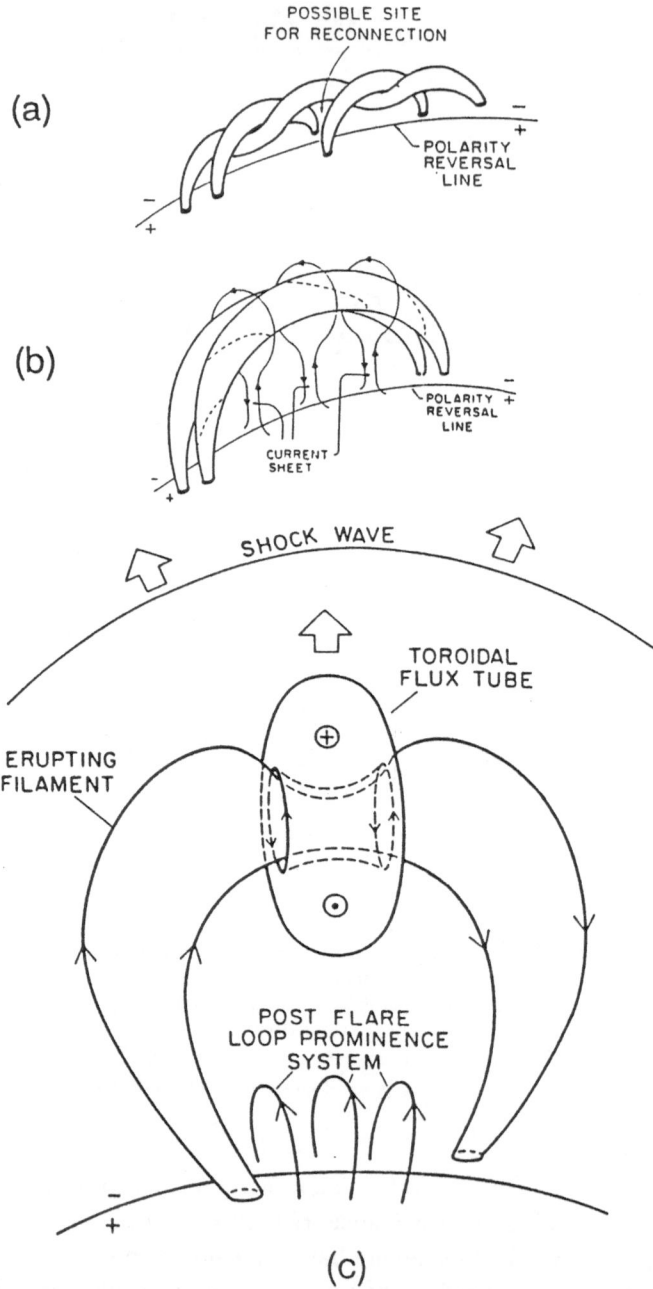

Fig. 3.   A possible interpretation of solar eruptions that involves both MHD and resistive processes. (a) shows a schematic representation of a possible magnetic configuration associated with a filament, that of a magnetic rope comprising a number of magnetic strands. Reconnection can occur where strands of opposite polarity are in contact with each other. (b) is a schematic representation of the evolution of the magnetic system after all connection with the photosphere has been broken, except the connection at the ends of the erupting flux tube. As a result of this eruption, magnetic field lines of the arcade form a new current sheet below the erupting flux tube. (c) is a schematic representation of the final form of the magnetic

If the flux tube is sufficiently long and sufficiently twisted, an eruption of the flux tube, corresponding to a rupture of the arcade, would be energetically favorable. This is the case since the tube can now move towards an untwisted state, so reducing the magnetic energy along its entire length. By making the tube long enough, and sufficiently highly twisted in its initial state, the energy so released can certainly exceed the finite energy of the open-field configuration that would be the limiting state after the eruption has occurred. An estimate of the critical condition for instability could be made either in terms of the small-amplitude energy theorem of Bernstein (1973a, b) or by calculating the initial and final magnetic field configurations and comparing their energies. (It should perhaps be emphasized that we are proposing that only *part* of the magnetic flux erupts into the open state. Hence there is no conflict between this proposal and the conjecture advanced by Aly (1984), that the completely open field is the state of maximum energy for a magnetic field with prescribed values of the normal field on the bounding surface.)

The above proposal represents a purely MHD instability. By contrast, it is possible that the instability responsible for eruptions involves both MHD and resistive processes. For instance, it may be more realistic to consider the initial state of a filament as being comprised of many individual strands, as shown in Figure 3(a). Reconnection can occur between adjacent strands of opposite polarity. This has two effects. One is that the reconnection would give rise to energy release that could lead to soft X-ray emission and Hα emission. The other consequence is that the remaining strands would be increasingly stressed. This may lead to runaway reconnection that then leads to the formation of a flux tube rooted in the photosphere only at its ends, as shown in Figure 3(b). This flux tube will be twisted. If the twist is sufficiently great, the tube would erupt towards an open configuration as shown in Figure 3(c).

3. Magnetic-Field Eruption and Particle Acceleration

As noted some time ago by Kopp and Pneuman (1976) and by Anzer and Pneuman (1982), the eruption of a filament will produce an important change in the topology of the magnetic arcade that overlays the filament before the eruption. This is shown schematically in Figure 3. The topology is that of an open bipolar magnetic-field configuration, that was proposed some time ago (Sturrock, 1968) as being a plausible explanation of two-ribbon flares. According to the preceding authors, reconnection at the current sheet separating the two parts of the open bipolar flux system results in energy release that can explain the long-lived soft X-ray emission and also the Hα emission. Cliver *et al.* (1986) point out that this process may, on occasion, also give rise to weak hard X-ray emission.

We have learned from observations of hard X-ray and gamma-ray emission from flares that particle acceleration, leading to nonthermal events, typically occurs during

field when the original field is sufficiently stressed that it results in ejection leading to an open configuration. Reconnection of the extended current sheet can now lead to the formation of a toroidal flux tube surrounding the magnetic field of the erupting filament.

the 'main' energy-release phase of solar flares (Bai, 1986b; Bai and Sturrock, 1989), a term that is used to include the impulsive and the gradual nonthermal phases of energy release. It has been proposed elsewhere (Sturrock et al., 1984) that this phase may be attributed to reconnection at the current sheets separating elementary flux tubes of an active region. It certainly is reasonable to look to reconnection as an explanation of energy conversion, since the release of free magnetic energy requires a change in the connectivity of magnetic-field lines. However, it is not clear that tearing-mode-type reconnection can accelerate sufficiently large numbers of particles sufficiently rapidly to explain the short time-scale of particle acceleration that has become apparent from SMM observations (Kane et al., 1986).

Another puzzle concerning flare behavior is the relationship between eruptions and the main phase of flares. It is well known that the eruption normally begins well before the main phase (Martin and Ramsey, 1972). This suggests that the eruption somehow leads to conditions that trigger an instability that is manifested as the main phase of a flare. We have suggested elsewhere (Sturrock, 1987, 1988) that the Kelvin–Helmholtz instability may develop in the neighborhood of an erupting filament when the speed attains a critical value, and that this instability may develop MHD turbulence that then speeds up the tearing-mode reconnection process in existing current-carrying magnetic-field systems. We now discuss another possible interpretation of the relationship between eruptions and the main phase of flares.

It was pointed out some time ago by Alfvén and Carlqvist (1967) that energy release in a flare may be related to current interruption. If, in a twisted magnetic flux tube, the resistivity were suddenly to increase in a localized region, a strong electric field must develop in that region to maintain the current. Such a current must persist until the field can unwind, a process that requires Alfvén-wave propagation, so that the acceleration must persist for the Alfvén-wave propagation time.

In their article, Alfvén and Carlqvist suggest that the relevant instability is the simple two-stream electron-ion instability. For the case of an isothermal electron-proton plasma, the condition for instability (Krall and Trivelpiece, 1973) is that

$$v_d > 1.3(kT/m_e)^{1/2} \approx 10^{5.7} T_e^{1/2} . \tag{3.1}$$

On noting that the relative drift velocity of the electrons and protons is given by

$$j = nc^{-1} ev_d , \tag{3.2}$$

and that the magnitude of the current is given by

$$|j| \approx \frac{1}{4\pi} Bb^{-1} , \tag{3.3}$$

where the magnitude of the curl of the magnetic field is taken to be determined by the length scale b, we find that the condition for instability becomes

$$nb < 10^{13.0} BT_e^{-1/2} . \tag{3.4}$$

For values typical of the coronal component of an active region $T = 10^{6.4}$, $n = 10^9$, and $B = 10^2$, we find that this instability will occur only if $b < 10^{2.8}$, i.e., if the scale of the current system is less than about 6 m. Even if this were to occur, clearly such a region would involve only a very small fraction of the volume of the magnetic-field configuration, and therefore lead to the release of only a very small fraction of the total free magnetic energy. It therefore seems unlikely that, in such conditions, current interruption will play a significant role in energy release.

We now suggest two changes in this model that make the idea more attractive. First, we may note that the ion-acoustic instability sets in at a lower critical value of the current density, provided the electron temperature is substantially higher than the ion temperature. If the electron temperature is substantially higher than the proton temperature, the condition for instability (Krall and Trivelpiece, 1973) is

$$v_d > (kT_i/m_i)^{1/2} \approx 10^{4.0} T_i^{1/2} . \qquad (3.5)$$

Hence the condition (3.4) is now replaced by

$$nb < 10^{14.7} BT_i^{-1/2} . \qquad (3.6)$$

The coronal conditions considered above now require that $b < 10^{4.2}$ for instability. This is still a very small value so that the objection made above is still applicable.

Our second change in the model requires a departure from the conventional interpretation of the mechanism that maintains the solar corona, that there is a *steady* heating mechanism that maintains the corona at a temperature of about 10^6 K. In recent years, some attention has been given to an alternative possibility that the corona is heated only by localized flare-like energy releases (see, for instance, Blake and Sturrock, 1985; Parker, 1988). For instance, it may be appropriate to regard the corona as composed of many small filaments, each filament comprising a small flux tube rooted at both ends in the photosphere. Sudden energy release in a small tube will lead to chromospheric evaporation, so that the tube is soon filled with plasma at a temperature of order one million degrees. However, this gas will cool by radiation and by conduction, tending to acquire the same temperature as the chromosphere.

This assumption has two consequences. The first is that the temperature to consider in Equations (3.4) and (3.6) may be nearer 10^4 K than 10^6 K. The second consequence is that the lower temperature of the gas leads to a lower scale height, so that the density may be very much lower than that we normally ascribe to the corona in an active region.

Consider, as an example, that the temperature in a loop is in fact of order 10^4 K, but T_e is higher than T_i. For instance, consider the case that $T_e = 10^{4.5}$ and $T_i = 10^{4.0}$. Also consider an elementary flux tube with radius of order 10^8 cm, corresponding to a flux of $10^{18.6}$ Mx if $B = 10^2$ G. Adopting $b = 10^8$, we now find that the ion-acoustic instability sets in if $n < 10^{6.7}$. However, for the temperatures quoted, the scale height is only $10^{8.0}$ cm, so that the density will drop from the chromospheric value of 10^{12} cm^{-3} to the required value of $10^{6.7}$ cm^{-3} in 12 scale heights, that is, in about $10^{9.1}$ cm or about 13 000 km.

We see that, with these revised assumptions, current interruption could very well occur in an active region so that it may indeed play a role in the main phase of solar flares, as suggested by Alfvén and Carlqvist.

We can now come back to the question posed in the Introduction: "Is there any physical reason why an eruption should lead to a high-energy event?" If the main phase may indeed be attributed to current interruption by the ion-acoustic instability, we need to inquire whether an eruption could lead to a sudden reduction in the plasma density and/or a sudden increase in the ratio of the electron temperature to the ion temperature. The answer to both these questions appears to be in the positive.

As we saw in Section 2, the magnetic arcade above an erupting filament will be displaced vertically as a result of the eruption. Hence the density at the top of an arcade loop will decrease. The density could therefore drop to the critical value for ion-acoustic instability. Note also that this interpretation provides a simple justification for the common assumption that acceleration occurs at the *top* of a loop.

However, in order for the ion-acoustic instability to occur, it is necessary that the electron temperature should be larger than the ion temperature. If a flux tube suddenly expands, the ions will tend to cool adiabatically whereas the electrons will tend to remain at the temperature of the boundaries, since the electron thermal conductivity is much higher than the ion-thermal conductivity. Hence it is plausible that, as the result of a sudden eruption, an overlying flux tube could be disturbed in such a way that the ion temperature drops more than the electron temperature.

4. Discussion

We have seen that the mechanism for eruption of a filament and/or a CME is most likely due to an MHD instability, or a combined MHD-resistive instability. We have also seen that the main phase may plausibly be interpreted in terms of current interruption by the ion-acoustic instability, if a current-carrying flux tube is suddenly expanded and raised as the result of filament eruption. However, the proposed model raises many questions that require investigation. For instance, it will be necessary to determine whether the ion-acoustic instability is the most likely mechanism of current interruption. Also, we need to study the thermodynamic evolution of plasma in a suddenly erupting flux tube, to determine whether the ratio of electron temperature to ion temperature will indeed exceed unity by a significant factor. We need to determine the DC electric field that would develop in such a system and the particle distribution that would result from the combined action of the DC and oscillatory electric fields.

In addition, we need to inquire whether it is indeed possible to explain the observed properties of the solar corona on the assumption that there is no steady coronal heating, but only impulsive flare-like heating. Assuming that the spatial scale of this impulsive behavior is sub-telescopic, we need to consider a complete cycle in the life-history of a small flux tube, and then determine whether the time-averaged radiation and other properties of such a flux tube are consistent with the observed radiation and other properties of the corona.

Acknowledgements

It is a pleasure to acknowledge helpful discussion on these topics with James Klimchuk and Taeil Bai and helpful comments on this article from B. C. Low. This work was supported by ONR through contract N00014–85–K–0111, and by NASA through grant NGL 05–020–272 and through the Solar-A collaboration under contract NAS8–37334 with Lockheed Palo Alto Research Laboratories.

References

Alfvén, H. and Carlqvist, P.: 1967, *Solar Phys.* **1**, 220.

Aly, J. J.: 1984, *Astrophys. J.* **283**, 349.

Anzer, U. and Pneuman, G. W.: 1982, *Solar Phys.* **79**, 129.

Bai, T.: 1986a, *Adv. Space Res.* **6**, 203.

Bai, T.: 1986b, *Astrophys. J.* **308**, 912.

Bai, T. and Sturrock, P. A.: 1989, *Ann. Rev. Astron. Astrophys.* **27** (in press).

Bernstein, I. B.: 1973a, in M. N. Rosenbluth and R. Z. Sagdeev (eds.), *Handbook of Plasma Physics*.

Bernstein, I. B.: 1973b, in A. A. Galeev and R. N. Sudan (eds.), *Basic Plasma Physics I*, p. 421.

Blake, M. L. and Sturrock, P. A.: 1985, *Astrophys. J.* **290**, 359.

Birn, J., Goldstein, H., and Schindler, K.: 1978, *Solar Phys.* **57**, 81.

Cliver, E. W., Dennis, B. R., Kiplinger, A. L., Kane, S. R., Neidig, D. F., Sheeley, N. R., and Koomen, M. J.: 1986, *Astrophys. J.* **305**, 920.

Kahler, S. W.: 1982, *J. Geophys. Res.* **87**, 3439.

Kane, S. R., Chupp, E. L., Forrest, K. J., Share, G. H., and Rieger, E.: 1986, *Astrophys. J.* **300**, L95.

Klimchuk, J. A. and Sturrock, P. A.: 1989, *Astrophys. J.* (in press).

Kopp, R. A. and Pneuman, G. W.: 1976, *Solar Phys.* **50**, 85.

Krall, N. A. and Trivelpiece, A. W.: 1973, *Principles of Plasma Physics*, McGraw-Hill, New York, p. 475.

Low, B. C.: 1977a, *Astrophys. J.* **212**, 234.

Low, B. C.: 1977b, *Astrophys. J.* **217**, 988.

Low, B. C.: 1980, *Astrophys. J.* **239**, 377.

Martin, S. F. and Ramsey, H. E.: 1972, in P. S. McIntosh and M. Dryer (eds.), *Solar Activity Observations and Predictions*, MIT Press, Cambridge, p. 371.

Parker, E. N.: 1988, *Astrophys. J.* **330**, 474.

Priest, E. R. and Milne, A. M.: 1980, *Solar Phys.* **65**, 315.

Sturrock, P. A.: 1968, in K. O. Kiepenheuer (ed.), 'Structure and Development of Solar Active Regions', *IAU Symp.* **35**, 471.

Sturrock, P. A.: 1987, *Solar Phys.* **113**, 13.

Sturrock, P. A.: 1988, *Outstanding Problems in Solar System Plasma Physics*, AGU Monograph (in press).

Sturrock, P. A., Kaufmann, P., Moore, R. L., and Smith, D. F.: 1984, *Solar Phys.* **94**, 341.

Yang, W.-H., Sturrock, P. A., and Antiochos, S. K.: 1986, *Astrophys. J.* **309**, 383.

SOLAR FLARES: THE GRADUAL PHASE

ZDENĚK ŠVESTKA

SRON Laboratory for Space Research Utrecht, The Netherlands

Abstract. One has to distinguish between two kinds of the gradual phase of flares: (1) a gradual phase during which no energy is released so that we see only cooling after the impulsive phase (*a confined flare*), and (2) a gradual phase during which energy release continues (*a dynamic flare*).

The simplest case of (1) is a *single-loop flare* which might provide an excellent opportunity for the study of cooling processes in coronal loops. But most confined flares are far more complicated: they may consist of sets of unresolved elementary loops, of conglomerates of loops, or they form arcades the components of which may be excited sequentially. Accelerated particles as well as hot and cold plasma can be ejected from the flare site (coronal 'tongues', flaring arches, sprays, bright and dark surges) and these ejecta may cool more slowly than the source flare itself.

However, the most important flares on the Sun are flares of type (2) in which a magnetic field opening is followed by subsequent reconnection of fieldlines that may continue for many hours after the impulsive phase. Therefore, the main attention in this review is paid to the gradual phase of this category of *long-decay* flares. The following items are discussed in particular: The wide energy range of dynamic flares: from eruptions of quiescent filaments to most powerful cosmic-ray flares. Energy release at the reconnection site and modelling of the reconnection process. The 'post-flare' loops: evidence for reconnection; observations at different wavelengths; energy deposit in the chromosphere, chromospheric ablation, and velocity fields; loops in emission; shrinking loops; magnetic modelling. The gradual phase in X-rays and on radio waves. Post-flare X-ray arches: observations, interpretation, and modelling; relation to metric radio events and mass ejections, multiple-ribbon flares and anomalous events, hybrid events, possible relations between confined and dynamic flares.

1. Introduction

One has to distinguish between two different kinds of the gradual phase of flares: there are flares, in which all the energy is released within a few seconds, or tens of seconds during the impulsive phase, and the gradual phase is simply the cooling of the heated flare plasma; but there are other flares, in which the release of energy continues during the gradual phase, in extreme cases for many hours. For simplicity, following the classification first proposed by Pallavicini, Serio, and Vaiana (1977), one can call the first class of events 'compact' or 'confined' flares and the second class 'two-ribbon', 'long-decay', or 'dynamic' flares. There are, of course, many hybrids of these two classes of flares, as we will discuss later on.

2. Confined Flares

The most simple case of a confined flare is a single-loop flare. Some people believe that such flares do not exist at all, because the basic mechanism of all flares is an interaction of loops (e.g., cf. Machado, 1987). On the other hand, theorists love single-loop flares, because they can be handled theoretically in a simple way (cf., e.g., Van Hoven, 1981). Well, as it seems, such flares do exist, but are probably very rare. As an example the

Solar Physics **121** (1989) 399–417.

reader is referred to a flare observed by the AS&E group on Skylab at 23:04 UT on 1 September, 1973 (Petrasso *et al.*, 1975): a young active region consisted of three loops; the central one flared for about seven minutes while the general configuration of the active region apparently did not change (cf. Petrasso and Krieger, 1976).

Most flares, however, are more complicated: more loops are involved in the brightening, or an arcade of loops is seen, in which the individual loops may brighten sequentially, one after another. Even in the Petrasso's flare one cannot be quite sure that it was just one simple loop that brightened: it may be a conglomerate of elementary loops, parallel or twisted, the structure of which was below the resolving power of 4 arc sec of the AS&E soft X-ray telescope.

2.1. SINGLE-LOOP FLARES

But let us suppose that in some cases we see a single-loop flare. Then, under the assumption of cooling through classical conduction and radiation, one can compute basic physical characteristics of the flaring loop. As an example, Pallavicini *et al.* (1983) computed the expected time evolution of maximum temperature and density in a single-loop compact flare, predicted the corresponding spectral line intensities, and compared them with available observations. The agreement with observations was found better for high-temperature lines than for low-temperature lines, which indicates that the cooling process somewhat deviates from expectations: some flares have decay times much longer than predicted. The authors suggested that this happened simply because the flaring structure in reality was not a single loop, but a conglomerate of shorter and longer loops: the shorter loops evolve more rapidly, the longer loops cool slower.

This may well be true, but Antiochos and Sturrock (1978) have shown that also the process of cooling need not be as simple as these and other authors had assumed. The hot thermal condensation near the top of the loop can cause continuing chromospheric evaporation even after the impulsive phase is over. The evaporation reduces the conductive heat flux into the chromosphere. Acton *et al.* (1982) applied these considerations to a compact flare observed by the SMM and actually found an evidence of continuous evaporation and improved agreement with observations: according to their results, about 7×10^{37} chromospheric atoms evaporated into the loop, and 3×10^{37} electrons were needed to produce the observed soft X-ray emission.

In a later paper, Antiochos and Sturrock (1982) proposed a model in which rapid radiative cooling at the flare loop base creates strong pressure gradients which, in turn, generate large downward flows. Hence, most of the thermal energy of the coronal flare plasma may be lost by mass motions rather than by conduction or radiation. In that case, the differential emission measure has a strong temperature dependence, and Schmahl, Kundu, and Erskine (1986), when comparing microwave and X-ray data, really found such a case in several flares. It seems that this 'condensation cooling' may be important after the evaporation phase is over and before radiation fully dominates the energy losses.

Even simple flares may greatly differ in their characteristics. Some flares studied by

Schmahl, Kundu, and Erskine (1986) showed an unusually high ratio between micro-wave and X-ray fluxes. On the other hand, Tsuneta (1987) describes compact flares in which this ratio is unusually low. He calls these flaring structures 'hot thermal flares' and suggests that density in these flares is abnormally high: in flares of low-density particle acceleration occurs first, and only after evaporation, i.e., density enhancement, the mode of energy release gradually changes into the plasma heating; in hot thermal flares the density is so high that the accelerated particles are thermalized since the flare onset.

2.2. ASSOCIATED PHENOMENA

Thus both the heating and cooling processes may widely differ even in the most simple confined flares. In addition to it, there are secondary phenomena that accompany some of these flares and eventually may become more impressive than the source flare itself: the cooling of these secondary features may be slower, and thus their lifetime longer than that of the confined flare. I will present here three examples.

2.2.1. *Coronal 'Tongues'*

First, I want to draw attention to De Jager's 'Queens' Flare'. Figures in De Jager *et al.* (1983) show how energetic electrons streamed from a low-lying confined limb flare into a previously existing magnetic loop system, thus forming a less bright but long-lived tongue above the limb. At 10 keV this tongue was 35 000 km long and could be seen for 90 min, while the lifetime of the source flare was less than 15 min. At 0.7 keV the tongue lived still longer. Its cooling was very slow and its structure must have been highly filamentary, with filling factor perhaps as low as 0.01.

2.2.2. *Flaring Arches*

As a second example, I refer to a recent paper by Martin and Švestka (1988), where we described the phenomenon of 'flaring arches'. In the brightest event we saw, a small confined subflare appeared at the primary footpoint. Within a few seconds streams of electrons enhanced the secondary footpoint at a distance of 57 000 km. The whole arch between the footpoints brightened in > 16 keV X-rays within one minute while the Hα flow needed 8 min to reach the secondary footpoint. In Hα the flow continued long after the subflare at the primary footpoint decayed. Another flaring arch we studied was 260 000 km long – I suppose Sara Martin will say more about this phenomenon in her review that follows. Similar events were also reported by Mouradian, Martres, and Soru-Escaut (1983), and by Rust, Simnett, and Smith (1985).

The fact that the Hα flow is seen in emission implies that the density of the cool (about 10 000 K) gas that fills the arch must be rather high: $\geq 10^{12}$ cm^{-3}. This follows from empirical considerations made years ago by Zirin, and from computations made recently by Fontenla and Machado (in Švestka *et al.*, 1987) or Heinzel and Karlický (1987). On the other hand, the X-ray energy spectrum requires temperatures in excess of 20 million Kelvin (or particle streams simulating such high temperatures). Thus the flaring arches are a mixture of a huge quantity of a cold dense material and hot, less dense plasma injected into preexisting coronal loops from confined flares or subflares.

2.2.3. *Bright Surges*

Sotirovski, Simon, and Rust (1986) observed a somewhat similar event: a bright surge, seen both in Hα emission and in X-rays, with the Hα flow delayed by about 5 min behind the X-ray enhancement. The authors also conclude that the surge was composed of both hot, X-ray emitting plasma and cold absorbing material. We have also seen such events when comparing Big Bear Observatory and HXIS data.

All this shows that even the 'simple' phenomenon of a compact or confined flare may be sometimes very complex and not at all easy to interpret and understand. It seems reasonable to consider the tongue, the arch, or the surge or spray, which often survive longer than the flare source itself, for a significant component of the gradual phase of these flaring events.

3. Dynamic Flares

However, the most important flares on the Sun belong to the other kind: flares, in which the energy release continues during the gradual phase. There is a wide energetic spectrum of these phenomena: from eruptions of quiescent filaments, the so called 'disparition brusque', when no Hα flare is observed, to powerful two-ribbon flares of X importance in X-rays and 4B importance in Hα, with a set of 'post'-flare loops connecting the ribbons. The common characteristic of all these events is soft X-ray brightening of the corona above the neutral line that lasts for many hours. Therefore, Kahler (1977) invented for them the name 'Long-Decay Events'.

Kopp and Pneuman (1976) suggested that these flares start with the opening of magnetic field lines and subsequent field line reconnection gives rise to the 'post'-flare loops. The resulting configuration is essentially the same that was earlier proposed by Sturrock (1968): energy release at the top of the loops and heating of Hα bright ribbons at their footpoints. Kopp and Pneuman added a dynamic development to this old Sturrock model: the plasma begins to flow upwards after the fieldline opening so that the gas pressure decreases, magnetic pressure prevails, and the process of fieldline reconnection sets in: first very fast, later with decreasing speed, while the neutral point is rising up. Therefore, these events are also called 'dynamic flares'.

3.1. KOPP AND PNEUMAN MODEL

While most people presently accept this model and some have contributed to its improvement, others are still skeptical, though nobody has been able so far to suggest anything better. Therefore, allow me to spend some time on arguments that support the Kopp–Pneuman interpretation.

3.1.1. *Energy Release during the Gradual Phase*

I think that there is now little doubt that in these flares energy continues to be released for a long time after the end of the impulsive phase. There are many observations that provide evidence for it – let me mention just a few (some more will become obvious in the following sections):

(1) Twenty years ago Bruzek (1969), during his stay at Sacramento Peak, discovered that the Hα post-flare loops do not expand, but that the growth of the loop prominence system is due to the generation of higher and higher loops while the lower ones fade. Each individual loop starts as a rapid brightening of a knot above existing loops which grows and eventually flows downward along the magnetic field lines.

(2) While this process can be usually followed in Hα for a relatively short period of one-and-half hour or less, X-ray observations show the growth of these loops for many more hours. On 29 July, 1973, for example, Skylab observed the growth of loops in a dynamic flare for 11 hours at least. The speed of growth, originally some 50 km s^{-1}, decreased then to 0.5 km s^{-1}, but the altitude of the X-ray loops was still increasing. Calculations showed that cooling time of the loops seen in X-rays was much shorter than the duration of the whole event; thus also in X-rays new loops had to be sequentially formed (Moore et al., 1980).

(3) In another dynamic flare observed by Skylab, MacCombie and Rust (1979) observed that for at least 8 hours bright X-ray flare loops were hotter at the tops than along the legs which implies that continuous heating must have been taking place at the loop tops.

(4) When studying X-ray images of the dynamic flare of 6 November, 1980, Švestka et al. (1987) have found a clear dependence of the loop top altitude on the temperature corresponding to the different images (Figure 1(a)): 20×10^6 K for HXIS and Fexxv, 10×10^6 K for Sixiii, 4×10^6 K for Mgxi, 2×10^6 K for Oviii, and 10 000 K for the Hα line. A plausible explanation is that we see here the newly formed loops in the hottest lines, and older loops at lower altitudes during the process of their cooling.

(5) Doyle and Raymond (1984) analyzed a major dynamic flare, taking spectra of the loops for more than 3 hours in the range 400–1335 Å. They found the total energy losses, integrated over the flare decay, far greater than the thermal energy content of the flare plasma at the beginning of observations, shortly after the flare maximum phase.

3.1.2. Source of the Energy Release

Thus the fact that energy is released during the gradual phase of dynamic flares has been established beyond any doubt. Another question is what kind of process causes this energy release and loop excitation. Kopp and Pneuman claim that it is reconnection of fieldlines that opened at the onset of the flare. The reconnection starts low in the corona, where the magnetic pressure is greatest, proceeds upwards, and gradually slows down in the upper layers. Still it can, slowly but persistently, continue for hours, as observed.

Really, I do not know any other interpretation that could explain the observed growth of the loop system. Which other process could release energy for 10 or more hours high in the corona, rising upwards with speeds of a few hundred meters per second? If there is no reconnection, i.e., if there are preexisting loops which are gradually excited by a rising agent, what is this agent, moving upwards for ten hours with a speed eventually as low as half-a-kilometer per second? Thus, really, the Kopp–Pneuman interpretation seems to be the only plausible explanation of the observed growth of loops in the dynamic flares.

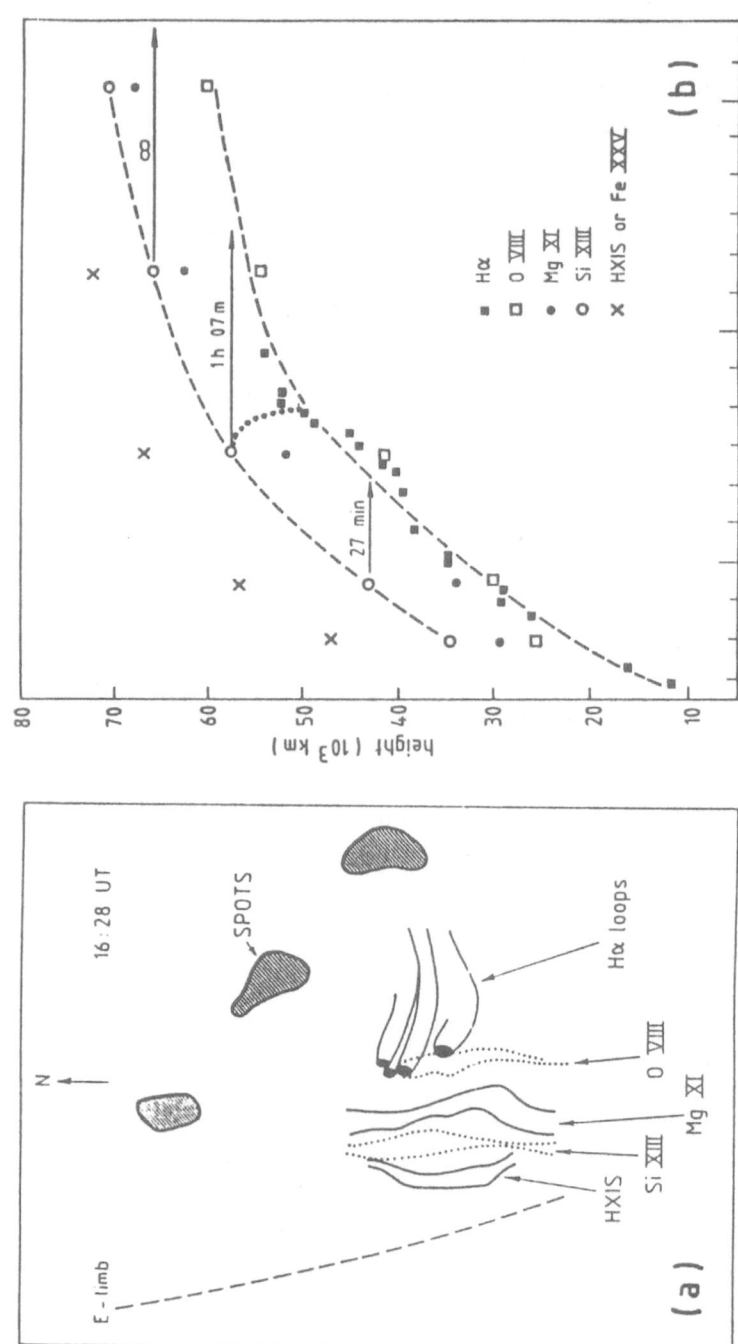

Fig. 1. (a) Positions of the tops of the loops in the dynamic flare of 6 November, 1980 in different X-ray lines of the SMM Flat Crystal Spectrometer and in the HXIS image (in X-rays above 3.5 keV), compared with the Hα loops. Time: 16:28 UT. (b) Time-variation of the measured altitude of the flare loops in Hα, in X-ray lines corresponding to different temperatures (cf. text), and in HXIS (or Fexxv which yields very similar images). Two dashed lines follow the growth of the loop system at low temperatures (10^4 K in Hα and 2×10^6 K in Oviii) and at 10^7 K (Sixiii), respectively. Arrows and times indicate the cooling of stationary loops. The real cooling path of the shrinking loops is indicated for the Sixiii loop at 16:28 UT by the dotted curve. (After Švestka et al., 1987.)

3.1.3. *Indirect Evidence for Reconnection*

With our present, limited spatial resolution, we do not have, and probably cannot have, any direct evidence of the reconnection process. But indirect evidence does exist:

Sakurai (1985), and before him Roy (1972), have shown that the loop-prominence systems of dynamic flares can be fit by potential-field modelling. In confined flares, according to Sakurai, potential-field modelling does not show any loops connecting the Hα footpoints or fitting the X-ray source. Thus the magnetic modelling indicates that the magnetic field structure is simplified in the case of a dynamic flare, possibly through reconnection of field lines which, prior to the field opening, were not potential.

Some 15 min after the maximum of the well-observed dynamic flare of 21 May, 1980 a new X-ray source appeared above the existing system of loops. This new source was first detected in the highest energy channels and only later on at lower energies (Švestka and Poletto, 1985). This indicates a source of small dimensions and high temperature in the corona.

Finally I may mention observations of post-flare loops made by Hanaoka, Kurokawa, and Saito (1986) during a solar eclipse. Their images indicated the existence of a cusp at the top of the loop system in the Fe XIV line, the hottest line they used.

3.2. MODIFICATIONS OF THE MODEL

3.2.1. *Additional Heating*

It was definitely established more than two decades ago by Kleczek (1964) that the plasma condensations seen in the post-flare loops cannot originate in the corona: there is simply not enough coronal material available to provide this plasma. Thus, the material seen in the loops must evaporate from below, from the chromosphere, and it must be heated to the observed temperatures, close to 20×10^6 K at the top of the loops. The original Kopp and Pneuman model assumed that a gas-dynamic shock propagates downward from the reconnection site and heats the upflowing plasma to $3–4 \times 10^6$ K. But this is insufficient when compared with the loop temperatures observed. Six years later, Pneuman (1982) added to it the global Ohmic heating due to reconnection. This might have raised the temperature by about a factor of two, but it was still far from the values actually observed at the tops of the newly formed loops.

But there are other ways to enhance the heating. Cargill and Priest (1983) suggested heating of post-flare loops by slow MHD shocks. Figure 2 illustrates their idea: the rising Y-type neutral point trails behind a pair of slow shocks which heat the upflowing plasma and bring it to rest. Another way to enhance the energy release during the reconnection process in the Kopp and Pneuman model was considered by Somov and Titov (1985): they demonstrated that a small transverse component of magnetic field in a turbulent current sheet could increase the energy output by several orders of magnitude. Kopp and Poletto (1984) applied the reconnection model to the well observed and relatively simple spotless dynamic flare of 29 July, 1973 and found good agreement with several parameters of the growing loop system during its entire lifetime starting soon after the flare onset.

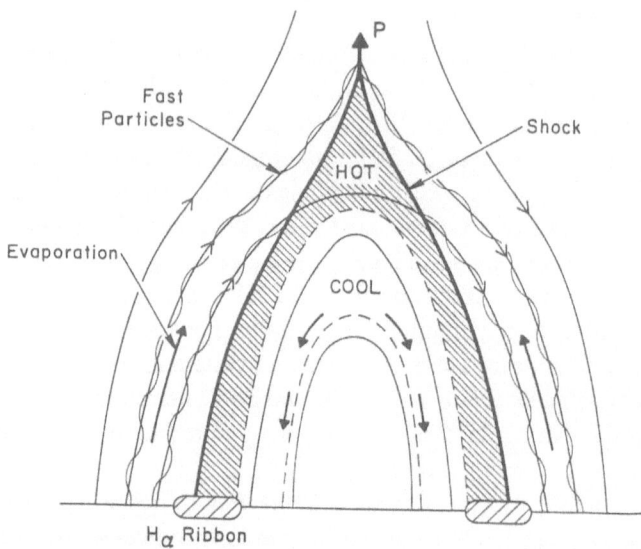

Fig. 2.   The system of dynamic-flare loops some time after the reconnection has started. The neutral point
P is rising vertically with decreasing velocity trailing behind it two slow MHD shocks (thick lines). These
shocks heat the plasma which evaporates from the chromosphere. Behind the shocks is then a series of hot
loops (shaded); the previously heated lower loops cool and one can eventually see the cooled material falling
in Hα. (After Cargill and Priest, 1983.)

3.2.2. *Chromospheric Evaporation*

According to Kleczek (1964), between 10^{16} and 10^{17} g of material must be supplied to
the top of post-flare loops during the gradual phase. Forbes and Priest (1983) have
considered direct particle acceleration as the source of this upward flow, but ended with
the result that an indirect, thermally-driven acceleration, such as ablation or evaporation,
is necessarily needed. Again, in the original Kopp and Pneuman model the amount of
evaporated material was only a fraction of that required, but there are ways to enhance
it.

Forbes and Malherbe (1986) have shown that in addition to the slow-mode MHD
shocks, considered by Cargill and Priest and shown in Figure 2, there also exists a
fast-mode shock generated by the reconnection process. This fast shock can contribute
to the chromospheric ablation and thus help to trigger a thermal condensation, that
means a loop prominence, provided the reconnecting magnetic fields are sufficiently
strong.

Indeed, Schmieder *et al.* (1987) found evidence for gentle chromospheric evaporation
during the gradual phase of major dynamic flares (Figure 3(a)): in three flares studied,
small blue shifts lasting for several hours were observed in the flare ribbons, which can
be interpreted as upward chromospheric flows with speeds of 0.5–10 km s^{-1}. These
upflows are sufficient to supply 10^{16} g or slightly more needed to maintain a dense
($n = 10^{12}$ cm^{-3}) Hα post-flare loop system in the corona. By contrast, the region
between the two ribbons exhibits large red shifts that are typical for falling material in

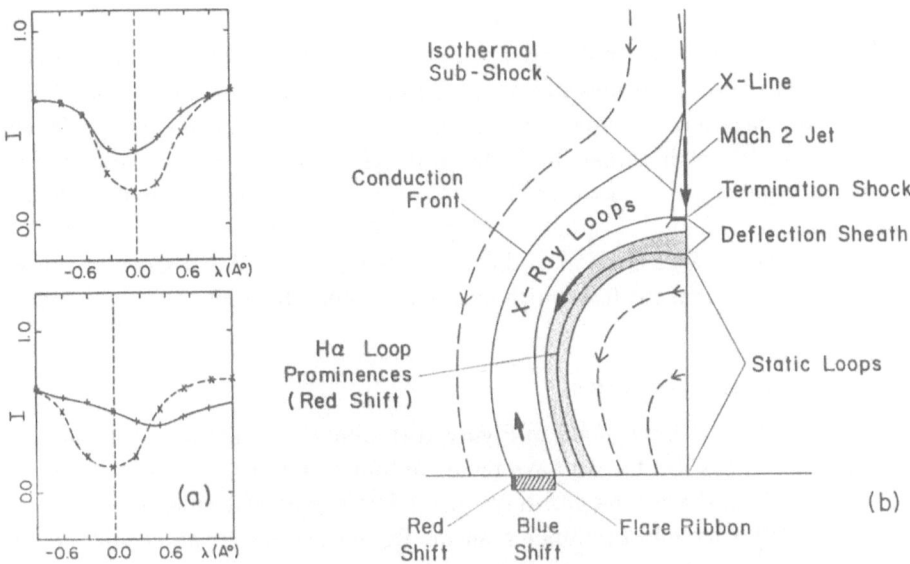

Fig. 3. (a) An example of observed Hα line profiles (solid lines) in the ribbons (*above*) and in the loops (*below*) of the dynamic flare of 16 May 1981. Dashed lines show the quiet-Sun reference profile. The blue shift in the ribbons corresponds to an upflow velocity of 4.5 km s^{-1}; the red shift in the loops corresponds to a downflow velocity of 24 km s^{-1} for optically thick, and 45 km s^{-1} for optically thin loops. (b) The inferred flow pattern for a reconnection model of the gradual phase (see text). (After Schmieder *et al.*, 1987.)

the cool Hα post-flare loops. The expected flow pattern is shown in Figure 3(b). In another analysis of chromospheric spectra of a dynamic flare by Wei-Qun and Cheng (1987) the authors deduced that about 9×10^{15} g of mass evaporated during the flare from the chromosphere to the corona.

One should realize that the density of 10^{12} cm^{-3} assumed by Schmieder *et al.* and also found in the study of the flare of 6 November, 1980, which I mentioned earlier (Figure 1), is the density in loops that, in projection on the solar disk, appeared in emission in the Hα line. Most post-flare loops seen on the disk are in absorption and their density is lower. For example, Shilova and Starkova (1987), who used Giovanelli's (1967) computations of hydrogen emission at various densities and temperatures, found the density in dark post-flare loops between 7×10^{10} and 10^{11} cm^{-3}.

The narrow red-shifted region at the outer edge of the Hα ribbon in Figure 3(b) can be explained as due to downward motion of the lower chromospheric level, which occurs when it is suddenly heated. MacNeice (1986) simulated the impact of a downward propagating coronal conduction front in a loop on the transition region and upper chromosphere. An accelerating upward moving ablation front and a downward moving compression is generated, just as the blue and red shifts in Figure 3 indicate.

The evaporation process has been studied by several authors, mainly by Canfield, Fisher, Gunkler, and their co-authors, but in most cases only evaporation during the impulsive phase has been considered. Fisher (1986) distinguishes two types of evaporation: an explosive one, if the chromosphere is heated by particle flows and the heating exceeds a certain threshold limit, and a non-explosive evaporation through conduction. The maximum upflow velocities during an explosive evaporation can reach 2.3 sound speeds as the maximum, whereas the non-explosive evaporation runs at speeds which are about 10 to 20% of this upper limit. During the gradual phase we encounter the lower speeds which generally agrees with Schmieder *et al.*'s values. A problem which still remains is the prevailance of radiative energy losses, demonstrated, e.g., by Doyle and Raymond (1984): a large fraction of the conducted flux is radiated away in the upper transition region so that the flux which eventually enters the upper chromosphere may be greatly reduced.

3.3. SHRINKING OF THE LOOPS

Let me draw attention once again to Figure 1(a): altitudes of the loop tops imaged in various spectral lines. So far we have tacitly assumed that a loop is newly formed at a certain altitude in the corona and stays there; first it is heated, thus becoming visible in X-rays, and then it cools, becoming sequentially visible in cooler and cooler lines until it eventually appears in Hα. During the Skylab workshop, Moore *et al.* (1980) even determined the cooling times from the time difference between the appearance of a loop at a certain altitude in X-rays and in Hα.

However, a study of the event of 6 November, 1980 reveals that a loop does not stay at the altitude at which it was formed (Švestka *et al.*, 1987). We have made more such images like Figure 1(a) and compared the time variation of the altitude of the loop tops in different lines. This is shown on Figure 1(b). Let us consider the observation of Si XIII at 16:28 UT. If the loop we see here stayed at the same altitude while cooling, it would appear after 1 hr and 7 min in the Hα line. For pure radiative cooling, this cooling time corresponds to a density of 4×10^9 cm^{-3} (and less if conduction also contributes to the cooling process). However, the Hα loops were in emission, hence their density must be 10^{12} cm^{-3} or more. Such a dense loop would cool purely by radiation, and the cooling time from the Si XIII temperature would not exceed 2 min. The implication is that the density must be increasing during the life of the loop, and the loop cannot stay at a constant altitude, but must gradually shrink. The real cooling path would then be similar to the dotted curve in Figure 1(b).

The whole process is schematically shown in Figure 4: a magnetic reconnection produces a loop; energy input from the reconnection point then evaporates chromospheric plasma into the loop which gradually cools and shrinks: the resulting O VIII and Hα loops are at much lower altitude than where the loop had been originally formed. According to Kopp and An (private communications), the chromospheric evaporation should first lead to an expansion of the loop. However, if the loop becomes very dense, and thus very heavy, gravitational forces will cause the loop to collapse and seek a different equilibrium configuration at a lower altitude.

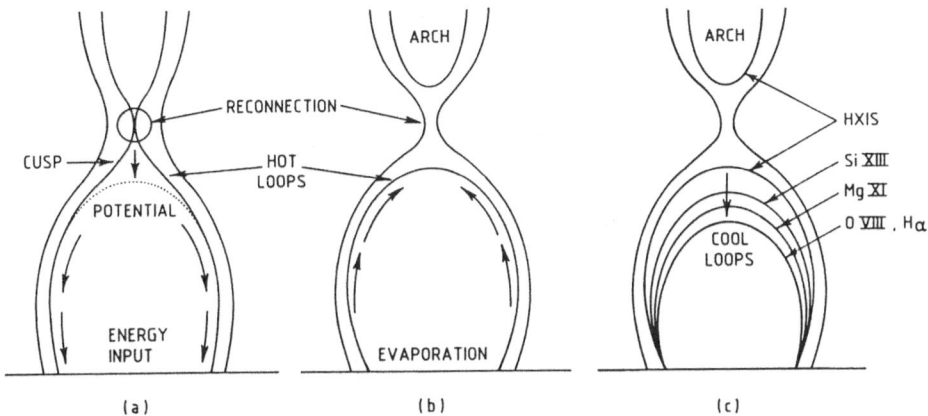

Fig. 4.   A tentative scenario of the loop formation (a), heating (b), and cooling (c), based on the observations
presented in Figure 1. (After Švestka *et al.*, 1987.)

3.4. GIANT POST-FLARE X-RAY ARCHES

The 'arch', shown in Figure 4, is the expected upper corona product of the reconnection
process. I suppose that its discussion should be included in a review of the gradual
phase, because, first, it may be considered as the gradual phase of a dynamic flare at
very high coronal altitudes and, second, it is probably related to some other observations
of the gradual phase which we will discuss in the next section.

The first arch of this kind was detected by HXIS in 4 keV X-rays on 22 May, 1980
(Švestka *et al.*, 1982a): the associated flare appeared at 21:00 the day before, and the
arch could be seen until 08 UT the next day, i.e., for 11 hours after the flare onset.
During the same period Culgoora observed a stationary type I noise storm above the
site of the arch. The flare of 6 November, to which Figures 1 and 4 refer, was associated
with another arch (Švestka *et al.*, 1982b). Again, Culgoora observed type IV continuum
above the source on the day of the flare, and a type I noise storm the day after. Figure 5
shows the time development of the arch in soft X-rays; the lower traces on the graph
show the time development of emission mensure and temperature. The temperature
peaks first, after approximately one hour, the flux peaks second, and the emission
measure last, 3.5 hours after the flare onset. This is, as under a magnifying glass, the
same development we see low in the solar atmosphere in flares. Thus the arch really
may be considered as the gradual phase of the dynamic flare at the altitude of some
100 000 km in the corona.

Švestka *et al.* (1982a) have suggested that such an arch is the upper product of the
reconnection process during dynamic flares. After the reconnection, the originally
sheared field lines produce much less sheared (quasi-potential, remember Sakurai's
(1985) modelling) post-flare loops below and the arch structure above. Though these
arches are created by some dynamic flares, their magnetic configuration then apparently
persists in the corona, because when another dynamic flare appears, essentially the same
structure reappears again in soft X-rays (Švestka, 1984).

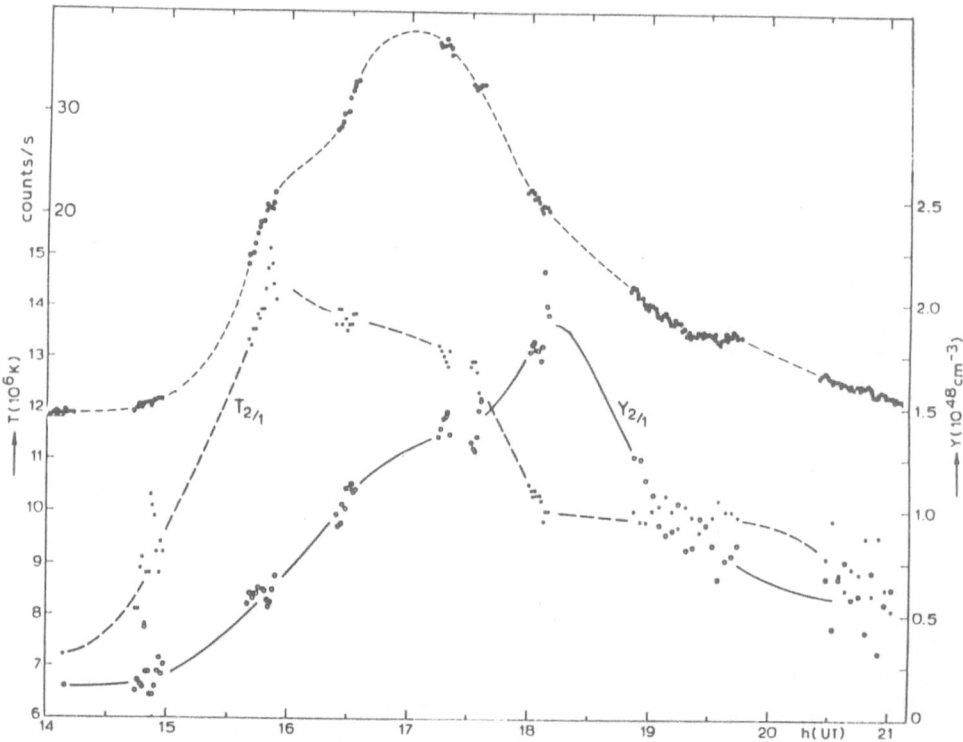

Fig. 5.   Time development of the giant post-flare arch that followed the dynamic flare of 14:44 UT on 6 November, 1980 (tentatively the 'arch' of Figure 4). All curves refer to an average altitude of 100 000 km. The upper curve shows the number of counts in 3.5–5.5 keV X-rays, the lower ones the temperature ($T_{2/1}$; left scale) and emission measure ($Y_{2/1}$; right scale). (After Švestka, 1984.)

3.5. RADIO TYPE IV EVENTS AND NOISE STORMS

Similar structures were actually seen long ago on radio waves. As early as 1961 Pick showed that metric type IV bursts, which as we know are associated with dynamic flares, consist of two parts: the first part sets in at the onset of the flare and the second part later on, during the first-part decay. In November 1968 Wild (1969) imaged at 80 MHz the radio event shown in Figure 6. The components A, B, C are components of a moving type IV burst and correspond to the first part. The component D is a stationary burst located above the site of the flare (the cross) and corresponds to the second part. There is little doubt that what Wild observed here as the source D is the same structure which HXIS observed as the arch in X-rays.

Klein *et al.* (1983) have found that the first part (the moving type IV) is associated with hard X-rays; in the second part the hard X-rays are missing, but the burst is associated with soft X-ray emission. Klein *et al.* conclude that during this second phase the energy release is less efficient, but still continues, and that the simultaneous occurrence of X-rays and metric type IV indicates existence of large-scale magnetic arches.

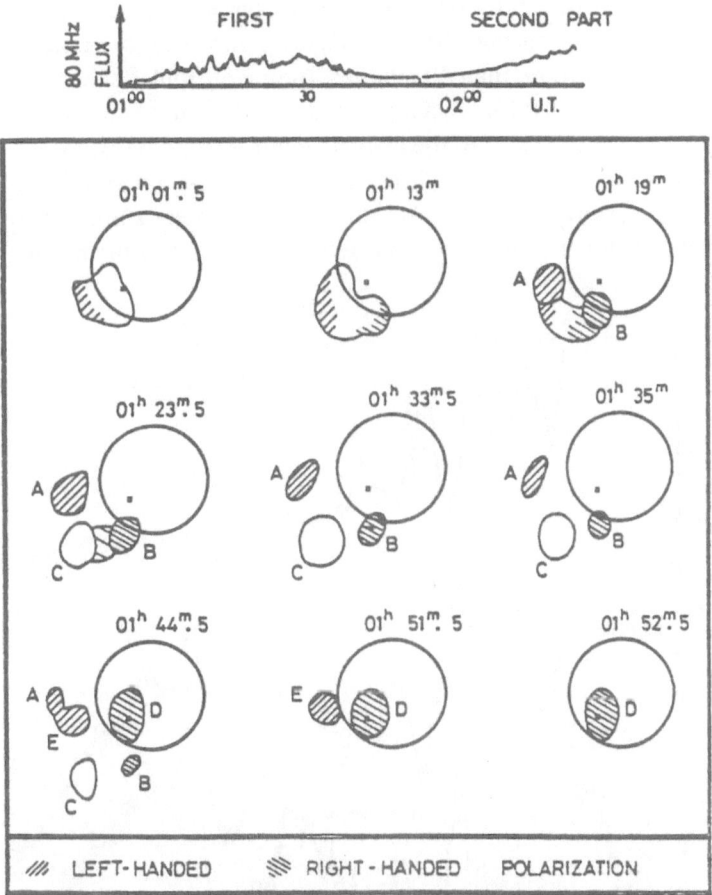

Fig. 6. Sources of the type IV burst on 22 November, 1968, as recorded by the 80 MHz radioheliograph at Culgoora. Position of the associated flare is marked with a cross. *Above*: record of the integrated 80 MHz flux showing the two parts of the type IV, corresponding to the moving (A, B, C) and stationary (D) sources, respectively. (After Wild, 1969.)

In another study Kai *et al.* (1986) observed in several dynamic flares delayed bursts which appeared 0.5 to 1 hour after the strong impulsive-phase bursts had faded. These delayed bursts were observed from microwaves to meter waves. The authors concluded that electrons were accelerated to the range of MeV even tens of minutes after the impulsive phase acceleration had ceased. The wide difference in wavelengths at which this delayed radio bursts appeared is again evidence that the delayed energy release occurs in a giant magnetic structure extending to at least 200 000 km altitude, and that is also the altitude where we see the tops of X-ray post-flare arches.

An intriguing problem, associated with long-lasting post-flare bursts is the hardening of the X-ray spectrum in some of them. Several such events were observed by Hinotori (e.g., Takakura *et al.*, 1984) and by the Hard X-Ray Burst Spectrometer on the SMM (e.g., Cliver *et al.*, 1986). Takakura *et al.* observed three successive peaks in a flare event which became progressively harder. Cliver *et al.* observed flare events in which the spectral index γ was decreasing with time. Figure 7 shows an example in which γ was steadily decreasing for more than 40 minutes. In one of similar sources, Vilmer, Kane, and Trottet (1982) estimated its density to at least 10^{10} cm^{-3}. In another case, when such a source was imaged close to the limb, its altitude could be estimated at 40 000 km and it did not change its position (during 14 min of imaging with spatial resolution of about 15 arc sec; Tsuneta *et al.*, 1984). Thus again these are stationary or very slowly moving bursts which are somehow connected with the tops of the newly formed loops below the post-flare arch.

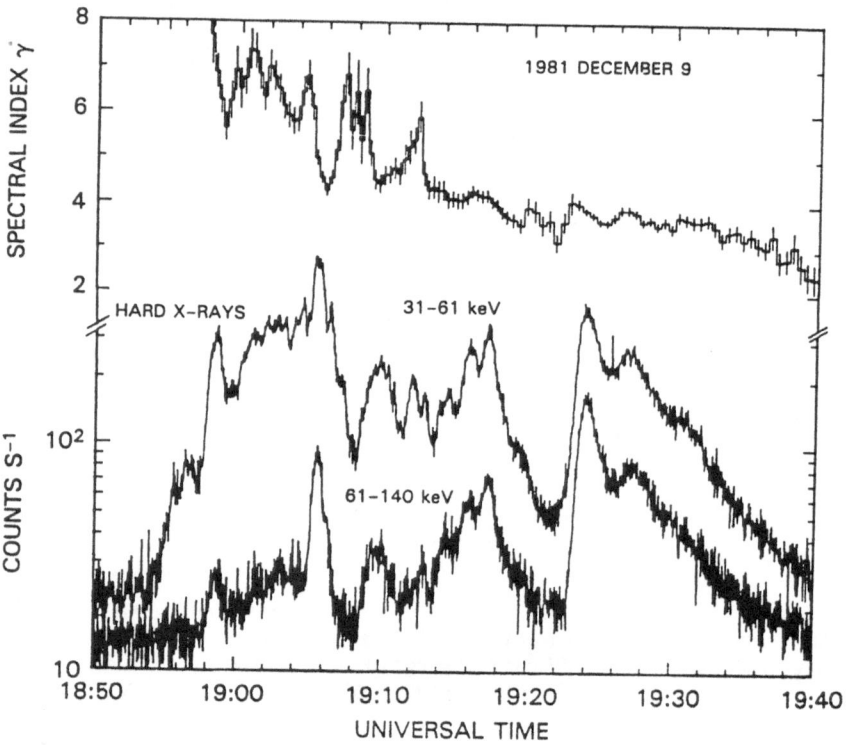

Fig. 7.   Hard X-ray time profiles (31–61 keV and 61–140 keV) of the dynamic flare on 9 December, 1981. The uppermost curve shows the gradual decrease of the spectral index γ, relatively unaffected by variations in the X-ray flux. (After Cliver *et al.*, 1986.)

As Cliver *et al.* (1986) have pointed out, there is multiple evidence that sources of this hard X-ray radiation are significantly higher than the sources we see during the impulsive phase: (1) There are very small related changes in Hα; (2) the frequency of maximum radio flux is low (which implies low magnetic field strength); and (3) the bursts are microwave rich; this implies less bremsstrahlung, i.e., low density. Tsuneta (1983) has interpreted the systematic hardening in terms of the energy-dependent collisional loss of electrons confined in a magnetic trap. Trapping, alone, however, without prolonged acceleration, in insufficient to account for the observed degree of spectral hardening, as Bai and Dennis (1845) pointed out.

3.6. Complex Structures of Dynamic Flares

Gopalswamy and Kundu (1987) have demonstrated how extremely diverse the positions of different radio bursts can be above a dynamic flare site. In the flare they studied, the various continua observed before, during, and after the flare were at different positions and altitudes, probably due to electron trapping in different arches. The type II burst, i.e., the flare-associated shock, moved in a completely separate direction. This is a particularly interesting observation, because it may explain why we sometimes see both the mass ejection associated with a dynamic flare and a quasi-stationary arch above the flare site.

Indeed, some dynamic flares may be extremely complex phenomena. Apart from the common two-ribbon flares, there are also three-ribbon and four-ribbon flares (cf. Tang, 1985). Ogir and Antalová (1986) suggested that a three-ribbon flare is a misleading image of two chromospheric ribbons and projected bright tops of the loops, but that, according to Tang, is not always the case. The third ribbon may really represent a complex structure of loops, like in the case shown in Figure 8, after Kundu, Schmahl, and Velusamy (1982). A four-ribbon flare is still more complex. Besides, one may encounter quite anomalous cases, like the 'circular two-ribbon flare' presented in her paper by Tang (1985).

3.7. Small Dynamic Flares

Tang (1985) has found in many small flares that what looks like ribbons in them does not usually show any measurable separation of the chromospheric bright structures. Therefore, these events are probably confined flare arcades. But some small flares do show the ribbon expansion which implies that not all dynamic flares are big long-duration events.

This seems to be confirmed by a study made by Lin, Lin, and Kane (1985). From an analysis of flares observed by ISEE-3 they deduced that in 40% of cases there are two components during the decay phase of solar flares: a usual flare plasma at about 10 million degrees, cooling, and a superhot one, with temperature of at least 30 million

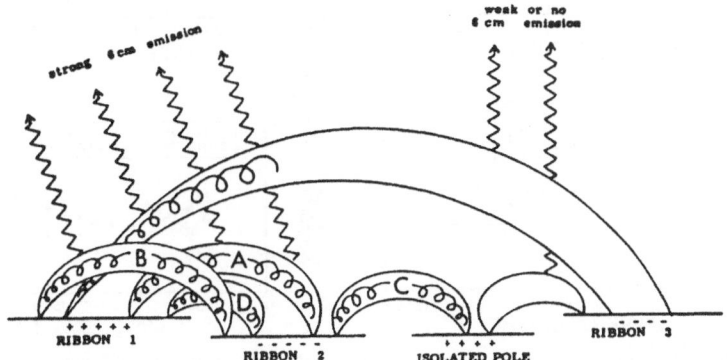

Fig. 8.   Sketch of the magnetic field structure of a three-ribbon flare on 25 June, 1980 deduced from radio, Hα, and magnetogram data. (After Kundu, Schmahl, and Velusamy, 1982.)

degrees, which requires an additional input of energy after the impulsive phase. In Figure 9 the right-hand flare event belongs to this category while the left-hand event does not: there a two-temperature fit does not produce any improvement. Note that the right-hand event, though moderately small and relatively short-lived, still has a more gradual time profile than the other one. Thus one can suppose that this was a relatively small dynamic flare, with continued release of energy in the late phase, whereas the left-hand event was a confined flare, with all energy released during the impulsive phase. Several other examples can be found in the paper by Lin, Lin and Kane.

4. Flare Hybrids

Some flares, however, are quite clearly hybrids of the two classes we discussed before. That dynamic flare to which Figures 1, 4, and 5 are related, on 6 November, 1980, started at 14:44 UT as a confined flare and the dynamic-flare loops began to develop only tens of minutes later. Harrison *et al.* (1983) presented an example of another flare, on 5 July, 1980, which started as a compact flare. De Jager (private communication) is

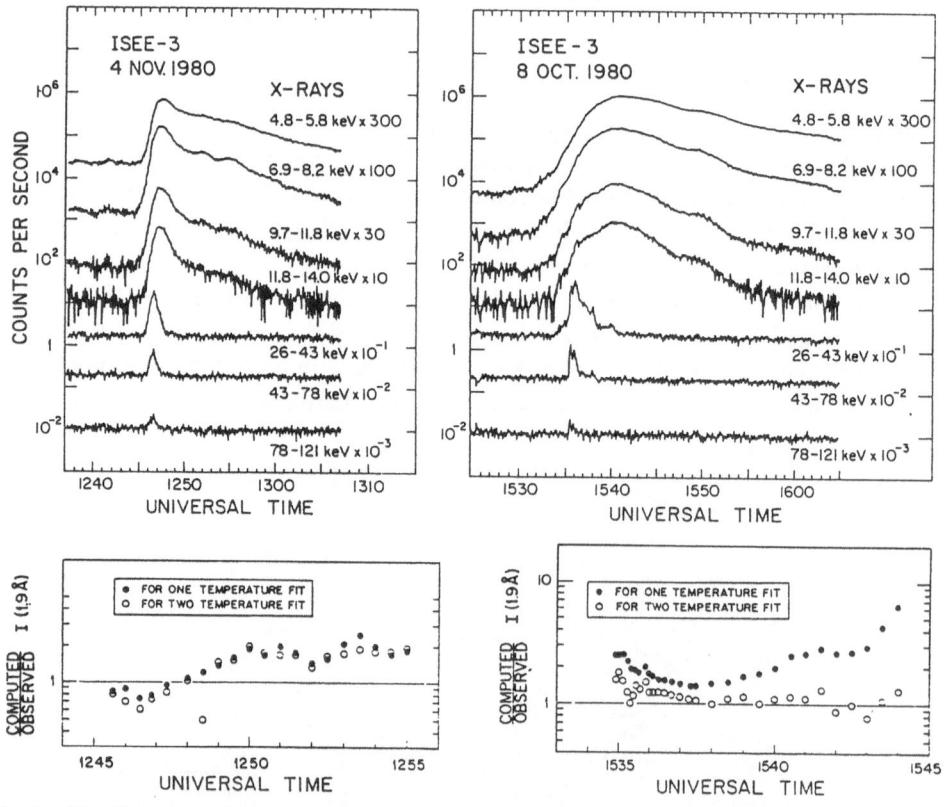

Fig. 9. The X-ray counting rate versus time profiles (*above*) and results of fitting procedure for one-temperature and two-temperature fit (*below*) for a tentatively confined (*left*) and dynamic (*right*) flare. (Examples taken from Lin, Lin, and Kane, 1985.)

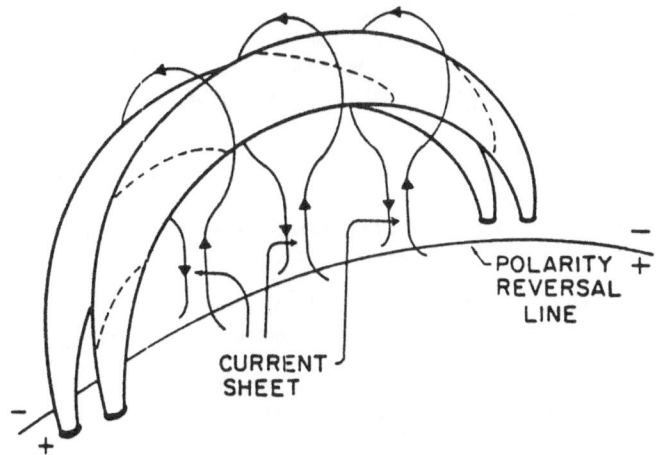

Fig. 10.   Schematic representation of the development of an extended current sheet beneath an erupting
filament. (After Sturrock *et al.*, 1984.)

of the opinion that quite often (and possibly always) a compact flare may serve as a
trigger for a dynamic flare; we do not know, actually, what causes the magnetic field
to open and thus start the dynamic flare process.

Sturrock *et al.* (1984) have suggested that the onset phase of a two-ribbon dynamic
flare may be due to reconnection of contiguous but oppositely directed magnetic flux
tubes linking a filament to the photosphere (Figure 10). A set of reconnections below
the filament causes its eruption and starts the flare. Until the filament erupts and the
field opens, and before the open field lines begin to reconnect, the set of reconnections
in the configuration of Figure 10 clearly can simulate the character of a confined flare.

As a matter of fact, there are still relatively few flares for which good images at
different wavelengths (X-rays, UV, optical, radio) are available. In the set of high-
resolution Hα data at the Big Bear Observatory, for example, one finds many strange
events, for which the classification is extremely difficult when no images of the coronal
X-ray or radio flare are available. Some flares seem to be hybrids of the two classes,
some others are either confined or dynamic flares accompanied by other associated
phenomena (surges, sprays, flaring arches, or loop connections of which only the remote
footpoints become visible in Hα). In the same way, the Hinotori flare classification (cf.,
e.g., Dennis, 1985) which is based solely on X-ray images with poor spatial resolution,
encounters many hybrids and events for which the classification is questionable. Hopefully, the rising solar cycle will provide more complete and more coordinated flare
observations which will help us to understand better all these varieties in the flaring
plasma configuration.

References

Acton, L. W., Canfield, R. C., Gunkler, T. A., Hudson, H. S., Kiplinger, A. L., and Leibacher, J. W.: 1982,
Astrophys. J. **263**, 409.

Antiochos, S. K. and Sturrock, P. A.: 1978, *Astrophys. J.* **220**, 1137.
Antiochos, S. K. and Sturrock, P. A.: 1982, *Astrophys. J.* **254**, 343.
Bai, T. and Dennis, B. R.: 1985, *Astrophys. J.* **292**, 699.
Bruzek, A.: 1969, in C. de Jager and Z. Švestka (eds.), *Solar Flares and Space Research*, North-Holland, Amsterdam, p. 61.
Cargill, P. J. and Priest, E. R.: 1983, *Astrophys. J.* **266**, 383.
Cliver, E. W., Dennis, B. R., Kiplinger, A. L., Kane, S. R., Neidig, D. F., Sheeley, N. R., and Koomen, M. J.: 1986, *Astrophys. J.* **305**, 920.
De Jager, C., Machado, M. E., Schadee, A., Strong, K. T., Švestka, Z., Woodgate, B. E., and van Tend, W.: 1983, *Solar Phys.* **84**, 205.
Dennis, B. R.: 1985, *Solar Phys.* **100**, 465.
Doyle, J. G. and Raymond, J. C.: 1984, *Solar Phys.* **90**, 97.
Fisher, G. H.: 1986, in D. Mihalas and K. H. Winkler (eds.), 'Radiation Hydrodynamics in Solar Flares', *Lecture Notes in Physics* **255**, 53.
Forbes, T. G. and Malherbe, J. M.: 1986, *Astrophys. J.* **302**, L67.
Forbes, T. G. and Priest, E. R.: 1983, *Solar Phys.* **88**, 211.
Giovanelli, R. G.: 1967, *Australian J. Phys.* **20**, 81.
Gopalswamy, N. and Kundu, M. R.: 1987, *Solar Phys.* **111**, 347.
Hanaoka, Y., Kurokawa, H., and Saito, S.: 1986, *Solar Phys.* **105**, 133.
Harrison, R. A., Simnett, G. M., Hoyng, P., Lafleur, H., and van Beek, H. F.: 1983, *Solar Phys.* **84**, 237.
Heinzel, P. and Karlický, M.: 1987, *Solar Phys.* **110**, 343.
Kahler, S. W.: 1977, *Astrophys. J.* **214**, 891.
Kai, K., Nakajima, H., Kosugi, T., Stewart, R. T., Nelson, G. J., and Kane, S. R.: 1986, *Solar Phys.* **105**, 383.
Kleczek, J.: 1964, in *AAS-NASA Symp. on Physics of Solar Flares*, NASA, Washington, D.C., p. 77.
Klein, L., Anderson, K., Pick, M., Trottet, G., Vilmer, N., and Kane, S.: 1983, *Solar Phys.* **84**, 295.
Kopp, R. A. and Pneuman, G. W.: 1976, *Solar Phys.* **50**, 85.
Kopp, R. A. and Poletto, G.: 1984, *Solar Phys.* **93**, 351.
Kundu, M. R., Schmahl, E. J., and Velusamy, T.: 1982, *Astrophys. J.* **253**, 963.
Lin, H. A., Lin, R. P., and Kane, S. R.: 1985, *Solar Phys.* **99**, 263.
MacCombie, W. J. and Rust, D. M.: 1979, *Solar Phys.* **61**, 69.
Machado, M. E.: 1987, *Solar Phys.* **113**, 57.
MacNeice, P.: 1986, *Solar Phys.* **103**, 47.
Martin, S. F. and Švestka, Z. F.: 1988, *Solar Phys.* **116**, 91.
Moore, R., McKenzie, D. L., Švestka, Z., Widing, K. G., and 12 co-authors: 1980, in P. A. Sturrock (ed.), *Solar Flares*, Colorado Associted University Press, p. 341.
Mouradian, Z., Martres, M. J., and Soru-Escaut, I.: 1983, *Solar Phys.* **87**, 309.
Ogir, M. B. and Antalová, A.: 1986, *Bull. Astron. Inst. Czech.* **37**, 344.
Pallavicini, R., Serio, S., and Vaiana, G. S.: 1977, *Astrophys. J.* **216**, 108.
Pallavicini, R., Peres, G., Serio, S., Vaiana, G., Acton, L., Leibacher, J., and Rosner, R.: 1983, *Astrophys. J.* **270**, 270.
Petrasso, R. D. and Krieger, A. S.: 1976, *Solar Phys.* **47**, 167.
Petrasso, R. D., Kahler, S. W., Krieger, A. S., Silk, J. K., and Vaiana, G. S.: 1975, *Astrophys. J.* **199**, L127.
Pick, M.: 1961, *Ann. Astrophys.* **24**, 183.
Pneuman, G. W.: 1982, *Solar Phys.* **78**, 229.
Roy, J. R.: 1972, *Solar Phys.* **26**, 418.
Rust, D. M., Simnett, G. M., and Smith, D. F.: 1985, *Astrophys. J.* **288**, 401.
Sakurai, T.: 1985, *Solar Phys.* **95**, 311.
Schmahl, E. J., Kundu, M. R., and Erskine, F. T.: 1986, *Solar Phys.* **105**, 87.
Schmieder, B., Forbes, T. G., Malherbe, J. M., and Machado, M. E.: 1987, *Astrophys. J.* **317**, 956.
Shilova, N. S. and Starkova, L. I.: 1987, *Kinematika i fizika nebeskich tel 3*, No. 6, 28.
Somov, B. V. and Titov, V. S.: 1985, *Solar Phys.* **102**, 79.
Sotirovski, P., Simon, G., and Rust, D. M.: 1986, in D. F. Neidig (ed.), *The Lower Atmosphere of Solar Flares*, National Solar Observatory, Sunspot, New Mexico, p. 71.
Sturrock, P. A.: 1968, in K. O. Kiepenheuer (ed.), 'Structure and Development of Solar Active Regions', *IAU Symp.* **35**, 471.

Sturrock, P., Kaufman, P., Moore, R. L., and Smith, D. F.: 1984, *Solar Phys.* **94**, 341.

Švestka, Z.: 1984, *Solar Phys.* **94**, 171.

Švestka, Z. and Poletto, G.: 1985, *Solar Phys.* **97**, 113.

Švestka, Z., Stewart, R., Hoyng, P., Van Tend, W., Acton, L. W., Gabriel, A. H., Rapley, C. G., and 8 co-authors: 1982a, *Solar Phys.* **75**, 305.

Švestka, Z., Dennis, B. R., Pick, M., Raoult, A., Rapley, C. G., Stewart, R. T., and Woodgate, B. E.: 1982b, *Solar Phys.* **80**, 143.

Švestka, Z. F., Fontenla, J. M., Machado, M. E., Martin, S. F., Neidig, D. F., and Poletto, G.: 1987, *Solar Phys.* **108**, 237.

Takakura, T., Ohki, K., Sakurai, T., Wang, J. L., Xuan, J. Y., Li, S. C., and Zhao, R. Y.: 1984, *Solar Phys.* **94**, 359.

Tang, F.: 1985, *Solar Phys.* **102**, 131.

Tsuneta, S.: 1983, Thesis, University of Tokyo.

Tsuneta, S.: 1987, *Solar Phys.* **113**, 35.

Tsuneta, S. Takakura, T., Nitta, N., Ohki, K., Tanaka, K., Makishima, K., Murakami, T., Oda, M., Ogawara, Y., and Kondo, I.: 1984, *Astrophys. J.* **280**, 887.

Van Hoven, G.: 1981, in E. R. Priest (ed.), *Solar Flare Magnetohydrodynamics*, Gordon and Breach, New York, p. 217.

Vilmer, N., Kane, S. R., and Trottet, G.: 1982, *Astron. Astrophys.* **108**, 306.

Wei-Qun, G. and Cheng, F.: 1987, *Solar Phys.* **107**, 311.

Wild, J. P.: 1969, *Solar Phys.* **9**, 260.

POLARIMETRY OF STELLAR ACTIVE REGIONS AND FLARES

I. TUOMINEN, J. HUOVELIN

Observatory and Astrophysics Laboratory, University of Helsinki, Tähtitorninmäki, SF-00130 Helsinki, Finland

and

YU. S. EFIMOV, N. M. SHAKHOVSKOY, and A. G. SHCHERBAKOV

Crimean Astrophysical Observatory, P/O Nauchny, SU-334413 Crimea, U.S.S.R.

Abstract. Observations of regular and irregular polarimetric variability in late-type stars are reviewed, and the related physical and geometrical effects are discussed. There are indications that the irregular part of the variability could be caused by transient events, possibly associated with flares. Polarimetric observations during flares are reviewed, and preliminary results of new observations of a well-known flare star, YY Geminorum, are presented. The results show that the small flare in YY Gem did not cause any significant variations in linear polarization, while the binary eclipse evidently causes an enhancement in the polarization. The reasons for the difficulties in stellar flare polarimetry are discussed. Finally, future prospects for the observations of flaring stars and for the utilization of linear polarimetry as a complementary method to other techniques of surface imaging of stellar activity and flares are presented.

1. Introduction

Magnetically active late-type dwarfs and giants have inhomogeneities on their surfaces that cause various observable effects in the spectral lines and light curves. These variations are partly caused by rotational modulation due to more or less permanent inhomogeneities on the stellar surface (i.e., magnetic spots, plages, gaseous or dust disks, gas streams, or 'starpatches'), or changes in the visibility or illumination of the atmosphere of a multiple star.

It is expected that the polarization of the integrated stellar light may change along with the above phenomena. RS CVn stars such as RS CVn (Pfeiffer, 1979), HR 5110 (Barbour and Kemp, 1981), HD 8357 and ER Vul (Liu and Tan, 1987) have shown variations in linear polarization, while most of these close binary stars seem to exhibit constant polarization (see, e.g., Weiler *et al.*, 1978; Liu and Tan, 1987). Most of the BY Dra-type spotted stars and dMe stars (i.e., 'flare stars') have not shown significant intrinsic polarimetric variability (see Efimov and Shakhovskoy, 1972; Pettersen and Hsu, 1981; Clayton and Martin, 1981), while BY Dra itself has been reported to exhibit marginally variable linear polarization (Koch and Pfeiffer, 1976; De Jager *et al.*, 1986).

Ordinary single late-type dwarfs seldom have significant irregular variations in their spectra and they usually show almost constant brightness in the optical region. RS CVn stars, BY Dra stars, FK Com stars, and dMe stars, on the contrary, often show various types of irregularities mixed with rotational modulation. The more irregular changes that sometimes dubiously resemble instrumental effects or variations due to telluric atmosphere are, in many cases, probably caused by transient phenomena qualitatively similar to those on the Sun. During these events, magnetic energy is suddenly released in the form of electromagnetic and particle radiation. The most intense of these phenomena on the Sun are flares that may cause brightening in the whole continuum from X-rays to radio waves, and intensified emission in many spectral lines, such as the chromospheric ultraviolet lines and the Hα line.

Since flares are closely related with magnetic activity and inhomogeneities on the stellar surface, one would expect to also see variations in the polarization during stellar flares, although the essential mechanism is not yet clear. In this report, we shall discuss suggested polarization mechanisms in late-type stars, review polarimetric observations of flaring stars, introduce new observations of YY Gem (Castor C), and finally discuss the interpretation and future prospects of stellar flare polarimetry.

2. Polarization in FK Comae, HD 199178 and Late-Type Dwarfs

The peculiar active GS giant FK Com and another chromospherically active G5 (sub)giant HD 199178 have shown significant variations in the broadband linear polarization (Piirola and Vilhu, 1982; Huovelin et al., 1987). The variations in HD 199178 are apparently close to phase locked polarization (see Figure 1(a)), matching the photometric period of 3.337 days determined by Bopp et al. (1983). The polarimetric variations in HD 199178 are, however, not stable, as shown by recent analysis by Jetsu et al. (1989). The changes are obviously connected with changing photometric variability. Jetsu et al. found significant changes in time-scales of few months in the light curves, as well as in the polarimetry.

FK Comae showed more irregular variability (see Figures 1(b–d)), without clear evidence of rotational modulation. Huovelin et al. (1987) suggest that the large polarization peak in FK Comae in the B band at phase 0.825 (see Figure 1(c)) could be due to a transient phenomenon, such as a flare. Simultaneous observations of other flare indicators were, however, not made.

G- and K-type dwarfs have linear polarimetric variations as well (e.g., Piirola, 1977; Tinbergen and Zwaan, 1981; Huovelin et al., 1985). A few of them also show evidence of low amplitude rotational modulation, as demonstrated by observations of Huovelin et al. (1986, 1989) and Huovelin, Saar, and Tuominen (1988). The most regular variations were found in such stars as the G0 dwarf HD 206860 (HN Pegasi), and the G2 dwarf HD 1835 (9 Ceti), both of which also show low amplitude photometric variability (Blanco, Catalano, and Marilli, 1979; Chugainov, 1980), and rotational modulation of the chromospheric Ca II H and K emission (Noyes et al., 1984). The photometric period of Chugainov for HD 1835 (7.655 days) matches well with the Ca II modulation period (7.7 days), while the photometry of Blanco, Catalano, and Marilli for HD 206860 implies a period of 24.9 days in the V band, contradicting the period of 4.7 days determined by Noyes et al. (1984). A reasonable physical explanation for the discrepancy in the HD 206860 period determinations is difficult to find, if both determinations are correct. The observed polarimetric variations of HD 206860 are demonstrated in Figures 2 and 3 (from Huovelin et al., 1989). The variations match well with the period 4.7 days (Figure 3).

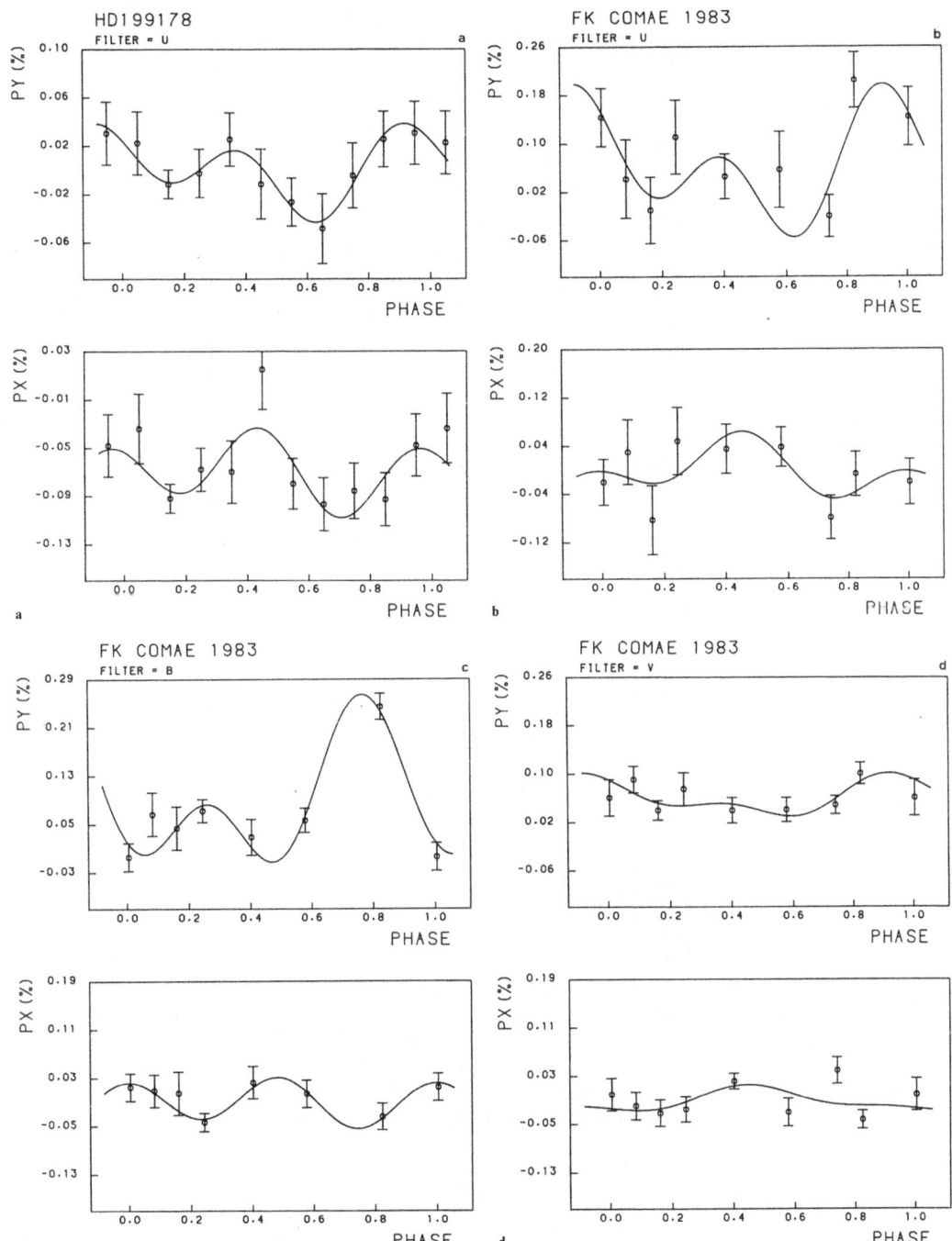

Fig. 1.   Normalized Stokes parameters $P_X = P\cos2\theta$ and $P_Y = P\sin2\theta$ vs photometric phase for HD 199178 in the U band, and for FK Comae in U, B, and V, from Huovelin *et al.* (1987). The continuous lines are second-order Fourier fits, and the error bars give the standard mean errors.

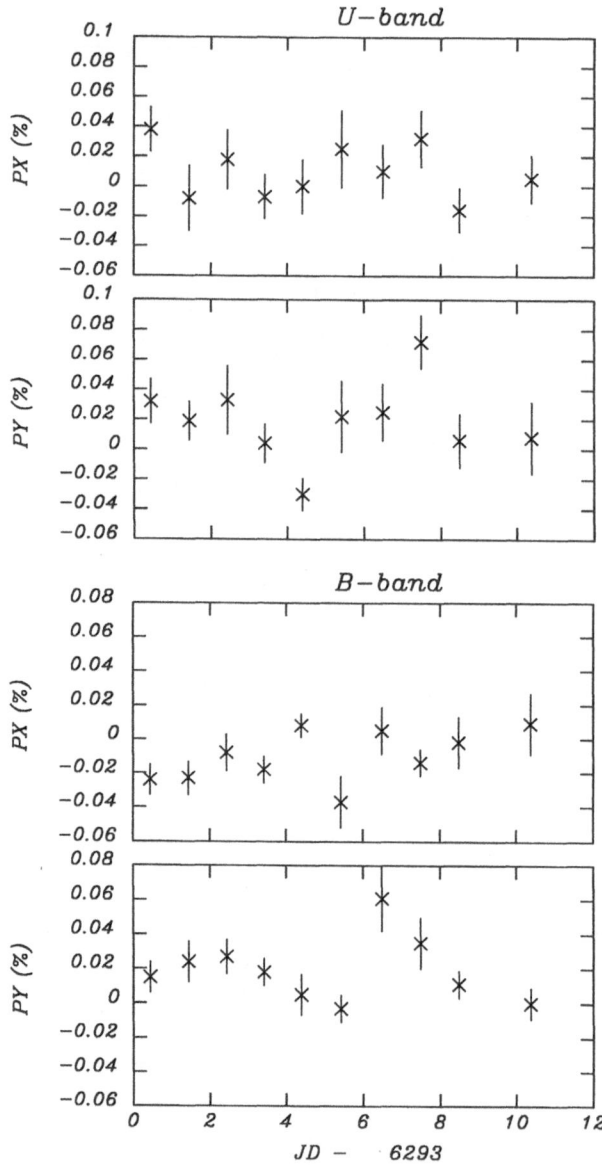

Fig. 2.   Nightly averages of P_X and P_Y vs Julian date ($2\,440\,000 +$) for HD 206860 in the U and B, from Huovelin *et al.* (1989). Errors as in Figure 1.

3. Sources and Mechanisms of Polarization

Linear polarization can generally be expected in stars with inhomogeneities or non-isotropic gas flows. These are (1) magnetic areas, i.e., star spots and plages (chromo-spherically active stars), (2) inhomogeneous circumstellar gas or dust envelopes or disks

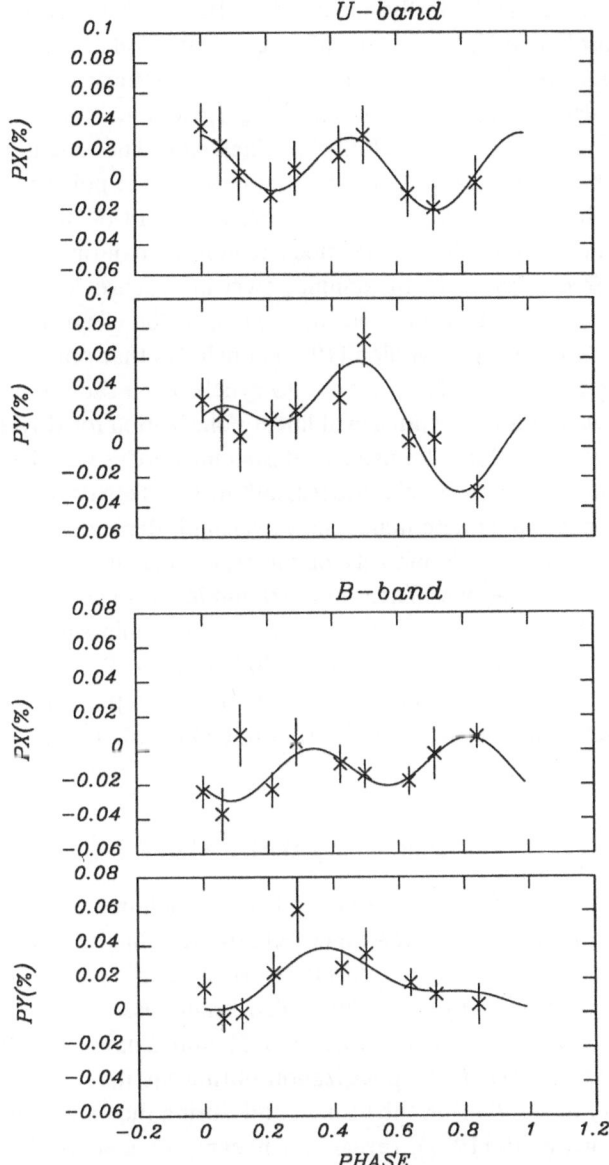

Fig. 3. Observations of Figure 2 as a function of rotational phase, calculated from period 4.7 days. Zero-phase arbitrary. The continuous line is the second-order Fourier fit.

(binaries, giants), (3) gas streams or non-isotropic stellar winds (interacting binaries), and (4) tidal and rotational deformations (rapidly rotating close binaries).

The most popular mechanisms that have been suggested as the source of linear polarization in the above cases are Rayleigh, Thomson, or Mie scattering in an *optically thin* medium (e.g., Shakhovskoy, 1965; Grinin and Domke, 1971; Cassinelli and

Haisch, 1974; Poeckert and Marlborough, 1976; Brown, Mclean, and Emslie, 1978; Pfeiffer, 1979; Piirola and Vilhu, 1982) and magnetic intensification, i.e., the net linear polarization in saturated Zeeman sensitive spectral lines (Warwick, 1951; Leroy, 1962; Kemp and Wolstencroft, 1974; Mullan and Bell, 1976; Tinbergen and Zwaan, 1981; Huovelin et al., 1985, 1986, 1989; Huovelin, Saar, and Tuominen, 1988). Also synchrotron radiation has been proposed as a possible source of polarization, but it would require magnetic fields or electron energies several orders of magnitude larger than observed in the Sun to be effective in the optical region, even during a post-flare outburst of electrons (the frequency of the maximum power in synchrotron radiation increases linearly with the magnetic field and with the square of the electron energy; Ginzburg and Syrovatskii, 1964). Hence, Mullan (1975) concludes that polarization in late-type Main-Sequence stars is most likely *not* due to synchrotron radiation.

The expected wavelength dependence of linear polarization for Rayleigh scattering in optically thin gas is λ^{-4}, while Thomson scattering introduces wavelength independent linear polarization. As for magnetic intensification, the density of saturated Zeeman lines in the given passbands determines the wavelength dependence. It is, therefore, a function of temperature and luminosity of the star. Saar and Huovelin (1989) have calculated models for broadband linear polarization in late-type stars, and their results indicate considerable changes with spectral type and luminosity from a power-law dependence $P \sim \lambda^{-8}$ ($\log g = 4$ and $T_{eff} = 4000$ K), to $P \sim \lambda^{-2.0}$ ($\log g = 3$ and $T_{eff} = 6500$ K) in the degree of linear polarization. A comparison with observations supports these predictions (Huovelin, Saar, and Tuominen, 1988).

4. Polarimetry During Flares

Linear polarization during flares has been monitored simultaneously with the usual flare indicators (Hα, photometry) only a few times. Efimov and Shakhovskoy (1972) observed linear polarization in the blue region during the rising and declining phases of a flare in EV Lac with integration times of 20 s. The brightness of EV Lac increased by 1.27 mag, and the duration of the flare was about 20 min. Efimov and Shakhovskoy did not find significant changes in the polarization during the flare.

The polarimetry of YZ CMi in the visual band during the short (about 4 min) flare, reported by Karpen et al. (1977), reveals no linear polarization above the 2σ level, although the magnitude in blue increased by about 0.8 mag. Also Eritsian (1978) observed a number of flares in EV Lac and AD Leo in the blue and the visual region, finding no significant differences in polarization between the active and the quiescent phases.

Pettersen and Hsu (1981) observed AD Leo before and during a flare in the ultraviolet region (≈ 360 nm), finding the average polarization of $P = 0.9 \pm 0.6\%$ during the preflare phase, and $P = 0.5 \pm 0.4\%$ during the high intensity phase of the flare (averages over about 10–15 min). The observed complex flare in the ultraviolet consisted of 3 maxima with about 5 min separation between the first and the last peak. Pettersen and

Hsu concluded their result as a nondetection of linear polarization during the flare of AD Leo.

De Jager *et al.* (1986) reported broadband *UBVRI* linear polarimetry observed during a flare of 20 min duration in BY Dra. The brightness of BY Dra increased by about 0.25 mag in the ultraviolet (360 nm), about 0.04 mag in the blue (440 nm), and was practically constant in *VRI*. In the ultraviolet, the average degree of polarization was $P = 0.40 \pm 0.12\%$ during the flare, compared with the average over the whole night $P = 0.056 \pm 0.040\%$. The polarization in *BVRI* was considerably smaller, following the decrease of the brightness with increasing wavelength. The 3σ detection in the ultraviolet can be taken as evidence for flare-induced polarization, although De Jager *et al.* carefully avoid calling it conclusive.

5. YY Geminorum

The spectroscopic double-lined binary YY Geminorum is the weakest component of the sextuple star α Geminorum (Castor). YY Geminorum consists of two almost equally massive stars (Struve, Herbig, and Horak, 1950; Struve and Zebergs, 1959; Bopp, 1974), both classified as dMle flare stars, with the Hα and Ca II H and K lines in emission (Moffett and Bopp, 1971; Joy and Abt, 1974). The distortions in the observed light curves suggest bright or dark spots on the surfaces of both components (Kron, 1952). The orbital period of this synchronously rotating system is 0.8142822 days (Van Gent, 1931), with very small eccentricity in the orbit (see Bopp, 1974), and almost equally deep eclipses at phases 0.0 and 0.5, indicating an inclination close to 90°.

The polarimetric observations of Pfeiffer and Koch (1973) showed variability with time, which may indicate the existence of a circumstellar cloud around the system, with scattering as the cause of polarization. The polarization has also shown variations in short time-scales (less than 50 min), which may be due to the spots, or changes in the circumstellar material, perhaps during flares (see Bopp, 1974).

New polarimetric observations of YY Geminorum by the present authors were obtained at the Crimean Astrophysical Observatory on March 5 and 6, 1988 with a five-channel (*UBVRI*) version of the double image chopping photometer-polarimeter described by Piirola (1973, 1975), connected with the 1.25 m optical telescope. The observations were made in five passbands, with effective wavelengths 0.36, 0.44, 0.53, 0.69, and 0.83 μm. The photometry was obtained with the same instrument, simultaneously with the polarimetry.

The Hα spectra were also observed simultaneously with the photopolarimetric observations, using the 2.6 m telescope Coudé spectrograph of the Crimean Observatory (10 spectra in the first night and 9 spectra in the second night).

Among other irregularities, a complex and comparatively small flare with two maxima was observed on March 6, at $\approx 18:00$ UT (phase 0.1 to 0.16). The photometric flare was strongest (about 0.2 mag) in the ultraviolet band, and was simultaneous with a brightness depression possibly caused by a dark spot crossing the stellar disk (see Figure 4(b)). The March 5 observations included the primary minimum (Figure 4(a))

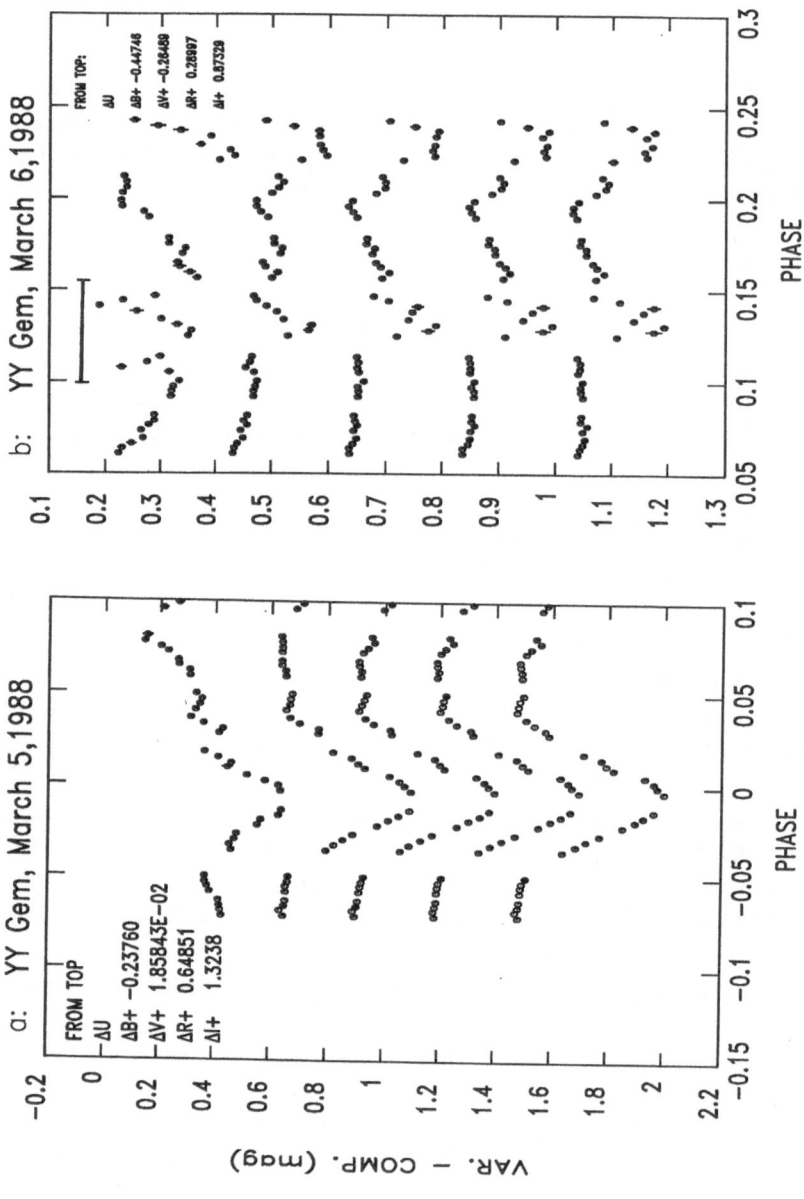

Fig. 4. The photometry of YY Germinorum in standard *UBVRI* system, in (a) March 5 and (b) March 6, 1988. The magnitudes are in the sense (variable − comparison + constant) = (A + const.), where the constant is the vertical shift of each light curve. The ephemeris for the phase is T_0 = JD 2424989.1169 and period 0.8142822 days (van Gent, 1931). The flare is seen in (b) between phases 0.1 and 0.16 (indicated by the horizontal bar).

which showed a small phase shift (≈ 5 min/≈ 60 years), indicating very high stability in the system, and high precision in Van Gent's (1931) period determination.

The simultaneous Hα spectrum showed slightly enhanced emission, as demonstrated by a comparison with another spectrum taken one orbital period earlier, when flares were not seen in the light curves. The equivalent width of the Hα emission during the flare was enhanced by about 250 mÅ (i.e., about 10% of the total emission).

The linear polarization is represented in Figure 5. The observations were grouped into independent phase bins with $\frac{1}{40}$ of a period per bin (about 0.5 hr). Each bin contained 5 to 8 observations, which is quite adequate for reliable error estimates. The polarization variations on March 6 (Figure 5(b)) were the most significant in the ultraviolet (U), while the other bands ($BVRI$) showed variations below 1σ. The degree of polarization during the flare was $P = 0.318 \pm 0.193\%$, which is not significantly different from the nightly average. A slight increase was observed during the post-flare phase 0.15 to 0.175 ($P = 0.634 \pm 0.235\%$), and the polarization was again lower during the phase 0.175 to 0.2 ($P = 0.396 \pm 0.193\%$). Thus, considering the error limits, the above results do not represent conclusive evidence for significant variations in the polarization due to the flare.

There was, however, significant difference in the average polarization on March 6 ($P = 0.335 \pm 0.082\%$) compared with the average of observations on March 5 ($P = 0.635 \pm 0.093\%$). The difference is probably connected with effects due to the

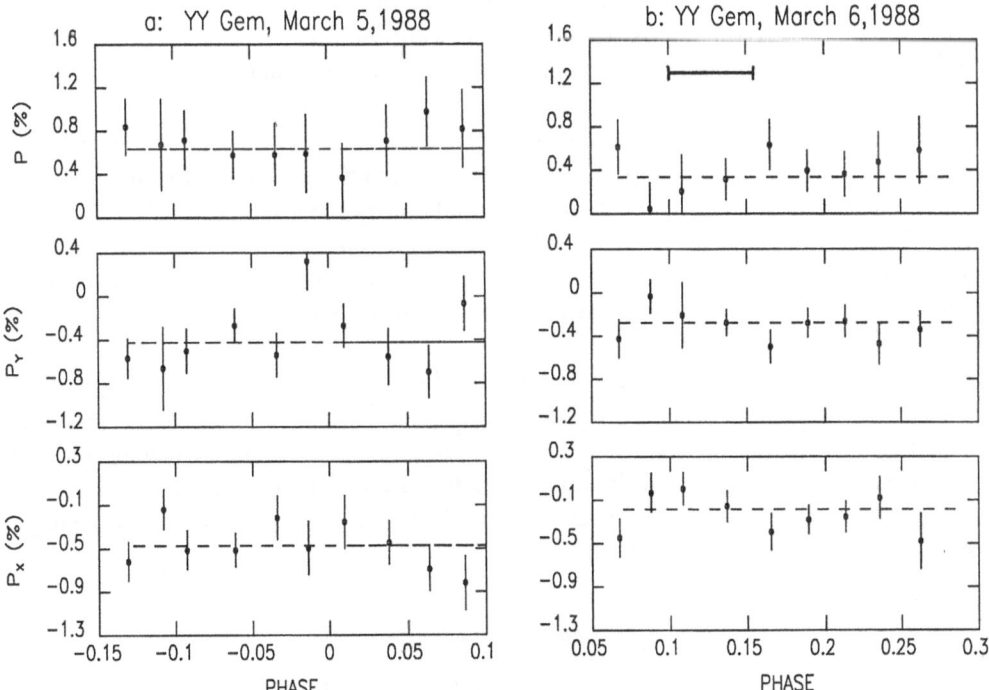

Fig. 5. The Stokes parameters and the degree of polarization in (a) March 5, and (b) March 6. The horizontal bar in (b) shows the time of the flare. The dashed lines show the nightly averages.

eclipse. Perhaps there is a circumstellar envelope or cloud that was asymmetrically illuminated near the time of the eclipse, causing the enhancement of the linear polarization.

6. Conclusions and Discussion

The polarization changes during and after stellar flares seem to be very difficult to detect, as shown by both previous and our new observations. Given the time resolution and observational accuracy, the observed polarization changes are only suggestive. The lack of observational evidence for a polarization enhancement during flares can be explained using the following arguments.

(1) Flares near the stellar disk center are the most intense and are the most easy to observe with photometry and spectroscopy. However, the expected linear polarization due to scattering or magnetic intensification will be small if the flare is observed face-on, since the linear polarization should be zero if the polarizing region is at the line-of-sight from the stellar disk center, while the maximum for simple scattering occurs at 90°, and for magnetic intensification at 45–55° from line-of-sight (Unno, 1955; Saar and Huovelin, 1989). We should, therefore, expect to see higher polarization with flares close to the stellar limb, instead of a hopeless search for a correlation between the flare intensity and the degree of linear polarization.

(2) Flares usually last only from a few to 10 min, causing large photometric variations. Large errors in polarimetric observations inevitably follow from rapidly changing brightness. Very short integration times, on the other hand, are followed by a low level of counts and poor photon statistics. A discussion of these effects is already presented by Efimov (1968), who estimated lower limits for observational errors in flare star polarimetry.

Flares thus require intense monitoring with a fast polarimeter, simultaneously with photometry and/or spectroscopy, to catch flares with various angles to the line-of-sight. The angular dependence of the polarization could then be determined, and a comparison with theoretical models as the source of polarization could be made. Once the mechanism is well determined, polarization observations will yield important complementary information on the geometry and structure of the flares, and on the magnetic fields in the flares.

The results of broadband linear polarimetry demonstrate, however, that it provides a useful complementary method to support spectroscopic surface imaging (Piskunov, Tuominen, and Vilhu, 1988) of spatial and temporal variations of active regions, i.e., the stellar butterfly diagram. Such simultaneous long term observations may give a method to compare the properties of stellar dynamos with those of the solar dynamo (e.g., Brandenburg and Tuominen, 1988).

Acknowledgements

This work has been supported by the Academy of Finland and the exchange program between the Academy of Finland and The Academy of Sciences of the U.S.S.R.

References

Barbour, M. S. and Kemp, J. S.: 1981, *Astrophys. J.* **246**, 203.

Blanco, C., Catalano, S., and Marilli, E.: 1979, *Astron. Astrophys. Suppl.* **36**, 297.

Brandenburg, A. and Tuominen, I.: 1988, *Adv. Space Res.* **7**, 185.

Bopp, W. B.: 1974, *Astrophys. J.* **193**, 389.

Bopp, W. B., Africano, J. L., Stencel, R. E., Noah, P. V., and Klimke, A.: 1983, *Astrophys. J.* **275**, 691.

Brown, J. C., Mclean, I. S., and Emslie, A. G.: 1978, *Astron. Astrophys.* **68**, 415.

Cassinelli, . P. and Haisch, B. M.: 1974, *Astrophys. J.* **188**, 101.

Chugainov, P. F.: 1980, *Izv. Krymsk. Astrofiz. Obs.* **61**, 124.

Clayton, G. C. and Martin, P. G.: 1981, *Astron. J.* **86**, 1518.

De Jager, C., Heise, J., Avgoloupis, S., Cutispoto, G., Kieboom, K., Herr, R. B. *et al.*: 1986, *Astron. Astrophys.* **156**, 95.

Efimov, Yu. S.: 1968, in L. Detre (ed.), 'Non-Periodic Phenomena in Variable Stars', *IAU Colloq.*, p. 169.

Efimov, Yu. S. and Shakhovskoy, N. M.: 1972, *Izv. Krymsk. Astrofiz. Obs.* **45**, 111.

Eritsian, M. A.: 1978, *Soobshch. Byurakansk. Obs. Akad. Nauk Arm. S.S.R.* **50**, 40.

Ginzburg, V. I. and Syrovatskii, S. I.: 1964, *The Origin of Cosmic Rays*, Mac Millan, New York.

Grinin, V. P. and Domke, H.: 1971, *Astrofizika* **7**, 211.

Huovelin, J., Saar, S. H., and Tuominen, I.: 1988, *Astrophys. J.* **329**, 882.

Huovelin, J., Linnaluoto, S., Piirola, V., and Tuominen, I.: 1985, *Astron. Astrophys.* **152**, 357.

Huovelin, J., Linnaluoto, S., Piirola, V., and Tuominen, I.: 1986, in M. Zeilik and D. M. Gibson (eds.), *Cool Stars, Stellar Systems and the Sun*, Springer-Verlag, New York, p. 333.

Huovelin, J., Piirola, V., Vilhu, O., Efimov, Yu. S., and Shakhovskoy, N. M.: 1987, *Astron. Astrophys.* **176**, 83.

Huovelin, J., Linnaluoto, S., Tuominen, I., and Virtanen, H.: 1989, *Astron. Astrophys. Suppl. Ser.* (in press).

Jetsu, L., Huovelin, J., Vilhu, O., Tuominen, I., Linnaluoto, S., Savanov, I., Shakhovskoy, N. M., and Efimov, Yu. S.: 1989, *Astron. Astrophys.* (submitted).

Joy, A. H. and Abt, H. A.: 1974, *Astrophys. J. Suppl.* **28**, 1.

Karpen, J. T., Crannell, C. J., Hobbs, R. W., Maran, S. P., Moffett, T. J., and Bardas, D. *et al.*: 1977, *Astrophys. J.* **216**, 479.

Kemp, J. C. and Wolstencroft, R. D.: 1974, *Monthly Notices Roy. Astron. Soc.* **166**, 1.

Koch, R. H. and Pfeiffer, R. J.: 1976, *Astrophys. J.* **204**, L47.

Kron, G. E.: 1952, *Astrophys. J.* **115**, 301.

Leroy, J.-L.: 1962, *Ann. Astrophys.* **25**, 127.

Liu, X. and Tan, H.: 1987, *Acta Astron. Sinica* **28**, 39.

Moffett, T. J. and Bopp, B. W.: 1971, *Astrophys. J.* **168**, L117.

Mullan, D. J.: 1975, *Astrophys. J.* **201**, 630.

Mullan, D. J. and Bell, R. A.: 1976, *Astrophys. J.* **204**, 818.

Noyes, R. W., Hartmann, L., Baliunas, S. L., Duncan, D. K., and Vaughan, A. H.: 1984, *Astrophys. J.* **279**, 793.

Pettersen, B. R. and Hsu, J.: 1981, *Astrophys. J.* **247**, 1013.

Pfeiffer, R. J.: 1979, *Astrophys. J.* **232**, 181.

Pfeiffer, R. J. and Koch, R. H.: 1973, *Inf. Bull. Var. Stars*, No. 780.

Piirola, V.: 1973, *Astron. Astrophys.* **27**, 383.

Piirola, V.: 1975, *Ann. Acad. Sci. Fennicae* **AVI**, No. 418.

Piirola, V.: 1977, *Astron. Astrophys. Suppl. Ser.* **30**, 213.

Piirola, V. and Vilhu, O.: 1982, *Astron. Astrophys.* **110**, 351.

Piskunov, N., Tuominen, I., and Vilhu, O.: 1989, *Astron. Astrophys.* (submitted).

Poeckert, R. and Marlborough, J. M.: 1976, *Astrophys. J.* **206**, 182.

Saar, S. H. and Huovelin, J.: 1989, *Astrophys. J.* (submitted).

Shakhovskoy, N. M.: 1965, *Soviet Astron. A.J.* **8**, 833.

Struve, O. and Zebergs, V.: 1959, *Astrophys. J.* **130**, 783.

Struve, O., Herbig, G., and Horak, H.: 1950, *Astrophys. J.* **112**, 216.

Tinbergen, N. and Zwaan, C.: 1981, *Astron. Astrophys.* **101**, 223.

Unno, W.: 1956, *Publ. Astron. Soc. Japan* **8**, 108.

Van Gent, H.: 1931, *Bull. Astron. Inst. Neth.* **6**, 99.

Warwick, J. W.: 1951, *Proc. Nat. Acad. Sci.* **35**, 196.

Weiler, E. J., Owen, F. N., Bopp, B. W., Schmitz, M., Hall, D. S., Fraquelli, D. A., Piirola, V., Ryle, M., and Gibson, D. M.: 1978, *Astrophys. J.* **225**, 919.

PARTICLE ACCELERATION IN SOLAR FLARES

LOUKAS VLAHOS

Department of Physics, University of Thessaloniki, 54006 Thessaloniki, Greece

Abstract. Particle acceleration during solar flares is a complex process where the main 'actors' (Direct (D.C.) or turbulent electric fields) are hidden from us. It is easy to construct a successful particle accelertion model if we are allowed to impose on the flaring region arbitrary conditions (e.g., strength and scale length of the D.C. or turbulent electric fields), but then we have not solved the acceleration problem; we have simply re-defined it. We outline in this review three recent observations which indicate that the following physical processes may happen during solar flares: (1) Release of energy in a large number of microflares; (2) short time-scales; (3) small length scales; and (4) coherent radiation and acceleration sources. We propose that these new findings force us to reformulate the acceleration process inside a flaring active region assuming that a large number of reconnection sites will burst almost simultaneously. All the well-known acceleration mechanisms (electric fields, turbulent fields, shock waves, etc.) reviewed briefly here, can be used in a statistical model where each particle is gaining energy through its interaction with many small reconnection sites.

1. Introduction

Particle acceleration (ions, electrons, and nuclei) is an essential part of the 'flare problem'. The other parts of the 'flare problem' are: (1) global structure of the active region and energy storage, (2) energy release (3) interaction of energized plasma with ambient plasma, (4) reaction of the atmosphere, and (5) radiation signatures. All these parts of the 'flare problem' are in fact one complex problem that is very hard to split into independent pieces. On the other hand solving the global problem is a task that goes beyond our instrumental and theoretical capabilities today.

Numerical simulations that are currently very popular 'suffer' from the fact that several aspects of the flare problem demand global MHD modeling (lengths $\geq 10^9$ cm and time-scales $\geq (L/V_A)$ s); others require 'kinetic' modeling ($L \approx c/\omega_{pe} \ll 1$ km, $r \approx$ many $(\omega_{pe}^{-1}, \Omega_e^{-1}) \ll 1$ s). The 'global modeling' imposes dynamic boundary conditions on the 'kinetic' part and the 'kinetic' modeling determines characteristic parameters (e.g., resistivity) for the global modeling. The comments made above are well known but in practice we ignore them and continue our work on the global or local level imposing artificial boundary or local conditions.

The physical process that control particle acceleration lie on the interface of the global and local phenomena. Several theoretical attempts made so far emphasize the global or the local aspect of the acceleration process. Let us mention a few characteristic examples. A number of articles have proposed a D.C. electric field as a possible mechanism for electron acceleration in solar flares (e.g., De Jager, 1986, and references therein). The strength of the electric field is estimated from the linear evolution of the tearing mode (no temporal or spatial dependence is considered). However, it is well known that the appearance of the electric field inside the plasma, as well as its temporal and spatial structure depends critically on the boundary conditions.

Another example of this approach is the acceleration of ions by MHD turbulence (Fermi acceleration) or shock waves. In this work a serious effort is made to match the spectrum of the observed data with the spectrum of the accelerated particles, but little attention is placed on the mechanism that excites the turbulence or drives the shocks (Forman, Ramaty, and Zweibel, 1985); Ramaty and Murphy, 1987, and references therein).

As a final example, consider the 'localized hot spots' or 'conduction fronts'. We have used these concepts for so long, almost a decade, even developing radiation models, but we know so little on the physics of their origin, stability and evolution (but see recent numerical simulation by Winglee, Pritchett, and Dulk, 1988).

A large number of review articles have appeared recently on particle acceleration in solar flares (Heyvaerts, 1981; Chupp, 1984; Vlahos *et al.*, 1986; De Jager, 1986; Forman, Ramaty, and Zweibel, 1985; Ramaty and Murphy, 1987; Sakai and Ohsawa, 1987; Scholer, 1988). These articles review the observed data and the mechanisms that can accelerate charged particles, but they omit the process of fitting to the existing data or the connection to the global energy release processes in solar flares.

In this review we will follow a different approach. First we will discuss a few *key* observational results, which seem to indicate a need for a new thinking on our research in particle acceleration. We will review a number of new theoretical ideas that indicates that the corona is probably full of small magnetic tubes (fibers). The acceleration mechanisms proposed so far will be reviewed *briefly* in Section 4. Finally in Section 5 we will discuss a new acceleration mechanism proposed especially for a fibrous corona and we will close our review with a summary of the current research.

2. Recent Observational Results

A detailed review of the recent observational results related to particle acceleration and recorded by the SMM or Hinotori satellites during the 1980 solar maximum was presented elsewhere (see Vlahos *et al.*, 1986). We will outline below three characteristic observations that, in our opinion, suggest the need for a new way of thinking on particle acceleration and transport in solar flares.

(1) The U.C. Berkeley balloon flight of June 27, 1980 was the first to observe the Sun with high-energy resolution (≤ 1 keV) and sensitivity in the energy range ≥ 20 keV (Lin *et al.*, 1983). They discovered the phenomenon of solar hard X-rays microflares which have peak fluxes $\simeq 10$–100 times less than the normal flares. These bursts occurred one every five minutes on average through the 141 min of solar observations. Although they are associated with small increases in soft X-rays, their spectra are best fit by power-laws which can extend up to ≥ 70 keV. These microflares are thus probably nonthermal in origin. The integral number of events varies roughly inversely with X-ray intensity (Figure 1), so that many more bursts may be occurring with peak fluxes below the limit

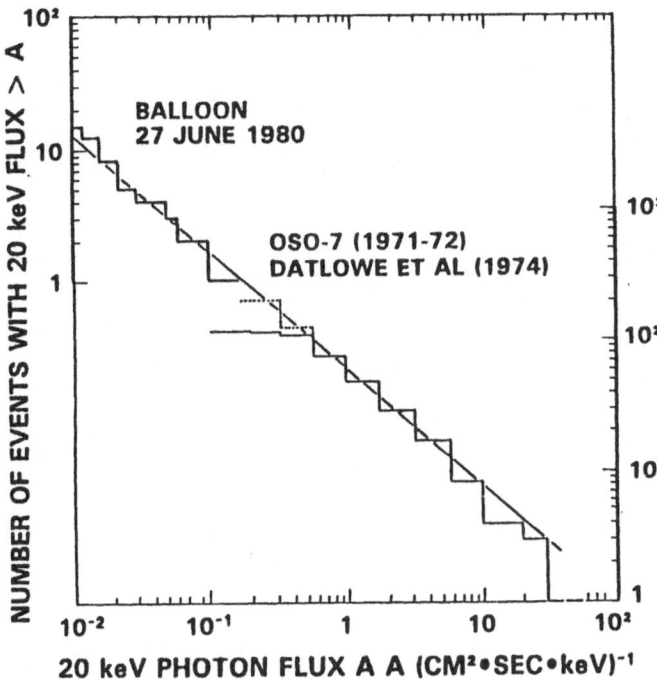

Fig. 1. The distribution of integral number of events versus peak 20 keV photon flux for the solar hard X-ray microflares observed in the balloon flight of June 27, 1980. Als shown for comparison is the distribution of solar flare X-ray bursts reported by Datlowe *et al.* (1974). The distributions have been arbitrarily moved vertically to show that their slopes are approximately the same (from Lin *et al.*, 1983).

of sensitivity. There is also some indication that these bursts may be made up of spikes of $\simeq 1$ s duration (Figure 2). Perhaps these are real 'elementary' bursts, a factor 10^2–10^3 smaller than the elementary flare bursts reported by De Jager and De Jonge (1978). Kaufmann *et al.* (1985) reached a similar conclusion regarding such 'microbursts' in the microwave domain. These hard X-ray 'microflares' and the 'microbursts' indicate the impulsive electron acceleration to above 20 keV energy is very common and may be the primary transient energy release mode in the solar corona.

(2) Spikes of duration less than 100 ms are well known in the 200–3000 MHz band. At meter wave lengths some have been reported near the starting frequency of type III bursts (Benz, Zlobec, and Jaeggi, 1982) at decimeter wave lengths as a part of type IV events (Droge, 1977) and at centimeter wave lengths superimposed on a gradual event (Slottje, 1978). In an analysis of 600 short decimetric events (excluding type IV's), Benz, Aschwanden, and Wiehl (1984) have found 36 events consisting only of spikes. An example of the data is presented in Figure 3 together with a hard X-ray time profile and a blow-up of some single spikes. Benz (1985) analyzed these data and reached the conclusion that the groups of spikes are always associated with groups of metric type III bursts. The spikes tend to occur in the early phase of the type III groups and predomi-

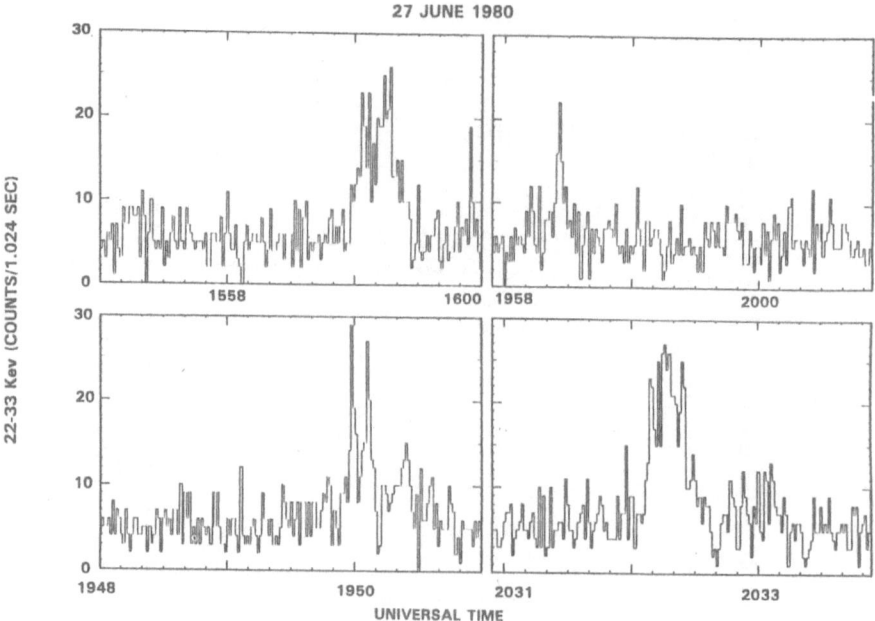

Fig. 2.   The four largest hard X-ray microflares are shown here at 1.024 s resolution (from Lin *et al.*, 1983).

nantly in the rising phase of hard X-rays. The half-power duration of the spikes is less than 100 ms, the time resolution of the instrument used. The spectrum of the spikes has been recorded and the typical half power widths are 3–10 MHz at 500 MHz. This puts a severe constraint on the spectral width of the radio emission and, therefore, on the generation mechanism. Assuming a locally homogeneous corona with a magnetic field scale length of 10 000 km, the source size in the direction of the field gradient must be equal to or less than 100 km. This is less than the limit imposed by time variations. Assuming this dimension for the lateral extent of the source, the lower limit of the brightness temperature is as high as 10^{15} K. The high brightness temperature of short duration (1–100 ms) spikes observed during the impulsive phase of some flares indicates that a *coherent radiation mechanism* is responsible.

(3) Observation of the time dependence of the gamma-ray fluxes from solar flares provided a great deal of information on the acceleration and interaction of the energetic particles. These time dependencies are determined by the temporal structure of the acceleration process, by the lag, due to propagation and trapping, between the acceleration and the interaction of the particles, and by the delay between the interaction of the particles and the emission of photons. Bremsstrahlung and most nuclear line emission are produced essentially instantaneously at the time of the interaction of the particles and, therefore, serve as the best tracers of the time dependencies of the acceleration and interaction processes. Timing studies based on these radiations define the total duration of particle interaction in flares, as well as the overall temporal structure of the emission.

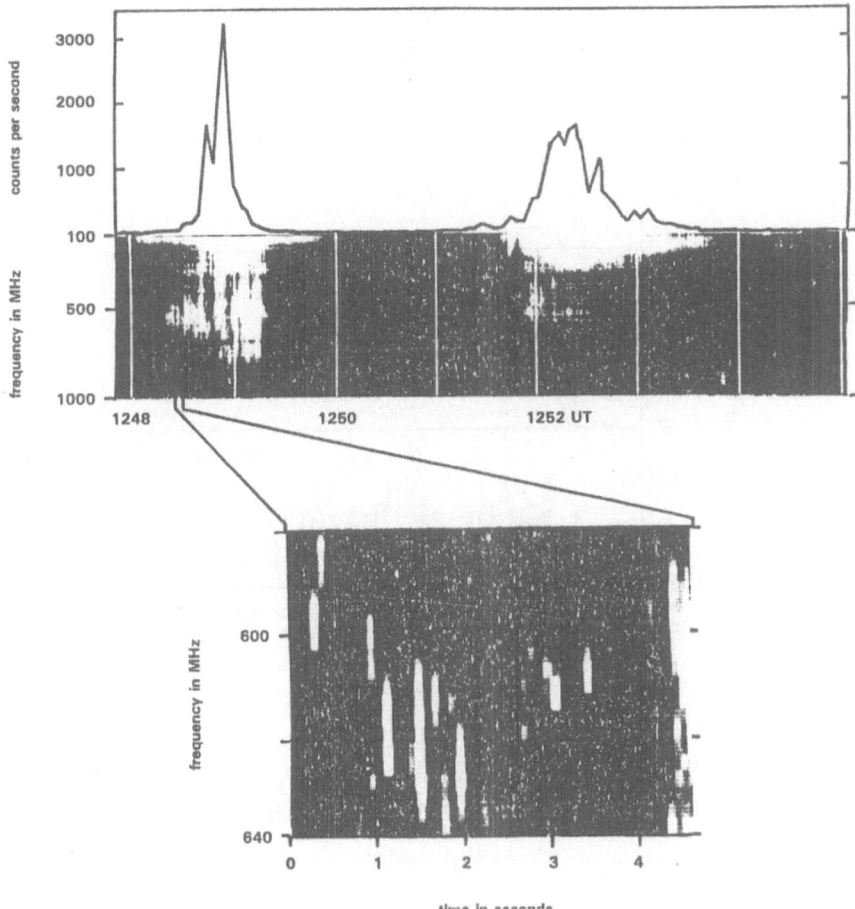

Fig. 3.   *Top*: Composed figure showing hard X-ray counts (> 30 keV, observed by HXRBS on board the Solar Maximum Mission) vs time, of the double flare of August 31, 1980 and radio spectrogram registered by the analog spectrograph at Bleien (Zürich). The spectrogram shows type III bursts at low frequency having starting frequency in correlation with the X-ray flux and spike activity above 300 MHz. *Bottom*: Blow-up of a small fraction of spectrogram produced from data of the digital spectrometer at Bleien (Zürich). The blow-up shows single spikes which are resolved in frequency (from Benz, 1985).

But of particular interest is the temporal relationship between the fluxes in the various energy channels, as these data provide information on the relationship between ion and electron acceleration. The gamma-ray spectrometer γ-ray observations > 0.3 MeV indicates a range of total flare duration from $\simeq 10$ s to over 1000 s. The total emission in the majority of these events consists of at least a few emission pulses. These separate emission pulses can be as short as $\simeq 10$ s and as long as 100 s. It is important to note that in a few events there are no delays detected between the emission in 40–120 keV and 10–25 MeV (Figure 4). Another aspect of the timing studies is the relationship between the starting times of the fluxes in different energy channels. Forrest and Chupp

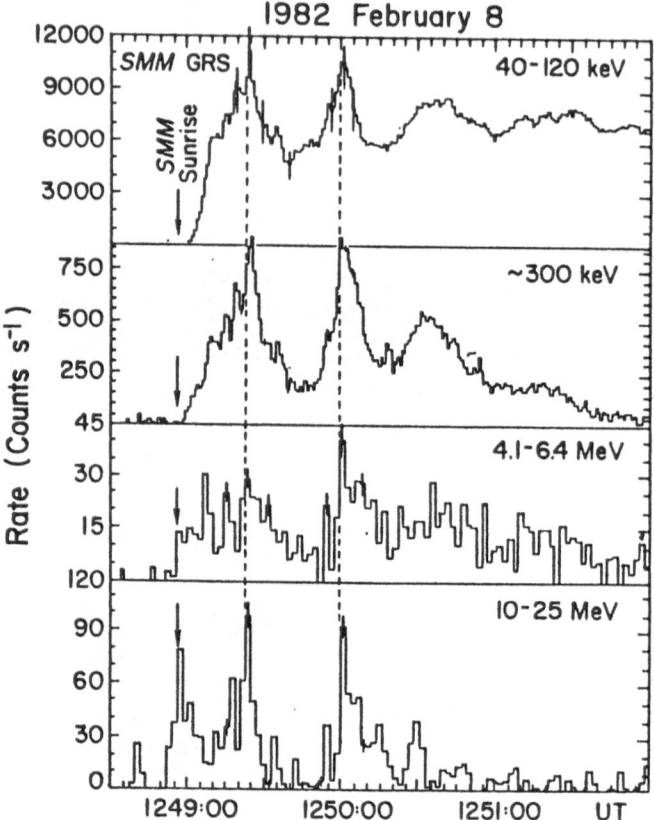

Fig. 4.   The observed time histories in 4 energy bands for a gamma-ray flare (from Chupp, 1984).

(1983) have studied this relationship for the 40–65 keV flux and the 4.1–6.4 MeV flux in two impulsive flares. They found that the starting time, defined as the time when flux above background was first detected, was the same in each energy band within ±2 s for the smaller flare and ±0.8 s for the larger flare, in spite the fact that these flares show evidence for a delay in the maximum of the fluxes of the same two energy bands.

The observations outlined above uncovered the presence of (1) very short time-scales, (2) small scale lengths, (3) large number of explosive phenomena, and (4) intense coherent radiation sources. We propose in this review that these observations are important new findings that should redirect our work on particle acceleration in solar flares.

3. A Model for the Fibrous Corona

Let us accept the idea proposed initially by Parker (1972) that the footpoints of the bipolar fields are subject to random shuffling and mixing, then tangential discontinuities (current sheets) are formed and the amplitude of each discontinuity increases with the passage of time. Eventually a point is reached where rapid reconnection of the magnetic field across the individual discontinuities destroys them as fast as they are created by the motions of the footpoints. We expect the bipolar fields above the surface of the Sun to be filled with small scale reconnection events, i.e., filled with nanoflares, microflares, or flares depending on the rate and total energy released. The spontaneous formation of tangential discontinuities is a peculiar consequence of the static equilibrium properties of the magnetic field imbedded in a infinitely conduction fluid. The discontinuities arise when the field is subjected to continuous but complex deformation, so that the magnetic lines of force are wound and wrapped about each other in complicated patterns (see discussion and references in Parker, 1988; Moffat, 1987; Low and Wolfson, 1988).

In this review we will assume that the active region is full of small (characteristic radius < 10–100 km) magnetic fibers randomly moving with a characteristic velocity $v \simeq 0.5 \text{ km s}^{-1}$. Nanoflares continuously heat the corona but under certain circumstances a large number of dissipation sites are present, increasing the total energy release by thousands or hundreds of thousands times, we will call this a 'flare' (see Figure 5). We propose that acceleration and transport should be re-examined in such an environment. In the next section we will briefly review the physics of the well known acceleration mechanisms and in Section 5 we will use many of these acceleration mechanisms inside a fibrous corona.

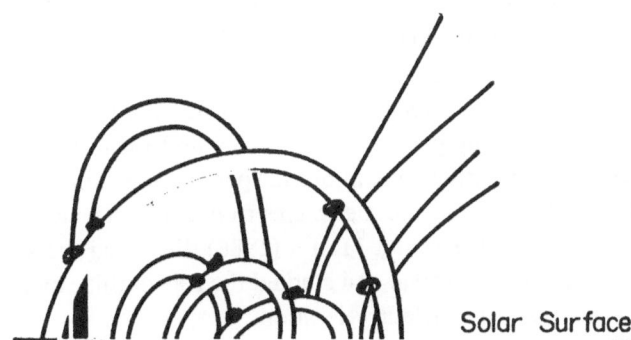

Fig. 5.   A catastrophic interaction of thousands of reconnection layers.

4. Main Acceleration Process

The general concepts of particle acceleration is illustrated by discussing briefly the three generic types of particle acceleration. (a) Coherent acceleration, (b) Fermi (or stochastic) acceleration, (c) shock wave acceleration. We will review briefly the main conclusions from these acceleration processes below.

4.1. COHERENT ACCELERATION

Coherent acceleration can be the result of a D.C. electric field (when the acceleration time is shorter than the time of change of the E-field), or a narrow-band electromagnetic wave. We will discuss these two physical processes separately.

The origin and strength of the electric field in solar flares is not well known. There are at least two possible ways that an electric field will appear in solar flares (i) magnetic reconnection or (ii) double layers.

There have been several attempts (Van Hoven, 1979; Smith, 1980; Heyvaerts, 1981) to estimate the electric field produced by the resistive tearing mode instability but the results disagree. The principal reason for the lack of agreement is that the induced E-field depends critically on the small scale structure of the magnetic field and the transport properties of the instability as it nears the point of saturation, and such nonlinear behavior is poorly known. There are two distinct mechanisms available in a reconnecting field, for accelerating particles, (a) the electric field $\mathbf{E}_0 = \eta J_0 \mathbf{e}_z$ (where η is the resistivity and J_0 is the current) in the tearing layer itself and (b) the $\mathbf{E} = -\mathbf{v} \times \mathbf{B}$ due to the flow velocity $\mathbf{v}$ outside the resistive layer which converts magnetic flux into the tearing layer. The strength of these fields is still an open question. The answer depends sensitively on the time development of the field structure at the reconnection point, which (in turn) depends on the local resistivity and on the external boundary conditions. A number of articles have calculated the detailed changes of local reconnection due to radiation losses and thermal conduction (Steinolfson and Van Hoven, 1984; Van Hoven, Tachi, and Steinoffson, 1984). Bulanov (1986) estimated the E-field from the rapid changes in a magnetic field structure in the course of the breaking of a current sheet, which gives rise to an induced electric field, $E \simeq (V_A/c)B$, where V_A is the Alfvén velocity, c is the speed of light. An *approximate estimate* of the dimensions of the current sheet is given and then the maximum energy gain by the particles and their spectrum is calculated (assuming conservation of particle flux in phase space). Depending on the structure of the magnetic field near the reversal the energy spectrum can be a *power-law or exponential form.* Buchner and Zeleny (1986) and Martin (1986) discussed the stochastization of orbits near the magnetic field reversal and studied the role of this stochastization to the reconnection efficiency and particle acceleration.

All these recent attempts are important steps towards our understanding of particle acceleration near the reconnection sheet, but as we mentioned above, depends so critically on the magnetic structure, the boundary conditions and the time evolution of the resistive instability. Thus we feel that it is not yet easy to construct detailed models based on this acceleration mechanism for solar flares.

Double layers were initially proposed almost twenty years ago by Alfvén and Carlqvist (1967). They assumed that in a current flowing through a plasma, a density depression may rise while the induction of the total circuit is large enough for the current to be maintained. A D.C. electric field must appear to adjust the velocity of the electron flow in such a way that the current density remains constant, $en(x)v(x) = J$. We can easily estimate the potential drop and the energy gained by the particles. It is possible to show rigorously the existence of self consistent solutions of the Vlasov equations able to sustain large potential drops, and able to accelerate electrons and ions to high energies (Block, 1978; Hubbard and Joyce, 1979). More informations for Double layers can be found in Smith (1983). Double layers too need a careful understanding of the large-scale structure circuits in solar flares, as well as the local conditions at the point that the circuit breaks down. Although it is an open question whether double layers are good candidates for particle acceleration, the whole subject should be re-examined for a fibrous corona since the conditions for double-layer formation are easier to achieve inside the fiber due to the stronger current localization.

The presence of an electric field inside the plasma (independent of its origin) is a subject that needs careful study. It is well-known that if the electric field is less than the Dreicer field, a small fraction of electrons $(n_r/n_0) \cong 0.5 \exp(-E_D/2E)$, where n_r is the number of density of the runaway particles, n_0 is the ambient density, and E_D is the Dreicer field, will run away. In the absence of a magnetic field (or if $\omega_{pe} \gg \Omega_e$) the energy gained by the runaway particles will be limited only by the scale length of the potential drop. In the presence of a magnetic field the scenario of the runaway particles changes since the electrons can excite an instability (the anomalous Doppler resonance instability) which scatters the electrons perpendicular to the magnetic field direction. The final result is that the tail will be isotropized and eventually thermalized. Moghaddam-Taaheri et al. (1985) and Moghaddam-Taaheri and Vlahos (1987) studied the evolution of the runaway tail as a function of electric field and found that for $E_\parallel < 0.2 E_D$ the anomalous Doppler resonance scattering is weak and the tail is possible to be accelerated to very high energies.

If the electric field exceeds the E_D inside the plasma the whole distribution will runaway and drive currents. Depending on the details of the ambient plasma parameters a number of current driven instabilities can be excited (see Heyvaerts, 1981, for a catalog of the potential instabilities and the necessary conditions for their excitation). Spicer (1983) and Holman (1985) have discussed the difficulties that arise when we attempt to accelerate all the necessary electrons for a hard X-ray burst from a single potential drop inside the flaring plasma.

We discussed earlier the discovery of intense narrow-band, highly polarized microwave spikes and their excellent correlation to X-rays and type III bursts. These observations have been interpreted as the signature of unstable loss-cone type electron distributions, formed inside a flaring magnetic loop (Holman, Kundu, and Eichler, 1983; Melrose and Dulk, 1982; Sharma, Vlahos, and Papadopoulos, 1982; Vlahos, Sharma, and Papadopoulos, 1983). Sprangle and Vlahos (1983) and Karimabadi et al. (1987) studied the interaction of coherent electromagnetic waves with the ambient plasma. It is well known that the relativistic cyclotron frequency and the wave phase change in such a way that the resonance between the electrons and the wave is maintained in a uniform magnetic field (Kolomenskii and Lebedev, 1963; Roberts and Buchsbaum, 1964) and the particles gain energy continuously. Karimabadi et al. (1987) extended those calculations to oblique em waves and showed that the fundamental and second harmonic will accelerate electrons to high energies and the acceleration is virtually unchanged when a wave with finite bandwidth is considered.

4.2. Fermi or (stochastic) acceleration

Stochastic acceleration of particles in turbulent fields is defined as the process that causes particles to change their energy in a random manner with many increases and decreases that lead finally to acceleration. Stochastic acceleration of ions or nuclei can result from Alfvén waves with wave lengths of the order of particle gyroradius. Alfvén waves will propagate parallel and antiparallel to the magnetic field directions. Electrons can also be accelerated stochastically by lower hybrid waves (Lampe and Papadopoulos, 1987; Benz and Smith, 1987). Formally stochastic acceleration is described as the solution of the diffusion equation in momentum space:

$$\frac{\partial f}{\partial t} = \frac{1}{p^2} \frac{\partial}{\partial p} p^2 D_{pp} \frac{\partial f}{\partial p} ,$$

where D_{pp} is the diffusion coefficient in the momentum space. The diffusion coefficient can be estimated from the wave spectrum. The solution of the diffusion equation for the particle distribution (Tverskoi, 1967) can be expressed in terms of modified Bessel functions for $E \ll mc^2$ (non-relativistic protons) or an exponential for $E \ll mc^2$ (a case relevant to electrons). Acceleration takes place only when the particles move with velocities equal or larger than the Alfvén velocity (the velocity of the moving scatterers). Assuming that the particles stay inside the acceleration volume for a finite time T, a series of characteristic spectra will be formed for the accelerated particles and can be used for fitting the observed data (Ramaty and Murphy, 1987, and references therein).

Melrose (1983) has demonstrated that magnetoacoustic turbulence with frequency $\omega \approx 30 \text{ s}^{-1}$ can accelerate ions from 100 keV to 30 Mev in ≈ 2 s if $(\delta B/B)^2 \approx 0.1$. Ambrosiano et al. (1989) followed particle inside a *turbulent reconnecting magnetic field* and found that particles are accelerated to very high energies. This turbulent neutral point mechanism includes both coherent and stochastic components of acceleration. Turbulence appears to influence the acceleration in several ways: it enhances the reconnection electric field while producing a stochastic electric field that gives rise to momentum diffusion; it also produces magnetic irregularities that trap test particles in the D.C. strong reconnection electric fields for times comparable to the magnetofluid characteristic time.

The main problem with stochastic acceleration in solar flares is that very little is known on the processes that generate the turbulent spectra used in the diffusion equation.

4.3. Shock wave acceleration

The theory of particle acceleration in shock waves combines coherent and stochastic elements. If we assume that there is no wave activity upstream and downstream of the shock then the main acceleration mechanism is the drift of ions and electrons along the $E = -V_s \times B$ field, where B is the value of the magnetic field and V_s is the upstream flow velocity as it is measured by an stationary on the shock frame. The shock frame is the frame moving with the shock discontinuity. Examples of ion motion in the shock

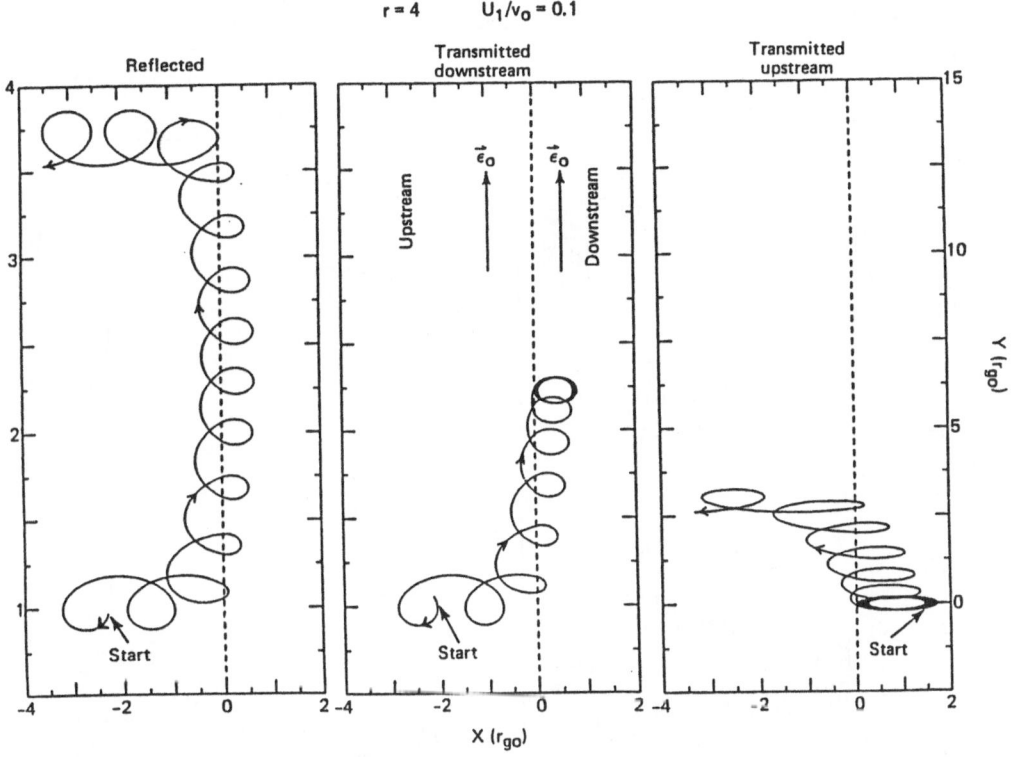

Fig. 6.   Sample particle trajectories in shock frame at quasi-perpendicular shock with $\theta_1 = 80°$ (Decker, 1989).

frame are shown in Figure 6. It is obvious that when $V_s \parallel B$ (parallel shock) the electric field approaches zero and drift acceleration is not important.

If the upstream and downstream plasma is turbulent, then ions and electrons are scattered. If we view the picture in the shock frame and assume that $V_s \parallel B$, then an energetic particle crossing the shock from upstream will be scattered back upstream since the randomly moving turbulence upstream have now a systematic velocity (the downstream fluid velocity) away from the shock. This scattering will change the energy (ε) of particle the by $(V_2/c)\varepsilon$ (V_2 is the downstream fluid velocity). The particle will then move backward cross the shock again and propagate upstream gaining $(V_1/c)\varepsilon$ (V_1 is the upstream fluid velocity), the total energy gain is $(\frac{3}{4})V_s\varepsilon$ if the upstream velocity is V_s and the downstream for a strong shock $(\frac{1}{4})V_s$. In other words the parallel shock organized the upstream and downstream turbulence such that the rate of energy is the first-order power of the turbulent velocity. This is in contrast with the stochastic acceleration which is proportional to the second-order power in V_s.

Shock wave acceleration includes elements of coherent and stohastic acceleration processes. This coupling of coherent and stochastic processes is more pronounced in oblique shocks (see Decker and Vlahos, 1986; Decker, 1989). Oblique shocks have $E \neq 0$ in the shock transition and turbulence scatter particles back and forth (see Figure 7).

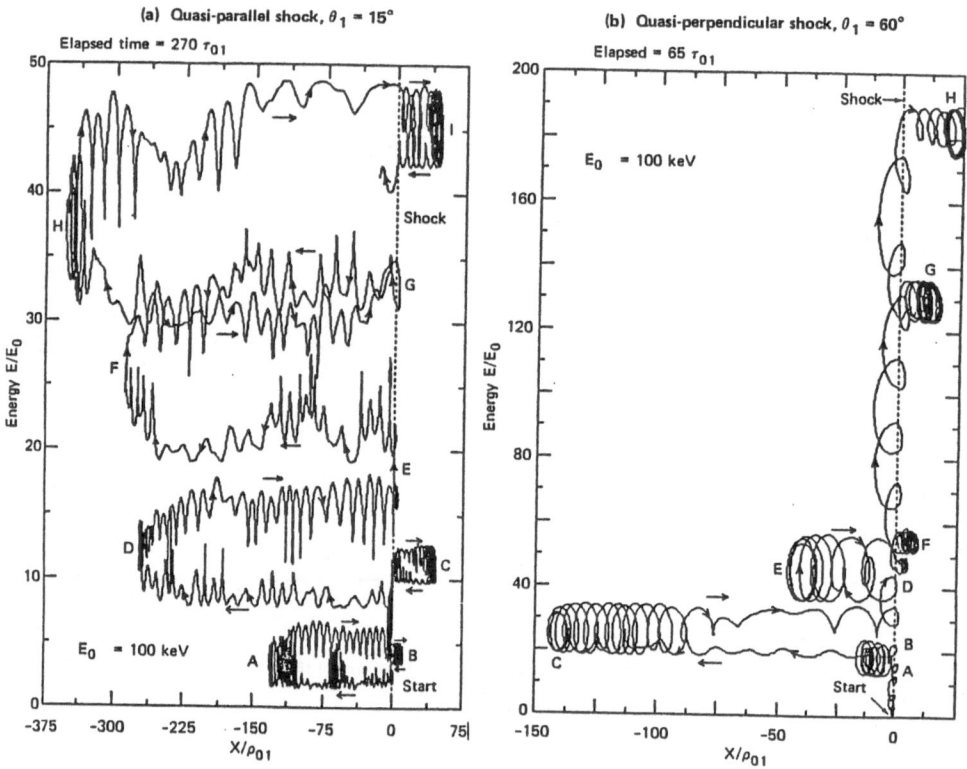

Fig. 7.   (a) Sample particle orbit showing multiple shock encounters at a quasi-parallel ($\theta_1 = 15°$) shock. Shown is the energy E/E_0 ($E_0 = 100$ keV) versus particle's x-coordinate in shock frame. Elapse time is 270 upstream gyroperiods. (b) Sample particle orbit at a quasi-perpendicular ($\theta_1 = 60°$) shock. Elapsed time is 65 upstream gyroperiods (from Decker and Vlahos, 1986).

Shock wave acceleration in solar flares have been used extensively (see Ramaty and Murphy, 1987) since it is possible to accelerate electrons and ions quickly and form the observed spectrum assuming that we know how to form shock waves and turbulence with the 'correct' characteristics (wave energy, wave spectrum, etc.). In solar flares shock waves have been associated with acceleration of particle in the upper corona since they were related to metric and decametric type II bursts.

In Section 6, acceleration in solar flares proceeds in coherent, stochastic or mixed mode inside reconnecting sheets, shock waves or double layers. Although here is an abundance of ideas on particle acceleration, none of them is sitting on firm ground yet. For this reason a number of flare models appeared recently using extensively 'black boxes' in places where we sunspect that particle acceleration is possible.

5. Acceleration in a Fibrous Corona

Assuming that an active region is a collection of small fibers then a 'flare' is the sudden release of energy in many small regions inside the energy release volume. We will show below that in that environment the coherent and stochastic elements of particle acceleration take another interesting twist.

Let us start with the assumption that inside an active region many sudden releases of energy appear almost simultaneously in many different spots (Figure 5). We also assume that the released energy is going primarily into heating.

Cargill, Goodrich, and Vlahos (1988) studied numerically the evolution of such a 'hot spot'. They used a hybrid numerical code (Winske, 1985) to follow the evolution of heated plasma. They found that once $\beta = 8\pi p/B^2$ exceeds unity a shock wave is formed moving away from the hot spot (Figure 8). The shock is formed in time scales $\ll 1$ s. Initially the formation of quasi-perpendicular shocks was studied, but recently Cargill (1989) extended these calculations in the formation of parallel shock waves. The formation time for parallel shocks is longer ($\simeq 100\ \Omega_i^{-1}$) but still much less than one second.

The formation of a large number of shock waves inside the active region initiates a number of important processes. (a) Each shock wave can be an efficient and fast accelerator (see Decker and Vlahos, 1987, and references therein). (b) Once a particle escapes from each shock, it is possible to continue gaining energy from a neighboring shock. Electrons and ions have now a 'mean-free path' for their interaction with a large number of shock waves (Toptyghin, 1980).

Even if the shock-particle interaction is coherent (e.g., shock drift) the N-shocks-particle interactions are stochastic. The important difference between this processes and the classical Fermi acceleration is the fact that the shock-particle interaction is much more efficient than wave-particle interaction. (c) Shock waves interact among themselves (Cargill et al., 1986; Cargill and Goodrich, 1987) and the result from their interaction is strong heating and acceleration of a small number of particles. Colliding shocks depart from their collision point with less energy and after many collisions will disappear.

Adding the two effects together, even if we start with localized 'heating' the N-shocks-particle interaction and the shock-shock interactions will heat and/or accelerate particles in a large volume.

Depending on the mean-free path for shock-particle or shock-shock collisions the energy release volume will end up as a large 'hot spot' (if shock-shock collisions are the dominate process) or 'a hot spot with a large number of accelerated particles' (if the N-shock-particle interactions are the dominant process). Thus we conclude, thermal or non-thermal flares can be produced in such environment, depending on the ratio of the characteristic mean-free paths mention above.

In summary heating, jets of fluid plasma and acceleration of a few electrons around the reconnecting sheet are coupled with the global heating and acceleration through the formation of many of shock waves. We feel that this approach, dictated by the

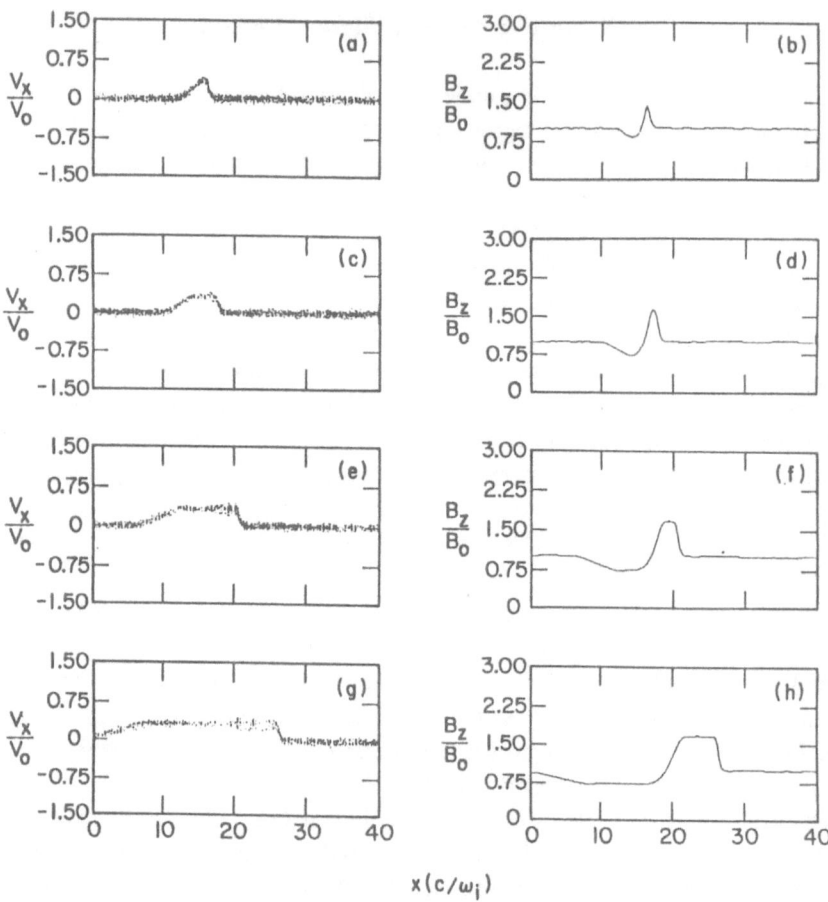

$$x(c/\omega_i)$$

Fig. 8.   The formation of a perpendicular shock due to a hot electron plasma. (8a, c, e, g) are (v_x, x) phase space at $t = 0.9, 1.8, 3.6$, and $7.2\ \Omega_i^{-1}$, respectively. (8b, d, f, h) is the magnetic field at the same times. The velocities in the phase space plots are normalized to v_0, the lengths to an ion inertial length (c/ω_i) of the initial plasma and the magnetic field to its initial value. In this case, v_0 is defined as $v_0^2 = 2kT_e/m_i$ (from Cargill, Goodrich, and Vlahos, 1988).

observational data (hard X-rays, microflares, microwave spikes, and fast acceleration of ions) and the current theoretical understanding of the evolution of active regions combine almost all the elements of particle acceleration processes mentioned in Section 4 but places them in different environment; the fibrous corona.

6. Summary

When computer companies start the development of a large software program (a new compiler or a new word processor) they split the construction of this program in tenths or sometimes hundreds of smaller jobs. When these partial jobs are finished they put them together in order to complete the initial software. Sometimes the problems rising during the final welding of all these individual efforts are as big as the construction of the initial program without splitting it. We have almost reached such a state in solar physics. Many different pieces of information processed locally, or developed as local models, are in fact part of the same global problem. Our effort to piece together models for energy release, acceleration, transport, radiation, etc., at present looks as difficult as the problem we initially started to work on. In my opinion it is important to realize that we are not going to make any real progress without a strategy that connects the analysis made of the local parts of this admittedly fundamental, but extremely hard problem, with the global problem.

Our main conclusions in this review are:

(1) Observations suggest that microbursts are possibly a fundamental process in solar flares.

(2) We have evidence for generation of coherent electromagnetic radiation in solar flares.

(3) We have evidence that particles are accelerated to all energies almost simultaneously.

(4) Global modeling of the solar active region suggest that bipolar fields above the surface of the Sun are filled with small scale reconnection events.

(5) A number of mechanisms exist that can accelerate electrons, ions or nuclei to high energies. The main problems are: how these mechanisms are connected to the energy release process and how they will accelerate the number of particles that are necessary.

(6) We have proposed that new observational and theoretical work demand now a different approach in solar flare physics. We must start deemphasizing the single loop model, or the huge single reconnection sheet model, etc., and move on to statistical ensembles for large numbers of energy releases and acceleration sites. We feel that a step in that direction is the proposal that N-shocks-particle and shock-shock interactions discussed briefly in this review. Although these statistical models have not been tested yet by observations, much information collected during the last solar maximum may prove helpful on this count.

Acknowledgements

I am in debt to Dr P. Cargill for presenting this talk on my behalf, reading and commenting on the first draft and collaborating with me in several of the topics presented in this review. Discussions I had had with Prof. K. Papadopoulos, Dr R. Decker, Dr P. Sprangle, Mr T. Anastasiadis and the members of the particle acceleration team during the SMM workshop, improved my understanding on particle acceleration.

References

Alfvén, H. and Carlqvist, P.: 1967, *Solar Phys.* **1**, 220.
Ambrosiano, J. J. Matthaeus, D., Goldstein, M. L., and Plante, D. R.: 1989, *J. Geophys. Res.* (in press).
Benz, A. O.: 1985, *Solar Phys.* **96**, 357.

Benz, A. O. and Smith, D.: 1987, *Solar Phys.* **107**, 299.

Benz, A. O., Zlobec, P., and Jaeggi, M.: 1982, *Astron. Astrophys.* **109**, 305.

Benz, A. O., Ashwanden, M. J., and Wiehl, J. J.: 1984, *Decimetric Radio Emission During Solar Flares*, Kunming Workshop Proceedings.

Block, L. P.: 1978, *Astrophys. Space Sci.* **55**, 59.

Buchner, J. and Zeleny, L. M.: 1986, *Proceedings of the Joint Varena–Abastumani International school and Workshop on 'Plasma Astrophysics'*, ESA SP-251, p. 195.

Bulanov, S. V.: 1986, *Proceedings of the Joint Varena–Abastumani International School and Workshop on 'Plasma Astrophysics'*, ESA SP-251, p. 185.

Cargill, P. J.: 1989, *J. Geophys. Res.* (submitted).

Cargill, P. J. and Goodrich, C. C.: 1987, *Phys. Fluids* **30**, 2504.

Cargill, P. J., Goodrich, C. C., and Vlahos, L.: 1988, *Astron. Astrophys.* **189**, 254.

Chupp, E. L.: 1984, *Ann. Rev. Astron. Astrophys.* **22**, 359.

Datlowe, D. W., Elcan, M. J., and Hudson, H. S.: 1974, *Solar Phys.* **39**, 127.

Decker, R. B.: 1989, *Space Sci Rev.* (in press).

Decker, R. D. and Vlahos, L.: 1986, *Astrophys. J.* **306**, 710.

De Jager, C.: 1986, *Space Sci. Rev.* **44**, 43.

De Jager and De Jonge, G.: 1978, *Solar Phys.* **58**, 127.

Droge, F.: 1977, *Astron. Astrophys.* **57**, 285.

Forrest, D. J. and Chupp, E. L.: 1983, *Nature* **305**, 291.

Forman, M. A., Ramaty, R., and Zweibel, E. G.: 1985, in P. A. Sturrock (ed.), *The Physics of the Sun*, D. Reidel Publ. Co., Dordrecht, Holland, p. 249.

Heyvaerts, J.: 1981, in E. R. Priest (ed.), 'Particle Acceleration in Solar Flares', *Solar Flare Magnetohydrodynamics*, Gordon and Breach, New York, p. 429.

Holman, G.: 1985, *Astrophys. J.* **293**, 584.

Holman, G., Kundu, M. R., and Eichler: 1983, in M. R. Kundu and T. E. Gergely (eds.), 'Radio Physics of the Sun', *IAU Symp.* **86**, 457.

Hubbard, R. F. and Joyce, C.: 1979, *J. Geophys. Res.* **84**, 4297.

Karimabadi, H., Menyuk, C. R., Sprangle, P., and Vlahos, L.: 1987, *Astrophys. J.* **316**, 462.

Kolomenskii, A. A. and Lebedev, A. N.: 1963, *Soviet Phys. Dokl.* **7**, 745.

Kaufmann, P. *et al.*: 1985, *Solar Phys.* **95**, 155.

Lampe, M. and Papadopoulos, K.: 1977, *Astrophys. J.* **212**, 886.

Lin, R. P., Schwartz, R. A., Kane, S. R., Pelling, R. M., and Harley, K. C.: 1983, *Astrophys. J.* **251**, L109.

Low, B. C. and Wolfson, R.: 1988, *Astrophys. J.* **324**, 574.

Martin, R. F.: 1986, *J. Geophys. Res.* **91**, 11985.

Melrose, D. B.: 1983, *Solar Phys.* **89**, 149.

Melrose, D. B. and Dulk, G. A.: 1982, *Astrophys. J.* **259**, 844.

Moffat, H. K., 1987, in G. Comte-Bellot and J. Mathieu (eds.), *Advances in Turbulence*, Springer-Verlag, p. 240.

Moghaddam-Taaheri, E. and Vlahos, L.: 1987, *Phys. Fluids* **30**, 3155.

Moghaddam-Taaheri, E., Vlahos, L., Rowland, H. L., and Papadopoulos, K.: 1985, *Phys. Fluids* **28**, 3034.

Parker, E.: 1972, *Astrophys. J.* **174**, 499.

Parker, E.: 1988, *Astrophys. J.* **330**, 474.

Ramaty, R and Murphy, R. J.: 1987, *Space Sci. Rev.* **45**, 213.

Roberts, C. S. and Buchsbaum, S. J.: 1964, *Phys. Rev.* **A135**, 381.

Sakai, Y. and Ohsawa, J.: 1987, *Space Sci. Rev.* **46**, 113.

Sharma, R. R., Vlahos, L., and Papadopoulos, K.: 1982, *Astron. Astrophys.* **112**, 377.

Scholer, M.: 1988, in O. Havnes *et al.* (eds.), *Activity in Cool Star Envelopes*, D. Reidel Publ. Co., Dordrecht, Holland, p. 195.

Slottje, C.: 1978, *Nature* **275**, 520.

Smith, D. F.: 1980, *Solar Phys.* **66**, 135.

Smith, R., 1983, in M. R. Kundu and G. Holman (eds.), 'Unstable Current Systems and Instabilities in Astrophysics', *IAU Symp.* **107**, 113.

Spicer, D. S.: 1983, *Adv. Space Res.* **2**, 135.

Sprangle and Vlahos, L.: 1983, *Astrophys. J.* **273**, L95.

Steinolfson, R. S. and Van Hoven, G.: 1984, *Astrophys. J.* **276**, 391.

Toptyghin, I. N.: 1980, *Space Sci Rev.* **26**, 157.
Tverskoi, B. A.: 1967, *Soviet Phys. JEPT* **25**, 317.
Van Hoven, G.: 1979, *Astrophys. J.* **232**, 572.
Van Hoven, G, Tachi, T., and Steinolfson, R. S.: 1984, *Astrophys. J.* **280**, 391.
Vlahos, L., Sharma, R. R., and Papadopoulos, K.: 1983, *Astrophys. J.* **275**, 374.
Vlahos, L. *et al.*: 1986, *Energetic Phenomena on the Sun*, NASA CP-2439, p. 2–1.
Winglee, R. M., Pritchett, P. L., and Dulk, G. A.: 1988, *Astrophys. J.* **327**, 968.

CYCLOTRON LINES IN THE SPECTRA OF SOLAR FLARES AND SOLAR ACTIVE REGIONS

V. V. ZHELEZNYAKOV and E. YA. ZLOTNIK

Institute of Applied Physics, Academy of Sciences of the USSR, Gorky, U.S.S.R.

Abstract. It was shown by Zheleznyakov and Zlotnik (1980a, b) that in complex configurations of solar magnetic fields (in hot loops above the active centres, in neutral current sheets in the preflare phase, in hot X-ray kernels in the initial flare phase) a system of cyclotron lines in the spectrum of microwave radiation is likely to be formed. Such a line was obtained by Willson (1985) in the VLA observations at harmonics of the electron gyrofrequency. This communication interprets these observations on the basis of an active region model in which thermal cyclotron radiation is produced by hot plasma filling the magnetic tube in the corona above a group of spots. In this model the frequency of the recorded 1658 MHz line corresponds to the third harmonic of electron gyrofrequency, which yields the magnetic field (196 ± 4) G along the magnetic tube axis. The linewidth $\Delta f/f \sim 0.1$ is determined by the 10% inhomogeneity of the magnetic field over the cross-section of the tube; the line profile indicates the kinetic temperature distribution of electrons over the tube cross-section with the maximum value 4×10^6 K. Analysis shows that study of cyclotron lines can serve as an efficient tool for diagnostics of magnetic fields and plasma in the solar active regions and flares.

1. Introduction

Solar microwave emission originating from bremsstrahlung and cyclotron emission of electrons in the inhomogeneous magnetoactive plasma of the solar corona usually has a rather flat frequency spectrum. However, in complicated systems of solar magnetic fields the spectra can have increased complexity, so that under definite conditions they can exhibit fine structure in the form of narrow-band features. A theoretical analysis by Zheleznyakov and Zlotnik (1980a, b) indicates that separate cyclotron lines and high-frequency cutoffs can be resolved in the microwave solar radio emission at cyclotron harmonics. Cyclotron features can form in the sources with different types of kinetic temperature and magnetic field distributions, such as neutral current sheets, regions where the magnetic field along the line-of-sight has a maximum at a certain point and in the magnetic flux tube filled with 'hot' electrons. The frequency spectrum and polarization are specific for each type of distribution. This permits diagnostics of active and preflare regions by observing the form of the fine structure of the microwave spectra.

Detection of cyclotron features and investigation of their source requires two-dimensional images at a number of closely-spaced wavelengths. Willson (1983) was the first who reported the possible detection of thermal cyclotron lines from solar active regions. Indirect evidence for the presence of a cyclotron line in the active region spectrum were obtained by Schmahl *et al.* (1984). Finally, a cyclotron line-like spectrum was observed by Willson (1985) using the VLA at 10 closely-spaced frequencies near 20 cm. Interpretation of this line and diagnostics of magnetic fields in the coronal plasma by its characteristics are proposed by Zheleznyakov and Zlotnik (1988) and in this communication.

Solar Physics **121** (1989) 449–456.

2. Inhomogeneous Magnetic Field Model

According to Willson (1985), a source was situated above a large bipolar group of spots with the linear dimension $2b = 8 \times 10^9$ cm ($\sim 100''$) and a magnetic field $B = 2000$ G at the photosphere. The radio source dimension was $50''$ to $100''$ (the interferometer beamsize was $3'' \times 4''$), the maximum brightness temperature, $T_{b_{max}}$, was approximately 4×10^6 K at the background temperature $T_c \sim (1.0-1.5) \times 10^6$ K. The observations were carried out at 10 closely-spaced frequencies between 1440 and 1720 MHz. The frequency spectrum at a chosen point on the source for two days is given in Figure 1. The cyclotron linewidth $\Delta f/f$ was about 0.1 and the line centre was at 1658 MHz. Polarization of the radiation was not observed (within an accuracy of 15%).

The observational data were interpreted by Willson (1985) on the basis of an optically thin source in the thermally-homogeneous model with a constant magnetic field. As will be shown below, the proposed model cannot explain simultaneously the observed line profile, the stability of the parameters and the absence of polarization. There are no such difficulties if the source is optically thick for ordinary and extraordinary modes. The brightness temperature is equal to the kinetic temperature of the plasma in the source, the polarization becomes weak and the line profile and the line width are determined

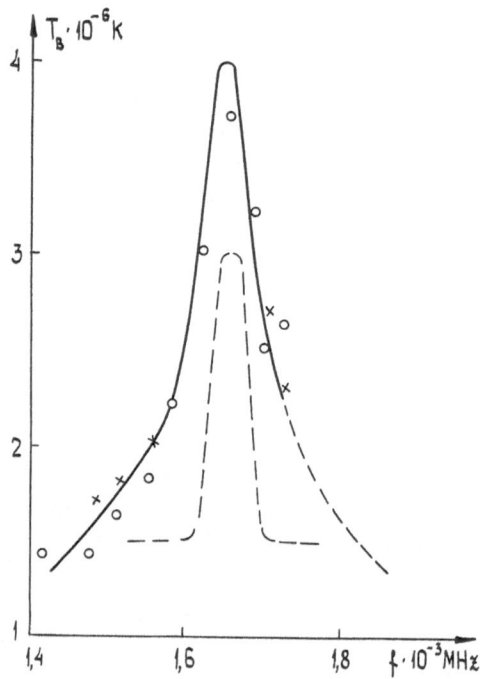

Fig. 1. Cyclotron line in the active region spectrum. The crosses and circles denote the brightness temperatures for two days of observation, recorded by Willson (1985). The solid line is the spectrum of cyclotron radiation of a hot loop with inhomogeneous magnetic field and temperature at the third cyclotron harmonic. The dashed line is the frequency spectrum of a homogeneous hot loop, which we calculated in the homogeneous model.

by the inhomogeneous distribution of plasma temperature and magnetic field in the source.

To analyze this version, we take a model shown in Figure 2. The source is a force tube of bipolar magnetic field, filled with a hot plasma (with the kinetic electron temperature T higher than the surrounding corona temperature T_c), i.e., a hot coronal loop. If the hot plasma pressure is small as compared to the magnetic pressure, then this plasma does not modify appreciably the magnetic field distribution in the plasma. The overall hot loop radiation shows no cyclotron lines because of the inhomogeneity of the field along the tube. Cyclotron lines can appear only if the antenna beam is narrow enough to cut out a region with a quasihomogeneous (in the plane perpendicular to the line-of-sight) magnetic field from the loop. Then the frequency spectrum of observed radio emission may contain cyclotron lines $f = s f_{B_0}$ ($s = 2, 3, 4$ are harmonic numbers, $f_{B_0} = eB_0/2\pi mc$ is the electron gyrofrequency, B_0 is the magnetic field at the centre of the loop). At the second harmonic the cyclotron line contains only an ordinary mode

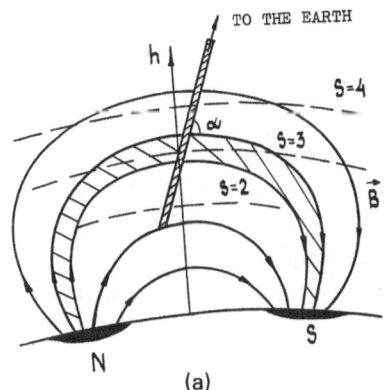

(a)

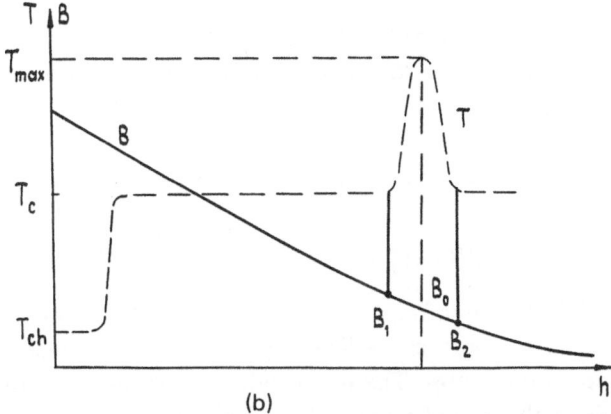

(b)

Fig. 2. Qualitative model of a source with hot electrons filling the magnetic flux tube (hot loop): (a) field lines, the loop (large hatching) and antenna pattern (small hatching); gyroresonant layers $s = 2, 3, 4$ (dashed lines); (b) magnetic field and temperature distribution versus height.

completely polarized, while at the fourth harmonic the extraordinary mode dominates. The third harmonic is partly polarized with an excess in the extraordinary mode if its optical depth $\tau_3 < 1$ and is not polarized if the radiative layer is optically thick ($\tau_3 \gg 1$) for both ordinary and extraordinary modes (for details see Zheleznyakov and Zlotnik, 1980a, b).

Since the antenna beam is much narrower than the angular dimension of the bipolar group of spots, the magnetic field variation across the pattern can be neglected. The magnetic field variation along the pattern (i.e., along the line-of-sight) within the hot loop is rather small: the field changes from B_1 to B_2, where $(B_1 - B_2)/B_1 \ll 1$ (Figure 2(b)). This variation is a key to interpretation of the observed cyclotron line. Following Willson (1985), we approximate the magnetic field B above the bipolar group of spots by the field of a horizontal dipole immersed at a depth of $b = 4 \times 10^9$ cm below the photosphere with a magnetic field of $B_{ph} = 2000$ G at the photosphere. Then the dependence of the field B on the height h above the centre of the group is given by

$$B = \frac{B_{ph}b^3}{(b + h)^3} . \tag{1}$$

The frequency at the centre of the cyclotron line, $f = 1658$ MHz at the third cyclotron harmonic ($f = 3f_{B_0}$), corresponds to a magnetic field $B = 196$ gauss, at an altitude $h_3 = 4.7 \times 10^9$ cm, i.e., at the centre of the loop with a maximum temperature $T = 4 \times 10^6$ K decreasing toward higher and lower values down to $T_c = (1-1.5) \times 10^6$ K, which correspond to the corona. The length of the loop along the line-of-sight is assumed to be of the order of the loop thickness Δh, i.e., approximately $(5-10) \times 10^8$ cm and the electron density within the layer is assumed to be $N \sim (10^9-10^{10})$ cm^{-3}.

Calculations show that the geometric thickness of an optically thick gyroresonance layer responsible for radiation at a frequency $f = 3f_B$ is much less than the transverse dimensions of the hot loop and its optical thickness is $\tau_3 \gg 1$ for both ordinary and extraordinary modes. In the range from $f_1 = 3f_{B_1}$ to $f_2 = 3f_{B_2}$ the brightness temperature at f must be equal to the plasma kinetic temperature at a height h_3 at which $f_B = f/3$:

$$T_b(f) = T(h_3) , \tag{2}$$

where the relationship between h_3 and f in the magnetic field model (1) follows from the condition

$$h_3 = b[(3f_{B_{ph}}/f)^{1/3} - 1] \tag{3}$$

($f_{B_{ph}} = eB_{ph}/2\pi mc$ is the gyrofrequency at the photospheric level). The frequency spectrum is expected to represent a line that repeats, in a sense, the temperature distribution in the coronal loop. The line width is determined by the magnetic field inhomogeneity along the line-of-sight in the loop. The distribution $T(h)$ across the coronal loop, reconstructed from the cyclotron line profile (the solid curve in Figure 1) is given in Figure 3 under the assumption that this line is the third cyclotron harmonic

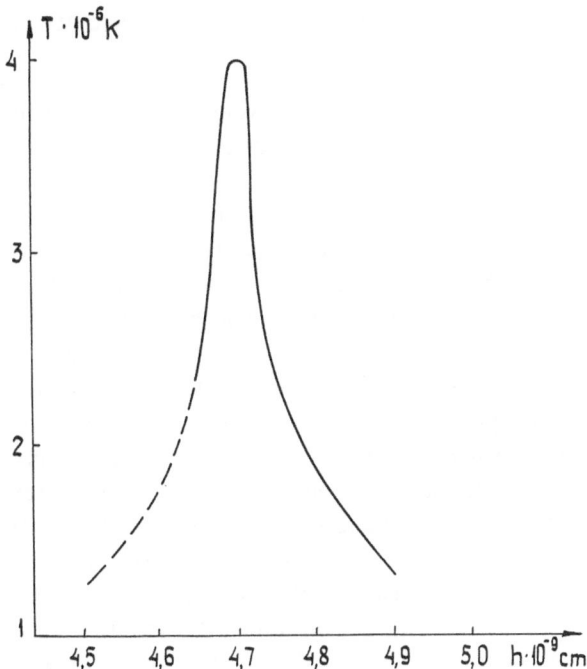

Fig. 3. Temperature distribution through a hot loop reconstructed from the observed cyclotron line profile (*h* is the height above the photosphere).

and the magnetic field varies with height as in expression (1). Since the assumption of optically thick gyroresonance layer $s = 3$ is valid when the plasma parameters range widely (for details see Zheleznyakov and Zlotnik, 1988), it should be natural to expect no polarization and relatively stable line characteristics for the two days of observation (the layers $s \geq 4$, radiation escaping the loop passes through in the corona, are too optically thin for extraordinary and ordinary modes to introduce a noticeable polarization).

Note that the relation between the observed line and other cyclotron harmonics is not so reliable. Indeed, the radiation in the cyclotron line at $s = 2$ passes through the layer $s = 3$ in the higher corona with a weaker magnetic field. Being optically thick for the extraordinary mode in a wide range of angles α between the magnetic field and the line-of-sight and optically thin for the ordinary mode, this layer can appreciably polarize the outgoing radiation (in contradiction to the observational data). The relationship between the observed line and the fourth harmonic is also hardly possible. The point is that at $N \sim 10^9$–10^{10} cm^{-3} the layer $f = 4f_B$ is optically thin in a wide range of angles α, the brightness temperature is much lower than the kinetic temperature and the radiation is polarized.

3. Homogeneous Magnetic Field Model

Taking into account the magnetic field inhomogeneity within the loop cross-section in our model permits one to explain the cyclotron line characteristics. Meanwhile the source with a homogeneous magnetic field considered by Willson does not provide such an opportunity. In his model the observed line width cannot be reconciled with the absence of noticeable polarization. According to Willson, the source has the optimal parameters: a depth $l = 10^8$ cm, a temperature $T = 3.8 \times 10^6$ K (at an ambient plasma temperature $T_c = 1.5 \times 10^6$ K), an electron density $N \sim 10^9$ cm^3 and a magnetic field $B = 145$ G or 119 G (in those fields the fourth and the fifth harmonics, respectively, yield a cyclotron line frequency 1650 MHz). The line calculated by Willson for $f = 4f_B$ and $\alpha = 70°$ is similar to the solid line in Figure 1. Our calculations show that the radiation of a homogeneous plasma layer with the above parameters has a different spectrum shown by a dashed line in Figure 1. That the observed cyclotron line would not be explained by the homogeneous source model is demonstrated by the dependence (calculated for the above parameters) of the brightness temperature T_b, the polarization degree ρ at the line centre (at $f = 4f_B = 1658$ MHz) and the relative line width $\Delta f/f \approx 2\sqrt{2}\,\beta_T \cos\alpha$ ($\beta_T = (\kappa T/mc^2)^{1/2}$; κ is a Boltzmann constant) on the angle α in Figure 4. It is assumed that the background is due to unpolarized coronal radiation with a brightness temperature $T_c = 1.5 \times 10^6$ K. The observed parameters of the cyclotron

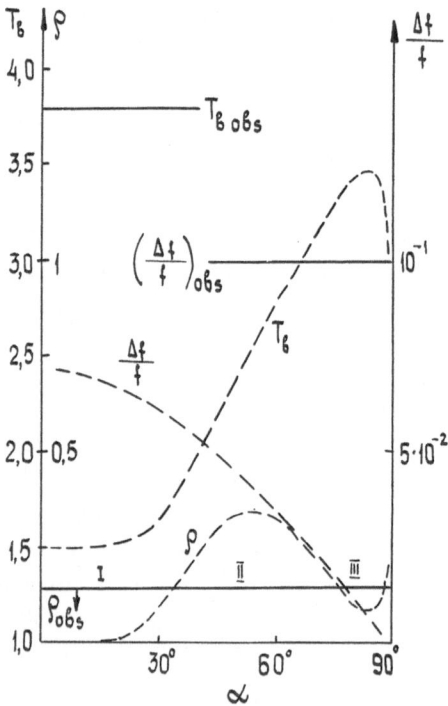

Fig. 4.   Dependence of brightness temperature T_b, polarization degree ρ and line width $\Delta f/f$ on angle α in the homogeneous model.

line are also indicated. The angles α can be divided into three intervals, those of small (I), middle (II), and large (III) angles. In the first interval the line width is closest to the observed one and the polarization degree is below 15%, in accord with the observations. Meanwhile this line becomes very weak and indistinguishable against the coronal radiation background ($T_b \approx T_c$). The second interval is inadequate because of the high degree of polarization ($\rho > 15\%$), the relative weakness of the line ($T_b < 3 \times 10^6$ K) and the reduced line width ($\Delta f/f < 5 \times 10^{-2}$). Finally, the third interval conforms with the polarizational observations and ensures a reasonable temperature at the line centre but does not explain the observed line width (in the theory the line is too narrow (see Figure 2)).

Strong dependence of the line characteristics on the angle α in the optically thin source does not make it possible to explain their constancy for two days of observation. For this time the angle α changed, due to the Sun's rotation, by approximately $10°$. If the angle α changed from $70°$ (the value adopted by Willson) to $80°$, then $\cos \alpha$ changed by almost a factor of two. In the model of a homogeneous source it should be natural to expect a two-fold variation of the Doppler line width, which, however, was not observed.

4. Conclusion

The proposed scheme of cyclotron line formation in an inhomogeneous optically thick source, a flux tube filled with hot electrons (a hot coronal loop) explains the observed properties of the line recorded by Willson (1985) such as its nearly stable maximum brightness temperature, profile and width, and low degree of polarization.

An investigation of cyclotron lines in the microwave spectrum is an effective tool for magnetic field and plasma diagnostics in solar active regions. It permits one to ascertain the parameters of the magnetic field and of the hot plasma in the magnetic flux tube above the bipolar group of spots. The magnetic field at the loop axis can be found from the relation $f_B = f_{max}/3$, where f_{max} is the frequency at the line centre. The kinetic temperature at the loop axis is equal to the brightness temperature at the line centre of the harmonic number s. The magnetic field inhomogeneity across the loop is determined by the line width: $\Delta B/B \sim \Delta f/f$. The dependence of the kinetic temperature on the magnetic field (or its distribution in the loop along the line-of-sight in the given model of magnetic field in the active corona is reconstructed from the line profile).

Since the line recorded by Willson (1985) probably refers to the third cyclotron harmonic, it appears in the region where the magnetic field is 196 G (to an accuracy of 2%, determined by the distance between the neighboring frequency channels) and where there is a hot loop with an axial temperature $T = 4 \times 10^6$ K and a temperature distribution similar to that shown in Figure 3.

Further progress in the study of cyclotron lines in the solar microwave spectrum would require spectral and polarization measurements at multiple frequencies corresponding to harmonics $s = 2, 3, 4$ (at VLA, for example, in the ranges 1100 MHz and 2200 MHz together with the operative range 1650 MHz). Detection of cyclotron features in these ranges will fully exclude the uncertainty in choosing the observed harmonic number and will permit reliable diagnostics of the physical parameters of the coronal plasma based on cyclotron lines.

References

Schmahl, E. J., Shevgaonkar, R. K., Kundu, M. R., and McConnel, D.: 1984, *Solar Phys.* **93**, 305.
Willson, R. F.: 1983, *Solar Phys.* **89**, 103.
Willson, R. F.: 1985, *Astrophys. J.* **298**, 911.
Zheleznyakov, V. V. and Zlotnik, E. Ya.: 1980a, in M. Kundu and T. Gergely (eds.), *Radio Physics of the Sun*, D. Reidel Publ. Co., Dordrecht, Holland, p. 87.
Zheleznyakov, V. V. and Zlotnik, E. Ya.: 1980b, *Astron. Zh.* **57**, 778.
Zheleznyakov, V. V. and Zlotnik, E. Ya.: 1988, *Pis'ma v Astron. Zh.* **14**, 461.

DISCUSSIONS

SPEAKER: ANTONUCCI

CHENG: Have you compared Ca XIX and Fe XXV Spectra from limb flare with those from disk flare?

ANTONUCCI: During the impulsive phase, limb flares show the same degree of non-thermal broadening in soft X-ray lines as disk flares, and do not show blue-shifted emission. That is, they are characterized by symmetric line profiles.

NITTA: Do you have to assume the "low-energy cutoff" of hard X-ray emitting electrons when you compare the evaporation velocity with the energy input? The value of the low-energy cutoff has hardly been known. Is it possible to obtain the temperature of hot components reliably with Fe XXV lines alone, e.g., by using the differential emission measure technique?

ANTONUCCI: Temperature measurements improve, in principle, by using spectral emission from ions with contribution fuctions in different ranges of temperature. Differential emission measure techniques, however, depend significantly on the sensitivity of the different channels and the abundances of the different elements. For what concerns the "hot" thermal component in flares, its temperature is best measured by combining Fe XXV and Fe XXVI spectra, as shown by the work of Tanaka and co-workers.

SMITH: The evaporated material should drive shock waves. For Ca XIX from the calculations of Fisher (1987, Ap.J.) we know that the shock should have a velocity $\simeq 3.1$ times that of the moving material. Now we have much higher velocities $\simeq 1000$ km s^{-1} in Fe XXV. Have you had time to figure out what velocity of shocks these velocities would imply?

ANTONUCCI: The evaporation velocities derived from the Fe XXV spectra can be expressed in units of a limiting velocity (which has been assumed equal 2.35 times the sound velocity in the evaporating plasma); maximum values in the velocity distributions reach up to 0.6 times the limiting velocity and are in the range from 0.3 to 0.6.

PALLAVICINI: If I understand correctly, the velocity of the upflowing plasma depends both on the energy input and on the density of the preflare loop. So, when you conclude that the velocities derived from Fe XXV are more consistent with the thick-target model than with the thermal model, have you compared models for the same preflare conditions in both cases?

ANTONUCCI: The comparison of observational velocities with those resulting from numerical simulations, which has been used to conclude that evaporation velocities are more consistent with thick-target models, has been made without discriminating for initial conditions. However, most of the simulations performed up to now consider initial densities of the order of 10^9 cm^{-3}.

Solar Physics **121**: 459–502, 1989.

P. B. BYRNE / Multi-Wavelength Observations of Stellar Flares: 61–74.

SPEAKER: BYRNE

BUTLER: I am no longer convinced that there was no X-ray counterpart to the flare observed in the U-band on YZ CMi. In fact an enhancement in the LE soft X-ray band was recorded by EXOSAT which peaked at 20h 06m UT; that is eleven minutes after the U-band peak. This, I think, was the corresponding X-ray event. The very short, impulsive, nature of the U-band flare made it appear that it was not connected to the later X-ray enhancement, but, if it had a more prolonged decay, I think there is no doubt that we would have associated the two enhancements as a single event.

BYRNE: Yes, this is so. A subsequent analysis of the ME data on this flare by R. Mewe (private communication) indicates that there is a weak flare close to the time of the U-band flare. A delayed soft (LE) X-ray flare 600-700 s after the U-band rise is not unreasonable as 1000 s delay between ME and LE was seen in the EQ Peg flare of Haisch et al. (1987, Astr. Ap. 181, 96). My main point in drawing attention to the lack of a strong LE flare was to establish the differences with the very similar optical flare on YZ CMi observed by Karpen et al. (1977, Ap.J. 216, 479) which was accompanied by a strong X-ray flare.

ACTON: It is interesting that both solar and stellar X-ray flares have electron densities in the 10^{11} to few times 10^{12} range. I suggest that this represents the density at which flare energy release is quenched.

BYRNE: Yes, I agree. In spite of the enormous range in energies from small solar, through dMe, to RS CVn star flares, the densities do not vary very much. I feel that this must be significant.

ACTON: Can you think of a solar analogue to the event discussed by Doyle of a strong optical "flare" without detectable X-ray emission?

BYRNE: It is difficult to make a direct comparison between optical stellar flares and solar flares. The latter do not have a strong broad band optical signature, so it is difficult to know what to compare. A second reservation is that we should bear in mind the limited sensitivity of the present generation of X-ray detectors. So it is possible that a soft X-ray flare occurred which was undetectable. I would point out, however, that the Kahler et al. (1977) flare on YZ CMi was very similar in optical energy and had a strong soft X-ray signature which we would have detected easily.

FALCIANI: The red asymmetric emission in Balmer lines can be explained with a Stark broadened profile, symmetric, roughly centered at the previous ($V_{//}$ = 0) line center, originating in the deepest, denser chromospheric layers, superimposed on a blue-shifted absorption, due to the cold material going up into the loop (evaporation, surge material). In some solar flares the velocity corresponding to the blue Doppler-shifted absorption fits very well the ascending velocity directly measured in the surge material that becomes visible later on. Do you think that this explanation might be applied to stellar flares, too?

BYRNE: Yes, it is possible to apply such a model but many others are also possible. It should be born in mind that the quality of the data are not sufficient to distinguish between the various possible models. Blueward absorption was recorded in the Mg II profiles of the II Peg 1983 February 2 flare (Doyle et al., 1989). I would caution that Stark broadened profiles do not fit the profiles of the Balmer lines of the YZ CMi 1985 March 5 flare (Doyle et al., 1988) at any phase of the flare.

PALLAVICINI: Just a short comment. Your statement that the EINSTEIN Observatory observed less energetic flare than EXOSAT is largely due to the fact that you have picked up the strongest ones among the flares observed by EXOSAT. Actually, in the complete EXOSAT data sample, there are X-ray flares which are weaker than those observed by EINSTEIN. And this simply reflects the fact that the energies of soft X-ray flares span more than 4 orders of magnitude and there are much less strong flares than small flares.

BYRNE: Yes, this is true. EINSTEIN, however, did not observe any flares as energetic as those observed by EXOSAT. There are systematic effects in the energy of flares as a function of their spectral type. So it may be that EINSTEIN spent more time on those stars which have less energetic flares.

MULLAN: dMe radio emission is often coherent whereas optical and X-rays are not. Therefore one does not expect to see a correlation in all cases between radio flares and flares in other spectral range.

BYRNE: Yes, this is true. But if we invoke a model which heats the lower atmosphere with a beam of highly relativistic electrons constrained by a magnetic field, it is difficult the see how these will not radiate in microwaves.

MULLAN: Are flares in RS CVn flares really an exact analogue of dMe flares? In the latter stars, radio emission is always coherent, whereas in RS CVn stars this is not (or hardly ever) true. Some of the energy release is RS CVn flares may be due to reconnection between the fields of one star and the other.

BYRNE: Again I agree with you that there are important differences between RS CVn and dMe flares. Not least of these differences are those of energy and time scale. Nevertheless, I have tried to show that there is evidence that the density, emission measure and, therefore, the flaring volumes are similar to dMe's. It is the duration of RS CVn flares which results in their having much greater energies. This would imply a very long-lived heating episode, lasting at least 7-12 hours and maybe longer.

RODONO': In comparing the relative timing and behaviour of radio and X-ray fluxes for stellar flares we must bear in mind that flare time evolutions are different in the different wavebands. For example, we may safely say that the impulsive 2-cm radio emission follows quite closely the hard X-ray enhancement, while in the 6-cm and soft X-ray bands the flare has a slower time evolution. Recent microwave observations of stellar flares with VLA have also shown that the flare peaks at 2-cm before that at 6-cm.

BRIAN R. DENNIS and RICHARD A. SCHWARTZ / Solar Flares: the Impulsive Phase: 75–94.

SPEAKER: DENNIS

CHENG: Is an impulsive HXR burst necessary for solar flares? My question is based on the previous OSO observations which show that 1/3 of all the flares OSO observed have no impulsive HXR bursts. Is this a sensitivity problem? A reverse question is: Are there impulsive HXR bursts observed from SMM with no associated soft X-ray bursts?

DENNIS: We have never seen a hard X-ray burst with no corresponding soft X-ray flare. There is a linear relation with about one order of magnitude scatter between the total number of counts seen in the hard X-ray burst spectrometer (HXRBS) on SMM integrated over the duration of the burst and the rate seen in the Ca XIX channel of the Bent Crystal Spectrometer also on SMM. Thus, for the large soft X-ray events >100 count/s in BCS, we always see a hard X-ray event significantly above the HXRBS background rate. Thus, I suspect that the absence of a hard X-ray event for some of the smaller soft X-ray events, particularly the events with the largest rise times, is a threshold effect connected to the HXRB sensitivity.

MULLAN: Is there a lower limit on the time-scale on which Hα varies in a solar flare impulsive phase (by analogy with stellar flares)?

DENNIS: The Hα observations with the best time resolution that have so far been reported are <1 s and significant fluctuations are seen during impulsive phases on these time scales. Thus, there is no evidence for a lower limit down to these time scales. These rapid fluctuations occur in individual kernels, however, and contribute very little to the integrated flux from the whole flare. Thus it might be said that the time profile of the Hα emission integrated over the full area of the flare has a rise time that is generally longer than 1 minute, similar to that for the soft X-ray emission.

RODONO': When simultaneous flare data in hard X-rays and UV transition region lines or continuum are available, do they always vary simultaneously (± 0.1 s), or are there exceptions?

DENNIS: In the few examples that we have, the UV transition zone lines and the UV continuum always vary nearly simultaneously with the hard X-rays (see Orwing and Wordgate 1986, and Cheng et al., 1988).

BAI: You showed a number of flares for which soft X-ray time profiles are similar to the time integral of hard X-ray time profiles. But this does not mean that all the energy of the soft X-ray emitting plasma comes from high-energy electrons. Direct heating is likely to have the same profile as the acceleration of high-energy electrons.

DENNIS: I agree. The uncertainty on the time integrated energy in the electrons that precipitate into the thick target is likely to be an order of magnitude because of our lack of information about the lower energy cutoff in the assumed power-law electron spectrum.

VIAL: Do you have many examples of the time lag of microwave radiation?

DENNIS: Yes, we have many examples of the time lag of the microwave radiation. A 2 second delay is about the longest that we see for an event that we would consider to be impulsive. Much longer delays up to $\approx$ 1 minute are seen for the more gradually varying, type C, flares.

GOPALSWAMY: The figure with footpoint HXR emission (Hinotori) shows that both footpoints have avoided the Hα ribbons. Why is this discrepancy in a disk event like this? Is this due to a projection effect?

DENNIS: I believe that the hard X-ray bright particles could be spatially coincident with the Hα ribbons to within the uncertainty in the relative alignment between the two images.

BHATNAGAR: In case of Hα kernels in a flare, all of them do not brighten simultaneously. Individual kernels are observed to brighten with delay times of several seconds to minutes. Hence, time correspondence with microwave and X-ray emission has to be taken with caution because one may consider good correspondence depending on which individual kernel is taken.

DENNIS: This is a good point and indicates the need for microwave and hard X-ray imaging with good time resolution. Nevertheless, Hα images taken with subsecond time resolution, as one now becoming available, can be usefully matched to hard X-ray and microwave time profiles with no spectral information. This is particularly true for the smaller, simpler flares, where the problem of multiple Hα kernels brightening at different times can be minimized.

MARTENS: I would like to argue that the case for the < 1 MeV proton beam model is stronger than you suggested: the proton beam model that Simnett, Henoux and myself propose (e.g. Ap. J. 330, L131) solves all these problems quite clearly. What is your comment on this?

DENNIS: The strengths and weaknesses of the < 1 MeV proton beam model will become evident with time as more work is carried out to explore the consequences of the model. While this model does suggest solutions to some problems with the electron beam model, there are many significant questions that must be answered such as, how are the hard X-rays and microwaves produced? Only when it has been subjected to the same scrutiny, as has the electron-beam model, we will be able to assess the strength of the case for the < 1 MeV beam model.

WOLFGANG DRÖGE, PETER MEYER, PAUL EVENSON, and DAN MOSES / Electron Acceleration in Solar Flares: 95–103.

SPEAKER: DROEGE

STURROCK: Do you have information on the ion abundances in energetic-ion events? How do the abundances compare for Class I and Class II events?

DROEGE: Class I events have a higher electron to proton ratio than Class II events. They also seem to be ^{3}He-rich.

SHAPIRO: Are proton data available for many of the flares you studied?

DROEGE: Proton spectra for 3 events included in our survey are published by McGuise and von Rosenvinge (1984, Adv. Space Res. 4, 117). In principle, more proton spectra from the IMP7/8 measurements should be available.

RIEGER: It is interesting to note, that for your short duration events the electron spectra get harder with increasing energy, in contrast to the proton spectra measured in space and inferred from the X-ray measurements, which have spectra that soften at high energies. For your long duration events the situation is the opposite. The question arises: Does stochastic acceleration create a harder electron spectrum than shock acceleration in contrast to the case for protons?

DROEGE: Diffusive shock acceleration without modifications produces power laws in momentum with the same spectral index for electrons and protons. In stochastic acceleration models, the spectral shape is determined by the rate at which particles gain energy from interacting with the plasma turbulence, and the rate at which particles escape from the acceleration region. Because electrons and protons might interact with different types of plasma curves, their spectra can be different, i.e. electron spectra can be harder at high energies than proton spectra.

A. GORDON EMSLIE / Models of Flaring Loops: 105–115

SPEAKER: EMSLIE

DOYLE: High time resolution (6 s) data on Ca XIX in the impulsive phase has been published for a few flares (Doyle & Bestlay). These show a complex picture, (although the signal-to-noise is poor) of several mass up-flows. We will have to wait for Solar A to obtain good data, as the sensitivity of the BCS is not sufficient for most C or M class flares to detect with high time resolution the mass up-flows.

EMSLIE: I believe that good signal-to-noise data, showing the changes in the Ca XIX spectra during the impulsive flare at, say, 3-5 s time resolution, should be extremely valuable in testing the type of models of which I spoke.

MOORE: Since your simple model works so well, what can you deduce about the electron acceleration mechanism?

EMSLIE: I can only say at this point that the data are consistent with the standard "black-box" picture in which the acceleration occurs near the loop of the flare loop. Questions regarding the acceleration mechanism are outside the scope of my talk, although it would be nice if someone could come up with a way of accelerating power-law electron spectra as required by the observations.

MACNEICE: I do not agree that excitation of EUV emission by a conduction front during the impulsive phase can be ruled out yet. Non-local heat flux may produce early pre-heating of the transition region ahead of the conduction front, preventing any initial dip in EUV emission, and reducing the lag time between hard X-ray and EUV emission. Also the temperature gradient in the transition region can be expected to be less severe. Finally even the numerical simulations using Spitzer-Hrm conductivity produce conflicting results regarding the initial EUV dip.

EMSLIE: Your comments are well taken. The conduction model used by Nagai and myself involved only local chemical conduction (with a replacement by a saturated heat flux where necessary), and it is possible that a more realistic treatement of thermal conduction, including global effects as you suggest, could produce a more satisfactory agreement with the observed behaviour of transition region lines during the impulsive phase. Notwithstanding this, however, the purpose of the comparison between conduction and electron heated models was to show that (i) the behaviour of EUV emission lines is not trivial to understand and (ii) agreement with observations can only be achieved for a spatially distributed form of energy inputs, such as collisional degradation of a nonthermal electron beam, or, perhaps, the global conduction physics to which you refer.

KOCHAROV: It seems to me that the observational data on solar neutrons and γ-rays are very informative. These data allow us to answer many questions concerning flare loops; the magnetic field and its gradient, the dimension of the arch, the number density and its gradient, so there is excellent possibility for solar plasma diagnostics in the acceleration and propagation region. New integrated experiments are needed. As far as theory is concerned, you have to consider together all the available data and try to explain them.

SERIO: How do your Ca XIX line profiles depend on the beam's low energy cutoff?

EMSLIE: For higher low-energy cutoffs, a larger portion of the beam energy resides in electrons which deposit their energy in deeper layers of the atmosphere. These layers can efficiently radiate away the deposited energy in UV and optical lines and continua, so that less energy is available to power the evaporation process that gives rise to the strong, blueshifted, Ca XIX profiles. Thus, for higher low-energy cutoff, the Ca XIX profiles are weaker.

NITTA: Isn't your argument that Ca IXX profiles depend on low-energy cutoff of electrons dependent upon the preflare atmospheric model you take?

EMSLIE: As long as the cutoff is such that the bulk of the electrons reach the preflare chromosphere (i.e., $E > 10$ KeV or so), then the results are insensitive to the preflare atmospheric model.

MARTENS: I have two questions regarding the relation between hard X-rays and micro-waves: 1) My impression from combined SMM-radio observations is that there is often a time delay of the micro-waves with regard to the hard X-rays: a time delay that is larger than the time resolution of the instruments; and, 2) Kundu showed yesterday an example that demonstrated that the number of electrons needed to generate the microwaves is much smaller than the number required for the hard X-rays. How do you reconcile these observations with a common electron population responsible for both types of radiation?

EMSLIE: Microwaves are radiated by much higher-energy electrons than these responsible for hard X-ray bremsstrahlung. Because of the steep spectra involved, the number of microwave-producing electrons is much less than those producing hard X-rays. The observed time delay may have its origins in the trapping and precipitation of the electrons, noting that a given trap may be collisionally thin for 100 KeV + electrons, yet collisionally thick for ≈ 20 keV electrons.

BAI: I would like to answer Dr. Martens' question regarding the electron number problem and the delay of microwaves with respect to hard X-rays. Energetic electrons which precipitate to the chromosphere produce hard X-rays while loosing energy quickly. Energetic electrons which have large pitch angles are trapped in the flare loop and produce microwaves. The number of trapped electrons is smaller than that of precipitating electrons. In addition, trapped electrons have longer lifetimes; therefore, microwaves are delayed a few seconds with respect to hard X-rays.

SMITH: There is a recent paper in Ap.J. by Spicer and Emslie which attempts to help the electron number problem by using a magnetic trap with an electrostatic field. With this greater confinement, can you get enough dumping from the trap to explain the other phenomena observed, like the impulsive EUV? In other word, is there really any energetic advantage?

EMSLIE: I thank you for pointing out this paper to me! The aims of that paper were to produce observed hard X-ray bursts using fewer electrons than in a standard thick target model; no attempts at investigating the physics of the precipitating component were made. I suspect, however, that the fewer number of electrons could make a reconciliation with, e.g., Ca XIX spectra, difficult.

BERNARD H. FOING / Stellar Flare Spectral Diagnostics: Present and Future: 117–133.

SPEAKER: FOING

LINSKY: Binarity in RS CVn systems can be responsible for the highly energetic phenomena observed because tidal forces induce rapid rotation and may change the differential rotation (with latitude) such that the dynamo generation of magnetic fields is enhanced greatly compared with slowly rotating single stars. Also, there is the possibility that magnetic fields of the two stars may interact and occasionally interconnect.

MACKINNON: I have to be pessimistic about the prospects for using radio observations, in the way hard X-rays have been used in the solar context, to develop a quantitative theory of energy transport in stellar flares. At least in dMe stars, radiation is coherent, and it is very difficult to say anything about total numbers, energy content, etc., of emitting particles. RS CVn may be better placed.

BYRNE: I would like to make a point concerning the scale of a stellar flare. In so far as we have information on the electron densities in stellar flares these would appear to be similar to solar values ($\simeq 10^{11}$–10^{12} cm^{-3}). The emission measures in both mid-transition region lines and in soft X-ray are also not very different from those in large solar flares. The critical difference appears to be in the duration of the flares and, therefore, in their total energies.

RODONO': At first glance, it might appear quite discouraging to study stellar flares with time- spatial- and spectral-resolutions much worse than in the solar case. However, we must always bear in mind the general idea underlying solar and stellar studies: i.e., the understanding of the basic physical mechanisms that produce solar flares may help in interpreting the global characteristics of seemingly similar phenomena occurring in the rather different stellar environments. Of course, we may expect very different behaviour and, eventually, substantially different mechanisms that are triggered during the course of the most energetic events on stars. We have still to learn a lot, beginning with trying to understand, for instance, why the surface differential rotation rate on very active RS CVn stars seems to be one or two orders of magnitude smaller than solar. Certainly, the binary nature of RS CVn systems, specifically the tidal interaction between the two stars, has some effect in modifying the "natural" differential rotation regime of the individual stars. This, on the other hand, provides one of the necessary conditions (e.g., high rotation-rate) for the development of activity phenomena. Certainly, RS CVn stars appear to have found a way of preserving their ability of producing high-energetic activity phenomena, possibly by developping rapidly spinning cores or less dramatic regimes of radial differential rotation, which at present escape detection.

STERN: Referring to Brendan Byrne's comment, I believe that the largest flare volumes derived using soft X-ray observations are in fact orders of magnitude larger than solar flares. On the other hand, the volumes measured using UU line fluxes may be comparable to the solar case.

V. GAIZAUSKAS / Preflare Activity: 135–152.

SPEAKER: GAIZAUSKAS

KOCHAROV: What is the dependence of flare-precursor events versus time to flare?

GAIZAUSKAS: The microwave precursors to which I alluded occur within 10 to 20 minutes before flare onset.

KOCHAROV: Systematic studies of X-ray precursors were carried out from 1975 to 1979 in experiments on Prognoz 4, 5, 6, and 7 led by my laboratory. Owing to the high sensitivity of the equipment we succeeded in revealing for a large number of flares (> 1000), slight changes in the soft X-ray intensity (in ≈80%) of the cases, which were considered X-ray precursors. On the average, a precursor appears 10-20 min. before the flare, which coincides in time with the beginning of activation of the dark filaments frequently observed before the flare onset. The magnitude and brightness distribution of flares with precursor was found to be the same as for all flares detected over the same time interval.

GAIZAUSKAS: I look forward to seeing your detailed case studies of X-ray precursors associated with filament eruptions.

BORNMANN: What are the ages of the active region complexes when they start producing flares?

GAIZAUSKAS: Major flares are usually produced during the first one or two rotations after a complex is formed. Smaller flares are produced thereafter, unless a major eruption of new flux rejuvenates the complex.

MARTENS: What is your opinion on the question of triggering of filament eruptions? How often do triggers occur, and of what kind are they?

GAIZAUSKAS: I do not believe that we can isolate one phenomenon as an inevitable flare trigger. So many things are going on at once in an evolving activity complex that you can imagine anyone of them to push a filament, already stressed by evolutionary processes, into an unstable regime: emerging flux, cancelling flux, footpoint motions, remote flares, etc.

KUNDU: I have a few comments to make on your discussion of microwave precursors. With regard to microwave flares, the precursor region is the same as the flaring region, as seen for example at 6 cm with the VLA using ≈ 1 arcsec resolution. Several tens of minutes before the flare, the precursor appears as a heating of the region (up to ≈ 15 x 10^6 °K). This preheating by itself is not a necessary condition for triggering the flare onset. A few minutes (< 10 min) before the impulsive onset of flares, one of the following manifestations must take place: 1) change of polarization of the flaring active region; 2) change in orientation of the plane of zero polarization; 3) physical appearance of new regions/structures; and (4) in some cases we see oppositely polarized bipolar regions or quadrupole structures. All these are manifestations of some emerging of newly "visible" microwave structure interacting with pre-existing structure, which produce a current sheet strongly suggestive or magnetic field reconnection, which ultimately is responsible for triggering the impulsive onset of flares.

GAIZAUSKAS: I agree with your statement except for the insistence that specific microwave phenomena must take place. But there is no doubt that the magnetic field is somehow restructuring rapidly just before flare and that microwave signatures contain important information on that process.

ACTON: Please give us your comments, from the point of view of an observer, on Gene Parker's conceptual model of magnetic stress building up in the corona from the random wandering of magnetic field footpoints in the photosphere.

GAIZAUSKAS: I have no trouble with Parker's concept. You can easily imagine that the chaotic fine structure in the quiet Sun originates in random local reconnections. There is obviously more order in space and in time for the magnetic field within active regions or we would not see homologous flares or flare kernels appearing adjacent to, but on opposite sides of, polarity inversion lines. But within this quasi-ordered state, the assumption that random reconnection triggers flares does not conflict with the lack for consistent pattern of pre-flare phenomena.

M. R. KUNDU, S. M. WHITE, and E. J. SCHMAHL / Simultaneous Multi-Frequency Imaging Observations of
Solar Microwave Bursts: 153–161.

SPEAKER: KUNDU

MULLAN: You observe not only at flare but also at quiescent
times. Can you detect non-thermal electrons during quiescent
times on the Sun or stars?

KUNDU: For the Sun, the active region producing a flare remains
heated (sometimes up to 10×10^6 °K) until the next flare, if it
occurs within a few hours. I'd like to interpret this heated
region as due to gyroresonance radiation. If non-thermal
electrons are involved in this region, they must show in the form
of rapid fluctuations of mini-flares. So far, no one has shown
conclusive evidence for this behaviour. As for the stellar case,
the question is really that of quiescent emission – whether or
not it is due an assemblage of many mini or microflares. We feel
that the quiescent emission consists of really quiet star
emission plus active region emission. Quiescent emission from UV
Ceti, for example, varies between 1 mJy and 3 mJy. We feel that
this emission is due to gyrosynchrotron radiations (producing
many miniflares) and therefore there is evidence for nonthermal
electrons.

JAN KUIJPERS / Radio Emission from Stellar Flares: 163–185.

SPEAKER: KUIJPERS

KRISHAN: I would like to inform you that I considered a "Free
Electron Laser" mechanism for type III solar radio bursts eight
years ago and more recently for the generation of nonthermal
continuum of quasars. The question that is asked all the time is:
" What are the observational signatures of a particular coherent
radiation process? " Polarization is one characteristic but there
are many ways of modifying it. Do you have any suggestions for
the observational signature of free electron laser type
mechanism?

KUIJPERS: The radiation process from double layers that I have in
mind is not identical to the free electron laser but can probably
be considered as a nonrelativistic extreme version. In view of
the multitude of theoretically existing coherent radiation
mechanisms, it is necessary to work out also the nonlinear
development in time before it can be applied to the observations.

JEFFREY L. LINSKY / Solar and Stellar Magnetic Fields and Structures: Observations: 187–196.

SPEAKER: LINSKY

KLIMCHUK: I have both a comment and a question. First, concerning the hydrodynamic models of Mariska, in order to get velocities of reasonable magnitude, it was necessary to assume loop pressures that are quite low ($nT \simeq 10^{14}$ °K cm^{-3}) compared to those observed in solar active regions. So, I feel that those results should be applied with some caution. My question is: What sorts of error bars do you get for the magnetic field measurements, both for the field strength and the filling factor?

LINSKY: I have called attention to some important systematic effects such as line blanketing, line saturation, and different thermal structures in the magnetic and nonmagnetic regions that can lead to systematic errors in the derived magnetic parameters. It is likely that systematic rather than random errors will dominate and the magnitude of the errors in the magnetic parameters probably depends more on the analysis technique than on the quality of the original data. I would estimate that the most careful analytic techniques now provide field strengths accurate to ± 30% and filling factors accurate to a factor of 2, although relative errors for data analyzed by the same techinque should be more reliable.

LIVI: Is there an explanation for the non uniform heating of magnetic loops in the Mariska model?

LINSKY: As I recall no explanation is provided in MARISKA's papers and asymmetric heating is strictly an ad hoc assumption that gives good answers. He places the heating in a 100 km wide region beginning 100 km above the base of the transition region in one footpoint of a loop. These numbers are arbitrary. One could speculate that the heating is localized to the footpoint in which there is a resonance between the local dissipation (controlled say by the particular geometry of the magnetic field) and the input of mechanical energy (say by convective motions or MHD turbulence).

PALLAVICINI: You have inferred magnetic fields of $\simeq 10^2$ G in the extended coronae of RS CVn stars (at typical distances of $\simeq 1$ R$_*$). Are these magnetic fields consistent with what you expect or measure at the surface of the stars?

LINSKY: This is an important question. For an isolated magnetic dipole, $B(\Omega) \simeq \Omega^{-3}$ and the deduced photospheric fields will probably be unreasonably large. However, the filling factor for fields in the photospheres of RS CVn stars is large, so $B(\Omega) \simeq \Omega^{-2}$ is more sensible. Also the coronal fields are probably twisted so that the fields may be much larger than even a $B(\Omega) \simeq \Omega^{-2}$ extrapolation.

SILVIA H. B. LIVI, SARA MARTIN, HAIMIN WANG, and GUOXIANG AI / The Association of Flares to Cancelling Magnetic Features on the Sun: 197–214.

SPEAKER: LIVI

POLETTO: Have you been able to establish any relation between the amount of flux which is cancelled and the importance of the flare?

LIVI: Not yet, but our goal is to do such quantitative analysis.

HENOUX: You use the word cancellation. Could you also interpret the magnetic decrease you observed as due to the submergence of a magnetic loop?

LIVI: Someone criticized our choice "cancellation", because the word had no physical meaning. That is exactly what we wanted: no preconceived model. The observations are not showing simple submergence of a magnetic loop, because ephemeral regions usually do not disappear by reversing their pattern of appearance and cancelling within themselves. Cancellation often occurs between one pole of the ephemeral region and a neighbouring magnetic fragment of opposite polarity. Besides, we also see brightenings and flares. I prefer to suggest reconnection as a related mechanism.

MOORE: I would like to point out that Dr. van Ballegoovijen has already a model (see his poster paper) for how such magnetic cancellation can lead to flares, and it fits your interpretation perfectly.

FALCIANI: Are you really sure that the phenomenon you are describing is well over the noise level of your videomagneto-graph?

LIVI: While playing the movie again, you can see a transient effect that is happening at flare time. I could include in the reference Severny's idea that magnetic fields suddenly decrease during a flare, but that was based on very few magnetograms taken hours apart and was not confirmed. Cancellation is a gradual process in our line-of-sight magnetograms, and the number of consistent images is more than enough. The magnetograms are increasing in sensitivity on 21 November 1987 as the cancelling feature is disappearing. On a previous study (Aust. J. Phys. 38, 855–73, 1985) we showed that isolated features changed less than 10% during 5.5 hours, while 16 closely-spaced opposite magnetic features were conspicuously cancelling.

SARA F. MARTIN / Mass Motions Associated with Solar Flares: 215–238.

SPEAKER:MARTIN

ZWAAN: Concerning your definitions of flaring arches and surges, is my understanding correct that the difference is mainly in the inclination: in surges matter is thrown up nearly vertically, and in flaring arches nearly horizontally? If that is so, what is the point in defining two separate categories?

MARTIN: The difference between most surges and most flaring arches is not their inclination but rather their brightness and apparent energy. Surges are usually defined or described as consisting of a narrow spike of mass that appears to be injected into the corona from the chromosphere, slows with increasing height, stops, and then falls back to the chromosphere along the same path that the mass had when moving upward. Most surges are seen only in absorption. Flaring arches begin as emitting mass which is thrown into the corona from a flare, follows the trajectory of a complete arch, and falls to the chromosphere at a peripheral site, which has brightened during the flare. Flaring arches are also seen in hard X-rays while most surges are absorption features, which are known not to be associated with hard X-rays. However, I have here illustrated examples of bright surges and surges that are partially in absorption and partially in emission to make the point that there is a spectrum of energies among surges; these bright energetic surges have many properties in common with flaring arches. However, we do not know if these bright surges are associated with hard X-rays.

BALLEGOOIJEN: How long does it take for a filament to reform after the flare?

MARTIN: The time of reformation seems to depend on the magnitude of the photospheric magnetic field around the filament channel. In many cases, especially in active regions with high magnetic flux density of the photospheric field, the filament will begin to reform before the end of the flare. However, if the magnetic flux density around a filament is low, such as around quiescent filaments, the reformation is very slow and can take several days or it might not reform at all.

SVESTKA: I can inform you that we have found, very recently, that the bright surge of 8 July 1980 you showed was seen in X-rays by HXIS. The X-ray emission preceded the Hα emission, along very much the same trajectory, by one or more minutes.

MARTIN: This is an additional evidence that flaring arches are just very energetic surges.

VERMA: In most cases solar surges do not produce hard X-ray bursts. In my study of about 50 solar surges, none was found to be associated with hard X-ray emission (Verma, 1985, Solar Phys. 97, 301).

KUIJPERS: Can you estimate if all of the gas in a surge falls back or if some of it escapes from the Sun?

MARTIN: There is no evidence that any of the mass of a surge escapes form the Sun. As far as we know, the surge mass flows into pre-existing closed coronal structures. Apparently, most of these coronal structures are large because the surges usually have a small amount of curvature. Only in a few cases, are surges observed to flow to the top of a closed arch and are then seen to flow down the other leg instead of returning to the source site.

D. J. MULLAN / Solar and Stellar Flares: Questions and Problems: 239–259.

SPEAKER: MULLAN

KUIJPERS: Concerning the extreme Alfven speed you derive on the assumption of the electron cyclotron maser, I should like to note that in the solar corona, the cyclotron maser is not the only coherent mechanism. Would you agree that a similar state in stellar coronae removes your problem of extreme Alfven speeds?

MULLAN: Some coherent mechanism is needed to explain strong dMe radio flares. I suggest ECM may not work, but there must be some mechanism at work. In that case, there may be no need for v_A > v_{A0}.

KUIJPERS: In the same context your large source length is derived on the assumption of loop pulsations. In microplasma physics there are, however, other ways of obtaining short pulsations (see, e.g., Aschwsanden, M.J.: 1987, Solar Phys.111, 113).

MULLAN: If other modes of maser are at work, then conversion from a time scale to a length scale will require the appropriate group velocity.

KRISHAN: You mentioned collisional losses of electron beam and consequently a very high column density. Electron beams are known to suffer collective losses, which being extremely efficient, require much smaller column densities. Has anyone looked at this?

MULLAN: Yes: Hamilton and Petrosian (Ap.J. 1987 or 1988) have looked at how electrons can be stopped also by plasma waves. See also Winglee, Pritchett, and Dulk (Ap.J. 1987 or 1988) who have analyzed how a beam of electrons propagates (including electrostatic and collective effects).

BYRNE: Recent studies of correlations between X-ray flares and VLA flares suggest that these are poorly correlated. So we may be observing different loops and even different events in the two frequency regimes.

MULLAN: A lack of correlation between radio and X-ray flares is not surprising because radio emission in dMe stars is coherent, whereas X-rays are not. Hence, a small change in physical conditions can drastically alter the ratio of radio to X-ray power. Therefore a large X-ray flare may be accompanied by either a huge radio flare or nothing detectable, even if both occur in exactly the same loop. Consequently, an underdetectable X-ray flare may give rise to an easily detectable radio signal if the physical conditions in the radio emitting gas are correct.

BYRNE: Using densities determined from the soft X-ray flares to determine preflare loop densities will overestimate this quantity.

MULLAN: Radio emission does not occur in a pre-flare loop, but in a flaring loop, although pre-flare loop density may be very high, much higher than "quiescent" loops (see Canfield, this conference). Electrons must stream down to the base of the loop and create a loss-cone distribution in order to create a velocity distribution favourable to coherent emission. Then radio emission can occur in the dense gas where the loss cone is created. Haisch's analysis gives mean densities for the whole loop: values in the dense lower gas will be higher than we have used, thereby exacerbating the problem. Admittedly, before the flare loop fills, the density will be lower overall, but this is offset by the higher densities in the radio emitting region. Moreover, in the pre-flare state the temperature will be small, so that the increase in density towards the feet of the loop will be very large. Hence, to calculate the column density, we have underestimated it by using the mean density. Even allowing for the effect you point out, electron beam penetration will be very difficult, especially since mean densities outside of flares are $\simeq 10^{10}$ cm^{-3}, anyway (Katsova, 1988).

BORNMANN: You mentioned the problem in explaining the Butterfly diagram which requires $(d\Omega/dr) < 0$. The recent thesis work of C. Morrow (BAAS 1988, Kansas City meeting) using Fourier Tachometer data relates to this problem. She concludes that the rotation rate both increases and decreases with depth in the convection zone, depending on the latitude:

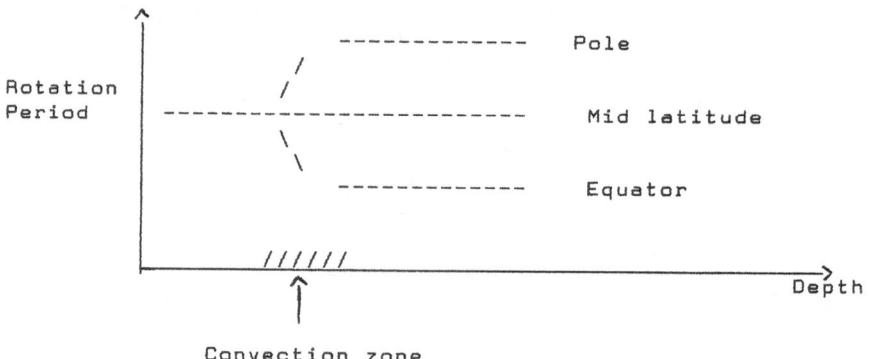

MULLAN: The data obtained by Harvey Duvall and Pomerantz at the South Pole Site in 1981 have been analyzed extensively and published in Nature (1986). The results show that the differential rotation, which is observed at the surface, persists inwards. That is, the pole continues to have the longest period, and the equator the shortest period, even at the base of the convection zone. This is consistent with the figure you have drawn, since the pole at the surface has a longer rotation than the equator. Hence, the existence of preferred longitudes suggests that the flare fields originate deeper than the convection zone. Also Duvall et al. (Nature, 1983) find that dΩ/dr is not negative, as required by dynamo theory, when averaged in latitude. It is not clear that this conclusion is necessarily inconsistent with the results you quote, depending on how much one weights the "equatorial" and "polar" regions in your plot.

VERMA: In addition to solar active longitudes, we have recently found for the first time that there are several active zones on the Sun, about six in each hemisphere (Verma et al.: 1987, Solar Phys. 112 , 341; Verma and Pande: 1988, Indian J. Radio & Sp. Phys. 17, 8).

DONALD F. NEIDIG / The Importance of Solar White-Light Flares: 261–269.

SPEAKER: NEIDIG

RODONO': Stellar flare observations show: a) pre-flare dips in a rather large fraction of flares; b) above 1 μm, in coincidence with the optical light increase, a "negative" flare (which is the mirror image of the optical one, but much fainter in relative intensity) is observed. Do you have evidence of similar behaviour in the solar case? Are IR observations of solar flares at wavelengths larger that 1 μm being done or planned?

NEIDIG: I know of no case where the solar white light flare shows pre-flare reductions in intensity. As far "negative" flares in the near infrared, observational coverage is extremely sparse and I know of no such examples. Presently we have no plans for flare patrols in the infrared, but perhaps we ought to.

KOJOIAN: In your talk, you have shown that the equation which characterizes the luminosity curves in both stellar and solar flares has the same functional form and is dependent on the same physical variables. This would appear to indicate that the same physical mechanism is acting in both cases but is just scaled. Equally, it would appear from other papers presented at this conference that the mechanism may be different. If you are also of this opinion, could you indicate how a different mechanism could account for the scaling which you have obtained and, further, what this other mechanism might be.

NEIDIG: The approximate equation given in the talk is of the form $L_{opt} \approx B_{6000\,Å}(T)\,\Delta\lambda\,A\,[1 - \exp(-\Delta\tau_{6000\,Å})]$, where A is the flare area and $\Delta\lambda \approx 10^4$ Å. This equation describes only the radiative losses, which, of themselves, imply nothing about the flare heating mechanism. If the temperatures in solar and stellar optical flares are sufficiently similar, then the emission mechanism should be the same, in which case the optical thickness, $\Delta\tau$, would be scaled by the same atmospheric density parameter.

E. N. PARKER / Solar and Stellar Magnetic Fields and Atmospheric Structures: Theory: 271–288.

SPEAKER: PARKER

LINSKY: Despite your general skepticism concerning the physics responsible for phenomena on the Sun, you have expressed some confidence that the physics of solar flares can be extrapolated to stellar fleres. The energy involved in flares on dMe stars can be 10^3–10^5 times larger. Does not this qualitative difference indicate a very different heating mechanism?

PARKER: Perhaps I have overstated my optimism. You are correct that the extreme flares on certain other solitary stars are so powerful as to suggest a situation that might be qualitatively different in some essential way from the solar flare. The enormous starspots occurring on some dM dwarfs are grossly different from sunspots too. And we have no right to assume that the solar repertoire includes every stellar phenomenon in the universe. Therefore, I should restrict my most intense optimism to the stars whose X-ray coronae are not qualitatively different from the Sun. On the other hand, we must not loose sight of the fact that present ignorance provides no evident objection to a considerable upward scaling of the magnetic active regions seen on the Sun. For instance, there is no objection of which I am aware to supposing that some stars have more vigorous photospheric convection, mixing the footpoints of bipolar magnetic regions five times larger in size, with mean fields five times greater than the usual 50–100 gauss observed on the Sun. The total magnetic energy of such a field would be $5^5 = 3125$ times greater then its solar couterpart. For a given rms misalignment of the local lines of force, and the same rate of shuffling of the footpoints, the rate of energy input per unit area is 25 times larger than in the Sun. The larger energy input and the larger dimensions allow higher coronal temperatures without an overwhelming increase in coronal density. So we can go a long way toward the extremes of hot bright X-ray coronae and enormous stellar flares solely on the basis of a simple and completely unimaginative upward scaling of the active corona of the Sun. The qualitative differences between the solar X-ray corona and flare and the known extreme stellar cases may, or may not, involve phenomena unknown on the Sun (and hence unknown to us). Intensive observational studies and further theoretical work will be essential to clarify the picture.

JORDAN: You envisage the active region heating arising from the reconnection of magnetic fields stressed by jostling at the footpoints. If magnetic flux emerges in an unstressed condition, how long would it take to produce a loop heated to 4–5×10^6 °K.

PARKER: With the estimate that the footpoint wandering proceeds at a velocity $u = 0.5$ km/s, the interweaving of the lines of force (in a magnetic bipole of length L) reaches a level where heating should begin in a time that is inversely proportional to L and which has a value of 1–2 hours for $L = 10^4$ km. Hence a freshly

emerging magnetic loop of a few thousand km should begin to show
heating in a fraction of an hour, assuming that it was unstressed
when it emerged.

STURROCK: I am dubious about your assertion that your proposed
mechanism for coronal heating will give a volume heating rate
that is independent of the overall linear dimensions. Uchida and
I considered a similar model in 1979, and we found that the
volume heating rate varies inversely as the square of the
overall linear dimensions.

PARKER: The proposed mechanism predicts that the heating rate per
unit area of the footpoints is independent of the length of the
magnetic field between the footpoints, all other things being
equal. Hence, the heating rate per unit volume is inversely
proportional to the length of the field. If the line of sight,
through a region filled with flux tubes, is proportional to the
length of the flux tubes, the energy input per unit area is then
independent of the length of the field.

MARTENS: There is one aspect in both the wave theory of coronal
heating and in your theory of spontaneous formation of current
sheets that does not satisfy me: Why does one see X-ray emetting
LOOP structures, and not the whole active region lighting up in
X-rays? Since the magnetic field probably fills the whole active
region, why is the heating confined to loops?

PARKER: I do not know the answer to your question. The heat input
needs not vary much between faint and bright flux bundles because
of the heat lost by downward thermal conduction. I can suggest
only that the shuffling and intermixing of the photospheric
footpoint of the field is patchy, so that in some places a flux
bundle is rapidly interwoven and in other places slowly
interwoven. It may be too that a flux bundle subject to a uniform
rate of intermixing tends to discharge its internal tangential
discontinuities intermittently, with dormant periods between each
phase of active dissipation. During the dormant time the
tangential discontinuities accumulate until they reach some
threshold and the active phase of heating begins again. But one
can only conjecture on these matters. Perhaps one day we will
observe the shuffling and mixing of the footpoints in the
photosphere to see how patchy is the effect. And perhaps
observation can determine if the dissipation of current sheets
involves cooperative effects and tends to procced in intermittent
phases.

GIOVANNI PERES / Hydrodynamic Models of Solar and Stellar Flares: 289–298.

SPEAKER: PERÉS

HAISCH: How does the magnetic field that your hydrodynamic simulation implies agree with the one I derived from the Alfven wave propagation time versus X-ray light curve argument?

PERES: Our lower limit of ≃100 Gauss is a factor of four smaller than the lower limit you derived.

MACNEICE: You justified the use of an implicitly collisional model of the coronal loop on the basis of the electron collisional mean free path being less than the loop dimension. In fact the use of Spitzer-Hrm heat flux demands that a much more rigourous condition (mean free path less than a few percent of the scale height) be satisfied. The degree to which Spitzer-Hrm heat flux fails in the "SMM Benchmark model" to which you referred, is illustrated in the poster paper by Ljepojevic and myself.

PERES: As for the bulk of thermal electrons, their mean free path is very short with respect to the loop dimensions. This ensures a hydrodynamic treatment. As for the high energy tail of the electron distribution, namely that responsible for considerable part of the heat conduction, the standard Spitzer-Hrm description is expected to fail. Better descriptions of thermal conduction, like those of Karpen et al. (1988), yourself and Peres et al. (1987), are available.

JORDAN: Were there any measurements of the electron density in the transition region of Proxima Cen from observations with IUE?

PERES: The IUE observations of this flare lacks the time resolution needed to study the rapid evolution of the flaring atmosphere.

PALLAVICINI: The flare you have modelled on Prox Cen is a rather modest flare by stellar standards. Suppose you want to model a much stronger stellar flare, you have to dump much more energy which will probably affect heavily the optically thick part at the loop footpoints. So, the first question is: Do you expect that the treatment of the optically thick part at the loop footpoints will affect critically the hydrodynamics of the coronal part? My second question is: Can your model predict the optical flare that presumably originates at the loop footpoints?

PERES: Stronger flares show larger amounts of evaporated material, therefore when we try to model those flares we tend to choose initial atmospheres with larger base pressures and, therefore, larger amounts of material in the chromosphere. This should reduce the effect of the treatment of the optically thick part on the corona evolution. Our model, however, does not try to fit, so far, the data of the chromosphere.

And coming to your second question we surely predict the presence of a shock propagating downward in chromosphere which has been connected with white light flares; and study other general hydrodynamic features of the chromosphere. Therefore we can, so far, make general predictions on this aspect.

CARGILL: Your simulations create very hot dense plasmas in the corona and transition zone. What magnetic field strengths do you need to confine the plasma to avoid vioalating the I-D assumption on the Sun?

PERES: The magnetic fields which would ensure confinement are not unreasonable for the solar cases studied so far.

CARGILL: What fields are required in the stellar flares you discussed?

PERES: The more demanding stellar case, I just presented, implies a coronal magnetic field of the order of 100 Gauss, which could easily be accepted even for a solar compact flare.

B. R. PETTERSEN / A Review of Stellar Flares and Their Characteristics: 299–312.

SPEAKER: PETTERSEN

PALLAVICINI: I would like to add a word of caution about the X-ray flare detected from the A-type star α Gem (Castor). The EXOSAT flare is certainly real; however, we have to take into account that both components of α Gem (A1V + A5 Vm) are spectroscopic binaries. The mass function leads to a broad range of spectral types for the secondary (from M to A-type). Moreover, the properties of the X-ray flare (time-scale, total energy, X-ray temperature) are consistent with those of flares from M dwarf flare stars. Therefore, we should not disregard the possibility that the flare on Castor may have originated from an unseen late-type component.

PETTERSEN: That may be the case also for other early type stars. Some of the flares have timescales and lightcurve shapes that are reminiscent of dMe flares. Only more detailed observations will show if any of these cases can be explained by the presence of convective companion stars. Until that has been thoroughly investigated, I would like to regard the situation as open for debate.

RODONO': If we are dealing with dynamo-activated flares, we would expect a dependence on the Rossby number rather than simply on the mean density or convection zone volume. Did you explore this possibility? In fact, in addition to convection, rotation rate is another important parameter to be taken into account for activity correlation studies.

PETTERSEN: No, I have not plotted flare activity versus Rossby number. For very active (i.e. saturated) flare stars the actual value of the rotation rate seems not to affect the situation. All of them rotate faster than about 5 km/s at equator and those few that are really rapid (>25 km/s) still seem to produce about the same level of flare activity. Rotation, as a determining parameter becomes apparent only when slower rotation rates are considered, as in K and M dwarf. For instance, for dK and dM stars with no emission lines the flare activity is dramatically smaller than for saturated stars.

POLETTO: Can you give a figure about the relative frequency of huge energetic flares with respect to small ones for stars of any given class?

PETTERSEN: The distribution of flare frequency with energy is well studied for dKe-dMe stars, the Sun, and some cluster flare stars. Shakhovskaya showed a diagram related to this problem in her talk. In a zeroth approximation, you can say that the cumulative distributions of flare energy vs.frequency have slopes near unity. Therefore if you move down by one order of magnitude in flare energy, flare frequency increases by a factor of 10. This is not true in every detail because some stars have slopes

different from unity. In fact, several observers have claimed
that the slope changes systematically with the spectral type for
solar neighbourhood stars.

SCHMITT: I wonder about the case of the activity being
proportional to the volume of the convection zone. In a sphere
most of the volume is concentrated near the surface, so in going
from a solar-like G star (with $T_{conv} = 0.7 R_*$) to a fully
convective M star, most of the volume change is due to the radius
change. Could one claim - using the same data - that activity is
proportional to the star's surface area as to its volume?

PETTERSEN: Your point is orrect, of course , from a geometrical
point of view. However, if the stellar radius were the key
parameter, the relationship would predict very high levels of
atmospheric activity (flares, X-rays, etc.) in early type
main sequence stars. Since that does not seem to be supported by
observations, I believe that the radius of the convection zone
gives a better parameter to describe the situation. Of course,
that immediately has implications for the physics. It allows
solar-like activity in stars with outer convection zones, such as
FGKM main sequence dwarfs, rapidly rotating subgiants, and fully
convective, rapidly rotating young contracting stars. Some forms
of activity seen, (e.g., in early type main-sequence stars) would
then have a different origin, or their outer structure would have
to allow convection.

KUIJPERS: Do you think it important to follow up the
first detection of variability in a Wolf Rayet star with a
monitoring programm for flares?

PETTERSEN: The only flare reported in the literature implies a
large energy release since it was seen against the background of
a bright Wolf Rayet binary. I don't think such flares are
abundant and a follow-up program could be very time-consuming.
However, flare-like phenomena in early type stars are not well
documented and I think it would be important to find out if these
outbursts have anything to do with the magnetic-related event on
the Sun and other convective cool stars.

GIANNINA POLETTO / Long-Duration Solar and Stellar Flares: 313–322.

SPEAKER: POLETTO

KUIJPERS: Why do you start with a magnetically open configuration? Should that not be explained precisely by a flare?

POLETTO: The reconnection model deals only with phenomena which occur after the field configuration has been torn open. To the model this is the initial field configuration and no attempt is made to account for events which occurred prior to this phase. While a comprehensive flare theory may consider mechanisms which lead to the field disruption, investigation of this problem was beyond our present purpose.

KUIJPERS: What is the physical basis that makes the field relax to precisely such a value that it can contain the X-ray emitting plasma?

POLETTO: If the field strength were smaller than required for magnetic confinement of the hot loop plasma, reconnection would not occur. On the other hand, if the field strength were larger than required for magnetic confinement of the plasma, reconnection would occur at a faster rate. Empirical evidence favouring this interpretation comes from the analysis of the 29 July 1973 flare on the Sun. From high resolution observation we may derive the electron density of the X-ray loops, as well as their temperature and their altitude, and determine the gas pressure at the tops of the loops. This value turns out to be in excellent agreement with the magnetic pressure predicted by the model at that height.

STURROCK: The results of your model depend on your assumption about how the " Y-type " point, separating closed from open field lines, increases with time. What assumption do you make, and on what basis?

POLETTO: According to the reconnection model the enhanced ohmic heating associated with fieldline merging shows up as localized thermal X-ray emission from the neutral point region. Therefore the X-ray source drift towards higher altitudes is supposed to trace the neutral point rise. In order to derive an analytical law representative of this upward motion, I made use of spatially resolved observations of X-ray loops in two-ribbon flares and each time I assumed the top of the newly formed X-ray loop to signal the position of the reconnecting region.

SCHMITT: Your model determines the rate of change of the magnetic energy whereas the observations relate to count rates in various passbands. Could you comment on your assumptions relating the change in magnetic energy to the observed radiative losses?

POLETTO: At this time one can only resort to the solar analogy and assume stellar flares to behave as solar flares. In the Sun radiative energy in the bands I used for the modelling amounts to about 1/10 of the total flare radiative losses. To be realistic, any model should provide at least as much energy. All the models I considered comply with this requirement.

ERICH RIEGER / Solar Flares: High-Energy Radiation and Particles: 323–345.

SPEAKER: RIEGER

SHAPIRO: Concerning particle beams: at least for very intense solar flares, energetic particles have been observed on earth. So at times, there are particle beams.

RIEGER: The beams you are referring to may have been produced later in the flare in the corona, or further out in interplanetary space. The question I am mainly concerned about is: Does the primary acceleration mechanism, acting for instance at the top of the loop, create beams or not? It appears to me that the detection of limb brightening can be explained by both: particle beam or omnidirectional distribution created, for instance, at the loop top.

MULLAN: Can you comment on the chemical composition of the target material from which the nuclear lines are emitted?

RIEGER: The shortness of time prevented me from talking about this subject. It is, however, extensively discussed by Ramaty and Murphy in 1987 (Space Sci. Rev. 45, 213).

BAI: You showed a very good correlation between X-ray fluences above 300keV and nuclear X-ray fluences. However, if you compare hard X-ray fluences > 30 keV and nuclear X-ray fluences, we do not find any good correlations. Therefore, your conclusion that one acceleration mechanism can accelerate all the particles cannot be applied to low-energy electrons.

RIEGER: What you say is certainly true. I think that at $\geq$ 30 keV in some events thermal emission, which does not reach to > 300 keV, may contribute to the radiation.

GRANDPIERRE: It seems to me that there is an energetic problem with the shock acceleration in the second phase. We know, that a large part of flare energy is involved in the particle flares. To be able to provide enough energy to these particles by shock waves, we would need much more energy in the shocks than in the flares, because of two reasons. The shocks propagates spherically and symmetrically, therefore their energy decreases with the square of the distance from their places of birth. Secondly, they can reach the particles only within a limited solid angle.

RIEGER: The total energy, which resides in the very high energy particles, about which I was talking, is only a minute part of the whole flare energy, so that I don't think, your energy argument holds.

TAKASHI SAKURAI / Magnetic Equilibria and Instabilities: 347–360.

SPEAKER: SAKURAI

KLIMCHUK: The method you have described for calculating sequences of magnetic equilibria, based on the so-called "generation function", places artificial constraints on the field. For example, in the case of an arcade that is uniform in the x-direction, the generation function dictates the B_x component of the field on each and every field line. This is an unphysical constraint that is not relevant to the Sun. What is relevant to the Sun is the connectivity of the field; that is, the locations of the footpoints in the photosphere. To my knowledge, none of the force-free equilibrium sequences calculated using connectivity boundary conditions, rather than the generating function, terminate with a loss of equilibrium. Regarding equilibrium sequences characterized by increasing pressure (with fixed shear), I share your concern that the pressures necessary for loss of equilibrium are unrealistically high. In the latest work of Zwingmann, for example, I believe that loss of equilibrium does not occur until the plasma beta is near 0.1, which is much too high for the corona. Finally, I believe that there is now a consensus on the issue of thermal stability. Most workers in the field, including Antiochos, agree that coronal loops are thermally stable, as long as they are not too low-lying.

MARTENS: I agree with your remark that the correct physical problem is done by describing the footpoint positions of the field lines. But I disagree with your statement that non-equilibrium has never been found. There is a poster here by Aad van Ballegooijen and myself that demonstrates the onset of non-equilibrium for just this situation.

HASSAM: From your presentation, it is apparent that you have studied the theory of ideal/non-ideal MHD instabilities in some detail. You have also studied vector magnetographs. To my knowledge, ideal MHD instabilities (or loss of equilibria) have growth quite disparate from non ideal instabilities. From your studies, would you associate solar flares with ideal instabilities, loss of equilibria or non-ideal instabilities?

SAKURAI: I would think that the primary driver for solar flares is either ideal MHD instabilities or a loss of equilibrium. These might lead to forced (or driven) reconnection, thereby releasing the magnetic energy in the form of heat and high energy particles.

CHEN: You stated that a force-free equilibrium (2-D arcade) can give information concerning pressure. It seems that the pressure may not be a physical quantity one can always prescribe in a coronal structure. The ability to prescribe pressure may depend on specific mechanisms. If so, a bifurcation with respect to specification of pressure need not be a physical one. This may be analogous to the point made by Jockers (in a 2-D arcade) versus specification of the toroidal field component (in a 2-D arcade) versus specification of footpoints. Do you have any comments regarding a 3-D structure such as a "toroidal" loop?

SAKURAI: I still feel it to be reasonable to suppose that we have the liberty of specifying the pressure as large as we like. For 3-D loops, some examples are presented by Low (1986) but much more remains to be done.

VELLI: In the 2-D calculations one finds instability when magnetic islands form above the photosphere. However it is not at all certain that instability would remain (linear or nonlinear), if the third, line-field dimension was included in the analysis. Would you comment on this point?

SAKURAI: I agree that the line tying in the third dimension (along the filament axis) is an important factor.

CARGILL: The growth rates for tearing modes are relevant for infinite or periodic media. The boundary condition of photo-spheric line tying has been shown (Hood and Priest, 1979; Einaudi and van Hoven, 1983, Mighredo and Cargill, 1983) to remove the possibility of tearing modes in such configurations because the singular surface present in infinite media is no longer present. This must be kept in mind when applying tearing mode theory to solar plasmas.

ANTIOCHOS: In your 3-D force-free fields calculations do you find any evidence for non-equilibrium? Do you find that your numerical method, i.e. code, works well for the force-free problem where the transverse field is specified at the photosphere?

SAKURAI: First, I encountered several cases where the iterative procedure did not converge. However this might well be a breakdown in the numerical scheme and may not be due to a non-equilibrium situation. Second, the force-free field may be determined by prescribing the normal component of the magnetic field (B_n) and the value of α, in either positive or negative polarity regions. The specification of all the three components of the magnetic field over-determines the solution.

J. H. M. M. SCHMITT, J. R. LEMEN, and D. ZARRO / A Solar Flare Observed with the SMM and *Einstein* Satellites: 361–373.

SPEAKER: SCHMITT

DOYLE: I have two comments concerning your estimate of Ne and V:

a) From the EINSTEIN data you only use the radiative cooling time: the derived N_e will therefore be a lower limit since you neglected thermal conduction and mass downflows, both of which are important for the coronal region.

b) The estimate of N_e based on the EM and the flare size from the FCS data may also be a lower limit since the FCS cannot resolve the flare and therefore the filling factor may be less than 1.

Hence your comparison between Ne and V derived from SMM and EINSTEIN should be treated with caution.

SCHMITT: To answer your first comment what has actually been determined - according to Moore et al. (1980) - is that radiative and conductive cooling time scales are about equal. With this the conduction length scale can be estimated, which turns out to be less than the derived length scales (cf. Schmitt, Harnden and Fink 1987). In other words, the derived picture is physically consistent.

As far as your second comment is concerned, I agree with you that the FCS filling may be less than one and that the desired densities are still lower limits. The filling factor problem is obviously aggravated in the stellar case; the point to keep in mind is that even the possibly conservative "stellar" estimates indicate very small filling factors for the flaring plasma.

SIMNETT: I wish to caution you about the accuracy of the parameters you might derive from full-star observations of soft X-rays. From the SMM-HXIS solar observations it is very clear that for many flares more than one structure is involved in the flare. At some point in the flare the X-ray emission from these structures becomes comparable, i.e. often the emission from one structure decays while it rises, temporarily, in another. Treating the situation as a single, unresolved event will merely result in some "average" value of the derived parameters, which gives a very misleading conclusion in the case of solar flares. Unfortunately, I cannot offer any practical suggestions as to how to treat stellar flares in a more realistic way.

SCHMITT: Your warnings are well taken, in fact, some of the observed stellar X-ray light curves have a very complex structure which may indicate what a number of spatial components participate in the flare. On the other hand, the very purpose of this work has been to investigate to what extent the "stellar" modelling of a solar flare can be verified (as falsified) by detailed observations. In the (only) case studied in detail the physical parameters derived from the "stellar" modelling agree with those derived from dedicated observations to within astronomical accuracy. This success I consider very encouraging; of course, this result does not imply that the physical parameters of all stellar flares have been correctly determined.

RODONO': I completely agree with Dermott Mullan's comment on the importance of quantifying the uncertainty in deriving physical parameters from spatially unresolved stellar flare observations. I should like also to stress that this beautiful piece of work by Jurgen Schmitt does not exhaust what can be done by using well planned and accurate stellar observations: I believe that we can learn a lot on stellar flares from multi-wavelength observations for the purpose of studying the effects of the global stellar parameters if any, and of the plasma physical state on the flare triggering and evolution, bearing always in mind what we have learned and might learn from the study of spatially-resolved solar flares.

JORDAN: You found that the peak emission measure from the Einstein measurements was significantly larger than that found from the Ca XIX lines with the SMM BCS instrument. Have you taken the emission measure distribution with temperature derived from SMM and folded it through the Einstein band responses to make a comparison that way?

SCHMITT: No. The IPC emission measure was derived taking the cooling function at $\log T = 7.2$. Since, as you said, lower temperature material does contribute to the IPC measurements, this procedure underestimates the effective cooling and hence overestimates the emission measure. Therefore including a DEM distribution would actually reduce the apparent discrepancy between IPC and SMM emission measures.

N. I. SHAKHOVSKAYA / Stellar Flare Statistics – Physical Consequences: 375–386.

SPEAKER: SHAKHOVSKAYA

FOING: How do you correct the bias in the diagram $\log E - \log \nu$ for (1) the detection threshold and (2) the time superposition of frequent small flares? If this statical correction is not made, any extrapolation or inferences about the role of small-scale flares for coronal heating may be incorrect.

SHAKHOVSKAYA: Only the upper parts of the flare energy spectra above the photometric detection threshold of flares are presented in Fig. 2. and were used to derive our statistical estimates.

RODONO': Have you tried to consider M_{bol} instead of M_V in your frequency energy correlations? The former might be more significant.

SHAKHOVSKAYA: No, I did not try.

RODONO': How did you compute the solar flare wide-band energy which is included in your plot of flare energy spectra?

SHAKHOVSKAYA: The solar flares were observed in $H\alpha$ line and the corresponding energy in B-band were calculated as in Gershberg, Mogilevsky, and Obridko (1987).

PETER A. STURROCK / The Role of Eruption in Solar Flares: 387–397.

SPEAKER STURROCK:

SERIO: What about small loops, say ≃10^3 km?

STURROCK: Flares occurring in loops that are only 10^3 km high could not be due to the current-interruption process I have suggested.

MOORE: What is the diameter of a typical current filament that would interrupt giving rise to a " single loop " flare? Is the total current carried in many narrow threads that fill a small fraction of the flare loops?

STURROCK: If each current filament arises from a magnetic knot, with a flux of about $10^{18.4}$ Mx, and if the field strength in the corona is about 100 gauss, the radius of a typical current filament would be 10^6 cm.

GAIZAUSKAS: Two observational comments on your description of an erupting filament. It is very common for active region filaments to exhibit downflow with a magnitude of tens of km/s for many hours preceding the eruption. The downflow is usually quite distinct at one footpoint of the filament, so much so that the location of that footpoint is easily identified with respect to photospheric features. The other footpoint of the same filament is often vague and difficult to identify as you have stated in your description of initial conditions. Can you safely ignore downflow as one of your initial conditions?

STURROCK: The point you make indicates that we really have an inadequate understanding of the structure and origin of filaments. The downflow suggests that a filament may be continually evolving, perhaps the result of the slow but steady emergence of new flux from below the photoshere. Even so, the motion can probably be ignored in considering the MHD stability of filaments.

CHEN: With respect to your statement that introduction of pressure can allow physical bifurcation in a 2-D arcade, it seems that pressure of an arcade is not a physical quantity that can be prescribed.

STURROCK: I agree that specifying the pressure at each point, as a given function of space and time, does not sound like a reasonable "thought experiment".

CHEN: As an alternative mechanism for magnetic energy release in the corona, you mention the possibility that a "mechanical" energy release mechanism may be operative. For example, Chen (1987, and poster paper at this meeting) describes a mechanism whereby a magnetic/current loop can become unstable and expand. The expansion velocity can range from very slow to ≃1200 km in the corona, with a correspondingly wide range of energy release rates (up to 10_{32} erg in 30 min) via drag heating. This may possibly be relevant to some motion-related phenomena.

STURROCK: In many flares, the kinetic energy of mass motion accounts for the biggest share of the energy released during the flare. Hence your proposal may well be relevant to an important component of the flare process.

CARGILL: What instability are you exactly proposing for the anomalous heating model of ion acoustic; this implies $T_e \gg T_i$ in the loop.

STURROCK: The instability I have considered so far is the ion-acoustic instability, and you are quite right that this requires $T_e \gg T_i$.

CARGILL: What exactly is responsible for the prompt acceleration of the protons?

STURROCK: The prompt acceleration mechanism would be simply direct acceleration by the field-aligned electric field that develops as a result of the attempted current interruption.

MULLAN: If there is no steady coronal heating, would you expect that, once instability is quenched, all loops should cool on time scales of $\simeq 10^2$ s and then reheat and re-cool repeatedly?

STURROCK: Yes, the picture is that coronal material would cool on a time-scale of order 10^4 s, then heat impulsively on a time-scale of perhaps 10^3 s, and so on.

HASSAM: I refer to your slide in which you listed four possible ways by which the reconnection rate could be enhanced over the Furth-Killeen-Rosenbluth rate. I would like to point out that one or both of two factors apply to each of the four mechanisms proposed: (a) the fact that resistivity, η, still acts as a nozzle in that the growth rate goes as some power of η; (b) the fact what some of these mechanisms have not yet been backed up by "hard" calculations. I would therefore submit that none of the four proposed mechanisms have convincingly shown us that the reconnection rate is sufficiently enhanced so that reconnection can be clearly considered as the process underlying flare phenomena.

STURROCK: I would submit that the work of Carreras and his colleagues, and of Sakai and his colleagues, are pretty "hard" calculations. However, I do agree that much work remains to be done, and that the precise role of reconnection in flares remains to be pinned down.

DING: Have you found any evidence of current sheet in active regions before solar flares?

STURROCK: Unfortunately we still have not discovered a way to detect current sheets in the corona.

SMITH: One of the best studied examples of coronal heating is the large loop of the 1980, November 5 flare (Martens et al. 1985, Solar Phys.). Martens et al. found that only 10^{-3} of the loop was heated and proposed ion tearing as a mechanism. Have you calculated in your sporadic heating model what fraction of a loop is being heated at any one time?

STURROCK: My very preliminary estimates are that the "filling factor" would be in the range 0.01 to 0.1 .

ACTON: Please comment on the observability of the current interruption event? The density is so low that there must be very little emission measure.

STURROCK: As long as the plasma density is low, the principal result of the interruption event is particle acceleration. But as soon as bombardment of the chromosphere leads to evaporation that fills the loop with hot, dense gas, the principal output would be X-ray emission.

SHAPIRO: Would you please elaborate on the nature of the shocks that might accelerate higher-energy ($\approx$ GeV) particles?

STURROCK: I have in mind the shocks that are responsible for Type II radio bursts. These may be either blast waves, caused by the sudden energy conversion related to the MHD instability that leads to filament eruption, or possibly bow shocks that run ahead of coronal mass ejections.

SVESTKA: You showed simultaneous flux peaks from 40 keV through 40 MeV, but that was at the very onset of the flare, during its impulsive phase. Later on, the energy range is much smaller. Do you accept reconnection as the source of energy release later in the flare development?

STURROCK: Yes. Certainly one must expect that reconnection plays a role in energy release. For instance, the morphology of two-ribbon flares is strongly suggestive of field line reconnection in an overlying current sheet.

CHENG: Observations in soft X-ray and EUV show that the flaring plasma is heated up 10-20 minutes before the onset of the impulsive phase. This means that before the flare instability occurs, the flaring loop already shows appreciable and obser- vable emission. I wonder how your cool and low density loop could reconcile with the observed preflare heating.

STURROCK: One possibility, that I mentioned, is that the filament eruption is coupled with field line reconnection. If this is the case, that stage of reconnection could be responsible for the soft X-ray emission that you mentioned.

ZDENĚK ŠVESTKA / Solar Flares: The Gradual Phase: 399–417.

SPEAKER: SVESTKA

SIMNETT: I have a question regarding the energy supply for long-duration soft X-ray events. It was pointed out by Sheeley et al. (Ap. J., 1985) that soft X-ray events lasting more than 6 hours were correlated with coronal mass ejections. The latter normally are associated with shocks, and it is believed that such shocks accelerate particles as they go out through the corona. A significant fraction of the particles go back to the Sun, where they deposit their energy, presumably near the base of the corona. Why do you not consider this mechanism as an energy source for the long duration events?

SVESTKA: Because I do not see how the deposit of energy of these particles could give rise to the well-defined loops we observe, with impressive brightness maximum at their tops. The particle acceleration is a stochastic chaotic process: the particles should deposit their energy all throughout the active region; why just in the loops we see? And how can the deposition last for so many hours, selecting sequentially higher loops? Besides, the fact that we record particles in space does not necessarily mean that a similar amount of particles flows downwards (cf. Bai's results, e.g., comparying intensity of γ-rays and number of protons in interplanetary space).

VERMA: What are the basic conditions in the active region for a flare to be impulsive or gradual?

SVESTKA: The confined flare (which you call impulsive, I suppose) is a local instability in a preexisting loop or a system of preexisting loops in an active region due, for example, to twisting of field lines. The dynamic (gradual) flare reflects a global instability, probably due to excessive shear and a trigger, which we do not know.

MARTENS: I have a question regarding confined flares. I have heard different opinions regarding simple loop flares, mainly from theorists, whether single loop flares exist, or whether all events involve multiple loops. What is your opinion on that, and can you give a specific example of a simple loop flare?

SVESTKA: I think that simple loops flares do exist, but are very rare. The soft X-ray flare, I showed on the first slide, looks like a single-loop flare. But, of course, better spatial resolution ($\approx$ 4 arcsec in Skylab) might have shown even in this case that the "single loop" was actually composed of a conglomerate of loops. I think that Steve Kahler and Mukul Kundu could show you other examples of such simple flares.

I. TUOMINEN, J. HUOVELIN, YU. S. EFIMOV, N. M. SHAKHOVSKOY, and A. G. SHCHERBAKOV /
Polarimetry of Stellar Active Regions and Flares: 419–429.

SPEAKER: TUOMINEN

MULLAN: Synchrotron emission may explain quiescent polarization. In 1975, I thought this was very difficult to understand, but since then, the connection between flaring and coronal heating suggests that synchrotron may now be considered as a viable source of polarization in M darfs even outside flares. IRAS data on flare stars show that 20% of flare stars are visible at 100 μm and 60 μm. Only synchrotron emission can explain the IRAS data. Hence, coronal heating apparently supplies the atmosphere of M dwarfs with lots of MeV electrons.

HENOUX: First a comment: During solar flares we observed linear polarization in Hα line, which seems to be due to particle bombardment of the solar chromosphere. My question is: What are the relative contributions of lines and continuum to the observed linear polarization for stellar flares?

TUOMINEN: It is difficult to say. Polarization for flares is small and difficult to measure. The magnetic intensification in lines seems to be the most probable mechanism in active regions of solar type stars.

V. V. ZHELEZNYAKOV and E. YA. ZLOTNIK / Cyclotron Lines in the Spectra of Solar Flares and Solar Active
Regions: 449–456.

SPEAKER: ZHELEZNYAKOV

ACTON: Is the distribution of temperature with height derived
from the theory or assumed?

ZHELEZNYAKOV: The distribution of kinetic temperature with
magnetic field strength in loop cross-section was derived from
observed cyclotron line profiles. The temperature distribution
with height was then derived, assuming a definite configuration
of the magnetic field above a bipolar group.

STURROCK: What are the ideal instrumental requirements of a radio
telescope that could make the required observations to detect
cyclotron lines?

ZHELEZNYAKOV: The angular resolution of the antenna should be
better than 5-10 arcsec because the linear sizes of loops are
about 50-100 arcsec. The desired frequency resolution of a
spectrograph is determined by the linewidth; it must not be worse
than 30-60 MHz at the operating frequency of, for example, 1800
MHz.

INDEX